Springer Theses

Recognizing Outstanding Ph.D. Research

Aims and Scope

The series "Springer Theses" brings together a selection of the very best Ph.D. theses from around the world and across the physical sciences. Nominated and endorsed by two recognized specialists, each published volume has been selected for its scientific excellence and the high impact of its contents for the pertinent field of research. For greater accessibility to non-specialists, the published versions include an extended introduction, as well as a foreword by the student's supervisor explaining the special relevance of the work for the field. As a whole, the series will provide a valuable resource both for newcomers to the research fields described, and for other scientists seeking detailed background information on special questions. Finally, it provides an accredited documentation of the valuable contributions made by today's younger generation of scientists.

Theses may be nominated for publication in this series by heads of department at internationally leading universities or institutes and should fulfill all of the following criteria

- They must be written in good English.
- The topic should fall within the confines of Chemistry, Physics, Earth Sciences, Engineering, and related interdisciplinary fields such as Materials, Nanoscience, Chemical Engineering, Complex Systems, and Biophysics.
- The work reported in the thesis must represent a significant scientific advance.
- If the thesis includes previously published material, permission to reproduce this must be gained from the respective copyright holder (a maximum 30% of the thesis should be a verbatim reproduction from the author's previous publications).
- They must have been examined and passed during the 12 months prior to nomination.
- Each thesis should include a foreword by the supervisor outlining the significance of its content.
- The theses should have a clearly defined structure including an introduction accessible to new Ph.D. students and scientists not experts in the relevant field.

Indexed by zbMATH.

Javier Rivera-Dean

Non-classical States of Light

Generation via Strong-Field Processes and Applications in Quantum Key Distribution

Doctoral Thesis accepted by
ICFO—The Institute of Photonic Sciences,
Barcelona, Spain

Foreword by Marcelo Ciappina

Springer

Author
Dr. Javier Rivera-Dean
Quantum Optics Theory Group
ICFO—The Institute of Photonic Sciences
Castelldefels, Barcelona, Spain

Supervisors
Prof. Marcelo Ciappina
GTIIT—Guangdong Technion—Israel
Institute of Technology
Shantou, China

Prof. Antonio Acín
ICFO—The Institute of Photonic Sciences
Castelldefels, Barcelona, Spain

ISSN 2190-5053 ISSN 2190-5061 (electronic)
Springer Theses
ISBN 978-3-031-73768-8 ISBN 978-3-031-73769-5 (eBook)
https://doi.org/10.1007/978-3-031-73769-5

This work was supported by Departament d'Innovació, Universitats i Empresa, Generalitat de Catalunya (2019_FI_B_01097).

This Springer imprint is published by the registered company Springer Nature Switzerland AG
The registered company address is: Gewerbestrasse 11, 6330 Cham, Switzerland

Supervisors' Foreword

It is a great pleasure to write the Foreword for this remarkable Ph.D. thesis, which represents a significant contribution to a new research field that marries attosecond physics and quantum information science. This thesis is a testament to the dedication, intellectual curiosity, hard work, and innovative spirit Javier has put into this endeavor. Specifically, his research delves into the complex and fascinating realm of non-classical states of light and their applications in quantum communication.

The thesis begins by contextualizing its research within the broader historical and scientific landscape. The early 20th century heralded the first quantum revolution, transforming our understanding of light and matter at the atomic and subatomic levels through groundbreaking discoveries within the then-incipient quantum mechanics. This period laid the groundwork for the second quantum revolution, which seeks to exploit quantum mechanics for advancements in information processing, communication, and computation.

At the core of the thesis is the exploration of non-classical states of light, that is, states exhibiting behaviors that defy classical physics and require a quantum mechanical framework for explanation. The manipulation and generation of these states are pivotal for advancing quantum technologies. Photonics, with its robustness and manageability, emerges as a promising platform in this context.

The thesis is divided into two primary parts: (i) Generation of Non-Classical States of Light through Strong-Field Processes and (ii) Non-Classical States of Light for Quantum Key Distribution.

The first part, (i), is devoted to investigating the interactions between high-intensity lasers and matter, demonstrating how strong-field phenomena can generate non-classical states of light. This part of the thesis covers: Above-Threshold Ionization (ATI): Examining how ATI processes could lead to entanglement between ionized electrons and the light field.—High-Harmonic Generation (HHG): Analyzing how non-classical light states, with potential applications in nonlinear optical processes, could be generated after HHG. - Complex Systems: Extending the previous analysis, mainly focused on atomic targets, to diatomic molecules and semiconductor materials, uncovering distinctive non-classical features in their quantum optical states compared to atomic systems.

In the second part, (ii), the application of non-classical light states in quantum communication, specifically focusing on quantum key distribution (QKD), is explored. This includes: - Device-Independent QKD: Investigations of photonic-based implementations for establishing secure communication without relying on the integrity of the devices used for key generation. - Dimensionality of Non-Classical States: A theoretical analysis of how varying the dimensionality of non-classical states impacts the key rate and security in QKD, providing insights into noise requirements and potential enhancements.

This thesis not only contributes to the theoretical understanding of non-classical states and their generation but also offers practical insights into their application in secure quantum communication. The meticulous research and innovative approaches presented in this work underscore its significance in the ongoing development of the symbiotic area of strong-field physics and quantum information science and technology. In conclusion, this thesis stands as a testament to Javier's exceptional ability to tackle complex scientific problems and contribute valuable knowledge to the field. It is with great respect and admiration that we present this foreword, acknowledging the profound impact of this research on the future of attosecond science, strong-field physics, and quantum technologies.

Last, but not least, we would like to write a few words to highlight the joy we have had with Javier on this journey. Besides being an excellent professional, he is an amazing person, and this is something we should clearly remark. Javier's kindness, dedication, and collaborative spirit have made this experience not only intellectually rewarding but also personally fulfilling. His ability to combine rigorous scientific inquiry with a warm and supportive presence is truly exceptional.

We are confident that Javier will shine brightly in the future, given his outstanding qualities, both professional and personal. His talent and character will undoubtedly lead to many successes and meaningful contributions to the field.

All the best in your future endeavors, Javier!

Shantou, China
Castelldefels (Barcelona), Spain
July 2024

Prof. Marcelo Ciappina
Prof. Antonio Acín

Abstract

The dawn of the last century marked the onset of the *first quantum revolution*, a period characterized by groundbreaking discoveries culminating in the establishment of quantum mechanics. Over time, the abstract concepts introduced by this new branch of Physics, evolved into indispensable practical devices shaping our daily lives. This technological evolution spurred our actual era, centered around information exchange and acquisition, laying the foundation for what is now termed the *second quantum revolution*. This phase aims to leverage quantum information science, which harnesses quantum mechanics' properties to propel advancements in information processing, communication, and computation, leading to revolutionary *quantum technologies*.

At the heart of advancing quantum technologies lies the exploration of what are known as *non-classical states*—physical manifestations exhibiting behaviors diverging from classical physics, necessitating the framework of quantum mechanics for explanation. Manipulating and generating these states delineates the frontier of progress in quantum technology. Therefore, it is crucial to devise methodologies for generating and controlling non-classical states. Photonics emerges as a promising platform within this context due to its robustness and exceptional manageability of this kind of states.

For the above reasons, this Thesis adopts a dual focus. Firstly, we delve into the generation of non-classical states of light through strong-field processes. These processes entail interactions between light and matter, where light intensities contend with the binding forces that keep electrons bound to their respective nuclei. Our exploration demonstrates the utility of strong-field phenomena in generating non-classical states of light, exhibiting intriguing features dependent on specific process dynamics and the materials involved in excitation. Secondly, we investigate the constraints and prerequisites of non-classical states of light sources—beyond those derived from the aforementioned strong-field processes—for the advancement of quantum communication. In particular, we analyze *quantum key distribution*, aiming to create a secret key exclusively known by the communicating parties for encrypting and decrypting messages.

Therefore, this Thesis can be understood as a zeroth step towards leveraging strong-field physics as a prospective tool for quantum information science applications, as well as an exploration about the advances and limitations of photonic-based setups for quantum key distribution.

List of Publications

This Ph.D. Thesis is based on the following articles and preprints:

- Maciej Lewenstein, Marcelo F. Ciappina, Emilio Pisanty, **Javier Rivera-Dean**, Philipp Stammer, Theocharis Lamprou and Paraskevas Tzallas, *Generation of optical Schrödinger cat states in intense laser–matter interactions*, Nature Physics **17**, 1104–1108 (2021). arXiv:2008.10221 (2020).
- **Javier Rivera-Dean**, Philipp Stammer, Emilio Pisanty, Theocharis Lamprou, Paraskevas Tzallas, Maciej Lewenstein and Marcelo F. Ciappina, *New schemes for creating large optical Schrodinger cat states using strong laser fields*, Journal of Computational Electronics **20**, 2111–2123 (2021). arXiv:2107.12811 (2021).
- **Javier Rivera-Dean**, Theocharis Lamprou, Emilio Pisanty, Philipp Stammer, Andrés F. Ordóñez, Andrew S. Maxwell, Marcelo F. Ciappina, Maciej Lewenstein and Paraskevas Tzallas, *Strong laser fields and their power to generate controllable high-photon-number coherent-state superpositions*, Physical Review A **105**, 033714 (Editor's suggestion) (2022). arXiv:2110.01032 (2021).
- Philipp Stammer, **Javier Rivera-Dean**, Theocharis Lamprou, Emilio Pisanty, Marcelo F. Ciappina, Paraskevas Tzallas and Maciej Lewenstein, *High Photon Number Entangled States and Coherent State Superposition from the Extreme Ultraviolet to the Far Infrared*, Physical Review Letters **128**, 123603 (2022). arXiv:2107.12887 (2021).
- **Javier Rivera-Dean**, Philipp Stammer, Andrew S. Maxwell, Theocharis Lamprou, Paraskevas Tzallas, Maciej Lewenstein, and Marcelo F. Ciappina, *Light-matter entanglement after above-threshold ionization process in atoms*, Physical Review A **106**, 063705 (2022). arXiv:2208.05245 (2022).
- Philipp Stammer, **Javier Rivera-Dean**, Andrew S. Maxwell, Theocharis Lamprou, Andrés F. Ordóñez, Marcelo F. Ciappina, Paraskevas Tzallas and Maciej Lewenstein, *Quantum Electrodynamics of Intense Laser-Matter Interactions: A Tool for Quantum State Engineering*, PRX Quantum **4**, 010201 (2023). arXiv: 2206.04308 (2022).
- Eva M. González-Ruíz, **Javier Rivera-Dean**, Anders S. Sørensen, Antonio Acín and Enky Oudot, *Device Independent Quantum Key Distribution with realistic single-photon source implementations*—Optics Express **32**, 13181–13196 (2024). arXiv:2211.16472 (2022).
- Utso Bhattacharya, Theocharis Lamprou, Andrew S. Maxwell, Andrés F. Ordóñez, Emilio Pisanty, **Javier Rivera-Dean**, Philipp Stammer, Marcelo F.

Ciappina, Maciej Lewenstein and Paraskevas Tzallas, *Strong laser physics, non-classical light states and quantum information science*, Reports on Progress in Physics, **86**, 9 (2023). arXiv:2302.04692 (2023).

- Theocharis Lamprou, **Javier Rivera-Dean**, Philipp Stammer, Maciej Lewenstein and Paraskevas Tzallas, *Nonlinear optics using intense optical Schrödinger "cat" states*—Now accecpted in Physical Review Letters. See: https://journals.aps.org/prl/accepted/0007fYe7J531e88b020c52178686c331e540b2006. arXiv: 2306.14480 (2023).
- **Javier Rivera-Dean**, Philipp Stammer, Andrew S. Maxwell, Theocharis Lamprou, Emilio Pisanty, Paraskevas Tzallas, Maciej Lewenstein and Marcelo F. Ciappina, *Quantum optical analysis of high-order harmonic generation in H_2^+ molecules*, Physical Review A **109**, 033706 (2024). arXiv:2307.12381 (2023).
- **Javier Rivera-Dean**, Philipp Stammer, Andrew S. Maxwell, Theocharis Lamprou, Andrés F. Ordóñez, Emilio Pisanty, Paraskevas Tzallas, Maciej Lewenstein and Marcelo F. Ciappina, *Non-classical states of light after high-harmonic generation in semiconductors: A Bloch-based perspective*, Physical Review B **109**, 035203. arXiv:2309.14435 (2023).
- Philipp Stammer, **Javier Rivera-Dean**, Andrew S. Maxwell, Theocharis Lamprou, Javier Argüello-Luengo, Paraskevas Tzallas, Marcelo F. Ciappina and Maciej Lewenstein, *Entanglement and squeezing of the optical field modes in high-harmonic generation*, Physical Review Letters **132**, 143603 (2024). arXiv: 2310.15030 (2023).
- **Javier Rivera-Dean**, Philipp Stammer, Andrew S. Maxwell, Theocharis Lamprou, Andrés F. Ordóñez, Emilio Pisanty, Paraskevas Tzallas, Maciej Lewenstein and Marcelo F. Ciappina, *Quantum optical analysis of high-harmonic generation in semiconductors*, in High-Order Harmonic Generation in Solids, eds. Marcelo Ciappina and Paraskevas Tzallas (World Scientific, Singapore, 2024).
- **Javier Rivera-Dean**, Anna Steffinlongo, Neil Parker-Sánchez, Antonio Acín and Enky Oudot, *Device-Independent Quantum Key Distribution beyond qubits*—to be submitted to New Journal of Physics. arXiv:2402.0016 (2024).

Acknowledgements

During the writing of this thesis, I engaged with various documents exploring the curiosities of renowned scientists. Among these, I encountered an anecdote that aptly mirrors my feelings as I face this blank page. It seems that at a certain point, Niels Bohr expressed his struggle to Paul Dirac about completing a sentence in a scientific article he was writing. Dirac's advice was simple—he had been taught at school never to start a sentence without knowing its end. Similarly, I find myself at a juncture where even commencing with this acknowledgments section seems daunting, for I am genuinely thankful to numerous individuals who have accompanied and supported me throughout this journey. For this reason, I will not follow Dirac's advice and just let my thoughts flow. I am sorry in advance if this takes too long.

First and foremost, this Ph.D. Thesis would not have been possible without the guidance of (which at least I consider to be) top-tier researchers with whom I had the fantastic opportunity to collaborate. Firstly, I am indebted to my supervisor, Antonio Acín, with whom I shared many football games on Wednesday afternoons. He gave me the freedom to explore topics I was interested in, even when they were not directly related to the main activities of the group. Secondly, I am extremely thankful to my co-supervisor, Marcelo Ciappina, who, despite the considerable distance between China and Barcelona, was always willing to meet whenever I needed help and guidance. His willingness and helpful character bridged the gap between these two far locations.

Although Maciej Lewenstein was my Master Thesis supervisor back in 2019, during my Ph.D. thesis, he also played this role even when he did not have to, offering his extremely knowledgeable guidance and immediate willingness to meet for any query. These characteristics were also shared by Paris Tzallas, with whom I engaged in numerous discussions, even during weekends. Without him, most of the results in this thesis would not be as interesting to a general audience as I believe they are.

In my perspective, earning a Ph.D. in a scientific field is a transformative journey, where individuals evolve in various dimensions: growing in independence in research, developing a critical mindset—both towards their own scientific results and those of others—and deepening knowledge in their field. If, to some (even infinitesimal) extent, these characteristics apply to me, it is because I had the opportunity to share my Ph.D. journey with individuals who are the quintessence of these qualities. One of the most significant influences on me was Philipp Stammer. Our countless discussions on various topics, a football match or two, and the wealth of knowledge I gained from him contributed immensely to my growth. Despite the geographical distance, Theocharis Lamprou played a fundamental role in my journey, and his experimental contributions became indispensable for my Ph.D. thesis. I eagerly look forward to the day we can meet in person. With Eva González Ruiz, I learned the basics of device-independent quantum key distribution and shared extensive discussions in this regard, learning from her expertise in single-photon sources. I tried to pay her back in the Secret Santa bringing her to the brighter side of the football history, but it seems Barça is not helping too much lately... Patrick Hümbeli, who was one of my teachers of the Machine Learning course at the Master, guided my initial steps in the Ph.D. He imparted me essential concepts of variational quantum algorithms and shared insights on efficient Python coding. Despite the progress, I acknowledge there is still a long journey ahead. With Anna Steffinlongo, I concluded the last steps of my Ph.D., but had the opportunity (together with Raja Yehia) to share some fresh strawberries in the office 372, as well as fruitful discussions about device-independent quantum key distribution.

I also had the privilege of working with an entire legion of postdocs who played a crucial role in enhancing not only my understanding of Physics but also my insights about the scientific world. In this context, Enky Oudot contributed significantly with endless discussions and humor, whether in my office, the meeting room, or during casual encounters in the corridor. He guided me through topics such as nonlocality, quantum key distribution, and entanglement characterization. I still owe him a game at League of Legends, but a bit of practice is in order first. Andy Maxwell, whom I initially met during my Master thesis when he came to ICFO as a visiting postdoc, generously shared his expertise on above-threshold ionization. While I may not accomplish a 5 km run in half an hour after writing this thesis, as I claimed during ICPEAC last summer, with a few weeks of practice, I believe I could manage. Emilio Pisanty laid the foundation for my (albeit brief, for now) scientific career. As the co-supervisor of my Master thesis, he introduced me to the world of strong-field physics, and with him I started my insights onto its quantum optical analysis. This Ph.D. Thesis owes him everything. Despite the brief time together at ICFO, Andrés Ordóñez's critical analysis has been an exceptional example to emulate. I look forward to the possibility of further collaboration in the future. Special thanks to Joe Bowles, my initial cosupervisor on this Ph.D. journey. I gained valuable insights from our discussions on variational quantum circuits and greatly appreciated his immense creativity.

I also want to express my gratitude to the entire Quantum Information Theory group at ICFO, including both current and past members, for the countless enjoyable moments, especially those spent at the *futbolín*. With the group now comprised of many members, it is challenging to mention everyone without the risk of inadvertently omitting someone. However, in addition to those already acknowledged, I extend special thanks to Cristian, Miqueleta (thanks again for revising the abstract in Catalan for this Thesis), Chung Yun, Vinicius, Enrique, Carlos, Dario, Matteo, Paolo, Jacopo,

Korbinian, Gabriel, Marcio, Moha, Felix, Gaël, Luke, Vicky, Máté, Marc-Olivier, Pedro, Donato, Fionnuala, Raffaele, Luca, Neil, Raja, Fede, Mariana and Leo.

Beyond QIT, I am also grateful to have made such good friends as Daniel Gonçalves, whom I have known since the Master in Photonics program, and with whom I have shared numerous travels on the ICFO bus; Javier Arrés, with whom I shared an office during the Master Thesis and engaged in spontaneous talks in the corridors throughout the Ph.D.; Rafael Luque, whom I already knew from our time in Tenerife during the Bachelor, met again during the Master, and once more at ICFO during the Ph.D.; Álvaro Moreno, with whom I shared an apartment during the SURF fellowship at IFISC in Palma de Mallorca in 2018, and later coincided with again at ICFO; and Javier Argüello, from whom I have also learned a lot about quantum simulators. I am also thankful to Francisco, Kasia, and Stefano, with whom I had the opportunity to participate in the Quantum Battles in Attoscience at University College London last summer. While preparing for the battles with them, I truly learned a lot about the attoscience of solids, and hopefully, I will continue to do so in the near future.

Oficialmente, mi doctorado comenzó el 1 de enero de 2020 (como *trainee*). Sin embargo, a mediados de diciembre de 2019, todavía me encontraba en Tenerife y no tenía un lugar donde quedarme en Barcelona. Así que, después de una búsqueda un tanto apresurada desde la isla, terminé encontrando el piso que compartí durante dos años con Henrry y Gilbert en el número 100 de Carrer de Vilamarí.Les agradezco enormemente el magnífico trato que tuvieron conmigo, tratándome como a un hermano pequeño y mostrándome en la práctica por qué Venezuela siempre será la octava isla canaria (con perdón a La Graciosa).

A pesar de no haber pasado mucho tiempo en Tenerife en los últimos años, mis amigos de allá se han encargado de hacer corta la distancia. Aunque quizás sin saberlo, me han brindado mucha ayuda a lo largo de estos años. En especial, me gustaría agradecer enormemente a Fer, con el que poco más y nos conocemos en el hospital al nacer; a Franchy, con el que muchas veces lloré de la risa en los regresos en tranvía de la facultad; y a Imo, por su increíble ingenio y carisma, pero con el que también me he desquiciado al verlo pegar más vueltas alrededor de balón que a Xavi Hernández, en aquellos (ahora un tanto lejanos) partidos que jugábamos todos en la cancha de al lado de la fábrica de Coca-Cola, en la del Sauzal, o en la del Barranco de Santos en Santa Cruz.

También me gustaría agradecer a aquellos profesores que me han contagiado (al menos parte) de su pasión a la Física, además de instruirme en ella. En especial,me gustaría agradecer a Daniel Alonso, el que fuera mi profesor de Física Estadística y tutor de Trabajo de Fin de Grado en la ULL; a Antonía Ruíz, profesora de Termodinámica y Física Estadística; y a Claudio Mirasso, mi supervisor en la beca SURF en el IFISC. Sin sus cartas de recomendación, igual no estaría escribiendo esta tesis.

Mi familia ha sido, desde siempre, un apoyo fundamental en todos los aspectos de mi vida, a pesar de que mi carácter, un tanto apático, haya dado lugar a contadas muestras afectivas y llamadas telefónicas una vez por semana, a lo sumo. Sin embargo, siempre he intentado estar presente en aquellos momentos en que más lo necesitaban, y espero seguir haciéndolo. Aspiro por lo tanto a que,por pequeña que sea, esta pueda considerarse una muestra de afecto por todo lo que me han dado.

Tengo una familia grande, y todos deberían sentirse aludidos, pero me gustaría expresar explicitamente mi gratitud a mi núcleo más cercano, conformado por mi madre Maruchi (doña Madre), mi (súper)tía Almudena, mi abuelo don Paco y mi abuela Mami (a pesar de los años que han pasado, *Mami no me ve cambiado, dice que sigo en mis quince*). En este núcleo también incluyo, como no, a mi hermano Manu y a su pareja Soraya (junto al resto de la tropa), que siempre me han acogido (y les garantizo que seguirán haciendo) en su casa. Quien les iba a decir que, a pesar de poner tanto esfuerzo en que no hubiese ni un pelo de gato en la habitación en la que dormía debido a mi (todavía no oficial pero con numerosos indicios) alergia, sus esfuerzos eran en vano pues los estados gato me los traía yo en la mochila.

En último lugar, aunque igual el más importante, quiero agradecer a mi pareja Nerea, de las mejores personas que he conocido en mi vida. Ella me ha apoyado en las duras y en las más duras, y ha tenido que convivir conmigo, esta tesis y mi adicción al trabajo. A pesar de ello, nos lo hemos pasado genial juntos y espero seguir haciéndolo por siempre. Por todo ello, esta vez los agradecimientos no van a una beca, sino a ti. Ojalá y pronto juntar a las dos *niñas*.

I acknowledge funding from the Secretaria d'Universitats i Recerca del Departament d'Empresa i Coneixement de la Generalitat de Catalunya, the European Social Fund (L'FSE inverteix en el teu futur)–FEDER, the Government of Spain (Severo Ochoa CEX2019-000910-S and TRANQI), Fundació Cellex, Fundació Mir-Puig, Generalitat de Catalunya (CERCA program) and the ERC AdG CERQUTE.

Contents

1 Introduction ... 1
References ... 6

2 Background ... 9
2.1 The Evolution of Light-Matter Interaction Along History ... 9
2.2 Light-Matter Interactions from a Semiclassical Perspective ... 14
2.2.1 A Brief Introduction to Classical Electromagnetism in Vacuum ... 14
2.2.2 Semiclassical Description of Light-Matter Interactions ... 16
2.2.3 The Length Gauge and the Dipole Approximation ... 18
2.2.4 Dynamics of a Free-Electron Interacting with an Electromagnetic Field ... 19
2.3 A Brief Introduction to Quantum Optics ... 21
2.3.1 Quantization of the Electromagnetic Field ... 21
2.3.2 Quantum Optical Observables for Probing Non-classical Light ... 25
2.3.3 A Summary of Different Quantum Optical States and Their Properties ... 35
2.3.4 Quantum Information Measures for Probing Non-classical Light ... 52
2.4 Light-Matter Interactions from a Quantum Optical Perspective ... 58
References ... 61

3 Non-classical States of Light After Strong-Laser Field Processes in Atoms ... 67
3.1 Semiclassical Analysis of Strongly Driven Laser-Atom Interactions ... 69
3.1.1 Dynamics of a Bound Electron Interacting with an Intense Electromagnetic Field ... 71

3.2 Quantum Optics Meets Strong-Field Physics: A State of the Art ... 87
3.2.1 Early (and Not So Early) Work ... 87
3.2.2 First Observations of Quantum Optical Signatures in the Field ... 89
3.2.3 Non-Classical Light in Strong-Field Physics ... 90
3.3 Quantum Optical Description of Above-Threshold Ionization and High-Harmonic Generation Processes in Atoms ... 93
3.3.1 Above-Threshold Ionization ... 94
3.3.2 High-Harmonic Generation ... 113
3.4 Discussion and Outlook ... 139
References ... 142

4 Non-classical States of Light After High-Harmonic Generation in Molecular and Solid Systems ... 151
4.1 High-Harmonic Generation Processes in Hydrogen Molecular Ions ... 153
4.1.1 A Linear Combination of Atomic Orbitals Picture of H_2^+ Molecules ... 154
4.1.2 A Quantum Optical Analysis ... 158
4.2 High-Harmonic Generation Processes in Semiconductor Materials ... 180
4.2.1 An Electron in a Periodic Potential ... 181
4.2.2 A Bloch-Based Semiclassical Description of High-Harmonic Generation ... 186
4.2.3 A Bloch-Based Quantum Optical Description of High-Har-Monic Generation ... 193
4.3 Discussion and Outlook ... 209
References ... 212

5 Non-Classical States of Light for Device-Independent Quantum Key Distribution ... 219
5.1 A Brief Introduction to Quantum Key Distribution ... 221
5.1.1 An Example of a Quantum Key Distribution Protocol: the Ekert 91 Protocol ... 222
5.1.2 Classical Postprocessing Employed on the *Raw* Key ... 225
5.1.3 Device Independent Quantum Key Distribution ... 226
5.2 DIQKD Using Quantum-Dot Based Single-Photon Sources ... 234
5.2.1 The Central Heralding Station Scheme ... 235
5.2.2 Imperfections on the Scheme When Using Quantum Dots as Single-Photon Sources ... 239
5.2.3 DIQKD in the Asymptotic Limit ... 242
5.2.4 DIQKD in the Finite Statistics Regime ... 247

5.3 DIQKD Beyond Qubits 251
5.3.1 Optimizing the Lower Bounds to the Key Rate 251
5.3.2 Upper Bounds on the Key Rate Based on the Convex-Com-Bination Attack 254
5.3.3 Comparison Between the Bounds 257
5.4 Discussion and Outlook 258
References 259

6 Conclusions 263
References 266

7 Additional Material of Chapter 2 267
7.1 Transformation of the Momentum Operator with Respect to $\hat{U}_{\mathrm{VG}}(t)$ 267
7.2 Poynting's Theorem and the Energy Carried by the Electromagnetic Field 268
Reference 269

8 Additional Material of Chapter 3 271
8.1 Numerical Computation of the ATI Wigner Functions 271
8.1.1 The Saddle-Point Approximation 271
8.1.2 Using a Numerical Integrator 272
8.2 Numerical Analysis for the BBO Interaction with Non-classical States of Light 274
References 277

9 Additional Material of Chapter 4 279
9.1 The Variational Principle 279
9.2 Properties of the Many-Body Left and Right-Like Localized States 280
9.3 About the ZnO Band Structure Parameters 282
9.4 A Wannier-Bloch Picture of HHG in Solids 282
9.4.1 Semiclassical Description 283
9.4.2 Quantum Optical Description 286
References 293

10 Additional Material of Chapter 5 295
10.1 Optimization of the Asymptotic Key Rate Using the Convergent Hierarchy on the Von Neumann Entropy 295
10.1.1 SDP Relaxation 295
10.1.2 Parallelization of the Optimization 296
10.2 Specifics About the Finite Size Analysis 296
10.2.1 Theorem 1 in E. Y.-Z. Tan Et Al, *Quantum* 5, 880 (2022) 296
10.3 Parametrization of the POVM Set 299
References 300

Abbreviations

ATI	Above-Threshold Ionization
BBO	Beta-Barium Borate
BCH	Baker-Campbell-Hausdorff
BSV	Bright Squeezed Vacuum
CC	Convex-Combination
CEP	Carrier-Envelope Phase
CGLMP	Collins-Gisin-Linden-Massar-Popescu
CHS	Central Heralding Station
CHSH	Clauser-Horne-Shimony-Holt
DIQKD	Device Independent Quantum Key Distribution
E91	Ekert-91
EAT	Entropy Accumulation Theorem
HATI	High-Order Above-Threshold Ionization
HHG	High-Harmonic Generation
HOM	Hong-Ou-Mandel
HWP	Half-Wave Plate
IR	Infrared
LCAO	Linear Combination of Atomic Orbitals
LIED	Laser-Induced Electron Diffraction
MIR	Mid-Infrared
NIR	Near-Infrared
NPA	Navascués-Pironio-Acín
NSDI	Non-Sequential Double Ionization
PBS	Polarized Beam Splitter
POVM	Positive-Operator Valued Measures
PPT	Positive Partial Transpose
PZW	Power-Zienau-Wolley
QED	Quantum Electrodynamics
QKD	Quantum Key Distribution
QS	Quantum Spectrometer
QWP	Quarter-Wave Plate

SAE	Single-Active-Electron
SATWAP	Salavrakos-Augusiak-Tura-Wittek-Acín-Pironio
SBEs	Semiconductor Bloch Equations
SDP	Semidefinite Programming
SFA	Strong-Field Approximation
SPDC	Spontaneous Parametric Down Conversion
XUV	Extreme-Ultraviolet

Chapter 1
Introduction

There is nothing like looking, if you want to find something. You certainly usually find something, if you look, but is not always quite the something you were after.

– J. R. R. Tolkien, *The Hobbit* (1937)

The early years of the 20th century witnessed the birth of quantum mechanics, in what it is referred nowadays as the *first quantum revolution* [1, 2]. During this period, a series of groundbreaking discoveries fundamentally changed the understanding of matter and energy at the atomic and subatomic levels, profoundly impacting various scientific disciplines and technological advances. Optics, in particular, underwent a significant transformation. During the 18th and 19th century, a vigorous debate took place regarding the corpuscular or wave-like nature of light, with the latter becoming firmly established thanks to the experimental results obtained by Thomas Young, which could not be explained by the corpuscular theory proposed by Sir Isaac Newton. However, in an attempt to describe the spectral radiation emitted by a black body, Max Planck proposed in 1901 that light-matter interactions involved the exchange of discrete packets of energy, rather than a continuous [3]. Later, in 1905, Albert Einstein used the same idea to explain the photoelectric effect [4]. This way, the concept of *photon* arose[1] and the wave-particle debate revived. However, in this case its scope widened to encompass matter as a whole through the famous concept of wave-particle duality, proposed by Louis-Victor de Broglie in 1924 [6].

Beyond Optics, quantum mechanics revolutionized the perception of the world and introduced new mathematical tools for describing it. Concepts such as *quan-*

[1] Although the concept of photon arose due to the contributions of Max Planck and Albert Einstein, the term *photon* was popularized by Gilbert N. Lewis in 1926 [5].

J. Rivera-Dean, *Non-classical States of Light*, Springer Theses,
https://doi.org/10.1007/978-3-031-73769-5_1

tum states and *wavefunctions* became crucial for explaining the properties of atoms, whose dynamics where well described by differential equations such as the well-known Schrödinger equation [7], or by matrix mechanics as proposed by Werner Heisenberg, Max Born and Pascual Jordan [8, 9]. However, like any new and revolutionary theory, quantum mechanics also faced several criticisms and sparked debates, some of which persist to the present. One of the most famous criticisms, which continues to influence the development of quantum theory, is the Einstein-Podolski-Rosen (EPR) paradox,[2] a thought experiment formulated by Albert Einstein, Boris Podolski and Nathan Rosen in 1935 in their paper *Can Quantum-Mechanical description of physical reality be considered complete?* [10]. In this work, Einstein, Podolski and Rosen demonstrated that the quantum formalism allowed for the existence of certain two-particle states—later on referred to as *entangled states* [11]—in which strong correlations could exist between non-commuting quantities, such as position and momentum [12], even when the particles are separated by considerable distances. Einstein famously referred to this phenomenon as "*spooky action at a distance*" expressing his discomfort with the implication in quantum mechanics that measurements on one entangled particle could affect the properties of another distant particle. His unease stemmed from the apparent nonlocal correlations and the departure from classical intuitions about the behavior of particles, challenging the classical notion that particle properties should have well-defined values independent of measurements. As the title of their work suggests, the EPR paradox questioned whether quantum mechanics provides a complete description of reality, or if there are additional *hidden variables* that determine the properties of particles independently of the measurements performed by the observers.

The question raised by the EPR paper remained on air until 1964, when John S. Bell presented what is now referred to as Bell's theorem. This theorem demonstrates the limitations of any theory attempting to explain the correlations observed in entangled systems using hidden variables while maintaining both locality—absence of faster-than-light influences—and realism—the assumption that physical properties exist independently of measurements [13]. In his work, John S. Bell considered a scenario similar to the one proposed by Einstein and his colleagues. He derived an inequality, now known as *Bell inequality*, that acts as a test: if the observed correlations between measurements on entangled particles violate this inequality, then the predictions of a hidden-variable theory are inconsistent with those of quantum mechanics.

Bell's discovery had profound practical consequences as it provided a means to design experiments, referred to as *Bell test experiments*, that can directly test the predictions of quantum mechanics involving entangled particles against those of classical hidden-variable theories. In this regard, the field of Photonics, the science and technology of generating, controlling and detecting photons, has played a fundamental role. This is due to the unique properties of light: great control and

[2] Although typically referred to as a paradox, the work by Einstein, Podolski and Rosen did not present any logical inconsistency, but raised important questions about the nature of reality and of measurement in quantum mechanics.

manipulation, suitability for entanglement generation, ease in detection by means of photodetectors and the possibility of performing experiments at long distances. By using light, Alain Aspect and his coworkers obtained strong violations of Bell's inequality in 1982 using photons entangled in polarization, which were emitted in nonlinear processes of radiative cascade of calcium [14, 15]. Although these experiments initially took place at distances of twelve meters, similar setups allowed for obtaining Bell inequality violations at separations of hundreds or even thousands of meters [16–18]. Moreover, these significant contributions to the scientific community were recently recognized with the 2022 Nobel Prize in Physics to Alain Aspect, John F. Clauser and Anton Zeilinger [19].

Understanding and harnessing the properties of entangled systems can lead to groundbreaking discoveries in information processing, communication and computation [20], which form the core of Quantum Information Science. In recent years, the applications of quantum information science have started to flourish, giving rise to what some refer as the *second quantum revolution* [1, 2]. One of its most prominent examples, which will be discussed in this Thesis, is quantum communication. Here, the unique properties of entangled states enable the establishment of provably secure communication between distant parties. And, as exemplified by the works of A. Aspect and his colleagues, photonic-based systems play a pivotal role in their implementation. However, to do so, *non-classical states* of light are needed, that is, states of light whose properties cannot be explained by classical theories, similar to the sources used in the previously mentioned Bell test experiments. Therefore, exploring methods to generate non-classical states of light is essential for the advancement of photonic-based quantum information science applications.

This Thesis is dedicated to the study of the generation of non-classical states of light and their use in quantum communication. On one hand, we investigate how interactions between high-intensity lasers and matter systems can result in the generation of non-classical states of light, which we analyze by treating light from a quantum mechanical perspective. On the other hand, we explore how non-classical states of light can be utilized for conducting device-independent quantum communication, meaning that the security of the protocol is not contingent on the specific manner in which the experiment is conducted.

Motivation and Results

After establishing a comprehensive background in Chap. 2, this Thesis is structured into two primary parts. The first encompasses Chaps. 3 and 4, delving into the creation of non-classical light states following intense laser-matter interactions. The second part, encapsulated in Chap. 5, centers on the scrutiny of (device-independent) quantum key distribution. It explores the potential advancements through photonic-based implementations in this field.

Non-classical States of Light After Strong-Laser Field Processes

Strong-laser field physics studies light-matter interactions under regimes of extremely high intensity—high enough to become comparable to the binding forces that keep electrons bound to their corresponding nuclei [21]. Beyond unveiling captivating Physics within exotic ionization regimes [22], strong-laser field physics has laid much of the groundwork for attosecond science [23, 24], a groundbreaking field of Physics whose merits have been recently recognized with the Nobel Prize in Physics 2023 [25]. However, much of the theoretical groundwork aimed at replicating and comprehending experimental outcomes has operated within a semiclassical framework, where the matter system is treated quantum mechanically while light is handled classically [26]. This leaves untapped the potential of strong-field physics within quantum optics, a rich domain ripe with the tools necessary for quantum information science applications. Bridging strong-field physics and quantum optics constitutes the main motivation of this first part of the Thesis, by analyzing strong-field processes within a quantum optical framework where both matter and light are approached from a quantum mechanical standpoint. The focus is to explore how strong-field physics contribute to generating non-classical states of light.

Our Contribution

We explore diverse strong-field processes and their potential for generating non-classical states. Chapter 3 delves into the interaction between strong-laser fields and atomic systems, focusing on two prominent processes: above-threshold ionization and high-harmonic generation. Above-threshold ionization involves the ionization of an electron by absorbing more energy than necessary to leave the atom, due to the strong-field interaction. Our analysis from a quantum optical perspective [27–29], reveals entanglement between the ionized electron and the light, stemming from the electron's post-ionization properties and their influence on the field.

Conversely, high-harmonic generation involves the laser-ionized electron recombining with its parent ion and emitting high-frequency radiation. We demonstrate, both theoretically and experimentally based on Refs. [27, 29–32], the generation of non-classical states of light with high photon numbers and along a wide range of frequencies through this process. We show theoretically and experimentally [33], that these states hold potential for driving other nonlinear optical processes. Moreover, we present theoretical techniques utilizing linear optical elements to generalize the obtained non-classical states [34].

In Chap. 4, we extend our analysis to the interaction of strong-fields with more complex systems, namely diatomic molecules [35] and semiconductor materials [36, 37]. From a semiclassical perspective, these exhibit richer electron dynamics that manifest themselves in the measured observables. We uncover, through theoretical analysis, how these dynamics contribute to diverse and distinctive non-classical features in the resulting quantum optical states, distinguishing them from atomic systems.

Non-classical States of Light for Quantum Key Distribution

When two parties seek to communicate confidentially, a strategy they can follow is to generate a secret object—a key—that they use to encrypt and decrypt the shared messages, ensuring the security of their exchange. Quantum key distribution leverages the principles of quantum mechanics to create keys that are provably secure [38]. However, conventional quantum key distribution relies on assumptions about the devices used by the communicating parties to generate the key. While these assumptions intend to safeguard the key's secrecy, they might be exploited by an eavesdropper to gain knowledge about the key [39]. To address this vulnerability, device-independent quantum key distribution emerges as a critical approach [40], operating under the assumption that both communicating parties distrust the devices employed for key generation [40]. Nevertheless, its implementation demands stringent practical requirements, pushing current technology to its limits [41, 42]. Understanding and seeking ways to minimize the experimental requirements for the implementation of device-independent quantum key distribution constitute the motivation of this second part of the Thesis.

Our Contribution

Firstly, based on Ref. [43], we delve into the feasibility of realistic photonic-based implementations utilizing quantum-dot-based non-classical state sources for establishing a device-independent quantum key distribution protocol. We scrutinize the secrecy of the generated key against various imperfections in the sources used by the communicating parties, assessing two distinct regimes: one where the parties generate infinitely large keys, and a more realistic scenario with finite-size keys.

Secondly, in a theoretical investigation abstracted from specific physical implementations that is based on Ref. [44], we analyze the potential advantage of varying the dimensionality of the non-classical states used by the parties to generate a secret key. This study involves analyzing both upper and lower bounds to the key rate—the rate at which the parties generate secret key information—and how these behave as the dimension increases. The differences between these bounds offer insights into noise requirements and potential enhancements feasible in device-independent quantum key distribution scenarios.

Notation Used in this Thesis

Throughout this thesis, the first-person plural pronoun *we* will be employed. This choice is motivated by two primary reasons: firstly, to recognize the collaborative essence inherent in scientific research; and secondly, to engage the reader as an active participant in this academic journey. By using *we*, the aim is to guide and involve the reader in this shared exploration, fostering a deeper understanding and connection with the research process.

Apart from this, other important notation to bear in mind is the following:

- Mathematical quantities expressed in **bold** represent vectorial quantities unless otherwise is explicitly stated.
- Mathematical quantities represented with a hat ($\hat{}$) denote operators.
- The use of SI units and atomic units (represented with a.u.) is used interchangeably and for convenience, specially in Chaps. 3 and 4. Atomic units is achieved by setting to one the main physical constants related to the electron dynamics. More specifically

$$\hbar = 1, \quad m_{\mathsf{e}} = 1, \quad |\mathsf{e}| = 1, \quad \text{and} \quad k_C = \frac{1}{4\pi\varepsilon_0} = 1,$$

where m_{e} represents the electron's mass, e its charge, k_C the Coulomb's constant and ε_0 the vacuum electric permittivity.
These units are used indistinctively mostly when referring to laser field intensities (denoted with I) and frequencies (denoted with ω). For these, the following conversion factors between SI units and atomic units become particularly useful

$$\omega = \frac{45.6 \text{ nm}}{\lambda} \text{ a.u.} \quad \text{and} \quad E = 0.053\sqrt{\frac{I}{10^{14}\text{W}/\text{cm}^2}} \text{ a.u.},$$

where λ denotes wavelength and E the laser electric field peak strength.

References

1. MacFarlane AGJ, Dowling JP, Milburn GJ (2003) Philos Trans R Soc London Ser Math Phys Eng Sci 361:1655. https://doi.org/10.1098/rsta.2003.1227
2. Aspect A (2023) Photonic Quantum Technologies. In: Benyoucef M (ed). Wiley-VCH GmbH, Weinheim, pp 9–30. Chap. 2. https://doi.org/10.1002/9783527837427.ch2
3. Planck M (1901) Annalen der Physik 309:553. https://doi.org/10.1002/andp.19013090310
4. Einstein A (1905) Annalen der Physik 322:891. https://doi.org/10.1002/andp.19053221004
5. Lewis GN (1926) Nature 118:874. https://doi.org/10.1038/118874a0
6. Broglie LD (1925) Annales de Physique 10:22. https://doi.org/10.1051/anphys/192510030022
7. Schrödinger E (1926) Phys Rev 28:1049. https://doi.org/10.1103/PhysRev.28.1049
8. Heisenberg W (1925) Zeitschrift für Physik 33:879. https://doi.org/10.1007/BF01328377

9. Born M, Jordan P (1925) Zeitschrift für Physik 34:858. https://doi.org/10.1007/BF01328531
10. Einstein A, Podolsky B, Rosen N (1935) Phys Rev 47:777. https://doi.org/10.1103/PhysRev.47.777
11. Schrödinger E (1935) Naturwissenschaften 23:844. https://doi.org/10.1007/BF01491987
12. Heisenberg W (1927) Zeitschrift für Physik 43:172. https://doi.org/10.1007/BF01397280
13. Bell JS (1964) Physics Physique Fizika 1:195. https://doi.org/10.1103/PhysicsPhysiqueFizika.1.195
14. Aspect A, Grangier P, Roger G (1982) Phys Rev Lett 49:91. https://doi.org/10.1103/PhysRevLett.49.91
15. Aspect A, Dalibard J, Roger G (1982) Phys Rev Lett 49:1804. https://doi.org/10.1103/PhysRevLett.49.1804
16. Tapster PR, Rarity JG, Owens PCM (1994) Phys Rev Lett 73:1923. https://doi.org/10.1103/PhysRevLett.73.1923
17. Weihs G, Jennewein T, Simon C, Weinfurter H, Zeilinger A (1998) Phys Rev Lett 81:5039. https://doi.org/10.1103/PhysRevLett.81.5039
18. Tittel W, Brendel J, Zbinden H, Gisin N (1998) Phys Rev Lett 81:3563. https://doi.org/10.1103/PhysRevLett.81.3563
19. Nobel Prize Outreach AB 2023 (2023) The Nobel prize in physics 2022. https://www.nobelprize.org/prizes/physics/2022/prize-announcement/. Accessed 1st Sept 2023
20. Nielsen MA, Chuang IL (2010) Quantum computation and quantum information: 10th, anniversary. Cambridge University Press
21. Keldysh LV (1965) Sov Phys JETP 20:1307. http://www.jetp.ac.ru/cgi-bin/e/index/e/20/5/p1307?a=list
22. Protopapas M, Keitel CH, Knight PL (1997) Rep Prog Phys 60:389. https://doi.org/10.1088/0034-4885/60/4/001
23. Corkum PB, Krausz F (2007) Nat Phys 3:381. https://doi.org/10.1038/nphys620
24. Krausz F, Ivanov M (2009) Rev Mod Phys 81:163. https://doi.org/10.1103/RevModPhys.81.163
25. Nobel Prize Outreach AB 2024 (2023) The Nobel prize in physics 2023. https://www.nobelprize.org/prizes/physics/2023/summary/. Accessed 4th Oct 2023
26. Amini K, Biegert J, Calegari F, Chacón A, Ciappina MF, Dauphin A, Efimov DK, de Morisson Faria CF, Giergiel K, Gniewek P, Landsman AS, Lesiuk M, Mandrysz M, Maxwell AS, Moszyński R, Ortmann L, Pérez-Hernández JA, Picón A, Pisanty E, Prauzner-Bechcicki J, Sacha K, Suárez N, Zaïr A, Zakrzewski J, Lewenstein M (2019) Rep Prog Phys 82:116001. https://doi.org/10.1088/1361-6633/ab2bb1
27. Rivera-Dean J, Lamprou T, Pisanty E, Stammer P, Ordóñez AF, Maxwell AS, Ciappina MF, Lewenstein M, Tzallas P (2022) Phys Rev A 105:033714. https://doi.org/10.1103/PhysRevA.105.033714
28. Rivera-Dean J, Stammer P, Maxwell AS, Lamprou T, Tzallas P, Lewenstein M, Ciappina MF (2022) Phys Rev A 106:063705. https://doi.org/10.1103/PhysRevA.106.063705
29. Stammer P, Rivera-Dean J, Maxwell A, Lamprou T, Ordóñez A, Ciappina MF, Tzallas P, Lewenstein M (2023) PRX Quantum 4:010201. https://doi.org/10.1103/PRXQuantum.4.010201
30. Lewenstein M, Ciappina MF, Pisanty E, Rivera-Dean J, Stammer P, Lamprou T, Tzallas P (2021) Nat Phys 17:1104. https://doi.org/10.1038/s41567-021-01317-w
31. Stammer P, Rivera-Dean J, Lamprou T, Pisanty E, Ciappina MF, Tzallas P, Lewenstein M (2022) Phys Rev Lett 128:123603. https://doi.org/10.1103/PhysRevLett.128.123603
32. Bhattacharya U, Lamprou T, Maxwell AS, Ordóñez A, Pisanty E, Rivera-Dean J, Stammer P, Ciappina MF, Lewenstein M, Tzallas P (2023) Rep Prog Phys 86:094401. https://doi.org/10.1088/1361-6633/acea31
33. Lamprou T, Rivera-Dean J, Stammer P, Lewenstein M, Tzallas P (2023) Nonlinear optics using intense optical Schrödinger "cat" states, arXiv:2306.14480 [physics, physics:quant-ph]. https://doi.org/10.48550/arXiv.2306.14480
34. Rivera-Dean J, Stammer P, Pisanty E, Lamprou T, Tzallas P, Lewenstein M, Ciappina MF (2021) J Comput Electron 20:2111. https://doi.org/10.1007/s10825-021-01789-2

35. Rivera-Dean J, Stammer P, Maxwell AS, Lamprou T, Pisanty E, Tzallas P, Lewenstein M, Ciappina MF (2024) Phys Rev A 109:033706. https://doi.org/10.1103/PhysRevA.109.033706
36. Rivera-Dean J, Stammer P, Maxwell AS, Lamprou T, Ordóñez AF, Pisanty E, Tzallas P, Lewenstein M, Ciappina MF (2023) Entanglement and non-classical states of light in a strong-laser driven solid-state system. arXiv:2211.00033 [physics, physics:quant-ph]. https://doi.org/10.48550/arXiv.2211.00033
37. Rivera-Dean J, Stammer P, Maxwell AS, Lamprou T, Ordóñez AF, Pisanty E, Tzallas P, Lewenstein M, Ciappina MF (2024) Nonclassical states of light after high-harmonic generation in semiconductors: a Bloch-based perspective. https://doi.org/10.1103/PhysRevB.109.035203
38. Wolf R (2021) Quantum key distribution: an introduction with exercises. In: Wolf R (ed) Lecture notes in physics. Springer International Publishing, Cham, pp 1–12. https://doi.org/10.1007/978-3-030-73991-1_1
39. Lydersen L, Wiechers C, Wittmann C, Elser D, Skaar J, Makarov V (2010) Nat Photonics 4:686. https://doi.org/10.1038/nphoton.2010.214
40. Acín A, Brunner N, Gisin N, Massar S, Pironio S, Scarani V (2007) Phys Rev Lett 98:230501. https://doi.org/10.1103/PhysRevLett.98.230501
41. Nadlinger DP, Drmota P, Nichol BC, Araneda G, Main D, Srinivas R, Lucas DM, Ballance CJ, Ivanov K, Tan EYZ, Sekatski P, Urbanke RL, Renner R, Sangouard N, Bancal JD (2022) Nature 607:682. https://doi.org/10.1038/s41586-022-04941-5
42. Zhang W, van Leent T, Redeker K, Garthoff R, Schwonnek R, Fertig F, Eppelt S, Rosenfeld W, Scarani V, Lim CCW, Weinfurter H (2022) Nature 607:687. https://doi.org/10.1038/s41586-022-04891-y
43. González-Ruiz EM, Rivera-Dean J, Cenni MFB, Sørensen AS, Acín A, Oudot E (2024) Opt Express 32:13181. https://doi.org/10.1364/OE.497935
44. Rivera-Dean J, Steffinlongo A, Parker-Sánchez N, Acín A, Oudot E (2024c) Device-independent quantum key distribution beyond qubits. arXiv:2402.00161 [quant-ph]. https://doi.org/10.48550/arXiv.2402.00161

Chapter 2
Background

"Journey before Destination."

Brandon Sanderson, The Way of Kings (2010).

This chapter delves into the concept of light-matter interactions, tracing their historical evolution from ancient civilizations to their modern mathematical description. It serves as a foundational cornerstone upon which the subsequent chapters of this Thesis are constructed. Within this chapter, we provide a comprehensive mathematical framework for understanding the interaction between light and matter, together with a brief introduction to quantum optics and some relevant quantum information concepts. Furthermore, the historical context provided in this chapter highlights the remarkable evolution of this field over the past century, particularly with the advent of quantum mechanics. Quantum mechanics has played a pivotal role in the development of many applications, some of which are discussed in subsequent chapters of this Thesis.

2.1 The Evolution of Light-Matter Interaction Along History

The concept of light-matter interaction has undergone significant evolution over history, closely intertwined with the conceptual understandings about the nature of light and matter. In ancient civilizations, this knowledge was practically oriented, with the macroscopic consequences of light-matter interaction being applied in daily

J. Rivera-Dean, *Non-classical States of Light*, Springer Theses,
https://doi.org/10.1007/978-3-031-73769-5_2

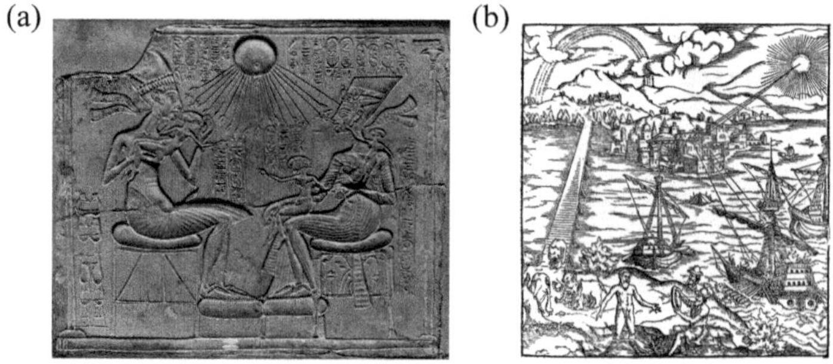

Fig. 2.1 Illustrations depicting the concept of light-matter interactions in Egyptian and Islamic civilizations. In **a**, a relief depicts Pharaoh Akhenaten and his wife Nefertiti, with three of their daughters basking the rays of Aten [1]. During Akhenaten's reign, Aten was elevated to the status of supreme god. In this relief, the rays of Aten bestow Ankh symbols upon them, symbolizing various aspects of life including physical life, eternal life, immortality, death and reincarnation. In **b**, an engraving from the title page of *Opticae Thesaurus*, a latin edition of the *Book of Optics* by Alhazen [2], illustrates different consequences of light-matter interactions from a macroscopic perspective. These include phenomena such as rainbows, the use of parabolic mirrors to focus and ignite ships, reflection effects on the buildings' windows and on mirrors, and refraction effects in water causing distorted images

life and serving as a focal point in many philosophical discussions. In the Egyptian culture, numerous examples of practical uses of light-matter interaction can be found in architecture and art. Egyptians utilized polished metal surfaces, such as copper and bronze, to create mirrors and other reflective surfaces. They also designed buildings, temples and tombs with specific orientations to make use of natural sunlight for illumination and create dramatic lightning effects. Moreover, in some of their artistic illustrations, the Sun was represented with rays, as shown in Fig. 2.1a. This concept later played a pivotal role in the development of geometrical optics. However, in their case, the symbolism of rays had a religious connotation, symbolizing the power of their god Aten, whose rays were believed to descend from the sky to the Earth, enabling the development of life.

The concept of rays was further developed by the ancient Greeks, leading to fruitful philosophical and speculative discussions about light and vision, which laid the groundwork for future work. For instance, Empedocles (c. 494–c. 434 BC) proposed that the eyes emitted rays which interacted with objects, allowing for the process of vision. These ideas were later incorporated by Euclid (c. 325–c. 265 BC) in his treatise *Optics*, where he elaborated geometrical discussions on optics and vision, including significant principles such as the equality between the angles of incidence and reflection. Ptolomy (c. 100–c. 170 AD) further expanded upon Euclid's mathematical treatment in his own book *Optics*, contributing to the understanding of colors, the description of optical instruments, and the concept of refraction.

While contrasted with empirical observations, much of the work developed by the ancient Greek scholars leaned toward deductive and theoretical reasoning. This approach underwent a significant transformation during the Islamic Golden Age, spanning roughly from the 8th to the 14th century. During this period, Islamic scholars placed great emphasis on empirical and detailed observation, experimentation, systematic recording of their results, as well as the development of empirically validated mathematical formulations. One of the most influential figures in the field of Optics during this period was Ḥasan Ibn al-Haytham (c. 965–c. 1040 AD), whose name was Latinized as Alhazen. In his monumental work *Kitāb al-Manāẓir* (Latinized as *Book of Optics*), Alhazen provided a comprehensive study of light, vision and optics, covering various phenomena such as reflection on surfaces, refraction when transitioning to media with different refractive indices, the properties of lenses and mirrors, and the formation of images within the eye. All of these phenomena are pictorially summarized in the book cover, shown in Fig. 2.1b. Another notable figure in the context of optics, and a precursor of Alhazen, was Ibn Sahl (c. 940–c. 1000 AD). In his work *Fī al-'āla al-muḥriqa* (Latinized as *On the burning instruments*), Ibn Sahl derived a general formula for the angles of incidence and refraction through different media, similar to the formula derived by Willebrord Snellious around six hundred years later, which is nowadays known as Snell's law.

The systematic recording during the Islamic Golden Age bore fruit during the Renaissance, which roughly spanned from the 14th to the 17th century. Access to translated versions of the ancient Greek and Islamic documentations became more widely available in Europe during this period.[1] As a result, the Renaissance marked a period of intense inquiry into optics, with scholars delving into the behavior of light, the nature of vision and the properties of lenses. Regarding this last point, Johannes Kepler (1571–1630) conducted research in optics and visual perception, laying the foundation for understanding how lenses focus light to form images. This work was crucial for the development and refinement of optical instruments like eyeglasses, microscopes and telescopes. Another important figure who significantly influenced the development of optics was René Descartes (1596–1650). In his work *Dioptrics*, the French philosopher suggested that light was composed of tiny particles that were emitted from a light source and traveled in straight lines, interacting with matter to produce sensations of color and brightness to the eyes, and undergoing the processes of reflection and refraction.

While Descartes' corpuscular theory faced challenges, it later initiated one of the most important debates in the field of optics: the debate between the corpuscular and the wave theory of light. In 1690, Christiaan Huygens (1629–1695) published his work *Traité de la lumière*, in which he proposed that light propagates through a medium in the form of waves, similar to the way sound travels through air. Here, Huygens introduced what is now known as Huygens-Fresnel principle, which states that each point in the wavefront acts as a source of secondary spherical waves, known as *wavelets*, collectively forming the new wavefront. This wave picture of light developed by Huygens not only allowed him to explain phenomena such as reflection and refraction, but also diffraction and, later on, interference effects. On the other side of the debate and as a supporter of a corpuscular theory of light,

[1] During the Middle Age in Europe, the access to these documents was controlled by religious institutions, and only accessible to a particular sector of the society.

Sir Isaac Newton (1642–1727) presented his ideas about the nature of light in his work *Opticks*, dated to the year 1704. Similar to Descartes, he suggested that light was composed of small corpuscles emitted from luminous bodies, which traveled in straight lines until they encountered a new medium, causing these corpuscles to change their velocity. This model allowed him to explain the phenomena of reflection and refraction. Furthermore, Newton attributed the different colors to corpuscles with varying properties, which made them behave differently under diffraction, explaining why a beam of white light split into different colors when traveling through a prism.

The debate about the nature of light continued throughout the 18th and 19th centuries, coinciding with a period when the use of reason, empiricism and the pursuit of knowledge was strongly emphasized. As a result, these centuries witnessed a notable increase in the publication of optics texts, contributing to the dissemination of knowledge about light and its interaction with matter. However, in the early years of the 19th century, the wave theory of light made a profound statement in support of this perspective. In 1803, Thomas Young (1773–1829) described what we now know as the double-slit experiment. He made light pass through two closely spaced slits, and observed a series of bright and dark bands on a screen placed beyond the two slits, known as an interference pattern. The presence of this pattern could not be described by Newton's corpuscular theory, and required a wave picture, where the bright and dark bands corresponded to the constructive and destructive interference of waves, respectively. Other experiments in 1809, following Rasmus Bartholin's (1625–1698) observation of double refraction in calcite crystals, lead Étienne-Louis Malus (1775–1812) to formulate what is currently known as Malus' law. This law quantitatively described how the intensity of one the rays obtained by birefringence gets modified when passing through a polarizing filter, accounting for the effects of polarization. Later on, between 1810 and 1820, Augustin-Jean Fresnel (1788–1827) made significant contributions to the wave picture of light. He specifically showed that the phenomena of polarization were indeed consistent with the wave picture whenever the periodic vibration of light was transverse to its propagation direction.

Meanwhile, as the balance of the debate between corpuscular and wave theory was shifting towards the latter, the field of electromagnetism began to flourish. In 1820, Hans Christian Ørsted (1777–1851) accidentally discovered that an electric current could deflect a magnetic needle from its original position, establishing one of the first connections between electricity and magnetism. In a similar direction, Michael Faraday's (1791–1867) experiments in 1831 with electromagnetic induction demonstrated that a changing magnetic field could induce an electric current in a nearby conductor. The mathematical framework for explaining these observations was provided between 1820 and 1840 by André-Marie Ampère (1775–1836), who described the behavior of electric currents and their interactions with magnetic fields. However, the major breakthrough in the field of electromagnetism was made in 1865 by James Clerk Maxwell (1831–1879) in his publication *A Dynamical Theory of the Electromagnetic Field*. In this work, Maxwell set forth the proposition that light was, in fact, a transverse electromagnetic wave, and characterized its propagation in space by using the now famous Maxwell's equations, unifying electricity and magnetism.

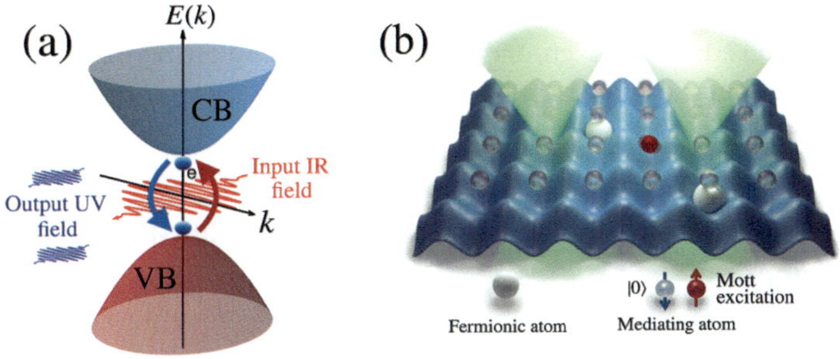

Fig. 2.2 Pictorial representations of modern applications of light-matter interaction. In **a**, an intense light pulse (in red) is employed to excite an electron between different energy levels in a solid sample to drive a highly nonlinear process known as high-harmonic generation. This process is used to generate light with shorter frequencies (in blue). In **b**, a figure adapted from Ref. [3] (reproduced with permission from Springer Nature and the authors), light serves as optical tweezers to precisely position atoms in specific locations, enabling the simulation of many-body quantum mechanical systems, applicable for quantum chemistry studies

These theoretical results by Maxwell were confirmed later on by Heinrich Rudolf Hertz (1857–1894) between 1880 and 1890, by generating and detecting radio waves (Fig. 2.2).

With the advent of quantum mechanics and special relativity in the early 20th century, the concept of light-matter interaction underwent a profound transformation. While earlier analyses had mainly focused on macroscopic aspects of this interaction, the introduction of concepts such as the photon (briefly discussed in Chap. 1) and Niels Bohr's (1885–1962) quantization of the atom [4] ushered in a new microscopic understanding of how light and matter interacted. Noteworthy discoveries by scientists like Pieter Zeeman (1865–1943) in 1896 and Johannes Stark (1874–1954) in 1913, demonstrated that electric and magnetic fields could have non trivial effects on the energy levels of atoms [5, 6]. Additionally, the development of groundbreaking technologies such as the laser [7] by Theodore H. Maiman (1927–2007) in 1960, and earlier the maser [8, 9], paved the way for unprecedented advancements in both technology and scientific understanding. These developments solidified light-matter interactions as an established branch of physics.

Today, light-matter interactions are actively used in a wide array of fields, particularly in physics. To name a few, they are instrumental in generating ultrafast high-energy sources [10, 11], probing and controlling electrons [12, 13], in the analysis of molecular [14, 15] and solid-state systems structure [16, 17] as well as for cooling and trapping atoms at low temperatures [18]. The latter application allows for the simulation of quantum many-body systems [19], useful for quantum technology applications [20].

Within the context of this Thesis, our focus is on the analysis of light-matter interactions for the purpose of generating non-classical states of light. As introduced in Chap. 1, non-classical states of light are fundamental for the development of photonic-based quantum information science applications. With this objective in mind, the following sections provide a comprehensive analysis of the theory of light-matter interactions, laying the groundwork and providing the necessary tools for the subsequent chapters.

2.2 Light-Matter Interactions from a Semiclassical Perspective

In this section, we initiate the analysis of light-matter interactions from a semi-classical perspective. By *semiclassical*, we are referring to a description where the electromagnetic field is characterized using classical electromagnetism, while the matter system is examined from a quantum mechanical standpoint. To achieve this objective, we commence by reviewing the fundamentlas of electromagnetism. Subsequently, we delve into the description of light-matter interactions, introducing certain approximations that are used in the upcoming chapters of this Thesis.

2.2.1 A Brief Introduction to Classical Electromagnetism in Vacuum

One of the primary approaches to commence the study of the electromagnetic field and its interactions with matter involves considering Maxwell's equations and the Lorentz force as a foundational set of equations. When examining the electric $\mathbf{E}(\mathbf{r}, t)$ and magnetic fields $\mathbf{B}(\mathbf{r}, t)$ in vacuum, i.e. in the absence of charges, the four Maxwell's equations can be expressed as

$$\nabla \mathbf{E}(\mathbf{r}, t) = 0, \tag{2.1}$$

$$\nabla \mathbf{B}(\mathbf{r}, t) = 0, \tag{2.2}$$

$$\nabla \times \mathbf{E}(\mathbf{r}, t) = -\frac{\partial \mathbf{B}(\mathbf{r}, t)}{\partial t}, \tag{2.3}$$

$$\nabla \times \mathbf{B}(\mathbf{r}, t) = \frac{1}{c^2}\frac{\partial \mathbf{E}(\mathbf{r}, t)}{\partial t}, \tag{2.4}$$

where in these expressions $\mathbf{r}$ denotes position, t stands for time and c represents the speed of light in vacuum. Under this scenario, we see from Eqs. (2.1) and (2.2) that there are no longitudinal components of the electric and magnetic fields, respectively.

For the description of electromagnetic field radiation, among other aspects, it is particularly suitable to introduce the scalar $\phi(\mathbf{r}, t)$ and vector $\mathbf{A}(\mathbf{r}, t)$ potentials. These are mathematical functions with no direct physical meaning, but that serve as a convenient way to express the electric and magnetic fields. They are related to these fields by

$$\mathbf{B}(\mathbf{r}, t) = \nabla \times \mathbf{A}(\mathbf{r}, t), \tag{2.5}$$

$$\mathbf{E}(\mathbf{r}, t) = -\frac{\partial \mathbf{A}(\mathbf{r}, t)}{\partial t} - \nabla \phi(\mathbf{r}, t). \tag{2.6}$$

An important point to highlight is that the introduced potentials jointly characterize the electric and magnetic field excitations, collectively introducing the concept of electromagnetic field. However, it is crucial to state that the definition of these potentials is not unique. For instance, given a function $f(\mathbf{r}, t)$ commonly referred to as *gauge function*, the substitutions $\mathbf{A}(\mathbf{r}, t) \to \mathbf{A}(\mathbf{r}, t) + \nabla f(\mathbf{r}, t)$ and $\phi(\mathbf{r}, t) \to \phi(\mathbf{r}, t) - \partial f(\mathbf{r}, t)/\partial t$ in Eqs. (2.5) and (2.6) leave the relations between the fields and the potentials unchanged. This invariance of the equations regarding the definition of the scalar and vector potentials is known as *gauge invariance*, whilst transformations previously mentioned are called *gauge transformations*. This freedom in the definition of scalar and vector potentials makes the approach particularly useful, as one can select them properly in order to simplify further equations. Here, we work with what is known as *Coulomb gauge* or *transverse gauge*, which imposes $\nabla \mathbf{A}(\mathbf{r}, t) = 0$, and leads to a particularly elegant form of the equation describing the propagation of the electromagnetic field. One of the implications of using this gauge in vacuum is that $\phi(\mathbf{r}, t) = 0$, which is obtained when combining the Coulomb gauge definition with Eqs. (2.1) and (2.6).

With these tools in place, we proceed to derive and solve the equation describing the propagation of the electromagnetic field in vacuum. To do so, we first introduce the definition of the electric and magnetic fields in terms of $\mathbf{A}(\mathbf{r}, t)$, as given in Eqs. (2.5) and (2.6), into Eq. (2.4)

$$\nabla \times \left[\nabla \times \mathbf{A}(\mathbf{r}, t)\right] = -\frac{1}{c^2}\frac{\partial \mathbf{A}(\mathbf{r}, t)}{\partial t}, \tag{2.7}$$

and taking into account that $\nabla \times [\nabla \times \mathbf{A}(\mathbf{r}, t)] = \nabla[\nabla \mathbf{A}(\mathbf{r}, t)] - \nabla^2 \mathbf{A}(\mathbf{r}, t)$, together with the Coulomb gauge condition, we arrive at

$$\nabla^2 \mathbf{A}(\mathbf{r}, t) - \frac{1}{c^2}\frac{\partial \mathbf{A}(\mathbf{r}, t)}{\partial t} = 0, \tag{2.8}$$

which indicates that the vector potential, and consequently the electromagnetic field due to Eqs. (2.5) and (2.6), propagates in free space as a wave, with its propagation direction perpendicular to the oscillation direction. To solve this equation, we model space as a cubic box of sizes of length L while applying periodic boundary conditions on the vector potential, i.e., $\mathbf{A}(\mathbf{r}, t) = \mathbf{A}(\mathbf{r} + \mathbf{L}, t)$. Under these conditions, we can obtain the solution to Eq. (2.8) by the separation of variables method, leading to

$$\mathbf{A}(\mathbf{r}, t) = \sum_{\mathbf{k},\mu} \epsilon_{\mathbf{k},\mu} \alpha_{\mathbf{k},\mu} e^{i(\mathbf{k}\cdot\mathbf{r} - \omega_{\mathbf{k}} t)} + \text{c.c.}, \tag{2.9}$$

where $\omega_{\mathbf{k}} = c\,|\mathbf{k}|$, $\boldsymbol{\epsilon}_{\mathbf{k},\mu}$ is a unit vector describing the polarization direction μ of wavemode $\mathbf{k}$, satisfying $\boldsymbol{\epsilon}_{\mathbf{k},\mu} \cdot \boldsymbol{\epsilon}_{\mathbf{k},\mu'} = \delta_{\mu,\mu'}$; and $\alpha_{\mathbf{k},\mu}$ represents the amplitude of the corresponding mode. Additionally, due to the periodic boundary conditions we have imposed, only certain modes $\mathbf{k}$ are allowed within our box. Specifically, these modes satisfy $k_i = \frac{2\pi}{L} m_i$, with $i = x, y, z$ and $m_i \in \mathbb{Z}$.

Finally, we can derive similar expressions for the electric and magnetic fields. To do so, we introduce the expression obtained for the vector potential, i.e. Eq. (2.9), into Eqs. (2.5) and (2.6) such that, after some algebraic operations, we arrive at

$$\mathbf{E}(\mathbf{r}, t) = \sum_{\mathbf{k},\mu} \boldsymbol{\epsilon}_{\mathbf{k},\mu} \big[(i\omega_{\mathbf{k}} \alpha_{\mathbf{k},\mu}) e^{i(\mathbf{k}\cdot\mathbf{r} - \omega_{\mathbf{k}} t)} + \text{c.c.} \big], \tag{2.10}$$

$$\mathbf{B}(\mathbf{r}, t) = \sum_{\mathbf{k},\mu} \big(\mathbf{k} \times \boldsymbol{\epsilon}_{\mathbf{k},\mu}\big) \big[(-i\alpha_{\mathbf{k},\mu}) e^{i(\mathbf{k}\cdot\mathbf{r} - \omega_{\mathbf{k}} t)} + \text{c.c.} \big]. \tag{2.11}$$

2.2.2 Semiclassical Description of Light-Matter Interactions

As mentioned in the previous subsection, along with Maxwell's equations, the Lorentz force serves as foundational equation to describe the interaction between light and matter from a semiclassical perspective. This equation, given by

$$\mathbf{F}(\mathbf{r}, \mathbf{v}, t) = Q\big[\mathbf{E}(\mathbf{r}, t) + \mathbf{v} \times \mathbf{B}(\mathbf{r}, t)\big], \tag{2.12}$$

describes the influence of an electric and magnetic field on an electric charge Q with mass m, which travels with velocity $\mathbf{v}$ through an otherwise charge-free medium. In terms of the vector potential, within the Coulomb gauge, this equation can be rewritten as

$$\begin{aligned} \mathbf{F}(\mathbf{r}, \mathbf{v}, t) &= Q\bigg[-\frac{\partial \mathbf{A}(\mathbf{r}, t)}{\partial t} + \mathbf{v} \times \big(\nabla \times \mathbf{A}(\mathbf{r}, t)\big)\bigg] \\ &= Q\bigg[-\frac{\mathrm{d}\mathbf{A}(\mathbf{r}, t)}{\mathrm{d}t} + \nabla\big(\mathbf{v} \cdot \mathbf{A}(\mathbf{r}, t)\big)\bigg], \end{aligned} \tag{2.13}$$

where in going from the first to the second equality, we considered the mathematical identity $\mathbf{v} \times (\nabla \times \mathbf{A}(\mathbf{r}, t)) = \nabla(\mathbf{v} \cdot \mathbf{A}(\mathbf{r}, t)) - (\mathbf{v} \cdot \nabla)\mathbf{A}(\mathbf{r}, t)$, and we introduced the concept of total derivative.[2]

The expression derived in Eq. (2.13) resembles the relation between a velocity-dependent potential $\mathcal{U}(\mathbf{r}, \mathbf{v}, t)$ and the force it generates. Specifically, this relation is given by

$$F_j(\mathbf{r}, \mathbf{v}, t) = -\frac{\partial \mathcal{U}(\mathbf{r}, \mathbf{v}, t)}{\partial r_j} + \frac{\mathrm{d}}{\mathrm{d}t}\bigg(\frac{\partial \mathcal{U}(\mathbf{r}, \mathbf{v}, t)}{\partial v_j}\bigg), \tag{2.14}$$

---

[2] $\mathrm{d}f(\mathbf{r}, t)/\mathrm{d}t = \partial f(\mathbf{q}, t)/\partial t + \sum_i (\partial f(\mathbf{q}, t)/\partial q_i)(\mathrm{d}q_i/\mathrm{d}t)$.

with F_j, r_j and v_j the force, position and velocity components along direction j, respectively. A direct comparison between this equation and Eq. (2.13) allows for the identification of the potential giving rise to the Lorentz force

$$\mathcal{U}(\mathbf{r}, \mathbf{v}, t) = -Q\mathbf{v} \cdot \mathbf{A}, \tag{2.15}$$

such that the Lagrangian describing the dynamics of an electric charge propagating within a region embedded by an electromagnetic field can be written as follows

$$\mathcal{L}(\mathbf{r}, \mathbf{v}, t) = \frac{1}{2}m\mathbf{v}^2 + Q\mathbf{v} \cdot \mathbf{A}(\mathbf{r}, t). \tag{2.16}$$

It is worth noting that, so far, we have not treated the electric charge degrees of freedom in terms of quantum operators. Up to this point, **r** and **v** are vectorial variables with scalar components. In order to perform a *quantization* of our system, we consider the procedure typically referred to as canonical quantization [21], where canonical variables are replaced by quantum operators, and their corresponding Poisson bracket by a commutation relation. In this case, the canonical variables are position **r** and canonical momentum **p**, the latter obtained as

$$p_i = \frac{\partial \mathcal{L}(\mathbf{r}, \mathbf{v}, t)}{\partial v_i} = mv_i + QA_i(\mathbf{r}, t), \tag{2.17}$$

which, together with the Lagrangian, leads through a Legendre transformation to

$$H(t) = \mathbf{p} \cdot \mathbf{v} - \mathcal{L}(\mathbf{r}, \mathbf{v}, t) = \frac{1}{2m}\Big[\mathbf{p} - Q\mathbf{A}(\mathbf{r}, t)\Big]^2, \tag{2.18}$$

commonly referred to as the minimal-coupling Hamiltonian.

At this point, we introduce the canonical quantization procedure mentioned earlier, where the pair of canonical variables $(\mathbf{r}, \mathbf{p})$ are substituted by the quantum operators $(\hat{\mathbf{r}}, \hat{\mathbf{p}})$, denoted with hats over the corresponding variables. These operators satisfy the commutation relations $[\hat{\mathbf{r}}, \hat{\mathbf{r}}] = [\hat{\mathbf{p}}, \hat{\mathbf{p}}] = \mathbf{0}$ and $[\hat{\mathbf{r}}, \hat{\mathbf{p}}] = i\hbar\mathbb{1}$, with $\hbar$ being the reduced Planck constant and $\mathbb{1}$ the identity operator. Introducing these canonical operators in Eq. (2.18), we obtain the Hamiltonian operator characterizing the interaction between a free-charge embedded in a *classical* field

$$\hat{H}_{\text{VG}}(t) = \frac{1}{2m}\Big[\hat{\mathbf{p}} - Q\mathbf{A}(\hat{\mathbf{r}}, t)\Big]^2, \tag{2.19}$$

such that the dynamics of this system are described by the Schrödinger equation

$$i\hbar\frac{\partial|\Psi_{\text{VG}}(t)\rangle}{\partial t} = \frac{1}{2m}\Big[\hat{\mathbf{p}} - Q\mathbf{A}(\hat{\mathbf{r}}, t)\Big]^2|\Psi_{\text{VG}}(t)\rangle, \tag{2.20}$$

with $|\Psi_{\text{VG}}(t)\rangle$ the quantum state of the charged particle.

2.2.3 The Length Gauge and the Dipole Approximation

The light-matter interaction Hamiltonian as given in Eq. (2.19), is commonly referred to as the *velocity gauge* Hamiltonian, wherein the coupling between light and matter is mediated by the canonical momentum operator $\hat{\mathbf{p}}$. However, it is also common practice to work with an alternative version of this Hamiltonian, where the coupling between light and matter occurs through the position operator $\hat{\mathbf{r}}$. This alternative version is typically referred to as the *length gauge* Hamiltonian. To transition between these two, we introduce the unitary transformation

$$|\Psi_{\text{VG}}(t)\rangle = \hat{U}_{\text{VG}}(t)\,|\Psi_{\text{LG}}(t)\rangle = e^{iQ\mathbf{A}(\hat{\mathbf{r}},t)\cdot\mathbf{r}/\hbar}\,|\Psi_{\text{LG}}(t)\rangle\,, \tag{2.21}$$

whose action on the momentum operator $\hat{\mathbf{p}}$ we write as

$$\hat{U}^{\dagger}_{\text{VG}}(t)\hat{\mathbf{p}}\hat{U}_{\text{VG}}(t) = \hat{\mathbf{p}} + Q\mathbf{A}(\hat{\mathbf{r}},t) + O\Big([\mathbf{A}(\hat{\mathbf{r}},t),\hat{\mathbf{p}}]\hat{\mathbf{r}}\Big), \tag{2.22}$$

where $O([\mathbf{A}(\hat{\mathbf{r}},t),\hat{\mathbf{p}}]\hat{\mathbf{r}})$ represents first and higher-order commutators explicitly involving $[\mathbf{A}(\hat{\mathbf{r}},t),\hat{\mathbf{p}}]\hat{\mathbf{r}}$ (see Appendix 7.1 for a detailed analysis). It is worth noting that, in this expression, the variable $\hat{\mathbf{p}}$ at the right-hand side of Eq. (2.22) denotes kinetic momentum, not the conserved canonical momentum, of the charged particle (note the similarity with Eq. (2.17)). By introducing Eq. (2.21) in the Schrödinger equation defined in Eq. (2.20), while taking into account Eq. (2.22), we obtain

$$i\hbar\frac{\partial|\Psi_{\text{LG}}(t)\rangle}{\partial t} = \left[\frac{\hat{\mathbf{p}}^2}{2m} - Q\mathbf{E}(\hat{\mathbf{r}},t)\cdot\hat{\mathbf{r}} + O\Big([\mathbf{A}(\hat{\mathbf{r}},t),\hat{\mathbf{p}}]\hat{\mathbf{r}}\Big)\right]|\Psi_{\text{LG}}(t)\rangle\,, \tag{2.23}$$

where we define the Hamiltonian characterizing the light-matter interaction in the length gauge as

$$\hat{H}_{\text{LG}} \equiv \frac{\hat{\mathbf{p}}^2}{2m} - Q\mathbf{E}(\hat{\mathbf{r}},t)\cdot\hat{\mathbf{r}} + O\Big([\mathbf{A}(\hat{\mathbf{r}},t),\hat{\mathbf{p}}]\hat{\mathbf{r}}\Big). \tag{2.24}$$

In many situations, as those analyzed in upcoming chapters of this Thesis, our interest lies in regimes where the wavelength of the applied laser field is significantly larger than the displacement undergone by the (non-relativistic) electron when accelerated by the field. In other words, we are interested in situations where $\mathbf{k}\cdot\mathbf{r} \ll 1$. This is precisely the regime where the dipole approximation becomes applicable, allowing us to approximate the term $e^{i\mathbf{k}\cdot\mathbf{r}}$ in Eq. (2.9) as

$$e^{i\mathbf{k}\cdot\mathbf{r}} = \sum_{n=0}^{\infty}\frac{(i\mathbf{k}\cdot\mathbf{r})^n}{n!} = 1 + i\mathbf{k}\cdot\mathbf{r} + \cdots \approx 1, \tag{2.25}$$

that is, we neglect the spatial dependence of the vector potential, as well as that of the electric and magnetic fields, that is, $\mathbf{A}(\hat{\mathbf{r}},t) \approx \mathbf{A}(t)$. Therefore, under the dipole approximation, we find that $[\mathbf{A}(t), \hat{\mathbf{p}}]\hat{\mathbf{r}} = 0$, which entails $\boldsymbol{O}([\mathbf{A}(t), \hat{\mathbf{p}}]\, \hat{\mathbf{r}}) = 0$ (see Appendix 7.1). Consequently, the length gauge Hamiltonian in Eq. (2.19) can be written as

$$\hat{H}_{\text{LG}}(t) \approx \frac{\hat{\mathbf{p}}^2}{2m} - Q\mathbf{E}(t) \cdot \hat{\mathbf{r}}. \tag{2.26}$$

2.2.4 Dynamics of a Free-Electron Interacting with an Electromagnetic Field

Before shifting our discussion to the quantum mechanical description of light and its interaction with matter, let us understand how the quantum state of the charged particle evolves under the length gauge Hamiltonian in Eq. (2.26). To achieve this, we proceed to solve the Schrödinger equation

$$i\hbar \frac{\partial |\Psi_{\text{LG}}(t)\rangle}{\partial t} = \left[\frac{\hat{\mathbf{p}}^2}{2m} - Q\mathbf{E}(t) \cdot \hat{\mathbf{r}}\right] |\Psi_{\text{LG}}(t)\rangle, \tag{2.27}$$

by first expanding the quantum state of the charged particle $|\Psi_{\text{LG}}(t)\rangle$ in the momentum representation as

$$|\Psi_{\text{LG}}(t)\rangle = \int \mathrm{d}\mathbf{v}' b(\mathbf{v}', t) \left|\mathbf{v}'\right\rangle, \tag{2.28}$$

with $b(\mathbf{v}, t)$ the probability amplitude of finding the electron with kinetic momentum $m\mathbf{v}$ at time t. Introducing this expansion in Eq. (2.27), and projecting the entire equation with respect to $|\mathbf{v}\rangle$, we can write the following

$$i\hbar\left[\frac{\partial b(\mathbf{v}, t)}{\partial t} + Q\mathbf{E}(t) \cdot \nabla_{\mathbf{v}} b(\mathbf{v}, t)\right] = \frac{1}{2} m\mathbf{v}^2 b(\mathbf{v}, t), \tag{2.29}$$

where we have taken into account that $\langle \mathbf{v}|\hat{\mathbf{r}}|\mathbf{v}'\rangle = i\hbar \nabla_{\mathbf{v}} \delta(\mathbf{v} - \mathbf{v}')$.

The left hand side in Eq. (2.29) corresponds to the total derivative of the complex function $b(\mathbf{v}, t)$ with respect to time. This becomes more apparent when considering the relationship between kinetic and canonical momentum, $m\mathbf{v} = \mathbf{p} - Q\mathbf{A}(t)$, and that $\mathbf{E}(t) = -\partial \mathbf{A}(t)/\partial t$, in agreement with Eq. (2.6). Thus, we are left with

$$i\hbar \frac{\mathrm{d}b(\mathbf{p}, t)}{\mathrm{d}t} = \frac{1}{2m}\left[\mathbf{p} - Q\mathbf{A}(t)\right]^2 b(\mathbf{p}, t), \tag{2.30}$$

which corresponds to the equation we would have obtained directly from the velocity gauge Hamiltonian, under the dipole approximation. It can be straightforwardly solved, leading to

$$b(\mathbf{p}, t) = \exp\left[-\frac{i}{2m\hbar}\int_{t_0}^{t} d\tau \left[\mathbf{p} - Q\mathbf{A}(\tau)\right]^2\right] b(\mathbf{p}, t_0), \tag{2.31}$$

with $b(\mathbf{p}, t_0)$ the initial condition.

We observe that, as a consequence of this evolution, the quantum state of the charged particle acquires an extra phase, corresponding to the kinetic energy accumulated by the particle while it propagates from an initial time t_0 to a final time t. In the context of the path-integral formulation, this phase is referred to as the action. Thus, if we aim to determine the probability of finding the particle in a given position $\mathbf{r}$ at a final time t, we can express it as a sum over all possible paths, i.e., as a sum over all possible probability amplitudes $b(\mathbf{p}, t)$, with the mean value of the obtained distribution as the most likely path to follow. The corresponding actions contribute to interference effects, suppressing trajectories that are not favorable from the classical standpoint. This can be better understood with an example. Let us assume that we want to determine the position of the charged particle at a specific final time t. To achieve this, we project Eq. (2.28) with respect to $|\mathbf{r}\rangle$. Thus, the probability of finding the particle in position $\mathbf{r}$ is given by

$$\begin{aligned} P(\mathbf{r}, t) &= |\langle \mathbf{r}|\Psi_{\mathrm{LG}}(t)\rangle|^2 \\ &= \left|\int d\mathbf{p} \exp\left[-\frac{i}{2m\hbar}\int_{t_0}^{t} d\tau \left[\mathbf{p} - Q\mathbf{A}(\tau)\right]^2 - \frac{i}{\hbar}\mathbf{p}\cdot\mathbf{r}\right] b(\mathbf{p}, t_0)\right|^2. \end{aligned} \tag{2.32}$$

From this expression, we see that this probability is given as the sum over all possible values of $\mathbf{p}$, encompassing all paths that result in the electron ending up at position $\mathbf{r}$ at time t. Each of these paths contributes with a certain complex-valued amplitude, and it is the interference between these amplitudes what causes the final probability distribution to spread in space, as illustrated in Fig. 2.3. In this figure, $P(x, t)$ is displayed as a function of the position for different final times during

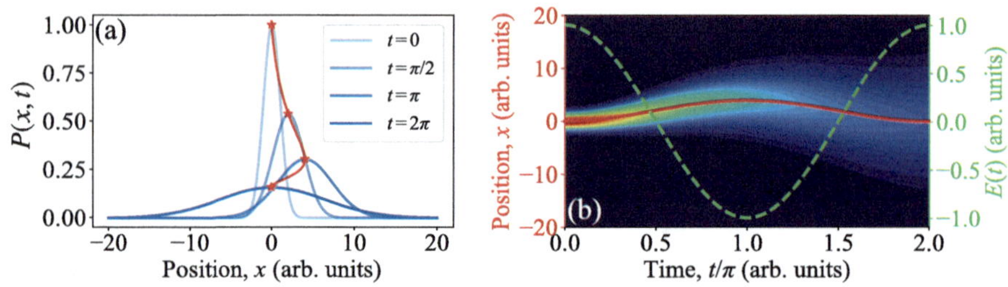

Fig. 2.3 Dynamics of a charged particle described quantum mechanically under the influence of a classical electromagnetic field. In **a**, the probability distribution of finding the electron at various positions x is depicted in blue at different time instances. The red curve passes through the maximum values (represente with the star markers) of these probability distributions. In **b**, a contour plot of these probability distributions is presented as a function of position and time, with the electric field indicated with the green dashed line. The red solid curve in this context represents the most probable trajectory, which aligns with the trajectory predicted by classical mechanics

one-dimensional motion. To begin, we assumed the initial momentum wavepacket to be a Gaussian centered around $p_x = 0$, i.e., $b(p_x, t_0) = \pi^{-1/2}e^{-p_x^2}$. Notably, as the final time t increases, the resulting Gaussian distribution becomes more spread out. However, when examining the behavior of its average value, the real-space trajectory of the charged particle aligns with classical physics expectations, as indicated by the red curves. In particular, at time t_0, the center of the wavepacket is situated at $x = 0$, and it undergoes acceleration driven by the field until reaching a maximum position around $x_{\text{max}} \approx 4$. Subsequently, when the sign of the applied field changes, the wavepacket center travels back to its original point, $x = 0$.

2.3 A Brief Introduction to Quantum Optics

In the previous section, light has been treated from a classical electromagnetism perspective, yielding to fascinating predictions about particles' behavior but leaving unexploited additional degrees of freedom within the optical domain. Quantum optics is a fundamental branch of physics that exploits these additional degrees of freedom. Specifically, it explores light's behavior and properties through the lens of quantum mechanics. In this section, we provide a concise introduction to the field, setting the stage for upcoming chapters of this Thesis. Our exploration covers key topics, including the quantization of the electromagnetic field and some properties of quantum states of light that we encounter in subsequent chapters. From these properties, we build the concept of *non-classical* states of light, as understood throughout this Thesis.

2.3.1 Quantization of the Electromagnetic Field

The quantization of the electromagnetic field presented in this text closely follows the approaches found in standard textbooks of quantum optics, such as Refs. [22–24], which are built upon the original formulation by Dirac [25, 26]. Our derivation begins by recognizing that the energy associated with an electromagnetic field can be expressed as a collection of uncoupled harmonic oscillators, each corresponding to a specific frequency mode. This insight can be intuitively grasped by considering the wave equation's solution, shown in Eq. (2.9). This reveals that the evolution in free space of the electromagnetic field bears a resemblance to that of a particle in a harmonic oscillator. In this analogy, each harmonic mode undergoes sinusoidal-like oscillations, transitioning between fixed maximum and minimum values.

In order to frame this statement within a mathematical context, let us begin by considering the energy carried by the electromagnetic field,[3] which we shall refer to

[3] This expression can be obtained by starting from Maxwell's equations leading to the Poynting's theorem, as presented in Appendix 7.2.

as H_{field} for convenience,

$$H_{\text{field}} = \frac{1}{2}\int d\mathbf{r}\left(\varepsilon_0\,|\mathbf{E}(\mathbf{r},t)|^2 + \frac{1}{\mu_0}\,|\mathbf{B}(\mathbf{r},t)|^2\right), \tag{2.33}$$

where ε_0 and μ_0 correspond to the vacuum electrical permittivity and magnetic permeability, respectively. By considering the expression of the electric field in Eq. (2.10), we obtain the following for the first term involving the norm of the electric field

$$\int d\mathbf{r}\,|\mathbf{E}(\mathbf{r},t)|^2 = \int d\mathbf{r}\left[\sum_{\mathbf{k},\mu}\sum_{\mathbf{k}',\mu'}\omega_{\mathbf{k}}\omega_{\mathbf{k}'}\boldsymbol{\epsilon}_{\mathbf{k},\mu}\cdot\boldsymbol{\epsilon}_{\mathbf{k}',\mu'}\right. \tag{2.34}$$

$$\left.\times\left(i\alpha_{\mathbf{k},\mu}e^{i(\mathbf{k}\cdot\mathbf{r}-\omega_{\mathbf{k}}t)} + \text{c.c.}\right)\left(i\alpha_{\mathbf{k}',\mu'}e^{i(\mathbf{k}'\cdot\mathbf{r}-\omega_{\mathbf{k}'}t)} + \text{c.c.}\right)\right]$$

$$= 2V\sum_{\mathbf{k},\mu}\omega_{\mathbf{k}}^2\alpha_{\mathbf{k},\mu}^*\alpha_{\mathbf{k},\mu} - V\sum_{\mathbf{k},\mu,\mu'}\omega_{\mathbf{k}}^2\boldsymbol{\epsilon}_{-\mathbf{k},\mu}\cdot\boldsymbol{\epsilon}_{\mathbf{k},\mu'}\left(\alpha_{-\mathbf{k},\mu}\alpha_{\mathbf{k},\mu'}e^{-i2\omega_{\mathbf{k}}t} + \text{c.c.}\right), \tag{2.35}$$

where $V = L^3$. In transitioning from the first to the second equality, we took into account that $\int d\mathbf{r}e^{i(\mathbf{k}-\mathbf{k}')\cdot\mathbf{r}} = \delta_{\mathbf{k},\mathbf{k}'}V$. Similarly, by considering the expression of the magnetic field in Eq. (2.11), we obtain for the second term

$$\int d\mathbf{r}\,|\mathbf{B}(\mathbf{r},t)|^2 = \int d\mathbf{r}\left[\sum_{\mathbf{k},\mu}\sum_{\mathbf{k}',\mu'}(\mathbf{k}\times\boldsymbol{\epsilon}_{\mathbf{k},\mu})(\mathbf{k}'\times\boldsymbol{\epsilon}_{\mathbf{k}',\mu'})\right. \tag{2.36}$$

$$\left.\times\left(i\alpha_{\mathbf{k},\mu}e^{-i(\mathbf{k}\cdot\mathbf{r}-\omega_{\mathbf{k}}t)} + \text{c.c.}\right)\left(i\alpha_{\mathbf{k}',\mu'}e^{-i(\mathbf{k}'\cdot\mathbf{r}-\omega_{\mathbf{k}'}t)} + \text{c.c.}\right)\right]$$

$$= \frac{2V}{c^2}\sum_{\mathbf{k},\mu}\omega_{\mathbf{k}}^2\alpha_{\mathbf{k},\mu}^*\alpha_{\mathbf{k},\mu} + V\sum_{\mathbf{k},\mu,\mu'}\omega_{\mathbf{k}}^2\boldsymbol{\epsilon}_{-\mathbf{k},\mu}\cdot\boldsymbol{\epsilon}_{\mathbf{k},\mu'}\left(\alpha_{-\mathbf{k},\mu}\alpha_{\mathbf{k},\mu'}e^{-i2\omega_{\mathbf{k}}t} + \text{c.c.}\right), \tag{2.37}$$

where $c = 1/\sqrt{\varepsilon_0\mu_0}$ represents the velocity of light in vacuum. In transitioning from the first to the second equality, we have taken into account the property $(\mathbf{a}\times\mathbf{b})\cdot(\mathbf{c}\times\mathbf{d}) = (\mathbf{a}\cdot\mathbf{c})(\mathbf{b}\cdot\mathbf{d}) - (\mathbf{a}\cdot\mathbf{d})(\mathbf{b}\cdot\mathbf{c})$. Finally, by combining these two expressions in Eq. (2.33), we arrive at the following form of the energy carried by the electromagnetic field

$$H_{\text{field}} = \sum_{\mathbf{k},\mu}\varepsilon_0 V\omega_{\mathbf{k}}^2\left(\alpha_{\mathbf{k},\mu}^*\alpha_{\mathbf{k},\mu} + \alpha_{\mathbf{k},\mu}\alpha_{\mathbf{k},\mu}^*\right), \tag{2.38}$$

and which corresponds to the energy of a set of uncoupled harmonic oscillators, each oscillating with frequency $\omega_{\mathbf{k}}$. This correspondence can be made clearer by introducing position-like and momentum-like variables, $X_{\mathbf{k},\mu}$ and $P_{\mathbf{k},\mu}$ respectively, which are related to $\alpha_{\mathbf{k},\mu}$ through the following (canonical) transformation

$$\alpha_{\mathbf{k},\mu} = \frac{1}{2\omega_{\mathbf{k}}\sqrt{\varepsilon_0 V}}\left[\omega_{\mathbf{k}} X_{\mathbf{k},\mu} + i P_{\mathbf{k},\mu}\right], \tag{2.39}$$

and when introduced in Eq. (2.38), leads to what we recognize as the Hamiltonian of a set of uncoupled harmonic oscillators

$$H_{\text{field}} = \frac{1}{2}\sum_{\mathbf{k},\mu}\left[P^2_{\mathbf{k},\mu} + \omega_{\mathbf{k}} X^2_{\mathbf{k},\mu}\right]. \tag{2.40}$$

The $(X_{\mathbf{k},\mu}, P_{\mathbf{k},\mu})$ variables are commonly referred to as the *quadratures* of the electromagnetic field. They form a pair of canonical variables similar to position and momentum variables, sharing the same Poisson bracket. Therefore, following the canonical quantization employed earlier in this chapter, we can replace this pair of canonical variables with the quantum operators $(\hat{X}_{\mathbf{k},\mu}, \hat{P}_{\mathbf{k},\mu})$. These operators satisfy the following set of commutation relations

$$[\hat{X}_{\mathbf{k},\mu}, \hat{P}_{\mathbf{k}',\mu'}] = i\hbar\delta_{\mathbf{k},\mathbf{k}'}\delta_{\mu,\mu'}, \quad [\hat{X}_{\mathbf{k},\mu}, \hat{X}_{\mathbf{k}',\mu'}] = [\hat{P}_{\mathbf{k},\mu}, \hat{P}_{\mathbf{k}',\mu'}] = 0, \tag{2.41}$$

such that the Hamiltonian operator characterizing the electromagnetic field in free space is given by

$$\hat{H}_{\text{field}} = \frac{1}{2}\sum_{\mathbf{k},\mu}\left[\hat{P}^2_{\mathbf{k},\mu} + \omega_{\mathbf{k}} \hat{X}^2_{\mathbf{k},\mu}\right]. \tag{2.42}$$

Therefore, in analogy to Eq. (2.40), we introduce the operator $\hat{a}_{\mathbf{k},\mu}$ and its Hermitian conjugate $\hat{a}^\dagger_{\mathbf{k},\mu}$ as

$$\hat{a}_{\mathbf{k},\mu} = \frac{1}{\sqrt{2\hbar\omega_{\mathbf{k}}}}\left[\omega_{\mathbf{k}}\hat{X}_{\mathbf{k},\mu} + i\hat{P}_{\mathbf{k},\mu}\right], \quad \hat{a}^\dagger_{\mathbf{k},\mu} = \frac{1}{\sqrt{2\hbar\omega_{\mathbf{k}}}}\left[\omega_{\mathbf{k}}\hat{X}_{\mathbf{k},\mu} - i\hat{P}_{\mathbf{k},\mu}\right], \tag{2.43}$$

which we refer to as annihilation and creation operators, respectively, for convenience. With these definitions and the relations in Eq. (2.41), we can derive the following set of commutation relations for these operators

$$[\hat{a}_{\mathbf{k},\mu}, \hat{a}^\dagger_{\mathbf{k}',\mu'}] = \delta_{\mathbf{k},\mathbf{k}'}\delta_{\mu,\mu'}, \quad [\hat{a}_{\mathbf{k},\mu}, \hat{a}_{\mathbf{k}',\mu'}] = [\hat{a}^\dagger_{\mathbf{k},\mu}, \hat{a}^\dagger_{\mathbf{k}',\mu'}] = 0, \tag{2.44}$$

and using them we can express the Hamiltonian operator in Eq. (2.42) as

$$\hat{H}_{\text{field}} \equiv \sum_{\mathbf{k},\mu} H^{(\mathbf{k},\mu)}_{\text{field}} = \sum_{\mathbf{k},\mu}\hbar\omega_{\mathbf{k}}\left[\hat{a}^\dagger_{\mathbf{k},\mu}\hat{a}_{\mathbf{k},\mu} + \frac{1}{2}\right], \tag{2.45}$$

while for the vector potential, electric and magnetic field operators, we can respectively write

$$\hat{\mathbf{A}}(\mathbf{r},t)=\sum_{\mathbf{k},\mu}\sqrt{\frac{\hbar}{2\omega_{\mathbf{k}}\varepsilon_0 V}}\boldsymbol{\epsilon}_{\mathbf{k},\mu}\Big[\hat{a}_{\mathbf{k},\mu}e^{i(\mathbf{k}\cdot\mathbf{r}-\omega_{\mathbf{k}}t)}+\text{h.c.}\Big], \tag{2.46}$$

$$\hat{\mathbf{E}}(\mathbf{r},t)=\sum_{\mathbf{k},\mu}\sqrt{\frac{\hbar\omega_{\mathbf{k}}}{2\varepsilon_0 V}}\boldsymbol{\epsilon}_{\mathbf{k},\mu}\Big[i\hat{a}_{\mathbf{k},\mu}e^{i(\mathbf{k}\cdot\mathbf{r}-\omega_{\mathbf{k}}t)}+\text{h.c.}\Big], \tag{2.47}$$

$$\hat{\mathbf{B}}(\mathbf{r},t)=\sum_{\mathbf{k},\mu}\sqrt{\frac{\hbar}{2\omega_{\mathbf{k}}\varepsilon_0 V}}(\mathbf{k}\times\boldsymbol{\epsilon}_{\mathbf{k},\mu})\Big[-i\hat{a}_{\mathbf{k},\mu}e^{i(\mathbf{k}\cdot\mathbf{r}-\omega_{\mathbf{k}}t)}+\text{h.c.}\Big], \tag{2.48}$$

obtained from applying the same canonical quantization procedure as schemed here to Eqs. (2.9), (2.10) and (2.11).

The quantum states for which the energies of the harmonic oscillator are well-defined, i.e. its eigenstates, are the so-called Fock states $\{|n_{\mathbf{k},\mu}\rangle : n \in \mathbb{W}\}$[4] [22–24], with n representing the number of excitations of the harmonic oscillator. These states satisfy

$$\hat{H}^{(\mathbf{k},\mu)}_{\text{field}}|n_{\mathbf{k},\mu}\rangle=\hbar\omega_{\mathbf{k}}\left(n+\frac{1}{2}\right)|n_{\mathbf{k},\mu}\rangle, \tag{2.49}$$

and the action of the creation and annihilation operators over them, respectively leads to

$$\hat{a}^{\dagger}_{\mathbf{k},\mu}|n_{\mathbf{k},\mu}\rangle=\sqrt{n+1}|(n+1)_{\mathbf{k},\mu}\rangle,\quad \hat{a}_{\mathbf{k},\mu}|n_{\mathbf{k},\mu}\rangle=\sqrt{n}|(n-1)_{\mathbf{k},\mu}\rangle, \tag{2.50}$$

that is, the creation operator *creates* an excitation on the harmonic oscillators, while the annihilation operator *destroys* an excitation.

According to this derivation, we observe that the energies obtained from the quantization of the harmonic oscillator are discrete. In our electromagnetic field perspective, this implies that the stable energies of a free electromagnetic field are discrete. This alignment with the observations made by Max Planck and Albert Einstein, as outlined in Chap. 1, reinforces the significance of Fock states and raises from a mathematical perspective the concept of photon. In the context of quantum optics, Fock states are often referred to as *photon-number* states and serve as a quintessential example of non-classical states of light. In the following section, we introduce observables used to distinguish between classical and non-classical states of light, which are relevant in the upcoming chapters of this Thesis.

---

[4] By $\mathbb{W}$ we denote the set of whole numbers, i.e., the set of natural numbers plus the zero element ($\mathcal{N}+\{0\}$).

2.3.2 Quantum Optical Observables for Probing Non-classical Light

The concept of non-classical light refers to those states of light that exhibit properties or behaviors that cannot be explained or described by classical electromagnetic theory. This kind of states are of fundamental importance for photonic-based quantum information science applications [27] such as quantum communication [28], quantum computation [29, 30], quantum simulation [31], quantum metrology [32] and quantum sensing [33], where their properties are directly exploited in order to find supra-classical advantages. Therefore, having methods to probe and characterize them becomes crucial for building reliable non-classical light sources that can be used in the aforementioned applications. In this section, we describe some of the main observables used for that purpose when considering single modes of light, i.e., modes of light with a well-defined polarization and frequency. To illustrate this further, we explore several examples where these observables lead to different, although non-exclusive, classifications between classical and non-classical states. Later in this chapter, we delve with the non-classical features when considering more than one mode of light.

2.3.2.1 Photon Statistics

Some of the inherent properties of non-classical states of light can be already observed by examining their photon statistics. Photon statistics involve studying the statistical distributions obtained through photon counting experiments. In typical experimental configurations [34, 35], a beam of light generated by a source impinges on a photomultiplier, where the incident light is converted into an electric current, in what is a practical implementation of the photoelectric effect [36], which is later postprocessed. The number of photons reaching the photodetector can vary depending on the light source used, leading to variations in the probability of ionizing a certain number of electrons in the photomultiplier. If various factors related to the experimental procedure and setup are met,[5] it is possible to retrieve the photon statistics of the considered light source (see e.g. Refs. [37–39] and references therein).

For a given quantum state of light defined in a specific mode $(\mathbf{k}, \mu)$, denoted as $|\Phi\rangle$, we can mathematically compute its photon statistics by projecting it onto a particular Fock state $|n_{\mathbf{k},\mu}\rangle$ and calculating the norm of this quantity, that is, $P_{\mathbf{k},\mu}(n) = |\langle n_{\mathbf{k},\mu}|\Phi\rangle|^2$. Typically, we can distinguish between three regimes depending on the properties of the states generated by the considered light source, leading to Poissonian, super-Poissonian and sub-Poissonian distributions. The differences among these regimes depend on the relationship between the mean photon number

[5] For example, accurately measuring chaotic light requires that the measurement duration is shorter than the coherence time of the source, or that the detector amplifiers function properly, allowing them to produce detectable current pulses when registering the arrival of single photons.

and the variance of the distribution, where the former is obtained straightforwardly by computing the mean value of the photon number operator $\hat{n}_{\mathbf{k},\mu} = \hat{a}^{\dagger}_{\mathbf{k},\mu}\hat{a}_{\mathbf{k},\mu}$, that is

$$\langle \hat{n}_{\mathbf{k},\mu} \rangle = \langle \Phi | \hat{n}_{\mathbf{k},\mu} | \Phi \rangle = \sum_{n=0}^{\infty} n P_{\mathbf{k},\mu}(n), \tag{2.51}$$

while the latter can be obtained as

$$(\Delta n_{\mathbf{k},\mu})^2 = \langle \hat{n}^2_{\mathbf{k},\mu} \rangle - \langle \hat{n}_{\mathbf{k},\mu} \rangle^2 = \sum_{n=0}^{\infty} n^2 P_{\mathbf{k},\mu}(n) - \left(\sum_{n=0}^{\infty} n P_{\mathbf{k},\mu}(n) \right)^2, \tag{2.52}$$

where in transitioning from the first to the second equality in both expressions we have introduced the identity in the photon number representation, i.e., $\mathbb{1} = \sum_{n=0}^{\infty} |n_{\mathbf{k},\mu}\rangle\langle \hat{n}_{\mathbf{k},\mu}|$.

The different cases we have just underlined are pictorially represented in Fig. 2.4. The central distribution shown in this plot represents the case of a Poissonian distribution, which is generally described by

$$P_{\mathbf{k},\mu}(n) = \frac{\langle \hat{n}_{\mathbf{k},\mu} \rangle^n}{n!} e^{-\langle \hat{n}_{\mathbf{k},\mu} \rangle}, \tag{2.53}$$

leading to a relationship between the variance and the mean photon number given by $(\Delta n_{\mathbf{k},\mu})^2 = \langle \hat{n}_{\mathbf{k},\mu} \rangle$. This type of probability distributions is typically found in the case of spatially and temporally coherent monochromatic fields, such as those

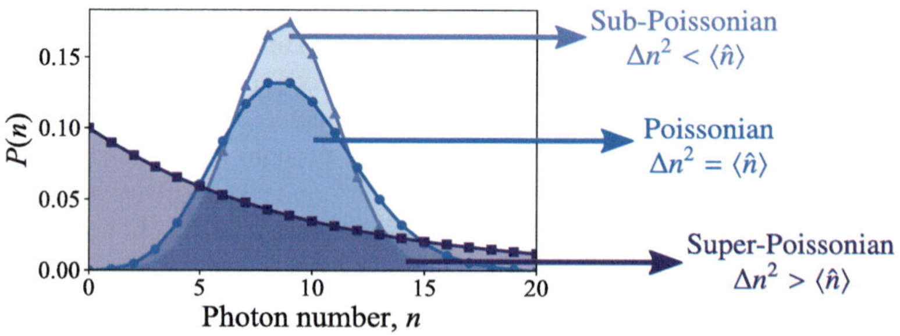

Fig. 2.4 A schematic representation of different types of photon statistics. The lowermost curve with squared markers illustrates the case of super-Poissonian behavior, where the variance in the photon number exceeds the mean value. This behavior is characteristic of thermal light, such as that produced by incandescent bulbs. The curve with circular markers positioned in the middle represents a Poissonian distribution, where the variance equals the mean photon number. This occurs with coherent light sources, which indeed satisfy Maxwell's equations in vacuum. Lastly, the uppermost curve with triangular markers demonstrates the case of sub-Poissonian statistics, where the variance is less than the mean photon number. Such distributions, categorized as non-classical, are generated by specific types of sources, including Fock state sources and certain kind of squeezed light sources

ideally generated by laser sources [34]. The distribution shown in the lowest part of the figure presents super-Poissonian statistics, which are defined as those for which $(\Delta n_{\mathbf{k},\mu})^2 > \langle \hat{n}_{\mathbf{k},\mu} \rangle$. This is indeed the case of thermal light sources [34], with incandescent bulbs being a primary example. In this particular case, the photon number distribution satisfies Bose-Einstein statistics [40], given by

$$P_{\mathbf{k},\mu}(n) = \frac{\langle \hat{n}_{\mathbf{k},\mu} \rangle^n}{\left(\langle \hat{n}_{\mathbf{k},\mu} \rangle + 1\right)^{n+1}}, \tag{2.54}$$

for which it can be seen that $(\Delta n_{\mathbf{k},\mu})^2 = \langle \hat{n}_{\mathbf{k},\mu} \rangle + \langle \hat{n}_{\mathbf{k},\mu} \rangle^2$.

The two cases we have just described can be properly characterized by using classical descriptions of light, as shown, for example, in Ref. [37] and references therein.[6] Thus, according to the photon statistics' criteria, if the photon statistics of a light source can be described by one of these two distributions, it is considered to generate classical states of light. On the other side of the spectrum, we have the case of sub-Poissonian distributions, as shown with the upper distribution in Fig. 2.4. In this case, we have that $(\Delta n_{\mathbf{k},\mu})^2 < \langle \hat{n}_{\mathbf{k},\mu} \rangle$, a feature that necessitates from quantum mechanical descriptions of light (see Ref. [38] and references therein). Light sources capable of producing photon statistics in this form are classified as non-classical. This includes sources capable of generating single photons, i.e. single-photon sources, or those producing specific kinds of squeezed light.

2.3.2.2 Second-Order Autocorrelation Function

The fact that certain sources exhibit varying levels of fluctuations in their photon number distribution compared to their mean photon number, has important implications for the probability of detecting radiation generated by the same source using different detectors at potentially different times. That is, when doing *autocorrelation* experiments. Before we explore the relationship between these aspects, let us motivate the concept of autocorrelation measurements within the context of non-classical light sources. To achieve this, we consider the setup depicted schematically in Fig. 2.5a, which corresponds to a specific arrangement of what is generally known as Hanbury-Brown and Twiss configuration. In this setup, a source emits a beam of light, which is subsequently divided into different optical paths by a 50:50 beam splitter. At the end of each optical path, two ideal photodetectors are positioned to perform a correlated measurement. When both photodetectors detect radiation simultaneously, the outcome is recorded as **1**; otherwise, it is recorded as **0**.

For **Case 1** in the figure, the input source is an ideal monochromatic laser field. As previously mentioned, such sources exhibit a Poissonian photon distribution and can be categorized as classical light sources, describable using classical electromagnetism. As a result, in accordance with the principles of classical optics [41], when

[6] In Ref. [37], it is shown that for thermal light sources, if the measurement time is above the coherent time of the source, then the obtained photon-number distribution becomes Poissonian.

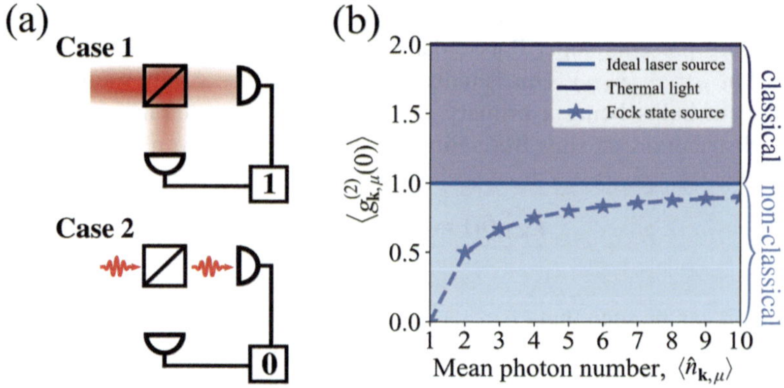

Fig. 2.5 Pictorial representation of second-order autocorrelation functions and their measurement outcomes for different light sources. In **a**, we illustrate second-order autocorrelation measurements using light from different sources. In both cases, light passes through a beam splitter and reaches two photodetectors. If these photodetectors detect light simultaneously, they produce a **1**; otherwise, they output a **0**. In **Case 1**, the input source is a coherent field, which evenly splits at the beam splitter, simultaneously reaching both photodetectors and resulting in a **1**. Conversely, in **Case 2**, a single photon that cannot be split by the beam splitter only reaches one photodetector and outputs a **0**. In **b**, the result of these measurements are depicted using the second-order autocorrelation function $g^{(2)}_{\mathbf{k},\mu}(0)$ for light sources with varying photon statistics. Light with $g^{(2)}_{\mathbf{k},\mu}(0) \geq 1$ is categorized as classical, while light with $g^{(2)}_{\mathbf{k},\mu}(0) < 1$ is considered non-classical

this radiation traverses the beam splitter, it separates into two beams of equal intensity. These beams reach their respective photodetectors simultaneously, yielding an outcome of **1**. In contrast, for **Case 2** in the figure, we consider a single-photon source that emits photons perfectly one at time, thus classifying it as a non-classical light source.[7] Since the energy of a single photon cannot be equally distributed between the two possible output paths of the beam splitter,[8] the photon will be detected by one of the two photodetectors but not both simultaneously. Therefore, in this scenario, the outcome of the correlated measurement is **0**.

The absence of correlation between the two photodetectors in the case of the single-photon source, in contrast to the correlation observed in the classical light source, underscores the potential significance of autocorrelation measurements in the field of quantum optics. The concept of autocorrelation functions and their relevance in quantum optics was initially introduced by Glauber [42], where he introduced a family of autocorrelation functions with the nth order autocorrelation function defined as follows

---

[7] In this case, we have $\langle \hat{n}_{\mathbf{k},\mu}\rangle = 1$ and $(\Delta n_{\mathbf{k},\mu})^2 = 0$.

[8] It is worth noting that for a free-field Hamiltonian, the energy difference between two consecutive excitations is given by $\hbar\omega_{\mathbf{k}}$, which corresponds to the energy of a single photon. In principle, the energy of a photon cannot be splitted into two photons of energy summing to $\hbar\omega_{\mathbf{k}}$ unless we are dealing with nonlinear processes, which we assume do not happen inside beam splitters.

$$G^{(n)}_{\mathbf{k},\mu}(\mathbf{r}_1,\cdots\mathbf{r}_n,\mathbf{r}_{n+1},\cdots\mathbf{r}_{2n};t_1,\cdots,t_n,t_{n+1},\cdots,t_{2n})$$
$$=\langle E^{(-)}_{\mathbf{k},\mu}(\mathbf{r}_1,t_1)\cdots E^{(-)}_{\mathbf{k},\mu}(\mathbf{r}_n,t_n)E^{(+)}_{\mathbf{k},\mu}(\mathbf{r}_{n+1},t_{n+1})\cdots E^{(+)}_{\mathbf{k},\mu}(\mathbf{r}_{2n},t_{2n})\rangle, \tag{2.55}$$

where $E^{(\pm)}_{\mathbf{k},\mu}(\mathbf{r_i},t_i)=\boldsymbol{\epsilon}_{\mathbf{k},\mu}\cdot\mathbf{E}^{(\pm)}(\mathbf{r}_i,t_i)$ with $\mathbf{E}^{(\pm)}(\mathbf{r},t)=\sum_{\mathbf{k},\mu}\boldsymbol{\epsilon}_{\mathbf{k},\mu}(i\omega_{\mathbf{k}}|\alpha_{\mathbf{k},\mu}|)$ $e^{\pm i(\mathbf{k}\cdot\mathbf{r}-\omega_{\mathbf{k}}t+\theta_{\mathbf{k},\mu})}$, where we have written $\alpha_{\mathbf{k},\mu}=|\alpha_{\mathbf{k},\mu}|e^{i\theta_{\mathbf{k},\mu}}$. From these functions, Glauber introduced the concept of *coherent* fields as those for which the normalized autocorrelation functions, denoted as $g^{(n)}(\mathbf{r}_1,\cdots\mathbf{r}_{2n};t_1,\cdots t_{2n})$, have unit magnitude, that is

$$\left|g^{(n)}_{\mathbf{k},\mu}(\mathbf{r}_1,\cdots\mathbf{r}_{2n};t_1,\cdots t_{2n})\right|=1,\forall n\in\mathbb{N}, \tag{2.56}$$

where the nth order normalized autocorrelation function is defined as

$$g^{(n)}_{\mathbf{k},\mu}(\mathbf{r}_1,\cdots,\mathbf{r}_{2n};t_1,\cdots,t_{2n})=\frac{G^{(n)}_{\mathbf{k},\mu}(\mathbf{r}_1,\cdots\mathbf{r}_{2n};t_1,\cdots t_{2n})}{\prod_{j=1}^{2n}\left[G^{(1)}_{\mathbf{k},\mu}(\mathbf{r}_j,\mathbf{r}_j;t_j,t_j)\right]^{1/2}}. \tag{2.57}$$

In the experiments presented in Fig. 2.5a, we were considering a particular case of a second-order autocorrelation function, specifically to $g^{(2)}_{\mathbf{k},\mu}(\mathbf{r}_1,\mathbf{r}_2;t)$, where two detectors were placed at positions $\mathbf{r}_1$ and $\mathbf{r}_2$, performing correlated measurement at the same instant of time. However, when characterizing non-classical states of light, it is often sufficient to examine single-point second-order autocorrelation functions, where $\mathbf{r}_1=\mathbf{r}$ and $\mathbf{r}_2=\mathbf{r}$, at times $t_1,t_3=t$ and $t_2,t_4=t+\tau$. In this particular case, the quantum optical version of the normalized second-order autocorrelation function takes the following simple form [43]

$$g^{(2)}_{\mathbf{k},\mu}(\tau)=\frac{\langle\hat{a}^\dagger_{\mathbf{k},\mu}(t)\hat{a}^\dagger_{\mathbf{k},\mu}(t+\tau)\hat{a}_{\mathbf{k},\mu}(t+\tau)\hat{a}_{\mathbf{k},\mu}(t)\rangle}{\langle\hat{a}^\dagger_{\mathbf{k},\mu}\hat{a}_{\mathbf{k},\mu}\rangle^2}, \tag{2.58}$$

where we have replaced $\mathbf{E}^{\pm}(\mathbf{r}_j,t_j)$ with the quantum operator $\hat{\mathbf{E}}^{\pm}(\mathbf{r}_j,t_j)$ as defined in Eq. (2.47). Specifically, for the case $\tau=0$, using Eq. (2.52) we can write[9]

$$g^{(2)}_{\mathbf{k},\mu}(0)=\frac{(\Delta n_{\mathbf{k},\mu})^2}{\langle\hat{n}_{\mathbf{k},\mu}\rangle^2}-\frac{1}{\langle\hat{n}_{\mathbf{k},\mu}\rangle}+1. \tag{2.59}$$

This expression indicates that the concept of second-order autocorrelation is inherently related to the fluctuations of the light source in use. According to Glauber's criteria, for spatially and temporally coherent monochromatic fields showing Poissonian distributions, we have $g^{(2)}_{\mathbf{k},\mu}(0)=1$. Conversely, for light sources with super-Poissonian statistics we obtain $g^{(2)}_{\mathbf{k},\mu}(0)>1$, as depicted in the upper part of Fig. 2.5b. In general, for statistically stationary fields such as thermal and ideal laser light,

[9] In fact, for a single-mode the time-dependent phase factors already cancel at the level of Eq. (2.58).

Eq. (2.58) yields[10] $g^{(2)}(\tau) \leq g^{(2)}(0)$. This is typically referred to as *photon bunching*, indicating that photons tend to form bunches rather than being distributed randomly.

On the other hand, non-classical fields exhibiting sub-Poissonian statistics result in $g^{(2)}(0) < 1$, as illustrated in the lower part of Fig. 2.5b for sources generating Fock states with a mean photon number $\langle \hat{n}_{\mathbf{k},\mu} \rangle$. Consequently, this result serves as a method for classifying non-classical states of light based on the second-order autocorrelation function, which consistently remains greater than or equal to one for classical sources. Moreover, in certain cases, such as single-photon sources, it is observed that $g^{(2)}(\tau) > g^{(2)}(0)$. This is often referred to as *photon antibunching*, signifying that fewer photons pairs are detected close together than further apart.

2.3.2.3 Wigner Functions

The behavior of individual modes of the electromagnetic field as harmonic oscillators enables a phase-space-like representation of the electromagnetic field using the set of optical quadratures $(X_{\mathbf{k},\mu}, P_{\mathbf{k},\mu})$. This representation bears resemblance to that of classical mechanics, where all conceivable values of a particle's position $\mathbf{r}$ and momentum $\mathbf{p}$ are depicted. However, owing to the inherent principles of quantum mechanics and the non-commutative nature of position-like and momentum-like operators, precise simultaneous measurements of non-commutative variables becomes unattainable. This circumstance establishes fundamental distinctions from phase-space representations in classical mechanics. In particular, these distinctions allow for establishing a delineation of classical and non-classical states of light, that is inherently distinct from the photon number distribution or the second-order autocorrelation function, which rely on the photon statistics originating from the considered source.

In classical mechanics, the Liouville density is a probability density function that enables the quantification of the likelihood of finding a dynamical system at a specific point within phase space, and at a given time t. This concept assumes particular significance when addressing scenarios characterized by imprecisely known, bounded or uncertain initial conditions, as it provides a probabilistic framework for the systematic description of a system's behavior and evolution over time [44]. The quantum mechanical analogue of the Liouville density is the Wigner function, first introduced by E. Wigner in 1932 within the context of quantum corrections to thermodynamic equilibrium [45], although it later found wide applicability in quantum optics (see e.g. Ref. [46] and references therein).

Given a density matrix $\hat{\rho} = \sum_i p_i \, |\psi_i \rangle\langle \psi_i|$ characterizing the state of a system, its Wigner function is defined as

[10] This can be obtained by applying the Schwartz inequality over the classical definition of the $g^{(2)}(\tau)$ function, with the obtained right-hand side being exact for statistically stationary fields. More details can be found in Ref. [43].

$$W(x, p) = \frac{1}{\pi\hbar}\int_{-\infty}^{\infty} \mathrm{d}y \langle x + y|\hat{\rho}|x - y\rangle e^{-i2py/\hbar}, \tag{2.60}$$

where in this context x and p represent points in phase space, extending beyond typical definitions of position and momentum, as they can also characterize the field quadratures. To elucidate the primary differences from the classical Liouville probability density, and underscore the importance of the Wigner function in quantum optics, let us explore some of its properties [46]:

1. *When integrating the Wigner function with respect to the momentum-like variable p, the position-like distribution of the corresponding state $\hat{\rho}$ is obtained.*
   This property can be straightforwardly demonstrated by integrating over p on both sides of Eq. (2.60). Specifically, we obtain

$$\begin{aligned}\int_{-\infty}^{\infty} \mathrm{d}p W(x, p) &= \frac{1}{\pi\hbar}\int_{-\infty}^{\infty} \mathrm{d}p \int_{-\infty}^{\infty} \mathrm{d}y \langle x + y|\hat{\rho}|x - y\rangle e^{-i2py/\hbar} \\ &= \int_{-\infty}^{\infty} \mathrm{d}y \langle x + y|\hat{\rho}|x - y\rangle \delta(y) \\ &= \langle x|\hat{\rho}|x\rangle,\end{aligned} \tag{2.61}$$

   where we introduced the identity $(1/\pi\hbar)\int_{-\infty}^{\infty} \mathrm{d}p\, e^{-i2py/\hbar} = \delta(y)$ when transitioning from the first to the second equality.
2. *When integrating the Wigner function with respect to the position-like variable x, the momentum-like distribution of the corresponding state $\hat{\rho}$ is obtained.*
   Similarly to what was done before, we integrate over x on both sides of Eq. (2.60)

$$\begin{aligned}\int_{-\infty}^{\infty} \mathrm{d}x W(x, p) &= \frac{1}{\pi\hbar}\int_{-\infty}^{\infty} \mathrm{d}x \int_{-\infty}^{\infty} \mathrm{d}y \langle x + y|\hat{\rho}|x - y\rangle e^{-i2py/\hbar} \\ &= \frac{1}{\pi\hbar}\int_{-\infty}^{\infty} \mathrm{d}x \int_{-\infty}^{\infty} \mathrm{d}y \langle x + y|\hat{\rho}|x - y\rangle e^{-ip(x+y)/\hbar} e^{ip(x-y)/\hbar} \\ &= \frac{1}{2}\int_{-\infty}^{\infty} \mathrm{d}x \int_{-\infty}^{\infty} \mathrm{d}y \langle p|x + y\rangle\langle x + y|\hat{\rho}|x - y\rangle\langle x - y|p >,\end{aligned} \tag{2.62}$$

   where we have taken into account that $\langle p|x\rangle = (1/\sqrt{2\pi\hbar})e^{-ipx/\hbar}$. At this point, we note that

$$\hat{\rho} = \frac{1}{2}\int_{-\infty}^{\infty} \mathrm{d}x \int_{-\infty}^{\infty} \mathrm{d}y \langle x + y|\hat{\rho}|x - y\rangle\, |x + y\rangle\langle x - y|\,, \tag{2.63}$$

   is a representation in the position basis of the density matrix operator, where the $1/2$ term appears from counting the number of elements twice. Therefore, we arrive to

$$\int_{-\infty}^{\infty} \mathrm{d}x\; W(x, p) = \langle p|\hat{\rho}|p\rangle. \tag{2.64}$$

3. *The norm of the Wigner function is bounded by* $1/(\pi\hbar)$.
   This is something that can be easily proven for pure states, where $\hat{\rho} = |\psi\rangle\langle\psi|$. In this case, the Wigner function can be expressed as

$$
\begin{aligned}
W(x, p) &= \frac{1}{\pi\hbar}\int_{-\infty}^{\infty} \mathrm{d}y \langle x + y|\psi\rangle\langle\psi|x - y\rangle e^{-i2py/\hbar} \\
&= \frac{1}{\pi\hbar}\int_{-\infty}^{\infty} \mathrm{d}y\ \tilde{\psi}(x + y)\tilde{\psi}^*(x - y),
\end{aligned}
\tag{2.65}
$$

   where we defined $\tilde{\psi}(x + y) = \langle x + y|\psi\rangle e^{-i(x+y)/\hbar}$. However, it is important to note that this expression represents the overlap between two quantum states $|\tilde{\psi}_1\rangle$ and $|\tilde{\psi}_2\rangle$, both of which satisfy the relationships $\langle y|\tilde{\psi}_1\rangle = \tilde{\psi}(x + y)$ and $\langle y|\tilde{\psi}_2\rangle = \tilde{\psi}(x - y)$. Therefore, we can write

$$
W(x, p) = \frac{1}{\pi\hbar}\langle\tilde{\psi}_2|\tilde{\psi}_1\rangle, \tag{2.66}
$$

   which leads us to

$$
|W(x, p)| = \frac{1}{\pi\hbar}\left|\langle\tilde{\psi}_2|\tilde{\psi}_1\rangle\right| \leq \frac{1}{\pi\hbar}. \tag{2.67}
$$

4. *The Wigner function can adopt negative values.*
   This property can be deduced from the previously discussed property, as the scalar product between two quantum states, denoted as $\langle\tilde{\psi}_2|\tilde{\psi}_1\rangle$, is generally a complex number. Consequently, in some cases, it can specifically take on negative values. However, it is essential to emphasize that the Wigner function is a real-valued function. This can be verified by considering the complex conjugate of Eq. (2.60)

$$
\begin{aligned}
W^*(x, p) &= \frac{1}{\pi\hbar}\int_{-\infty}^{\infty} \mathrm{d}y \langle x - y|\hat{\rho}|x + y\rangle e^{i2py/\hbar} \\
&= \frac{1}{\pi\hbar}\int_{-\infty}^{\infty} \mathrm{d}y' \langle x + y'|\hat{\rho}|x - y'\rangle e^{-i2py'/\hbar} = W(x, p),
\end{aligned}
\tag{2.68}
$$

   where in going from the second to the third equality, we introduced the change of variables $y = -y'$.

From the first and second properties, it follows that the Wigner distribution provides a comprehensive description of the quantum state of a system in phase-space, as it fully characterizes the state along both momentum-like and position-like coordinates. In fact, the Wigner function establishes a one-to-one correspondence with the quantum state $\hat{\rho}$, allowing either representation to be used interchangeably. However, the fourth property establishes a fundamental distinction from classical probability densities, which are constrained to positive values. For this reason, the Wigner function is often referred to as a *quasiprobability* distribution, as it may not satisfy the positivity requirement of classical probability densities for certain quantum states.

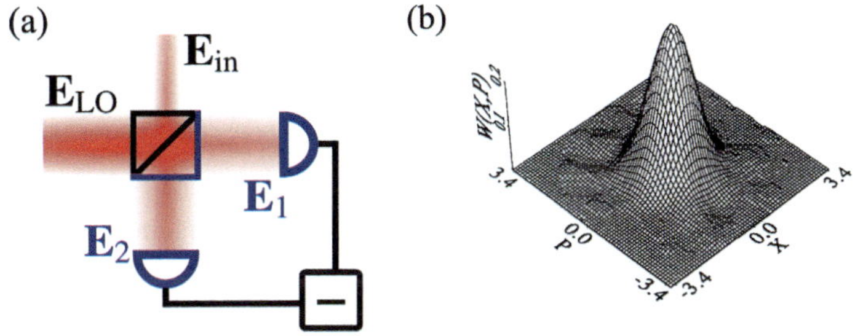

Fig. 2.6 Homodyne detection schemes and their role in computing the Wigner function of a light source. In **a**, we present a schematic representation of a homodyne detection setup, where an input light beam $\mathbf{E}_{\text{in}}$ is combined with a well-characterized local oscillator $\mathbf{E}_{\text{LO}}$ using a beam splitter. The resulting output beams $\mathbf{E}_1$ and $\mathbf{E}_2$ are measured by two photodetectors placed at the beam splitter's exits and the results subtracted. By adjusting the phase of the local oscillator and performing post-processing of the recorded signals, one can ultimately construct the Wigner function of the input beam $\mathbf{E}_{\text{in}}$. In **b**, we display the recorded Wigner function of a vacuum state obtained in Ref. [48], which we have appropriately adapted to this text (reprinted figure with permission of D. T. Smithey et al., *Phys. Rev. Lett.* **70** 9, pp. 1244–1247 (2023) Copyright (2023) by the American Physical Society)

Consequently, the Wigner function introduces an alternative criterion for distinguishing classical from non-classical states of light: states for which negativities exist in their Wigner function representation are classified as non-classical states of light. Specifically, Gaussian states, i.e. those for which the wavefunction can be expressed in terms of a Gaussian function, have Wigner functions which are positive allover the phase-space, while for non-Gaussian states this is not necessarily the case [47].

Interestingly, the Wigner function is not merely a mathematical construct for characterizing the quantum state of a system; it can also be experimentally reconstructed through homodyne detection [48]. In typical homodyne detection setups [38, 49], two light beams $E_{\text{in}}(t)$ and $E_{\text{LO}}(t)$, which we assume to be composed of single modes of light, are introduced in a beam splitter.[11] In this configuration, $E_{\text{in}}(t)$ represents the beam of light whose quantum state we intend to characterize, while $E_{\text{LO}}(t)$ is commonly referred to as the local oscillator. The local oscillator usually consists of an intense laser field with controllable relative phase with respect to $E_{\text{in}}(t)$. These two beams are combined in the beam splitter, resulting in the fields $E_1(t)$ and $E_2(t)$ whose intensities are subsequently measured, as illustrated in Fig. 2.6a. For a 50:50 beam splitter, the amplitudes of these two fields are related to those of $E_{\text{in}}(t)$ and $E_{\text{LO}}(t)$ by

[11] Here, we omit position dependence $\mathbf{r}$ as the beams are subsequently measured at specific locations. The difference in the path length taken by each beam after the beam splitter contributes to the relative phase between the two beams.

$$E_1(t) = \frac{1}{\sqrt{2}}\big[E_{\rm LO}(t) + E_{\rm in}(t)\big], \tag{2.69}$$

$$E_2(t) = \frac{1}{\sqrt{2}}\big[E_{\rm LO}(t) - E_{\rm in}(t)\big], \tag{2.70}$$

such that the powers P_1 and P_2 detected by the photodetectors measuring these fields is given by

$$P_1 \propto \langle E_1(t)^2\rangle = \frac{1}{2}\Big[\langle E_{\rm LO}(t)^2\rangle + \langle E_{\rm in}(t)^2\rangle + 2\langle E_{\rm LO}(t)E_{\rm in}(t)\rangle\Big], \tag{2.71}$$

$$P_2 \propto \langle E_2(t)^2\rangle = \frac{1}{2}\Big[\langle E_{\rm LO}(t)^2\rangle + \langle E_{\rm in}(t)^2\rangle - 2\langle E_{\rm LO}(t)E_{\rm in}(t)\rangle\Big], \tag{2.72}$$

where the $\langle\cdot\rangle$ notation represents here time-averaging over a certain number of cycles. Thus, finally subtracting these two powers, we end up getting

$$\Delta I = I_1 - I_2 \propto \langle E_{\rm LO}(t)E_{\rm in}(t)\rangle. \tag{2.73}$$

On the other hand, by combining Eqs. (2.10) and (2.39), for a single-mode of the electromagnetic field we can write

$$E_{\mathbf{k},\mu}(t) = \frac{1}{\sqrt{\varepsilon_0 V}}\Big[-\omega_{\mathbf{k}}X_{\mathbf{k},\mu}\sin(\omega_{\mathbf{k}}t) + P_{\mathbf{k},\mu}\cos(\omega_{\mathbf{k}}t)\Big], \tag{2.74}$$

where we have omitted out the dependence with the position $\mathbf{r}$ for simplicity. By substituting this into Eq. (2.73), and integrating over a certain number of optical cycles,[12] we obtain

$$\Delta I \propto \omega_{\mathbf{k}}^2 X_{\mathbf{k},\mu}^{(\rm LO)} X_{\mathbf{k},\mu}^{(\rm in)} + P_{\mathbf{k},\mu}^{(\rm LO)} P_{\mathbf{k},\mu}^{(\rm in)}. \tag{2.75}$$

This result is fundamental because, as mentioned earlier, the local oscillator is considered a well-controlled field with perfectly known parameters. By changing these, we can obtain any linear combination of the quadrature operators, from which the momentum-like and position-like distributions can be deduced. According to the properties of Wigner functions, these distributions are indeed the marginals of the associated Wigner functions. Consequently, an inverse transformation[13] of these measured marginals allows us to experimentally obtain the Wigner distribution of the $E_{\rm in}(t)$ field. In Fig. 2.6b, we present an example of an experimental reconstruction of a vacuum state $|n_{\mathbf{k},\mu} = 0_{\mathbf{k},\mu}\rangle$, as described in Ref. [48], where the Wigner distribution corresponds to a Gaussian.

[12] $\int_0^{2n\pi} dx \sin(x)\cos(x) = 0$ and $\int_0^{2n\pi} dx \sin[2](x) = \int_0^{2n\pi} dx \cos[2](x) = \pi$ when $n \in \mathbb{N}$.

[13] One of the most commonly used methods for this purpose is the inverse Radon transformation (see e.g. Ref. [50] and references therein).

Finally, we conclude this section about the Wigner function by providing an alternative, but equivalent, definition to the one provided in Eq. (2.60). According to A. Royer, the Wigner function of a given density matrix $\hat{\rho}$ can be also computed as [51]

$$W(\beta) = \frac{2}{\pi}\text{tr}(\hat{\rho}\hat{D}_{\mathbf{k},\mu}(\beta)\Pi\hat{D}_{\mathbf{k},\mu}(-\beta)), \tag{2.76}$$

where $X_{\mathbf{k},\mu} \equiv \text{Re}(\beta)$ and $P_{\mathbf{k},\mu} \equiv \text{Im}(\beta)$. In this expression, $\hat{D}_{\mathbf{k},\mu}(\beta)$ is the commonly known as displacement operator acting on the field mode $(\mathbf{k}, \mu)$, and which is given by

$$\hat{D}_{\mathbf{k},\mu}(\beta) = \exp\left[\beta\hat{a}^{\dagger}_{\mathbf{k},\mu} - \beta^{*}\hat{a}_{\mathbf{k},\mu}\right], \tag{2.77}$$

with $\beta \in \mathbb{C}$. Here, β is related to the optical quadratures by $X_{\mathbf{k},\mu} \equiv \text{Re}(\beta)$ and $P_{\mathbf{k},\mu} \equiv \text{Im}(\beta)$. In the remaining of this Thesis, we stick to this formulation of the Wigner function.

2.3.3 *A Summary of Different Quantum Optical States and Their Properties*

So far, we have introduced observables that can be used to discern between classical and non-classical states of light. In the following, we present some states relevant for this Thesis and its context, classifying them in classical or non-classical according to the aforementioned observables.

2.3.3.1 Fock States

As mentioned earlier, the set of Fock states or photon number states $\{|n_{\mathbf{k},\mu}\rangle : n \in \mathbb{W}\}$ are eigenstates of the free-field Hamiltonian obtained in Eq. (2.45) with eigenvalues $\hbar\omega_{\mathbf{k}}(n + 1/2)$. Interestingly, even though the energy for Fock states is non-zero, including the vacuum state $|n_{\mathbf{k},\mu} = 0_{\mathbf{k},\mu}\rangle$, the mean value of the electric field for these states is zero for all values of n

$$\begin{aligned}\langle n_{\mathbf{k},\mu}|\hat{E}_{\mathbf{k},\mu}(t)|n_{\mathbf{k},\mu}\rangle = i\sqrt{\frac{\hbar\omega_{\mathbf{k}}}{2\varepsilon_0 V}}\Big[&\sqrt{n}\langle n_{\mathbf{k},\mu}|n_{\mathbf{k},\mu} - 1\rangle e^{-i\omega_{\mathbf{k}}t} \\ &- \sqrt{n+1}\langle n_{\mathbf{k},\mu}|n_{\mathbf{k},\mu} + 1\rangle e^{-i\omega_{\mathbf{k}}t}\Big] = 0,\end{aligned} \tag{2.78}$$

while they have non-vanishing field fluctuations, which linearly increase with the number of photons n

$$\Delta E^2_{\mathbf{k},\mu}(t) = \frac{\hbar\omega_{\mathbf{k}}}{2\varepsilon_0 V}\langle n_{\mathbf{k},\mu}|\hat{E}^2_{\mathbf{k},\mu}(t)|n_{\mathbf{k},\mu}\rangle = \frac{\hbar\omega_{\mathbf{k}}}{2\varepsilon_0 V}(2n+1), \tag{2.79}$$

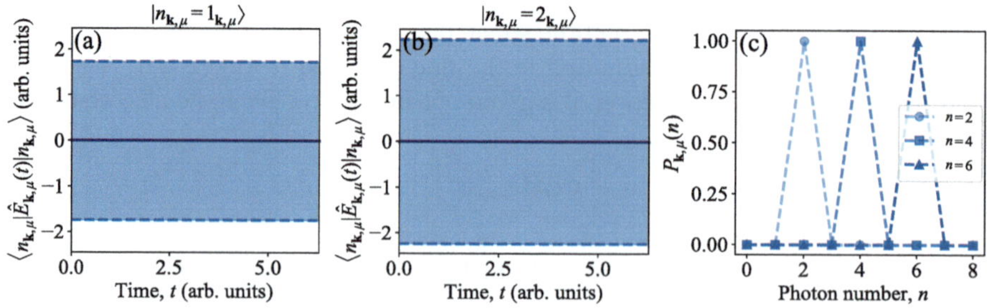

Fig. 2.7 **Mean value and variance of the electric field operator for different Fock states alongside their photon statistics.** In panels **a** and **b**, we depict the mean value (represented by the dark blue curve) and the variance (illustrated as the lighter blue region) of the electric field operator for $|n_{\mathbf{k},\mu} = 1_{\mathbf{k},\mu}\rangle$ and $|n_{\mathbf{k},\mu} = 2_{\mathbf{k},\mu}\rangle$, respectively. In **c**, the photon number statistics of various Fock states are displayed as a function of the photon number

as shown in Fig. 2.7a and b, where the mean value of the field (dark blue curve) along with the corresponding uncertainties (light blue region delimited by the dashed lines) are displayed for $|n_{\mathbf{k},\mu} = 1_{\mathbf{k},\mu}\rangle$ and $|n_{\mathbf{k},\mu} = 2_{\mathbf{k},\mu}\rangle$, respectively. From these results, we can conclude that Fock states do not satisfactorily recover the notion of oscillating electric fields that is expected from classical descriptions of light. Furthermore, considering the uncertainties in the field quadratures we obtain

$$\left.\begin{aligned} \Delta X^2_{\mathbf{k},\mu} &= \frac{\hbar}{2\omega_{\mathbf{k}}}(2n+1) \\ \Delta P^2_{\mathbf{k},\mu} &= \frac{\hbar\omega_{\mathbf{k}}}{2}(2n+1) \end{aligned}\right\} \Rightarrow \Delta X_{\mathbf{k},\mu}\Delta P_{\mathbf{k},\mu} = \frac{\hbar}{2}(2n+1) \geq \frac{\hbar}{2}, \tag{2.80}$$

that is, they exceed the minimum uncertainty of $\hbar/2$ according to Heisenberg's principle [52], except when $n = 0$ for which this minimum value saturates.

Regarding their photon number probability distribution, Fock states have a well-defined photon number, which results in deterministic photon-number probability distributions

$$P_{\mathbf{k},\mu}(n) = \left|\langle n_{\mathbf{k},\mu}|n'_{\mathbf{k},\mu}\rangle\right|^2 = \delta_{n,n'}, \tag{2.81}$$

with mean value $\langle\hat{n}_{\mathbf{k},\mu}\rangle = n$ and variance $(\Delta n_{\mathbf{k},\mu})^2 = 0$. This corresponds to an extremely sub-Poissonian scenario where $(\Delta n^2_{\mathbf{k},\mu}) < \langle\hat{n}_{\mathbf{k},\mu}\rangle$ for $n_{\mathbf{k},\mu} > 0$, leading to $g^{(2)}_{\mathbf{k},\mu}(0) = 1 - 1/\langle\hat{n}_{\mathbf{k},\mu}\rangle \leq 1$ according to Eq. (2.59). Therefore, based on their sub-Poissonian statistics and their time-zero second-order autocorrelation function, Fock states can be regarded as non-classical states of light whenever $n > 0$.

The presence of non-classical behaviors also extends to their Wigner function representation. For a Fock state $|n_{\mathbf{k},\mu}\rangle$, this function is given by

$$W(\beta) = \langle n_{\mathbf{k},\mu}|\hat{D}(\beta)\Pi\hat{D}(-\beta)|n_{\mathbf{k},\mu}\rangle = (-1)^n e^{-\frac{1}{2}|2\beta|^2} L_n(|2\beta|^2), \tag{2.82}$$

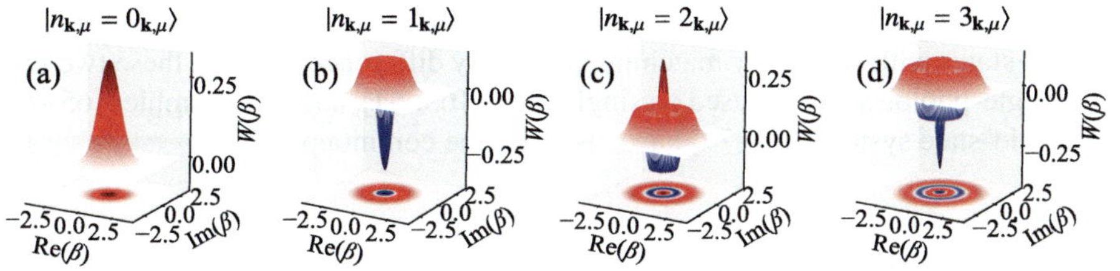

Fig. 2.8 Wigner functions distribution of different Fock states. In panels **a** through **d**, we present the Wigner distributions of Fock states, spanning from $n = 0$ to $n = 3$. With the exception of the first case in the left, i.e. a vacuum state, all the others exhibit negative regions along the z axis. The central peak's sign—if positive or negative—depends on whether n is even or odd, respectively

where $L_n(|2\beta|^2)$ is a Laguerre polynomial,[14] which is obtained from the fact that the matrix elements of the displacement operator in the Fock basis are given by $\langle n_{\mathbf{k},\mu}|\hat{D}(\beta)|n_{\mathbf{k},\mu}\rangle = e^{-|\beta|^2/2}L_n(|\beta|^2)$. In Fig. 2.8, we present the Wigner function for various Fock states, each characterized by a different photon number n. As observed, the behavior of the Wigner function varies depending on the considered photon number. Specifically, when $n = 0$, the Wigner distribution is described by a Gaussian, which is in agreement with the first experimental observations presented in Fig. 2.6b. However, for $n > 0$, the Wigner function oscillates between positive and negative regions, maintaining a symmetrical distribution around the origin ($\text{Re}(\beta) = 0, \text{Im}(\beta) = 0$). Interestingly, when n assumes even values, the Wigner distribution exhibits a global maximum at the center. Conversely, for n being an odd number, it manifests a global minimum.

These distinctive features exhibited by the Wigner function of Fock states lead us to classify them as non-classical states of light. It is important to emphasize that, for Fock states, the three criteria outlined earlier seamlessly converge, unequivocally categorizing Fock states as either classical (for $n = 0$) or as non-classical (for $n > 0$) states. Nevertheless, this unified classification does not necessarily hold true for other quantum states, as we illustrate in various forthcoming examples.

Lastly, the experimental generation of pure Fock states presents a significant challenge, with many methods primarily focused on approximating single-photon states. Among the most common methods employed for this purpose are Spontaneous Parametric Down Conversion (SPDC) sources [53–56]. In SPDC, a powerful laser beam interacts with a nonlinear crystal yielding, with some probability, pairs of photons that have orthogonal polarizations. From these events, the generation of single-photon states with a specified polarization can be achieved when a photon with the orthogonal polarization is measured, leading to single-photon sources that could be subsequently utilized in quantum information science experiments [57–62]. However,

[14] $L_n(x) \equiv \sum_{k=0}^{n} \binom{n}{k} \frac{(-1)^k}{k!} x^k$.

given that many transitions between excited and ground states are mediated by single-photon states with frequency matching the energy difference between these two levels, single-photon sources based on single atoms [63, 64], atomic ensembles [65–73] and solid-state systems [74–79] have also become commonplace.

2.3.3.2 Coherent States

In the process of characterizing the properties of Fock states, we reached the conclusion that these states fall short in faithfully recovering the classical concept of the electromagnetic field, typically envisioned as an oscillating field, in stark contrast with the results of Fig. 2.7a and b. At this juncture, a fundamental question arises: *What constitutes the optimal quantum optical description of classical oscillating fields, precisely as they manifest when solving Maxwell's equations in vacuum?* This question stands as well-posed, given that, in numerous practical scenarios, we find ourselves tasked with describing interactions between light and matter where the input source corresponds to one of these classical oscillating fields. A prominent example is provided by the SPDC sources previously discussed. Consequently, a comprehensive quantum optical characterization of such interactions necessitates a meticulously characterized initial state of light.

The answer to the question posed in the previous paragraph was notably provided by Roy J. Glauber in his seminal works Refs. [80, 81], where he introduced the concept of *coherent states of light*. Before delving into the properties of these states, let us first gain an intuitive understanding on how they naturally emerge. Experimental evidence, most notably from the pioneering experiments by Heinrich Hertz, firmly establishes that an oscillating charge current $\mathbf{J}(\mathbf{r}, t) \in \mathbb{R}^3$ possesses the capability to generate an electromagnetic field in free space that behaves in accordance with the predictions of Maxwell's equations. As such, if coherent states of light truly reproduce the properties of these classical electromagnetic fields, they should naturally arise when we consider a quantum optical description of the interaction between this classical current and the electromagnetic field. The Hamiltonian characterizing this interaction is described by[15]

$$\hat{H} = \hat{H}_{\text{field}} + \int \mathrm{d}\mathbf{r}\, \mathbf{J}(\mathbf{r}, t) \cdot \hat{\mathbf{A}}(\mathbf{r}), \tag{2.83}$$

which in the interaction picture with respect to the free-field Hamiltonian can be expressed as[16]

---

[15] This Hamiltonian is analogous to the one presented in Eq. (2.20). However, in this context, it is necessary to consider a charge density rather than a single current. Moreover, for simplicity we neglect the contribution of the diamagnetic term $\hat{\mathbf{A}}(\mathbf{r})^2$.

[16] In the interaction picture with respect to $\hat{H}_{\text{field}}$ the annihilation operator acquires a time dependence $\hat{a}_{\mathbf{k},\mu} \to \hat{a}_{\mathbf{k},\mu} e^{-i\omega_{\mathbf{k}} t}$.

$$\hat{H}_{\text{int}} = \int \mathrm{d}\mathbf{r}\, \mathbf{J}(\mathbf{r}, t) \cdot \hat{\mathbf{A}}(\mathbf{r}, t), \tag{2.84}$$

with the vector potential operator defined in Eq. (2.46). Thus, the Schrödinger equation describing the dynamics of this system is given by

$$\begin{aligned} i\hbar\frac{\partial|\Phi(t)\rangle}{\partial t} &= \int \mathrm{d}\mathbf{r}\, \mathbf{J}(\mathbf{r}, t) \cdot \hat{\mathbf{A}}(\mathbf{r}, t)\, |\Phi(t)\rangle \\ &= \sum_{\mathbf{k},\mu} \sqrt{\frac{\hbar}{2\omega_{\mathbf{k}}\varepsilon_0 V}} \int \mathrm{d}\mathbf{r}\, \mathbf{J}(\mathbf{r}, t) \cdot \boldsymbol{\epsilon}_{\mathbf{k},\mu}\Big[\hat{a}_{\mathbf{k},\mu} e^{i(\mathbf{k}\cdot\mathbf{r}-\omega_{\mathbf{k}}t)} + \text{h.c.}\Big]|\Phi(t)\rangle \end{aligned} \tag{2.85}$$

where $|\Phi(t)\rangle$ represents the quantum optical state, which we consider to be initially in a vacuum state in all field modes $|\Phi(t_0)\rangle = \bigotimes_{\mathbf{k},\mu}|0_{\mathbf{k},\mu}\rangle$.

From Eq. (2.85), we observe that the classical current couples independently and simultaneously to all the electromagnetic field modes. However, the field modes do not couple with each other. This implies that the time-evolution operator $\hat{U}(t)$ describing the quantum optical dynamics can be expressed as the product of operators acting independently in each mode $(\mathbf{k}, \mu)$, that is

$$\begin{aligned} \hat{U}(t, t_0) &= \prod_{\mathbf{k},\mu} \hat{U}_{\mathbf{k},\mu}(t, t_0) \\ &= \prod_{\mathbf{k},\mu} \hat{\mathcal{T}} \exp\left\{-\frac{i}{\hbar}\sqrt{\frac{\hbar}{2\omega_{\mathbf{k}}\varepsilon_0 V}} \int_{t_0}^{t} \mathrm{d}\tau \int \mathrm{d}\mathbf{r}\, \mathbf{J}(\mathbf{r}, \tau) \cdot \boldsymbol{\epsilon}_{\mathbf{k},\mu}\Big[\hat{a}_{\mathbf{k},\mu} e^{i(\mathbf{k}\cdot\mathbf{r}-\omega_{\mathbf{k}}\tau)} + \text{h.c.}\Big]\right\}, \end{aligned} \tag{2.86}$$

where $\hat{U}_{\mathbf{k},\mu}(t)$ is the time-evolution operator acting on mode $(\mathbf{k}, \mu)$,[17] and $\hat{\mathcal{T}}$ represents the time-ordering operator. For simplicity, let us focus on a single mode $(\mathbf{k}, \mu)$, and discretize the time interval $[t_0, t]$ into N steps of size Δt. In the limit $N \to \infty$, we can express $\hat{U}_{\mathbf{k},\mu}(t, t_0)$ as

$$\hat{U}_{\mathbf{k},\mu}(t, t_0) = \lim_{N\to\infty} \prod_{i=0}^{N-1} \hat{U}_{\mathbf{k},\mu}(t_{i+1}, t_i), \tag{2.87}$$

where we identify $t_N = t$. For each of the time steps of size Δt, we can express the corresponding time-evolution operator as

$$\hat{U}_{\mathbf{k},\mu}(t_{i+1}, t_i) = \exp\Big[\hat{O}_{\mathbf{k},\mu}(t_{i+1})\Delta t\Big], \tag{2.88}$$

with $\hat{O}_{\mathbf{k},\mu}(t_{i+1})$ defined as

[17] Formally, this operator is defined as $\hat{U}_{\mathbf{k},\mu}(t) \equiv \hat{U}_{\mathbf{k},\mu}(t) \bigotimes_{\mathbf{k}'\neq\mathbf{k},\mu'\neq\mu} \mathbb{1}_{\mathbf{k}',\mu'}$.

$$\hat{O}_{\mathbf{k},\mu}(t_{i+1}) \equiv -\frac{i}{\hbar}\sqrt{\frac{\hbar}{2\omega_{\mathbf{k}}\varepsilon_0 V}}\int \mathrm{d}\mathbf{r}\, \mathbf{J}(\mathbf{r},\tau)\cdot\boldsymbol{\epsilon}_{\mathbf{k},\mu}\Big[\hat{a}_{\mathbf{k},\mu}e^{i(\mathbf{k}\cdot\mathbf{r}-\omega_{\mathbf{k}}\tau)} + \text{h.c.}\Big] \\ = f(t)\hat{a}^\dagger_{\mathbf{k},\mu} - f^*(t)\hat{a}_{\mathbf{k},\mu}. \tag{2.89}$$

In this last expression, $f(t)$ is a function containing the spatial integral and all the complex-valued functions multiplying the annihilation operator. This form proves highly convenient, as the commutator of the operator $\hat{O}(t_i)$ with itself, evaluated at different instants of time, is proportional to the identity. To be more precise,

$$[\hat{O}(t_i), \hat{O}(t_j)] = \big(f^*(t_i)f(t_j) - f(t_i)f^*(t_j)\big)\mathbb{1}, \tag{2.90}$$

which facilitates the implementation of the Baker-Campbell-Hausdorff (BCH) formula (see e.g. Refs. [24, 82] and references therein) for combining the operators that act during the different and infinitesimal time intervals $(t_i, t_{i+1}]$ in Eq. (2.87). This convenience arises because all higher-order commutators that appear in the BCH formula's definition[18] vanish. Consequently, by applying this relation to Eq. (2.87) and taking the limit $N \to \infty$, we arrive at

$$\hat{U}_{\mathbf{k},\mu}(t, t_0) = e^{i\varphi_{\mathbf{k},\mu}(t)} \exp\Big[\alpha_{\mathbf{k},\mu}(t)\hat{a}^\dagger_{\mathbf{k},\mu} - \alpha^*_{\mathbf{k},\mu}(t)\hat{a}_{\mathbf{k},\mu}\Big], \tag{2.91}$$

where $\varphi_{\mathbf{k},t}(t)$ is a phase factor arising from the BCH expansion, and is given by

$$\varphi_{\mathbf{k},\mu}(t) = \frac{1}{\hbar\omega_{\mathbf{k}}\varepsilon_0 V}\int \mathrm{d}\mathbf{r}\int_{t_0}^{t}\mathrm{d}t_1\int_{t_0}^{t_1}\mathrm{d}t_2\,\big[\mathbf{J}(\mathbf{r},t_1)\cdot\boldsymbol{\epsilon}_{\mathbf{k},\mu}\big]\big[\mathbf{J}(\mathbf{r},t_2)\cdot\boldsymbol{\epsilon}_{\mathbf{k},\mu}\big]\sin[\omega_{\mathbf{k}}(t_1-t_2)], \tag{2.92}$$

while for $\alpha_{\mathbf{k},\mu}(t)$ we get

$$\alpha_{\mathbf{k},\mu}(t) = \frac{1}{i\hbar}\frac{\hbar}{2\omega_{\mathbf{k}}\varepsilon_0 V}\int_{t_0}^{t}\mathrm{d}\tau\int \mathrm{d}\mathbf{r}\,\big[\mathbf{J}(\mathbf{r},\tau)\cdot\boldsymbol{\epsilon}_{\mathbf{k},\mu}\big]e^{-i(\mathbf{k}\cdot\mathbf{r}-\omega_{\mathbf{k}}\tau)}. \tag{2.93}$$

Comparing the result obtained in Eq. (2.91) with the displacement operator given in Eq. (2.77) we can conclude that, as a consequence of the coupling of the classical electric current to the vector potential operator, the initial vacuum state undergoes a displacement by a quantity $\alpha_{\mathbf{k},\mu}(t)$. Notably, this displacement corresponds to the Fourier transform of the vectorial components of the charge current. For instance, if the charge density is moving in a specific spatial direction, the displacement operator would exclusively excite electromagnetic field modes polarized along that particular direction. In general, we can express the quantum state of the system after the time-evolution as

$$|\Phi(t)\rangle = \bigotimes_{\mathbf{k},\mu}\hat{D}_{\mathbf{k},\mu}\big(\alpha_{\mathbf{k},\mu}(t)\big)|0_{\mathbf{k},\mu}\rangle \equiv \bigotimes_{\mathbf{k},\mu}|\alpha_{\mathbf{k},\mu}(t)\rangle, \tag{2.94}$$

[18] The BCH formula is given by $e^{\hat{X}}e^{\hat{Y}} = e^{\hat{Z}}$ with $\hat{Z} = \hat{X} + \hat{Y} + (1/2)[\hat{X},\hat{Y}] + (1/12)[\hat{X},[\hat{X},\hat{Y}]] + \cdots$.

where we define $|\alpha_{\mathbf{k},\mu}(t)\rangle \equiv \hat{D}_{\mathbf{k},\mu}(\alpha_{\mathbf{k},\mu}(t))|0_{\mathbf{k},\mu}\rangle$, representing them as displaced vacuum states. According to our derivation, we designate the set $\{|\alpha_{\mathbf{k},\mu}\rangle : \alpha_{\mathbf{k},\mu} \in \mathbb{C}\}$[19] as coherent states of light defined on mode ($\mathbf{k}$, μ), since they emerge similarly to classical electromagnetic fields from oscillating charge currents. In the remainder of this section, we substantiate this nomenclature by computing various quantities to demonstrate that these states closely resemble the behavior of classical electromagnetic waves. Furthermore, we establish that they satisfy the criteria of *coherence* as defined by Roy J. Glauber.

To proceed, let us first introduce some properties of this set of coherent states[20]:

1. *Coherent states are eigenvectors of the annihilation operator, with eigenvalues equal to the (generally complex) amplitudes of the corresponding coherent states.*

   This can be straightforwardly demonstrated by considering the following property of displacement operators

$$\hat{D}^\dagger(\alpha)\hat{a}\hat{D}(\alpha) = \hat{a} + \alpha, \tag{2.95}$$

   together with the fact that they are unitary, i.e., $\hat{D}^\dagger(\alpha)\hat{D}(\alpha) = \hat{D}(\alpha)\hat{D}^\dagger(\alpha) = \mathbb{1}$. Specifically, when applying the annihilation operator to a coherent state, we can express it as

$$\hat{a}\,|\alpha\rangle = \hat{D}(\alpha)\hat{D}^\dagger(\alpha)\hat{a}\hat{D}(\alpha)\,|0\rangle = \hat{D}(\alpha)(\hat{a} + \alpha)\,|0\rangle = \alpha\,|\alpha\rangle\,. \tag{2.96}$$

2. *Coherent states can be expanded in the Fock basis as*

$$|\alpha\rangle = e^{-|\alpha|^2/2}\sum_{n=0}^{\infty}\frac{\alpha^n}{\sqrt{n!}}\,|n\rangle\,. \tag{2.97}$$

   To prove this property, let us project a coherent state $|\alpha\rangle$ onto a given Fock state $|n\rangle$. By using the property introduced in Eq. (2.95), we can express it as

$$\begin{aligned}\langle n|\hat{D}(\alpha)|0\rangle &= \langle 0|\,\frac{\hat{a}^n}{\sqrt{n!}}\hat{D}(\alpha)\,|0\rangle = \langle 0|\,\hat{D}(\alpha)\frac{(\hat{a}+\alpha)^n}{\sqrt{n!}}D(\alpha)\,|0\rangle \\ &= \frac{\alpha^n}{\sqrt{n!}}\langle 0|\hat{D}(\alpha)|0\rangle = \frac{\alpha^n}{\sqrt{n!}}e^{-|\alpha|^2/2}.\end{aligned} \tag{2.98}$$

[19] When referring to coherent states, we keep the subscript in the definition of the complex number α, unlike what we did for the case of Fock states, for convenience later on in this Thesis.

[20] In this context, we omit the field modes subindexes, as these properties hold true for any quantum harmonic oscillator, regardless on their physical interpretation.

3. *The set of coherent states form an overcomplete basis, with closeness relation given by*

$$\mathbb{1} = \frac{1}{\pi}\int d^2\alpha\, |\alpha\rangle\langle\alpha|\,, \tag{2.99}$$

the integral performed over the complex plane, i.e., $d^2\alpha \equiv d\mathrm{Re}(\alpha)d\mathrm{Im}(\alpha)$.
To verify this property, let us project the right-hand side of the equation onto Fock states $|m\rangle$ and $|n\rangle$

$$\frac{1}{\pi}\int d^2\alpha \langle n|\alpha\rangle\langle\alpha|m\rangle = \frac{1}{\pi\sqrt{n!m!}}\int d^2\alpha e^{-|\alpha|^2}\alpha^n(\alpha^*)^m, \tag{2.100}$$

and by expressing the complex number α as $\alpha = |\alpha|\, e^{i\theta}$, we obtain

$$\begin{aligned}\frac{1}{\pi}\int d^2\alpha \langle n|\alpha\rangle\langle\alpha|m\rangle &= \frac{1}{\pi\sqrt{n!m!}}\int d^2\alpha e^{-|\alpha|^2}\,|\alpha|^{n+m}\,e^{i\theta(n-m)} \\ &= \frac{1}{\pi\sqrt{n!m!}}\int_0^{2\pi} d\theta e^{i\theta(n-m)}\int_0^{\infty} d\, e^{-|\alpha|^2}\,|\alpha|^{n+m+1} \\ &= \frac{2}{n!}\delta_{n,m}\int_0^{\infty} d\,|\alpha|\; e^{-|\alpha|^2}\,|\alpha|^{2n+1} = \delta_{n,m} = \langle n|\mathbb{1}|m\rangle\,,\end{aligned} \tag{2.101}$$

where in transitioning from the first equality to the second, we employed polar coordinates.

4. *The overlap between two coherent states* $|\alpha\rangle$ *and* $|\beta\rangle$ *is non-vanishing and given by*

$$\langle\beta|\alpha\rangle = e^{-\frac{1}{2}(|\beta|^2+|\alpha|^2-2\beta^*\alpha)}. \tag{2.102}$$

This property can be straightforwardly demonstrated by considering that, given two displacement operators $\hat{D}(\beta)$ and $\hat{D}(\alpha)$, their combined action can be expressed as

$$\hat{D}(\beta)\hat{D}(\alpha) = e^{\frac{1}{2}(\beta\alpha^*-\beta^*\alpha)}D(\alpha+\beta), \tag{2.103}$$

which enables us to rewrite the inner product between coherent states as

$$\begin{aligned}\langle\beta|\alpha\rangle &= \langle 0|\,D^\dagger(\beta)D(\alpha)\,|0\rangle = e^{-\frac{1}{2}(\beta\alpha^*-\beta^*\alpha)}\langle 0|\alpha-\beta\rangle \\ &= e^{-\frac{1}{2}(\beta\alpha^*-\beta^*\alpha)}e^{-\frac{1}{2}(|\alpha|^2+|\beta|^2-\beta^*\alpha-\beta\alpha^*)} = e^{-\frac{1}{2}(|\beta|^2+|\alpha|^2-2\beta^*\alpha)},\end{aligned} \tag{2.104}$$

where we took into account that $\hat{D}(-\beta) = \hat{D}^\dagger(\beta)$.

With these properties in mind, let us first proceed to demonstrate how these states capture the concept of classical electromagnetic fields. To do so, we calculate the expected value of the electric field operator for a specific mode $(\mathbf{k}, \mu)$

$$\begin{aligned}\langle\alpha_{\mathbf{k},\mu}|\hat{E}_{\mathbf{k},\mu}(t)|\alpha_{\mathbf{k},\mu}\rangle &= \sqrt{\frac{\hbar\omega_{\mathbf{k}}}{2\varepsilon_0 V}}\boldsymbol{\epsilon}_{\mathbf{k},\mu}\big[i\alpha_{\mathbf{k},\mu}e^{-i\omega_{\mathbf{k}}t}+\text{c.c.}\big]\\ &= 2\sqrt{\frac{\hbar\omega_{\mathbf{k}}|\alpha_{\mathbf{k},\mu}|^2}{2\varepsilon_0 V}}\sin(\omega_{\mathbf{k}}t-\theta_{\mathbf{k},\mu}),\end{aligned} \tag{2.105}$$

where we represent the coherent state amplitude as $\alpha_{\mathbf{k},\mu} = |\alpha_{\mathbf{k},\mu}|e^{i\theta_{\mathbf{k},\mu}}$. Additionally, for the variance of the electric field we obtain

$$\Delta\hat{E}^2_{\mathbf{k},\mu}(t) = \langle\alpha_{\mathbf{k},\mu}|\hat{E}^2_{\mathbf{k},\mu}(t)|\alpha_{\mathbf{k},\mu}\rangle - \langle\alpha_{\mathbf{k},\mu}|\hat{E}_{\mathbf{k},\mu}(t)|^2\alpha_{\mathbf{k},\mu}\rangle = \frac{\hbar\omega_{\mathbf{k}}}{2\varepsilon_0 V}, \tag{2.106}$$

that is, it remains constant regardless of the magnitude of the mean photon number[21] $\langle\hat{a}^\dagger_{\mathbf{k},\mu}\hat{a}_{\mathbf{k},\mu}\rangle = |\alpha_{\mathbf{k},\mu}|^2$. This stands in contrasts with what we observed for Fock states, where the field's variance increased linearly with the number of photons. This distinctive behavior stems from the fact that, in the case of coherent states, the uncertainty relation between the field quadratures reaches its minimum possible value. More precisely, we have

$$\left.\begin{aligned}\Delta X^2_{\mathbf{k},\mu} &= \frac{\hbar}{2\omega_{\mathbf{k}}}\\ \Delta P^2_{\mathbf{k},\mu} &= \frac{\hbar\omega_{\mathbf{k}}}{2}\end{aligned}\right\} \Rightarrow \Delta X_{\mathbf{k},\mu}\Delta P_{\mathbf{k},\mu} = \frac{\hbar}{2}. \tag{2.107}$$

Therefore, based on Eqs. (2.105) and (2.106), we observe that the mean value of the field operator with respect to the set of coherent states recovers the solution of Maxwell's equations in vacuum[22] (illustrated by the dark blue curve in Fig. 2.9a), and exhibits an amplitude-independent field variance (depicted by the light blue region in Fig. 2.9a). Thus far, according to Eq. (2.107), these states appear to provide an optimal description of the classical field. Yet, there is a fundamental property they must satisfy in order to be classified as coherent states, which is the property of *coherence* itself. As previously mentioned, Glauber's definition of a coherent source is one for which all normalized autocorrelation functions, defined in Eq. (2.57), have magnitude equal to unity. Here, to compute the quantum optical version of these autocorrelation functions, we make the replacement $\hat{E}^+_{\mathbf{k},\mu}(\mathbf{r},t)\propto\boldsymbol{\epsilon}_{\mathbf{k},\mu}\hat{a}_{\mathbf{k},\mu}e^{i(\mathbf{k}\cdot\mathbf{r}-\omega_{\mathbf{k}}t)}$ and $\hat{E}^-_{\mathbf{k},\mu}(\mathbf{r},t)\propto\boldsymbol{\epsilon}_{\mathbf{k},\mu}\hat{a}^\dagger_{\mathbf{k},\mu}e^{-i(\mathbf{k}\cdot\mathbf{r}-\omega_{\mathbf{k}}t)}$ in that expression. Considering that $\langle\alpha_{\mathbf{k},\mu}|\hat{E}^+_{\mathbf{k},\mu}(\mathbf{r},t)|\alpha_{\mathbf{k},\mu}\rangle \propto \boldsymbol{\epsilon}_{\mathbf{k},\mu}\alpha_{\mathbf{k},\mu}e^{i(\mathbf{k}\cdot\mathbf{r}-\omega_{\mathbf{k}}t)}$ and that $\langle\alpha_{\mathbf{k},\mu}|\hat{E}^-_{\mathbf{k},\mu}(\mathbf{r},t)|\alpha_{\mathbf{k},\mu}\rangle \propto \boldsymbol{\epsilon}_{\mathbf{k},\mu}\alpha^*_{\mathbf{k},\mu}e^{-i(\mathbf{k}\cdot\mathbf{r}-\omega_{\mathbf{k}}t)}$, the evaluation of the nth-order autocorrelation function for these coherent states leads to

[21] Note that, for a coherent state $|\alpha_{\mathbf{k},\mu}\rangle$, its mean photon number is given by $\langle\alpha_{\mathbf{k},\mu}|\hat{a}^\dagger_{\mathbf{k},\mu}\hat{a}_{\mathbf{k},\mu}|\alpha_{\mathbf{k},\mu}\rangle = |\alpha_{\mathbf{k},\mu}|^2$.

[22] It is worth noting that in our calculations we have restricted to $\mathbf{r} = \mathbf{0}$. However, similar results are obtained for arbitrary values of $\mathbf{r}$.

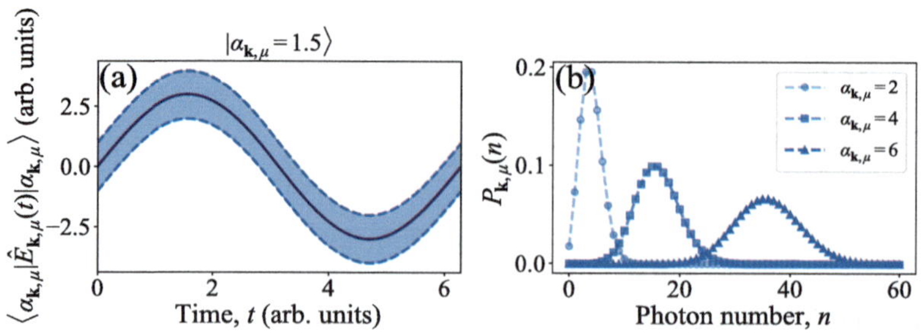

Fig. 2.9 Mean value and variance of the electric field operator for coherent states alongside their photon statistics. In **a**, we depict the mean value (indicated by the dark blue curve) and the variance (illustrated as the lighter blue region) of the electric field operator for $|\alpha_{\mathbf{k},\mu} = 1.5\rangle$. In **b**, the photon number statistics of various coherent states are displayed as a function of the photon number. These distributions exhibit a Poissonian behavior with their peak values located at $|\alpha_{\mathbf{k},\mu}|^2$, and the variance of the distribution increasing proportionally with this quantity

$$\left|g^{(n)}(\mathbf{r}_1, \cdots, \mathbf{r}_{2n}; t_1, \cdots, t_{2n})\right| = \left| \frac{\prod_{j=1}^{2n}\left[G^{(1)}_{\mathbf{k},\mu}(\mathbf{r}_j, \mathbf{r}_j; t_j, t_j)\right]^{1/2}}{\prod_{j=1}^{2n}\left[G^{(1)}_{\mathbf{k},\mu}(\mathbf{r}_j, \mathbf{r}_j; t_j, t_j)\right]^{1/2}} \right| = 1 \quad \forall n \in \mathbb{N}, \tag{2.108}$$

which justifies the name given to these states.

There are two other important properties that can be deduced from Eq. (2.108). The first property is that coherent states, when subjected to free evolution, consistently satisfy Eq. (2.108) at all times. This property is implicitly evident in the preceding expression, assessed within the interaction picture with respect to free-field evolution. The second property is that if this expression holds for all values of n, it will hold in particular for $g^{(2)}_{\mathbf{k},\mu}(0)$. Therefore, in accordance with Eq. (2.59), this implies that the photon statistics of coherent states follow a Poissonian distribution. Specifically, the probability of having $n_{\mathbf{k},\mu}$ photons in a coherent state $|\alpha_{\mathbf{k},\mu}\rangle$ is given by

$$P_{\mathbf{k},\mu}(n) = \left|\langle n_{\mathbf{k},\mu}|\alpha_{\mathbf{k},\mu}\rangle\right|^2 = \frac{|\alpha_{\mathbf{k},\mu}|^{2n}}{n!} e^{-\frac{|\alpha_{\mathbf{k},\mu}|^2}{2}}, \tag{2.109}$$

which corresponds to a Poissonian distribution with a mean value $\langle \hat{n}_{\mathbf{k},\mu}\rangle = |\alpha_{\mathbf{k},\mu}|^2$. In Fig. 2.9b, we illustrate this distribution for different values of $\alpha_{\mathbf{k},\mu}$.

Based on the observables mentioned earlier for probing the presence of non-classical light, both photon statistics and the autocorrelation functions classify coherent states of light as classical states. Not surprisingly, the same conclusion is reached when considering their Wigner function, which does not exhibit negativities. As shown in Fig. 2.8a, where the Wigner function of the vaccum state is shown, the resulting distribution shows a Gaussian-like behavior. Specifically, for any coherent state with a non-vanishing amplitude, its Wigner function is identical to that of a vacuum state but displaced in phase space. This can be demonstrated easily by substituting the definition of coherent states as displaced vacuum states into Eq. (2.76)

$$\begin{aligned}W_{\mathbf{k},\mu}(\beta) &= \langle 0_{\mathbf{k},\mu}|\hat{D}_{\mathbf{k},\mu}(-\alpha)\hat{D}_{\mathbf{k},\mu}(\beta)\hat{\Pi}\hat{D}_{\mathbf{k},\mu}(-\beta)\hat{D}_{\mathbf{k},\mu}(\alpha)|0_{\mathbf{k},\mu}\rangle\\ &= \langle 0_{\mathbf{k},\mu}|\hat{D}_{\mathbf{k},\mu}(\beta-\alpha)\hat{\Pi}\hat{D}_{\mathbf{k},\mu}(-\beta+\alpha)|0_{\mathbf{k},\mu}\rangle\\ &= \langle 0_{\mathbf{k},\mu}|\hat{D}_{\mathbf{k},\mu}(\tilde{\beta})\hat{\Pi}\hat{D}_{\mathbf{k},\mu}(-\tilde{\beta})|0_{\mathbf{k},\mu}\rangle,\end{aligned} \tag{2.110}$$

where in the last equality we defined $\tilde{\beta} \equiv \beta - \alpha$.

2.3.3.3 Squeezed Coherent States

In our previous derivation, we observed that coherent states emerge when a displacement operator, i.e. a unitary operator dependent on a linear combination of creation and annihilation operators, is applied to a vacuum state. However, in the field of nonlinear optics, it is common to encounter scenarios where creation and annihilation operators raised to second-order powers or higher do indeed appear, which are related to nonlinear effects. For instance, let us consider the case of second harmonic generation processes, where a photon with frequency of 2ω is utilized to generate two photons of frequency ω. This *light-light* interaction process is described by the Hamiltonian[23] [83]

$$\hat{H} = \hbar\kappa\Big[\hat{a}_{2\omega}(\hat{a}^\dagger_\omega)^2 + \hat{a}^\dagger_{2\omega}\hat{a}^2_\omega\Big], \tag{2.111}$$

where κ is a coupling constant. In this expression, we have omitted the polarization subscript and replaced the wavevector $\mathbf{k}$ with the associated frequency of the corresponding mode for simplicity.

In processes of this kind, the *pump* field of frequency 2ω, which drives the interaction, is typically characterized by a coherent state $|\alpha_{2\omega}\rangle$ of very high intensity, while the field of frequency ω is in a vacuum state. A common practice in perturbative nonlinear optical processes is to employ the parametric approximation [83]. This approximation involves treating certain aspects of the interaction as parameters. Specifically, we treat the pump field classically and neglect its depletion, allowing us to approximate Eq. (2.111) as follows

$$\hat{H} \approx \hbar\kappa\,|\alpha_{2\omega}|\,\big[(\hat{a}^\dagger_\omega)^2 e^{i\theta} + \hat{a}^2_\omega e^{-i\theta}\big], \tag{2.112}$$

where we have expressed $\alpha_{2\omega} = |\alpha_{2\omega}|\,e^{i\theta}$. It is important to note that this approximated Hamiltonian holds in regimes where $\kappa\Delta t \to 0$, $\alpha_{2\omega} \to \infty$ and $\kappa\alpha_{2\omega}\Delta t =$ constant [83], where $\Delta t = t - t_0$ represents the interaction time between the 2ω-field and the corresponding medium. From this Hamiltonian, it straightforwardly follows that the time-evolution describing the dynamics of this system within the time window Δt is given by

[23] The Hamiltonian in Eq. (2.111) is given in the interaction picture with respect to the free-field Hamiltonian.

$$\hat{S}_\omega(\varkappa) = \exp\left[\frac{1}{2}\left(\varkappa^* \hat{a}_\omega^2 - \varkappa \hat{a}_\omega^{\dagger 2}\right)\right], \tag{2.113}$$

where $\varkappa = i2\kappa\, |\alpha_{2\omega}|\, e^{i\theta}$. Thus, the state of the ω-field after the interaction is given by $|\varkappa, 0_{\mathbf{k},\mu}\rangle = \hat{S}(\varkappa)\, |0_\omega\rangle$.

The above example is common in quantum optics, illustrating how unitary operators can naturally depend on linear combinations of second-order powers of creation and annihilation operators, individually. However, this particular example is fundamental [84] as it gives rise to what is typically referred to as *squeezed coherent* states [85–87], defined as the set $\{|\varkappa, \alpha_{\mathbf{k},\mu}\rangle = \hat{S}_{\mathbf{k},\mu}(\varkappa)\hat{D}_{\mathbf{k},\mu}(\alpha)|0_{\mathbf{k},\mu}\rangle : \varkappa, \alpha \in \mathbb{C}\}$ for a single field mode $(\mathbf{k}, \mu)$, where the operator in Eq. (2.113) is commonly known as the squeezing operator [88]. Due to their properties, these states have found significant applications in improving the measurement sensitivity in interferometric setups [89–93]. What makes them particularly important for this application is their ability to exhibit reduced variances for specific quadrature operators when compared to coherent states. In particular, for real values of $\varkappa$, we obtain

$$\left.\begin{aligned} \Delta X_{\mathbf{k},\mu}^2 &= \frac{\hbar}{2\omega_{\mathbf{k}}} e^{-2\varkappa} \\ \Delta P_{\mathbf{k},\mu}^2 &= \frac{\hbar\omega_{\mathbf{k}}}{2} e^{2\varkappa} \end{aligned}\right\} \Rightarrow \Delta X_{\mathbf{k},\mu} \Delta P_{\mathbf{k},\mu} = \frac{\hbar}{2}, \tag{2.114}$$

from which it can be seen that squeezed coherent states saturate the lowest Heisenberg uncertainty. Nevertheless, in this case the variance along one of the quadratures becomes smaller than for coherent states while along the other becomes greater. This means that measurements involving the corresponding quadrature operator would be less noisy when using squeezed coherent states than when using coherent states.

The reduced variances concerning specific quadratures, which are clearly observable in the Wigner function representation of these states (as shown in Fig. 2.10c and f for $\varkappa > 0$ and $\varkappa < 0$, respectively) have implications on the electric field's behavior. As depicted in panels (a) and (d), they can result in a reduced uncertainty in the electric field's amplitude or phase, depending on the values of $\varkappa$ considered previously. Consequently, the precision with which the electric field's amplitude is known has consequences on the photon statistics. These effects are evident in panel (b), where a sub-Poissonian behavior is observed for $\varkappa > 0$. Conversely, if the phase is known with greater accuracy instead, the resulting photon statistics exhibit a super-Poissonian behavior, as illustrated in panel (e).

Although squeezed coherent states do not exhibit Wigner negativities and may not necessarily display sub-Poissonian statistics, they are generally categorized as non-classical states of light. This classification mainly arises from the behavior of uncertainties along the different quadratures we have just discussed about, an effect that cannot be explained by classical electromagnetism.

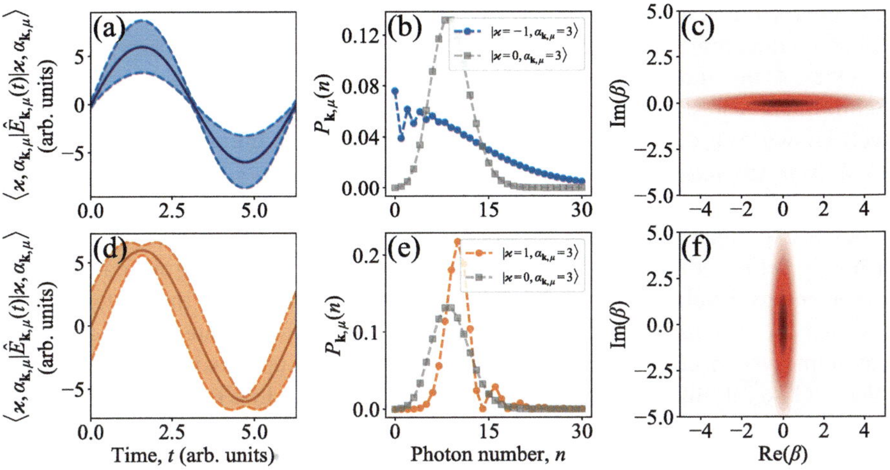

Fig. 2.10 Mean value and variance of the electric field operator for various squeezed states alongside their photon statistics and Wigner functions. In panels **a** and **d**, we illustrate the mean value (represented by the central darker curve) and the variance (illustrated as the lighter region) of two displaced squeezed states with $\alpha_{\mathbf{k},\mu} = 3.0$ and $\kappa = -1.0$ (phase-squeezed states) and $\kappa = 1.0$ (amplitude-squeezed states), respectively denoted with blue and orange colors. In panels **b** and **e**, the photon number statistics of these states are compared with that of a coherent state having the same amplitude (in gray color with squared markers), revealing super-Poissonian and sub-Poissonian statistics, respectively. Finally, in panels **c** and **f**, the Wigner function of these two states for $\alpha_{\mathbf{k},\mu} = 0.0$ and $\varkappa = -1.0$ and $\varkappa = 1.0$, respectively, is presented, highlighting squeezing features along perpendicular directions

2.3.3.4 Coherent State Superpositions

While coherent states are classical states of light, a linear combination of coherent states with varying complex-valued amplitudes $\alpha_{\mathbf{k},\mu}$, can exhibit strong non-classical characteristics. In mathematical terms, a coherent state superposition is typically represented as

$$|\Phi\rangle = \sum_{i=1}^{N} a_i |\alpha^{(i)}_{\mathbf{k},\mu}\rangle \quad \text{with } \alpha^{(i)}_{\mathbf{k},\mu} \neq \alpha^{(j)}_{\mathbf{k},\mu} \ \forall i \neq j, \tag{2.115}$$

where $N \geq 2$ and a_i is the probability amplitude weighting each coherent state $|\alpha^{(i)}_{\mathbf{k},\mu}\rangle$. Due to quantum interference effects among the various terms within this linear combination, these states may exhibit distinctive features such as Wigner negativities or photon statistics that deviate from the expected Poissonian distribution of individual coherent states. Notably, these characteristics are highly dependent on the differences in amplitude between the coherent states composing the superposition. The larger this difference, the lower the overlap between the individual states within the superposition, as described in Eq. (2.102) discussed earlier.[24] This lower overlap enhances quantum interference effects between the terms of the superposition.

One of the most characteristic examples in this context is that of optical Schrödinger cat states, which represent the quantum optical counterpart to the well-known Schrödinger's cat state. The concept was initially introduced as a *Gedankenexperiment* by Erwin Schrödinger in Ref. [94], where he contemplated an hypothetical scenario permitted by the principles of quantum mechanics. In this thought experiment, Schrödinger's cat is confined within a sealed box alongside a highly lethal poison that is triggered by the decay of an atomic particle to a lower energy level. When no measurements are conducted on the system meaning that the box remains closed and unobserved, the total state of the combined cat-atom system can be described as a macroscopic superposition of the form[25] $|\Psi\rangle = (1/\sqrt{2})[|\text{alive}\rangle \otimes |\text{excited}\rangle + |\text{dead}\rangle \otimes |\text{ground}\rangle]$, that is, in a coherent combination of the cat being both alive and dead, with the atom in an excited and ground state respectively, until a measurement is performed. The term *macroscopic* here underscores the possibility that the two states in this superposition could potentially manifest observable effects at a macroscopic scale, such as the cat displaying signs of life, movement or other macroscopic behaviors. Remarkably, it has been recently contemplated that superpositions of this form can be achieved with certain biological systems[26] such as bacteria [95, 96] or tardigrades [97]. In the context of quantum optics, the analog to Schrödinger's cat is given by coherent state superpositions. Consequently, optical Schrödinger cat states are defined as $|\Phi\rangle = a_1|\alpha^{(1)}_{\mathbf{k},\mu}\rangle + a_2|\alpha^{(2)}_{\mathbf{k},\mu}\rangle$, where the amplitude or phase of the different coherent states within the superposition could potentially be distinguished at a macroscopic scale.

The two most characteristic examples of optical Schrödinger cat states are commonly referred to as *even* (denoted as $|\Phi_+\rangle$) and *odd* (denoted as $|\Phi_-\rangle$) cat states, defined as follows

$$|\Phi_+\rangle = \frac{1}{\sqrt{\mathcal{N}_+}}\Big[|\alpha_{\mathbf{k},\mu}\rangle + |-\alpha_{\mathbf{k},\mu}\rangle\Big] \quad \text{with} \quad \mathcal{N}_+ = 2 + 2e^{-2|\alpha_{\mathbf{k},\mu}|^2}, \tag{2.116}$$

$$|\Phi_-\rangle = \frac{1}{\sqrt{\mathcal{N}_-}}\Big[|\alpha_{\mathbf{k},\mu}\rangle - |-\alpha_{\mathbf{k},\mu}\rangle\Big] \quad \text{with} \quad \mathcal{N}_- = 2 - 2e^{-2|\alpha_{\mathbf{k},\mu}|^2}, \tag{2.117}$$

and show photon statistics consisting of either even or odd Fock states, respectively, as a consequence of the quantum interference between the two distinct coherent states. This becomes evident when we incorporate the relation in Eq. (2.97) into the

---

[24] It is worth noting that, as per the properties of coherent states discussed earlier, $|\langle\alpha^{(i)}_{\mathbf{k},\mu}|\alpha^{(j)}_{\mathbf{k},\mu}\rangle| \to 0$ as $|\alpha^{(j)}_{\mathbf{k},\mu} - \alpha^{(i)}_{\mathbf{k},\mu}| \to \infty$.

[25] Specifically, one would have to wait a time equal to one radioactive half-time of the atom in order to get such a superposition.

[26] Unlike the original Schrödinger cat experiment, biological systems subjected to such superpositions do not ultimately survive or succumb, but rather their behavior is examined in terms of light emission.

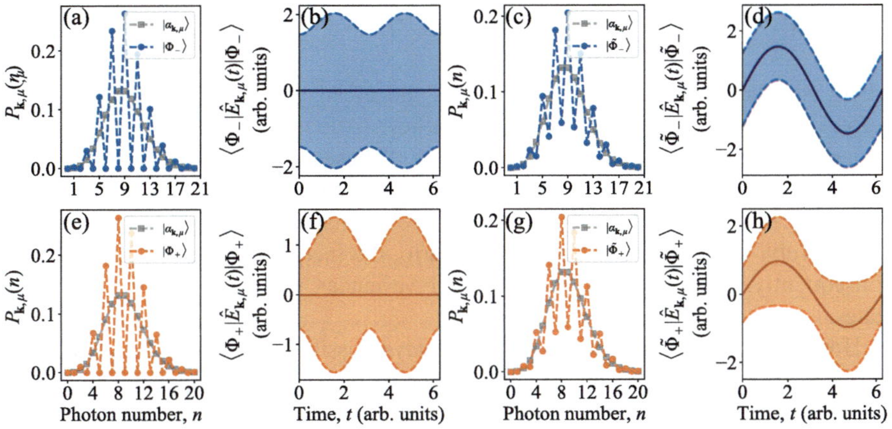

Fig. 2.11 Mean value and variance of the electric field operator for various coherent state superpositions alongside their photon statistics. In panels **a** and **e**, we illustrate the photon number statistics of odd and even cat states, comparing it with that of a coherent state of amplitude $\alpha_{\mathbf{k},\mu} = 3.0$ state (represented by the gray curve with squared markers). The same is done in panels (c) and (g) for unbalanced coherent states superpositions of the form $|\tilde{\Phi}\rangle = 0.7|\alpha_{\mathbf{k},\mu}\rangle \pm 0.3|-\alpha_{\mathbf{k},\mu}\rangle$. Conversely, in panels **b**, **f**, **d** and **h** we illustrate the mean value (represented by the central darker curve) and its variance (illustrated as the lighter region) for perfectly balanced (**b** for odd **f** for even) and unbalanced (**d** for odd and **h** for even) coherent state superpositions

definitions of these coherent state superpositions. Specifically, for even cat states, we obtain

$$\begin{aligned}|\Phi_+\rangle &= \frac{e^{-|\alpha_{\mathbf{k},\mu}|^2/2}}{\mathcal{N}_+}\left[\sum_{n=0}^{\infty}\big(1+(-1)^n\big)\frac{\alpha_{\mathbf{k},\mu}^n}{\sqrt{n!}}\,|n_{\mathbf{k},\mu}\rangle\right]\\ &= \frac{e^{-|\alpha_{\mathbf{k},\mu}|^2/2}}{\mathcal{N}_+}\sum_{n=0}^{\infty}\frac{\alpha_{\mathbf{k},\mu}^{2n}}{\sqrt{(2n)!}}\,\big|(2n)_{\mathbf{k},\mu}\big\rangle,\end{aligned} \tag{2.118}$$

while following the same procedure we get for the odd version of these states

$$\begin{aligned}|\Phi_-\rangle &= \frac{e^{-|\alpha_{\mathbf{k},\mu}|^2/2}}{\mathcal{N}_-}\left[\sum_{n=0}^{\infty}\big(1+(-1)^{n+1}\big)\frac{\alpha_{\mathbf{k},\mu}^n}{\sqrt{n!}}\,|n_{\mathbf{k},\mu}\rangle\right]\\ &= \frac{e^{-|\alpha_{\mathbf{k},\mu}|^2/2}}{\mathcal{N}_-}\sum_{n=0}^{\infty}\frac{\alpha_{\mathbf{k},\mu}^{2n+1}}{\sqrt{(2n+1)!}}\,\big|(2n+1)_{\mathbf{k},\mu}\big\rangle.\end{aligned} \tag{2.119}$$

The corresponding photon statistics of these states are depicted in Fig. 2.11a and e, for odd and even cat states respectively. In these plots, it is evident that peaks in the photon number probability distribution are obtained for odd and even values of n, respectively. Moving on the electric field's mean value and variances, we obtain[27]

[27] In deriving this expression, we assumed $\alpha_{\mathbf{k},\mu} \in \mathbb{R}$ for simplicity.

$$\langle\Phi_\pm|\hat{E}_{\mathbf{k},\mu}(t)|\Phi_\pm\rangle = 0, \quad \Delta E^2_{\mathbf{k},\mu}(t) = \langle\Phi_\pm|\hat{E}^2_{\mathbf{k},\mu}(t)|\Phi_\pm\rangle = 2\alpha^2_{\mathbf{k},\mu}\left(\frac{\mathcal{N}_\mp}{\mathcal{N}_\pm} - \cos(\omega_{\mathbf{k}}t)\right) + 1, \tag{2.120}$$

where, similar to the case of Fock states, the mean value of the electric field is zero even when $\alpha_{\mathbf{k},\mu} \neq 0$. Nevertheless, a key distinction lies in the electric field variance, which exhibits associated oscillations. These oscillations become negligible as $\alpha_{\mathbf{k},\mu} \to \infty$, ultimately resulting in $\mathcal{N}_\mp = \mathcal{N}_\pm$. In panels (b) and (f), the mean value of the electric field as well as its variance are displayed for odd and even states, respectively, with $\alpha_{\mathbf{k},\mu} = 0.7$. In this scenario, the inequality $\mathcal{N}_+ > \mathcal{N}_-$ holds, and consequently, odd cat states exhibit larger variances when compared to even cat states.

However, in many situations, it is not always possible to generate these perfectly balanced even and odd superpositions, where both coherent states appear in the superposition with equal probability. More generally, we may encounter regimes where the superpositions take the form $|\Phi_\pm\rangle = a_+|\alpha_{\mathbf{k},\mu}\rangle + a_-|-\alpha_{\mathbf{k},\mu}\rangle$ with $|a_+| \neq |a_-|$. We refer to these states as *unbalanced* coherent state superpositions, more specifically as unbalanced even cat states $|\tilde{\Phi}_+\rangle$, and unbalanced odd cat states $|\tilde{\Phi}_-\rangle$. In these instances, the dominance of one of the coherent states within the superposition leads to photon statistics that exhibit behavior closer to a Poissonian distribution, as shown in Fig. 2.11c and g, while still retaining peaks at odd and even values which are characteristic of odd and even cat states. This imbalance has consequences for the mean value of the field, which now displays the expected oscillating behavior due to the larger contribution of one coherent state over the other, as depicted in panels (d) and (h).

Another key characteristic of even and odd cat states is their Wigner function representation, where the quantum interference between the two coherent states results in a pattern of oscillating negative and positive regions. This characteristic is illustrated in Fig. 2.12 for two different values of $\alpha_{\mathbf{k},\mu}$. Specifically, when $\alpha_{\mathbf{k},\mu} \gtrsim 2$, it becomes possible to distinguish the two coherent states in the superposition, and an interference pattern emerges due to quantum interference. When the two coherent states in the superposition have equal probability amplitudes, both maxima representing the coherent states within the superposition are equal, as shown in panels (a) and (e). In contrast, for unbalanced superpositions, one maximum becomes greater than the other, as depicted in panels (b) and (f). Conversely, as $\alpha_{\mathbf{k},\mu}$ decreases, there is a limit where, for balanced cat states, the two coherent states in the superposition cannot be distinguished. This limit becomes evident in panels (c) and (g). In these figures, we observe that, for odd cat states, a homogeneous doughnut-like shape similar to that of a single-photon state emerges.[28] On the other hand, even cat states yield a squeezed-like state with a maximum centered at the origin.[29] When the coherent

[28] It is worth noting that in the regime $\alpha_{\mathbf{k},\mu} \ll 1$, odd cat states can be well approximated by single-photon states.

[29] In the regime $\alpha_{\mathbf{k},\mu} \ll 1$, even cat states can be well approximated by vacuum states.

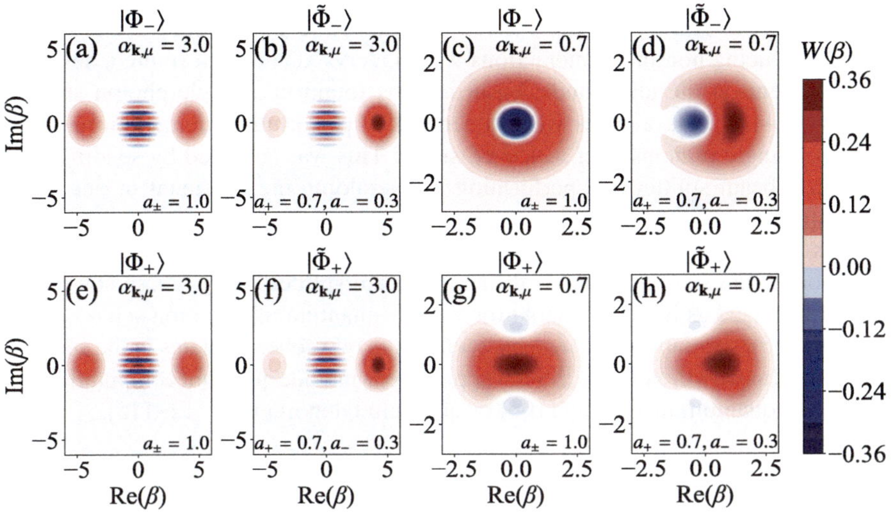

Fig. 2.12 **Wigner functions of different coherent state superpositions.** In panels **a** through **d**, the Wigner function of odd cat states is presented, where **a** and **c** correspond to balanced coherent state superpositions with $\alpha_{\mathbf{k},\mu} = 3.0$ and $\alpha_{\mathbf{k},\mu} = 0.7$ respectively, while panels **b** and **d** correspond to unbalanced superpositions of the form $|\tilde{\Phi}_{-}\rangle = 0.7|\alpha_{\mathbf{k},\mu}\rangle - 0.3|-\alpha_{\mathbf{k},\mu}\rangle$. Conversely, in panels **e** through **h**, the Wigner distribution of even cat states is shown, where panels **e** and **g** correspond to balanced coherent state superpositions with $\alpha_{\mathbf{k},\mu} = 3.0$ and $\alpha_{\mathbf{k},\mu} = 0.7$ respectively, while panels **f** and **h** to unbalanced superpositions of the form $|\tilde{\Phi}_{+}\rangle = 0.7|\alpha_{\mathbf{k},\mu}\rangle + 0.3|-\alpha_{\mathbf{k},\mu}\rangle$

state superposition is unbalanced, one of the maxima becomes more pronounced than the other, and the resulting Wigner functions become less symmetric compared to the balanced cases, as illustrated in panels (d) and (h).

In practice, various methods have been developed to generate these states, with many of them relying on heralding measurements. Heralding measurements involve performing a measurement on an additional system, typically known as an ancilla, that is entangled[30] with the system in which we aim to generate the desired state. The measurement on the ancilla projects the state of the other part of the system into a specific one, with the result obtained from the ancilla's measurement providing information about which state has been generated. Specific outcomes from this measurement correspond to the desired state, in our case to specific coherent state superpositions.

Some of the early experimental implementations for generating Schrödinger optical cat states were based on ion traps or cavity Quantum Electrodynamics (QED) systems (see e.g. Refs. [98, 99] and references therein). In these setups, the final state of ions trapped in the ion trap or atoms inside the cavity was utilized as the heralding system to generate even or odd cat states on the optical state. However, in recent years, full optical based methods have been employed for this purpose.

[30] The concept of entanglement is further discussed in Sect. 2.3.4.

For instance, in Ref. [100], unbalanced coherent state superpositions were generated by inducing nonlinear interactions within a crystal and performing a heralding measurement on the idler mode, ensuring it is found in a single-photon state. In Refs. [101–103], these states were generated by performing subtraction measurements of single photons over squeezed states. This was achieved by sending them through a beam splitter and performing the heralding measurement in one of the beam splitter outputs. Similar methods have been proposed to generate these states, such as squeezing a single-photon [104] or performing an homodyne detection on a portion of the state that passes through a beam splitter [105]. These states have been recognized as highly valuable for various quantum information science applications [106], specially when $\alpha_{\mathbf{k},\mu} > 2$ where the two coherent states in the superposition are almost orthonormal.[31] Such applications include quantum computation [29, 109–111], quantum metrology [106] or quantum teleportation [112–115].

2.3.4 Quantum Information Measures for Probing Non-classical Light

Up to this point, our discussions have mostly revolved around scenarios involving single field modes $(\mathbf{k}, \mu)$, and we have introduced several quantum optical observables that can be employed to distinguish classical from non-classical states of light. However, the postulates of quantum mechanics permit the description of composite systems [116], where the total Hilbert space can be expressed as the tensor product of individual Hilbert spaces for each constituent system, i.e., $\mathcal{H}_T \equiv \bigotimes_{i=1}^N \mathcal{H}_i$, with N the total number of individual systems. This fundamental postulate also paves the way for one of the most profound concepts in quantum mechanics with no classical counterpart—entanglement. The idea of entanglement has been a subject of intense scrutiny over the years, contributing to a deeper understanding of quantum mechanics [94, 117].

Throughout this chapter, we have briefly described how the presence of entanglement plays a pivotal role in generating certain non-classical states of light, such as single-photon states or optical Schrödinger cat states. Beyond these specific applications, the concept of entanglement holds a central position in the realm of quantum information science [118]. It finds practical utility in a multitude of applications [119], including quantum computation [120–124], and quantum communication [125–127].

Entanglement is associated with the presence of quantum states that cannot be expressed as a tensor product of their individual component systems [116], that is

[31] It is important to note that, the experimental observations mentioned thus far, and some more recent ones such as the one in Ref. [107], have been able to obtain cat states with $\alpha_{\mathbf{k},\mu} < 2$. However, recent theoretical methods based on free-electron-photon interaction could potentially extend this range [108].

$$|\Psi\rangle \in \mathcal{H}_T \text{ is entangled if } |\Psi\rangle \neq \bigotimes_{i=1}^{N} |\psi_{\mathcal{H}_i}\rangle, \tag{2.121}$$

or more generally for mixed states as

$$\hat{\rho} \in \mathrm{B}(\mathcal{H}_T) \text{ is entangled if } \hat{\rho} \neq \sum_{i=1} p_i \bigotimes_{j=1}^{N} \hat{\rho}^{(i)}_{\mathcal{H}_j}, \tag{2.122}$$

with $\mathrm{B}(\mathcal{H}_T)$ the set of bounded operators. Since this concept does not possess a classical counterpart, entangled states are said to be non-classical. However, the concept of *non-classicality* differs here compared to the previous section in that, in term of light states, more than one single field mode is involved.

In general, determining whether a quantum state $\hat{\rho}$ in arbitrary dimensions is entangled or not is known to be NP-hard[32] [128]. However, in some cases, specially when dealing with bipartite systems, one can define mathematical functions that provide a quantitative characterization of entanglement, known as entanglement measures. Here, we briefly introduce some of the measures that are used in the incoming chapters of this Thesis.

2.3.4.1 Measures of Entanglement

Let us begin by considering pure, bipartite states, denoted as $|\psi_{AB}\rangle \in \mathcal{H}_A \otimes \mathcal{H}_B$. In the context of light, the subscripts A and B represent different field modes. These modes can differ in frequency, potentially yielding frequency-entangled states, or they can differ in polarization, potentially leading to polarized-entangled states, or a mix of both. In this context, the von Neumann entropy [129] provides a readily computable entanglement measure, often referred to as entanglement entropy [130, 131], applicable for pure states $\hat{\rho}^2 = \hat{\rho}$. This quantity is defined as

$$\mathcal{S}(\hat{\rho}) := -\mathrm{tr}(\hat{\rho}_A \log_d(\hat{\rho}_A)), \tag{2.123}$$

where $\hat{\rho}_A = \mathrm{tr}_B(\hat{\rho})$, representing the partial trace of the joint state $\hat{\rho} = |\psi_{AB}\rangle\langle\psi_{AB}|$. with respect to one of the bipartitions. This entanglement measure is bounded as $0 \leq \mathcal{S}(\hat{\rho}) \leq 1$, with the lower bound being satisfied for separable states, and the upper bound corresponding to what are typically referred to as maximally entangled states. This can be observed by considering the Schmidt decomposition [116] of the joint state $|\psi_{AB}\rangle$, which is

[32] NP is the set of all decision problems for which instances with an affirmative answer, i.e. a *yes*, can have their proofs verified in polynomial time. NP-complete is the complexity class that includes all problems X in NP to which any other NP problem Y can be reduced to X in polynomial time. Finally, if a problem is classified as NP-hard, it is at least as hard as NP-complete problems, although it may not necessarily belong to NP, nor be a decision problem.

$$|\psi_{AB}\rangle = \sum_{i=0}^{d-1} \lambda_i \, |i_A\rangle \otimes |i_B\rangle \,, \tag{2.124}$$

where λ_i are non-negative coefficients satisfying $\sum_{i=0}^{d-1} \lambda_i^2 = 1$, and $\{|i_A\rangle\}$ and $\{|i_B\rangle\}$ are sets of orthonormal states for $\mathcal{H}_A$ and $\mathcal{H}_B$, respectively, typically referred to as Schmidt basis. From this expression, it follows that the density matrix $\hat{\rho}_A$ can be expressed as

$$\hat{\rho}_A = \sum_{i=0}^{d-1} \lambda_i^2 \, |i_A\rangle\langle i_A| \,, \tag{2.125}$$

and consequently the entanglement entropy reads

$$\mathcal{S}(\hat{\rho}) = -\sum_{i=0}^{d-1} \lambda_i^2 \log_d(\lambda_i^2), \tag{2.126}$$

from which it can be seen that if $\lambda_i = \delta_{0,i}$, then the state is separable resulting in $\mathcal{S}(\hat{\rho}) = 0$. On the other hand, if $|\psi_{AB}\rangle$ is maximally entangled, then $\lambda_i = (1/\sqrt{d})$ and consequently $\mathcal{S}(\hat{\rho}) = 1$.

However, in many cases the Hilbert spaces $\mathcal{H}_A$ and $\mathcal{H}_B$ have extremely large dimensions, making it numerically challenging to find the Schmidt decomposition of these states. In such contexts, the *linear entropy* could be a suitable alternative, which is defined as

$$\mathcal{S}_{\text{lin}}(\hat{\rho}) := 1 - \text{tr}(\hat{\rho}_A^2), \tag{2.127}$$

where $\text{tr}(\hat{\rho}_A^2)$ represents the purity,[33] and therefore this entanglement measure satisfies $0 \leq \mathcal{S}_{\text{lin}}(\hat{\rho}) \leq 1 - (1/d)$ with $\dim(\mathcal{H}_A) = d$. In fact, this entanglement measure can be obtained as a first-order approximation of the von Neumann entropy. Specifically, by expanding the logarithm in polynomial series[34] in Eq. (2.123), we obtain

$$\begin{aligned} \mathcal{S}(\hat{\rho}) &= -\text{tr}(\hat{\rho}_A \log_d(\hat{\rho}_A)) = -\text{tr}\Big(\hat{\rho}_A \frac{1}{\log(d)} \sum_{n=1}^{\infty} \frac{(-1)^{n+1}}{n} (\hat{\rho} - 1)^n\Big) \\ &\approx \frac{1}{\log(d)} \Big[\text{tr}(\hat{\rho}_A) - \text{tr}(\hat{\rho}_A^2)\Big] = \frac{1}{\log(d)} \Big[1 - \text{tr}(\hat{\rho}_A^2)\Big]. \end{aligned} \tag{2.128}$$

Because the linear entropy is directly obtained from the entanglement entropy, it logically follows that it is only applicable for bipartite pure states. However, in many situations, the joint state of the system is characterized by a mixed state $\hat{\rho}$, for which

[33] For pure states, we have $\hat{\rho} = |\psi\rangle\langle\psi|$, and therefore $\text{tr}(\hat{\rho}^2) = 1$. For maximally mixed states, $\hat{\rho} = \sum_{i=0}^{d-1} \frac{1}{d} \, |\psi_i\rangle\langle\psi_i|$ which leads to $\text{tr}(\hat{\rho}^2) = 1/d$.

[34] $\log(1 + x) = \sum_{n=1}^{\infty} [-(-x)^n/n]$.

these entanglement measures are not valid.[35] In this scenario, other entanglement measures are necessary, such as the logarithmic negativity, defined as

$$E_N(\hat{\rho}) := \log(2N_{AB} + 1), \tag{2.129}$$

where N_{AB} is the negativity. The negativity is given as the absolute value of the sum of all negative eigenvalues of $\hat{\rho}^{T_B}$, where $\hat{\rho}^{T_B}$ is the partial transpose with respect to one of the subsystems. For instance, if

$$\hat{\rho} = \sum_{ijkl} p_{k,l}^{i,j} \left|i_A\middle\rangle\!\middle\langle j_A\right| \otimes \left|k_B\middle\rangle\!\middle\langle l_B\right|, \tag{2.130}$$

then for the partial transpose we get

$$\hat{\rho}^{T_B} = \sum_{ijkl} p_{k,l}^{i,j} \left|i_A\middle\rangle\!\middle\langle j_A\right| \otimes (\left|k_B\middle\rangle\!\middle\langle l_B\right|)^{T_B} = \sum_{ijkl} p_{k,l}^{i,j} \left|i_A\middle\rangle\!\middle\langle j_A\right| \otimes \left|l_B\middle\rangle\!\middle\langle k_B\right|. \tag{2.131}$$

This entanglement measure is based in the Positive Partial Transpose (PPT) criterion, also known as the Peres–Horodecki criterion [132, 133]. The PPT criterion serves as a necessary condition for a joint density matrix to be separable. Notably, for Hilbert spaces with dimensions $2 \otimes 2$ and $2 \otimes 3$, the PPT criterion is not only necessary but also sufficient. This means that, in the context of a Hilbert space $\mathcal{H}$, the set of states satisfying the PPT criterion and those that are separable coincide. However, for dimensions greater than these, the absence of non-negative eigenvalues after performing the partial transpose operation does not guarantee that the state is separable, as some entangled states may not show negative eigenvalues. In this case, the set of PPT states becomes greater than that of separable states, as shown in Fig. 2.13a.

In Fig. 2.13b, the various entanglement measures we have just discussed are presented for a qubit pure state of the form $|\psi_{AB}\rangle = \sqrt{p}\,|00\rangle + \sqrt{1-p}\,|11\rangle$. As evident, in all three cases, the degree of entanglement only reaches zero when $p = 0$ or $p = 1$, indicating that the state is separable. Conversely, it reaches its maximum when $p = 1/2$, representing maximally entangled states.

2.3.4.2 Bell Nonlocality

One of the most intriguing consequences of entanglement is its connection to the concept of Bell nonlocality [134], a crucial tool for the development of secure quantum communication protocols as we will discuss in Chap. 5. In typical Bell experiments, there are two distant parties, usually referred to as Alice and Bob, each possessing a system that may have interacted previously. In the case of light, these systems

---

[35] It is worth noting that for the separable mixed state $\hat{\rho} = \frac{1}{2}\left|0_A\middle\rangle\!\middle\langle 0_A\right| \otimes \left|1_B\middle\rangle\!\middle\langle 1_B\right| + \frac{1}{2}\left|1_A\middle\rangle\!\middle\langle 1_A\right| \otimes \left|0_B\middle\rangle\!\middle\langle 0_B\right|$, the entropy of entanglement is $\mathcal{S}(\hat{\rho}) = 1$, even though it is separable.

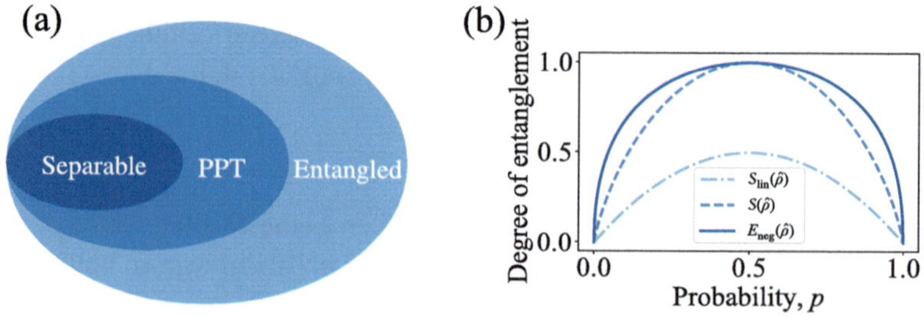

Fig. 2.13 Set of entangled states and entanglement measures to quantify the amount of entanglement. In **a**, a pictorial representation of the set of entangled states is shown, with the innermost subset corresponding to the set of separable states. Enclosing this subset we have the set of PPT states, i.e., those for which their partial transpose is positive. As mentioned in the main text, for composite systems of dimensions $2 \otimes 2$ and $2 \otimes 3$, the sets of PPT and separable states coincide. In **b**, we present the different entanglement measures that have been discussed, computed for states of the form $|\psi\rangle = \sqrt{p}\,|00\rangle + \sqrt{1-p}\,|11\rangle$

could be different field modes $(\mathbf{k}, \mu)$ that come from a common source. In their respective systems, Alice and Bob perform measurements denoted as x and y, generating outcomes a and b, respectively. These outcomes are obtained for a single run of the experiment, and even if Alice and Bob repeat their measurement choices in other runs, the outcomes may vary. However, after conducting the experiment numerous times, Alice and Bob can approximate $P(a, b|x, y)$, which represents the probability distribution of obtaining outcomes a and b when measurements x and y are performed. In what follows, we denote $\boldsymbol{P} = \{P(a, b|x, y)\}$, that is the set of probability distributions for different measurement settings and outcomes, as a *behavior*.

A set of behaviors is termed *local* if their corresponding probability distributions satisfy the relation [134]

$$P(a, b|x, y) = \int_{\Lambda} \mathrm{d}\lambda\, q(\lambda) P(a|x, \lambda) P(b|y, \lambda), \tag{2.132}$$

where λ is typically referred to as *hidden variables*. These variables follow a probability distribution $q(\lambda)$, and represent a set of past factors that may have locally influenced the outcomes obtained by Alice and Bob. However, it is assumed here that these hidden variables do not affect the choice of measurements both parties perform. In other words, and from a classical mechanics point of view, this expression assumes that events occurring in the space-time region where Alice is situated should not have effects in the space-time region where Bob is located. Alternatively, any set of behaviors not belonging to the local set $\mathcal{L}$ is considered *nonlocal*.

Let us now transition to the quantum mechanical framework, where the systems shared by Alice and Bob are generally described by the quantum state $\hat{\rho}$. Accordingly, and in agreement with the postulates of quantum mechanics [116], the measurements of Alice and Bob are defined by quantum operators $\hat{M}_x^A$ and $\hat{M}_y^B$, which can in turn be expressed as a set of operators $\{\hat{M}_{a|x}\}$ and $\{\hat{M}_{b|y}\}$ satisfying the following properties

$$\langle\hat{M}_{a|x}\rangle \geq 0 \,\forall a, x \quad \text{and} \quad \sum_{a=0}^{d_A} \hat{M}_{a|x} = \mathbb{1}, \tag{2.133}$$

$$\langle\hat{M}_{b|y}\rangle \geq 0 \,\forall b, y \quad \text{and} \quad \sum_{b=0}^{d_B} \hat{M}_{b|y} = \mathbb{1}, \tag{2.134}$$

where d_A and d_B are the number of outcomes that Alice and Bob can respectively obtain. The sets of operators satisfying these properties are usually known as Positive-Operator Valued Measures (POVM) [116]. Thus, and according to the Born rule [135], the corresponding probability distributions are given by

$$P(a, b|x, y) = \text{tr}(\hat{\rho}\hat{M}_{a|x} \otimes \hat{M}_{b|y}), \tag{2.135}$$

and the set of behaviors satisfying this kind of distributions is called the *quantum* set $\mathcal{Q}$.

The sets $\mathcal{L}$ and $\mathcal{Q}$ are actually closed, bounded and convex [134]. The last property implies that if behaviors $\boldsymbol{P}_1$ and $\boldsymbol{P}_2$ belong to one of these sets, then a convex combination of the form $\mu\boldsymbol{P}_1 + (1-\mu)\boldsymbol{P}_2$, with $0 \leq \mu \leq 1$, also belongs to the same set. In these cases, the hyperplane separation theorem becomes useful, which establishes that for each behavior $\boldsymbol{P}$ not belonging to a specific convex set, there exists a hyperplane that separates this behavior from the corresponding set. The main consequence this has, from a mathematical viewpoint, is that we can define inequalities of the form

$$\sum_{a,b,x,y} s_{x,y}^{a,b} P(a, b|x, y) \leq S. \tag{2.136}$$

In virtue of this, it was shown in Ref. [136] that the local set actually belongs to the quantum set, i.e. $\mathcal{L} \subset \mathcal{Q}$, and the set of inequalities separating these two sets are the famous Bell inequalities, as shown in Fig. 2.14.[36] The violation of one of these inequalities indicates that there are certain sets of correlations predicted by quantum mechanics that cannot be reproduced by local hidden-variable models. However, thus far, we have not discussed the role of entanglement in this context. In fact, entanglement is a necessary, although generally not sufficient, condition to find nonlocal correlations. For instance, let us consider a separable state of the form $\hat{\rho} = \sum_i p_i \hat{\rho}_A \otimes \hat{\rho}_B$. According to the Born rule, we can write

[36] In this pictorial representation, the non-signaling set $\mathcal{NS}$ [137, 138] is also shown. This set imposes that the marginal probabilities of Alice are independent of Bob's measurement choices, i.e., that Bob cannot affect Alice by his choice of input.

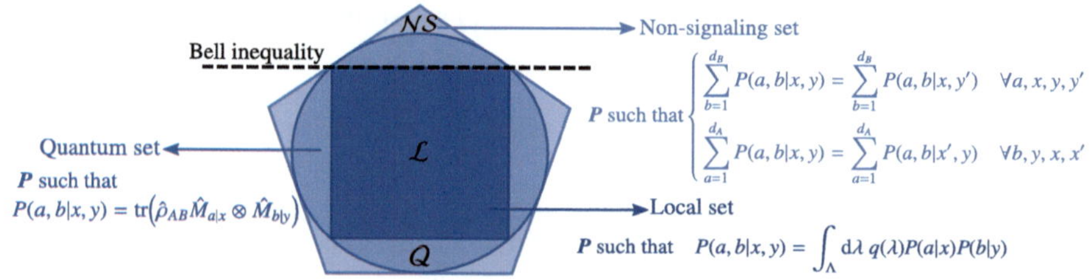

Fig. 2.14 Pictorial representation of the non-signaling, quantum and local sets. The innermost set corresponds to the local ($\mathcal{L}$) set, consisting of behaviors $\boldsymbol{P}$ that admit a local hidden-variable model. Enclosing this set, we find the set of quantum correlations ($\mathcal{Q}$), representing behaviors that satisfy the Born rule. The set of hyperplanes distinguishing behaviors belonging to the local set are the Bell inequalities (black dashed line). Finally, the non-signaling set ($\mathcal{NS}$) is that which only requires that Alice's measurements cannot signal Bob's measures and vice versa, meaning that Alice's measurement choices cannot affect Bob's results

$$
\begin{aligned}
P(a,b|x,y) &= \sum_i p_i \mathrm{tr}[(\hat{\rho}_A \otimes \hat{\rho}_B)(\hat{M}_{a|x} \otimes \hat{M}_{b|y})] = \sum_i p_i \mathrm{tr}(\hat{\rho}_A \hat{M}_{a|x})\mathrm{tr}(\hat{\rho}_B \hat{M}_{b|y}) \\
&= \sum_i p_i P(a|x)P(b|y),
\end{aligned}
\tag{2.137}
$$

which corresponds to a discretized version of the local hidden-variable model in Eq. (2.132). More generally, for pure entangled states it is possible to find local measurements that result in correlations violating a Bell inequality [139–141]. However, this is not the case for mixed states, as demonstrated in Refs. [142, 143], where it was found that certain states, known as Werner states, could exhibit under certain circumstances entanglement features while their correlations could be described using local hidden-variable models.

2.4 Light-Matter Interactions from a Quantum Optical Perspective

Up to this point, we have presented a semiclassical analysis of light-matter interactions, and provided a brief introduction to quantum optics and certain aspects of non-classical states of light. However, the majority of the analysis in this Thesis, specially in Chaps. 3 and 4, revolves around the examination of light-matter interactions from a quantum optical perspective. Therefore, similar to the analysis done in Sect. 2.2, we now introduce the fundamental concepts in this context by considering the interaction of a free electron with an electromagnetic field.

With that purpose in mind, we begin by considering the minimal-coupling Hamiltonian in Eq. (2.18). In terms of the canonical quantization procedure, it can be written as

$$\hat{H}_{\mathrm{VG}} = \frac{1}{2m}\Big[\hat{\mathbf{p}} - Q\hat{\mathbf{A}}(\hat{\mathbf{r}})\Big]^2 + \frac{1}{2}\int \mathrm{d}\mathbf{r}\Big(\varepsilon_0|\hat{\mathbf{E}}(\mathbf{r})|^2 + \frac{1}{\mu_0}\varepsilon_0|\hat{\mathbf{B}}(\mathbf{r})|^2\Big), \tag{2.138}$$

where we have replaced the canonical momentum and the position by their corresponding quantum operators, as well as for the vector potential and the electric and magnetic fields. Similarly to the case of Eq. (2.18), this Hamiltonian is given in its velocity-gauge form. Here, we aim to find a length gauge form similar to that of Eq. (2.27), which will serve as a basis for the analysis in Chaps. 3 and 4. To achieve this, we implement the Power-Zienau-Wolley (PZW) transformation given by

$$\hat{U}_{\mathrm{PZW}} = \exp\left[\frac{i}{\hbar}\int \mathrm{d}\mathbf{r}\ \hat{\mathbf{P}}(\mathbf{r})\cdot\hat{\mathbf{A}}(\mathbf{r})\right]. \tag{2.139}$$

In this expression, $\hat{\mathbf{P}}(\mathbf{r})$ is the operator characterizing the electric polarization. From the perspective of classical electromagnetism, the electric polarization satisfies the following equation

$$\nabla\mathbf{P}(\mathbf{r}) = -\rho(\mathbf{r}), \tag{2.140}$$

where $\rho(\mathbf{r})$ is the charge density. For the case of a single particle of charge Q located at position $\mathbf{r}_\alpha$, the charge density can be expressed as $\rho(\mathbf{r}) = Q\delta(\mathbf{r}-\mathbf{r}_\alpha)$. In this particular case, a solution to the differential equation in Eq. (2.140) can be expressed using a Fourier-integral representation of the Dirac delta function [144]. This expression is given by

$$\mathbf{P}(\mathbf{r}) = Q\int_0^1 \mathrm{d}s\ \mathbf{r}_\alpha\delta(\mathbf{r}-s\mathbf{r}_\alpha), \tag{2.141}$$

where, in order to obtain the associated operator, we apply the canonical quantization rules and replace the electron's position $\mathbf{r}_\alpha$ with the position operator $\hat{\mathbf{r}}$.

By definition, the PZW transformation leaves the position $\hat{\mathbf{r}}$ and the vector potential $\hat{\mathbf{A}}(\mathbf{r})$ operators unchanged, while for the electric field operator we find[37]

$$\hat{\hat{\mathbf{E}}}(\hat{\mathbf{r}}) = \hat{U}^\dagger_{\mathrm{PZW}}\hat{\mathbf{E}}(\hat{\mathbf{r}})\hat{U}_{\mathrm{PZW}} = \hat{\mathbf{E}}(\hat{\mathbf{r}}) - \frac{1}{\varepsilon_0}\hat{\mathbf{P}}(\mathbf{r}), \tag{2.142}$$

while for the canonical momentum operator we can write [144]

$$\hat{\hat{\mathbf{p}}} = \hat{U}^\dagger_{\mathrm{PZW}}\hat{\mathbf{p}}\hat{U}_{\mathrm{PZW}} = \hat{\mathbf{p}} - Q\hat{\mathbf{A}}(\hat{\mathbf{r}}) + \hat{\mathcal{O}}, \tag{2.143}$$

[37] It is worth noting that this operator is equivalent to the displacement operator, so the transformation it yields on the electric field is equivalent.

similar to Eq. (2.22). It is important to note that, in this case, there is a sign mismatch between both results, which arises as a consequence of a difference in the sign convention used in the definitions of the length gauge and the PZW transformations. In this context, the operator $\hat{O}$ includes the additional terms resulting from the transformation.

Applying these transformations to the velocity-gauge version of the Hamiltonian $\hat{H}_{\text{VG}}$, we obtain

$$\begin{aligned}\hat{\tilde{H}} &= \hat{U}^{\dagger}_{\text{PZW}}\hat{H}_{\text{VG}}\hat{U}_{\text{PZW}}\\ &= \frac{1}{2m}\left(\hat{\tilde{\mathbf{p}}} + Q\hat{O}\right)^2 + \frac{1}{2}\int \mathrm{d}\mathbf{r}\left[\varepsilon_0\left|\hat{\tilde{\mathbf{E}}}(\mathbf{r}) - \frac{1}{\varepsilon_0}\hat{\mathbf{P}}(\mathbf{r})\right|^2 + \frac{1}{\mu_0}\varepsilon_0|\hat{\mathbf{B}}(\mathbf{r})|^2\right],\end{aligned} \tag{2.144}$$

which after expanding the terms in this expression, leads to

$$\hat{\tilde{H}} = \frac{\hat{\tilde{\mathbf{p}}}^2}{2m} + \frac{1}{2\varepsilon_0}\int \mathrm{d}\mathbf{r}\,\left|\hat{\mathbf{P}}(\mathbf{r})\right|^2 - \int \mathrm{d}\mathbf{r}\,\hat{\tilde{\mathbf{E}}}(\mathbf{r})\cdot\hat{\mathbf{P}}(\mathbf{r}) + \hat{H}_{\text{field}} + f(\hat{O}), \tag{2.145}$$

where $\hat{H}_{\text{field}}$ represents the free-field Hamiltonian as in Eq. (2.45), and $f(\hat{O})$ is a function that includes all the terms involving the $\hat{O}$ operator. Specifically, the latter includes the diamagnetic energy and the interaction of the magnetic field with the magnetization defined by the oscillating charge [144]. Moreover, it can be shown that when applying the dipole approximation, these terms vanish as they are proportional to spatial derivatives of the vector potential, now a space-independent quantity. In light of this approximation, we can approximate the term involving the scalar product between the electric field and the polarization operators as

$$\begin{aligned}\int \mathrm{d}\mathbf{r}\,\hat{\tilde{\mathbf{E}}}(\mathbf{r})\cdot\hat{\mathbf{P}}(\mathbf{r}) &= Q\int_0^1 \mathrm{d}s\int \mathrm{d}\mathbf{r}\,\hat{\tilde{\mathbf{E}}}(\mathbf{r})\cdot\hat{\mathbf{r}}\delta(\mathbf{r}-s\hat{\mathbf{r}})\\ &= Q\int_0^1 \mathrm{d}s\,\hat{\tilde{\mathbf{E}}}(s\hat{\mathbf{r}})\cdot\hat{\mathbf{r}} \approx Q\hat{\tilde{\mathbf{E}}}\cdot\hat{\mathbf{r}}.\end{aligned} \tag{2.146}$$

Thus, within the dipole approximation, the length gauge Hamiltonian describing the interaction of light with a free charge under a quantum optical perspective is given by

$$\hat{H}_{\text{LG}} = \frac{\hat{\tilde{\mathbf{p}}}^2}{2m} - Q\hat{\tilde{\mathbf{E}}}\cdot\hat{\mathbf{r}} + \hat{H}_{\text{field}} + \frac{1}{2\varepsilon_0}\int \mathrm{d}\mathbf{r}\,\left|\hat{\mathbf{P}}(\mathbf{r})\right|^2, \tag{2.147}$$

which will serve as starting point for the quantum optical analysis presented in Chaps. 3 and 4.

References

1. Encylopæedia Britannica, Akhenaton and Nefertiti under the sun of god Aton, howpublished (2023). Accessed: Friday, 21st Sept 2023. https://www.britannica.com/biography/Nefertiti
2. Wikipedia commons, Book of Optics: title page engraving, howpublished (2023). Accessed Friday, 21st Sept 2023. https://commons.wikimedia.org/wiki/File:Thesaurus_opticus_Titelblatt.jpg
3. Argüello-Luengo J, González-Tudela A, Shi T, Zoller P, Cirac JI (2019) Nature 574:215. https://doi.org/10.1038/s41586-019-1614-4
4. Bohr N (1913) Lond Edinburgh Dublin Philosoph Mag J Sci 26:1. https://doi.org/10.1080/14786441308634955
5. Zeeman P (1896) Verslagen en Mededeelingen der Kon. Academie van Wetenschappen, Afd. Natuurkunde 5:181. https://ui.adsabs.harvard.edu/abs/1896VMKAN...5..181Z
6. Stark J (1913) Nature 92:401. https://doi.org/10.1038/092401b0
7. Maiman TH (1960) Phys Rev Lett 4(564). https://doi.org/10.1103/PhysRevLett.4.564
8. Gordon JP, Zeiger HJ, Townes CH (1954) Phys Rev 95:282. https://doi.org/10.1103/PhysRev.95.282
9. Gordon JP, Zeiger HJ, Townes CH (1955) Phys Rev 99:1264. https://doi.org/10.1103/PhysRev.99.1264
10. Papadogiannis NA, Witzel B, Kalpouzos C, Charalambidis D (1999) Phys Rev Lett 83:4289. https://doi.org/10.1103/PhysRevLett.83.4289
11. Paul PM, Toma ES, Breger P, Mullot G, Augé F, Balcou P, Muller HG, Agostini P (2001) Science 292:1689. https://doi.org/10.1126/science.1059413
12. Corkum PB, Krausz F (2007) Nat Phys 3:381. https://doi.org/10.1038/nphys620
13. Krausz F, Ivanov M (2009) Rev Mod Phys 81:163. https://doi.org/10.1103/RevModPhys.81.163
14. Zuo T, Bandrauk AD, Corkum PB (1996) Chem Phys Lett 259:313. https://doi.org/10.1016/0009-2614(96)00786-5
15. Lein M, Marangos JP, Knight PL (2002) Phys Rev A 66:051404. https://doi.org/10.1103/PhysRevA.66.051404
16. Geneaux R, Marroux HJB, Guggenmos A, Neumark DM, Leone SR (2019) Philosoph Trans R Soc A: Math Phys Eng Sci 377:20170463. https://doi.org/10.1098/rsta.2017.0463
17. Buades B, Picón A, Berger E, León I, Di Palo N, Cousin SL, Cocchi C, Pellegrin E, Martin JH, Mañas-Valero S, Coronado E, Danz T, Draxl C, Uemoto M, Yabana K, Schultze M, Wall S, Zürch M, Biegert J (2021) Appl Phys Rev 8:011408. https://doi.org/10.1063/5.0020649
18. Wineland DJ, Drullinger RE, Walls FL (1978) Phys Rev Lett 40:1639. https://doi.org/10.1103/PhysRevLett.40.1639
19. Lewenstein M, Sanpera A, Ahufiger V (2012) Ultracold atoms in optical lattices: simulating quantum many-body systems. Oxford University Press, Oxford, UK
20. Acín A, Bloch I, Buhrman H, Calarco T, Eichler C, Eisert J, Esteve D, Gisin N, Glaser SJ, Jelezko F, Kuhr S, Lewenstein M, Riedel MF, Schmidt PO, Thew R, Wallraff A, Walmsley I, Wilhelm FK (2018) New J Phys 20:080201. https://doi.org/10.1088/1367-2630/aad1ea
21. Dirac PAM, Fowler RH (1925) Proc R Soc Lond Ser A Contain Papers Math Phys Char 109:642. https://doi.org/10.1098/rspa.1925.0150
22. Loudon R (1973) The quantum theory of light (Oxford University Press, Oxford, UK, 1973) Chap. 3, pp 125–179
23. Scully MO, Zubairy MS (2001) Quantum theory of radiation. In: Quantum optics. Cambridge University Press, Cambridge, UK (Chap. 1), pp 1–45
24. Gerry C, Knight P (2005) Field quantization. In: Introductory quantum optics. Cambridge University Press, Cambridge, UK (Chap. 1), pp 10–42
25. Dirac PAM, Bohr NHD (1927) Proc R Soc Lond Ser A Contain Papers Math Phys Char 114:243. https://doi.org/10.1098/rspa.1927.0039
26. Fermi E (1932) Rev Mod Phys 4:87. https://doi.org/10.1103/RevModPhys.4.87

27. O'Brien JL, Furusawa A, Vučković J (2009) Nat Photonics 3:687. https://doi.org/10.1038/nphoton.2009.229
28. Gisin N, Thew R (2007) Nat Photonics 1:165. https://doi.org/10.1038/nphoton.2007.22
29. Ralph TC, Gilchrist A, Milburn GJ, Munro WJ, Glancy S (2003) Phys Rev A 68:042319. https://doi.org/10.1103/PhysRevA.68.042319
30. Kok P, Munro WJ, Nemoto K, Ralph TC, Dowling JP, Milburn GJ (2007) Rev Mod Phys 79:135. https://doi.org/10.1103/RevModPhys.79.135
31. Aspuru-Guzik A, Walther P (2012) Nat Phys 8:285. https://doi.org/10.1038/nphys2253
32. Giovannetti V, Lloyd S, Maccone L (2011) Nat Photonics 5:222. https://doi.org/10.1038/nphoton.2011.35
33. Pirandola S, Bardhan BR, Gehring T, Weedbrook C, Lloyd S (2018) Nat Photonics 12:724. https://doi.org/10.1038/s41566-018-0301-6
34. Arecchi FT (1965) Phys Rev Lett 15:912. https://doi.org/10.1103/PhysRevLett.15.912
35. Martínez Ricci ML, Mazzaferri J, Bragas AV, Martínez OE (2007) Amer J Phys 75(707). https://doi.org/10.1119/1.2742400
36. Einstein A (1905) Ann Phys 322:891. https://doi.org/10.1002/andp.19053221004
37. Loudon R (1973) The quantum theory of light. Oxford University Press, Oxford, UK (Chap. 3), pp 82–124
38. Loudon R (1973) The quantum theory of light. Oxford University Press, Oxford, UK (Chap. 6), pp 233–287
39. Vogel W, Welsch D-G (2006) Quantum optics. Wiley-VCH Verlag GmbH & Co. KGaA, Weinheim, Germany (Chap. 6), pp 173–236
40. Bose V (1924) Z Phys 26:178. https://doi.org/10.1007/BF01327326
41. Hecht E (2017) Optics, Global Edition. Pearson Education Limited, Harlow, England (Chap. 4), pp 96–158
42. Glauber RJ (1963) Phys Rev 130:2529. https://doi.org/10.1103/PhysRev.130.2529
43. Scully MO, Zubairy MS (2001) Field-field and photon-photon interferometry. In: Quantum optics. Cambridge University Press, Cambridge, UK (Chap. 4), pp 46–71
44. Goldstein H, Poole C, Safko J (2002) Classical mechanics, 3rd edn. Addison Wesley, Boston, US (Chap. 9), pp 368–429
45. Wigner E (1932) Phys Rev 40:749. https://doi.org/10.1103/PhysRev.40.749
46. Schleich WP (2001) Quantum optics in phase space. Wiley-VCH Verlag Berlin GmbH, Berlin, Germany (Chap. 3), pp 67–98
47. Hudson RL (1974) Rep Math Phys 6:249. https://doi.org/10.1016/0034-4877(74)90007-X
48. Smithey DT, Beck M, Raymer MG, Faridani A (1993) Phys Rev Lett 70:1244. https://doi.org/10.1103/PhysRevLett.70.1244
49. Mitchell MW (2007) Quantum optics for the impatient. Lecture notes for the course of advanced quantum optics given in the Master in Photonics at Universitat Politècnica de Catalunya (Chap. 4), pp 19–26. https://mitchellgroup.icfo.es/mg/pmwiki.php
50. Lvovsky AI, Raymer MG (2009) Rev Mod Phys 81:299. https://doi.org/10.1103/RevModPhys.81.299
51. Royer A (1977) Phys Rev A 15:449. https://doi.org/10.1103/PhysRevA.15.449
52. Heisenberg W (1927) Z Phys 43:172. https://doi.org/10.1007/BF01397280
53. Bao X-H, Qian Y, Yang J, Zhang H, Chen Z-B, Yang T, Pan J-W (2008) Phys Rev Lett 101:190501. https://doi.org/10.1103/PhysRevLett.101.190501
54. Haase A, Piro N, Eschner J, Mitchell MW (2009) Opt Lett 34:55. https://doi.org/10.1364/OL.34.000055
55. Fekete J, Rieländer D, Cristiani M, de Riedmatten H (2013) Phys Rev Lett 110:220502. https://doi.org/10.1103/PhysRevLett.110.220502
56. Magnitskiy S, Frolovtsev D, Firsov V, Gostev P, Protsenko I, Saygin M (2015) J Russ Laser Res 36:618. https://doi.org/10.1007/s10946-015-9540-x
57. Aspect A, Grangier P, Roger G (1982) Phys Rev Lett 49:91. https://doi.org/10.1103/PhysRevLett.49.91

58. Aspect A, Dalibard J, Roger G (1982) Phys Rev Lett 49:1804. https://doi.org/10.1103/PhysRevLett.49.1804
59. Marcikic I, de Riedmatten H, Tittel W, Zbinden H, Gisin N (2003) Nature 421:509. https://doi.org/10.1038/nature01376
60. de Riedmatten H, Marcikic I, Scarani V, Tittel W, Zbinden H, Gisin N (2004) Phys Rev A 69:050304. https://doi.org/10.1103/PhysRevA.69.050304
61. de Riedmatten H, Marcikic I, van Houwelingen JAW, Tittel W, Zbinden H, Gisin N (2005) Phys Rev A 71:050302. https://doi.org/10.1103/PhysRevA.71.050302
62. Vallés A, D'Ambrosio V, Hendrych M, Mičuda M, Marrucci L, Sciarrino F, Torres JP (2014) Phys Rev A 90:052326. https://doi.org/10.1103/PhysRevA.90.052326
63. McKeever J, Boca A, Boozer AD, Miller R, Buck JR, Kuzmich A, Kimble HJ (2004) Science 303:1992. https://doi.org/10.1126/science.1095232
64. Hijlkema M, Weber B, Specht HP, Webster SC, Kuhn A, Rempe G (2007) Nat Phys 3:253. https://doi.org/10.1038/nphys569
65. Chou CW, Polyakov SV, Kuzmich A, Kimble HJ (2004) Phys Rev Lett 92:213601. https://doi.org/10.1103/PhysRevLett.92.213601
66. Laurat J, Riedmatten HD, Felinto D, Chou C-W, Schomburg EW, Kimble HJ (2006) Optics Expr 14:6912. https://doi.org/10.1364/OE.14.006912
67. Thompson JK, Simon J, Loh H, Vuletić V (2006) Science 313:74. https://doi.org/10.1126/science.1127676
68. Matsukevich DN, Chanelière T, Jenkins SD, Lan S-Y, Kennedy TAB, Kuzmich A (2006) Phys Rev Lett 97:013601. https://doi.org/10.1103/PhysRevLett.97.013601
69. Chen S, Chen Y-A, Strassel T, Yuan Z-S, Zhao B, Schmiedmayer J, Pan J-W (2006) Phys Rev Lett 97:173004. https://doi.org/10.1103/PhysRevLett.97.173004
70. Du S, Kolchin P, Belthangady C, Yin GY, Harris SE (2008) Phys Rev Lett 100:183603. https://doi.org/10.1103/PhysRevLett.100.183603
71. Srivathsan B, Gulati GK, Chng B, Maslennikov G, Matsukevich D, Kurtsiefer C (2013) Phys Rev Lett 111:123602. https://doi.org/10.1103/PhysRevLett.111.123602
72. Zhao L, Guo X, Liu C, Sun Y, Loy MMT, Du S (2014) Optica 1:84. https://doi.org/10.1364/OPTICA.1.000084
73. Farrera P, Heinze G, Albrecht B, Ho M, Chávez M, Teo C, Sangouard N, de Riedmatten H (2016) Nat Commun 7:13556. https://doi.org/10.1038/ncomms13556
74. Zinoni C, Alloing B, Monat C, Zwiller V, Li LH, Fiore A, Lunghi L, Gerardino A, de Riedmatten H, Zbinden H, Gisin N (2006) Appl Phys Lett 88:131102. https://doi.org/10.1063/1.2190466
75. Buckley S, Rivoire K, Vučković J (2012) Rep Prog Phys 75:126503. https://doi.org/10.1088/0034-4885/75/12/126503
76. Kutluer K, Mazzera M, de Riedmatten H (2017) Phys Rev Lett 118:210502. https://doi.org/10.1103/PhysRevLett.118.210502
77. Uppu R, Pedersen FT, Wang Y, Olesen CT, Papon C, Zhou X, Midolo L, Scholz S, Wieck AD, Ludwig A, Lodahl P (2020) Sci Adv 6:eabc8268. https://doi.org/10.1126/sciadv.abc8268
78. Ollivier H, Maillette de Buy Wenniger I, Thomas S, Wein SC, Harouri A, Coppola G, Hilaire P, Millet C, Lemaître A, Sagnes I, Krebs O, Lanco L, Loredo JC, Antón C, Somaschi N, Senellart P (2020) ACS Photonics 7:1050. https://doi.org/10.1021/acsphotonics.9b01805
79. Tomm N, Javadi A, Antoniadis NO, Najer D, Löbl MC, Korsch AR, Schott R, Valentin SR, Wieck AD, Ludwig A, Warburton RJ (2021) Nat Nanotechnol 16:399. https://doi.org/10.1038/s41565-020-00831-x
80. Glauber RJ (1963) Phys Rev Lett 10:84. https://doi.org/10.1103/PhysRevLett.10.84
81. Glauber RJ (1963) Phys Rev 131:2766. Note publisher: American Physical Society. https://doi.org/10.1103/PhysRev.131.2766
82. Scully MO, Zubairy MS (2001) Coherent and squeezed states of the radiation fieldy. In: Quantum optics. Cambridge University Press, Cambridge, UK (Chap. 2), pp 1–45
83. Scully MO, Zubairy MS (2001) Squeezing via nonlinear optical processes. In: Quantum optics. Cambridge University Press, Cambridge, UK (Chap. 4), pp 460–486

84. Wu L-A, Kimble HJ, Hall JL, Wu H (1986) Phys Rev Lett 57:2520. https://doi.org/10.1103/PhysRevLett.57.2520
85. Kennard EH (1927) Z Phys 44:326. https://doi.org/10.1007/BF01391200
86. Slusher RE, Hollberg LW, Yurke B, Mertz JC, Valley JF (1985) Phys Rev Lett 55:2409. https://doi.org/10.1103/PhysRevLett.55.2409
87. Walls DF (1983) Nature 306:141. https://doi.org/10.1038/306141a0
88. Plebanski J (1956) Phys Rev 101:1825. https://doi.org/10.1103/PhysRev.101.1825
89. Caves CM (1981) Phys Rev D 23:1693. https://doi.org/10.1103/PhysRevD.23.1693
90. Xiao M, Wu L-A, Kimble HJ (1987) Phys Rev Lett 59:278. https://doi.org/10.1103/PhysRevLett.59.278
91. Jaekel MT, Reynaud S (1990) Europhys Lett 13:301. https://doi.org/10.1209/0295-5075/13/4/003
92. The LIGO Scientific Collaboration (2011) Nat Phys 7:962. https://doi.org/10.1038/nphys2083
93. Schnabel R (2017) Physics Reports series Squeezed states of light and their applications in laser interferometers 684:1. https://doi.org/10.1016/j.physrep.2017.04.001
94. Schrödinger E (1935) Naturwissenschaften 23:844. https://doi.org/10.1007/BF01491987
95. Coles D, Flatten LC, Sydney T, Hounslow E, Saikin SK, Aspuru-Guzik A, Vedral V, Tang JK-H, Taylor RA, Smith JM, Lidzey DG (2017) Small 13:1701777. https://doi.org/10.1002/smll.201701777
96. Marletto C, Coles DM, Farrow T, Vedral V (2018) J Phys Commun 2:101001. https://doi.org/10.1088/2399-6528/aae224
97. Lee KS, Tan YP, Nguyen LH, Budoyo RP, Park KH, Hufnagel C, Yap YS, Møbjerg N, Vedral V, Paterek T, Dumke R (2022) New J Phys 24:123024. https://doi.org/10.1088/1367-2630/aca81f
98. Wineland DJ (2013) Rev Mod Phys 85:1103. https://doi.org/10.1103/RevModPhys.85.1103
99. Haroche S (2013) Rev Mod Phys 85:1083. https://doi.org/10.1103/RevModPhys.85.1083
100. Zavatta A, Viciani S, Bellini M (2004) Science 306:660. https://doi.org/10.1126/science.1103190
101. Dakna M, Anhut T, Opatrný T, Knöll L, Welsch D-G (1997) Phys Rev A 55:3184. https://doi.org/10.1103/PhysRevA.55.3184
102. Ourjoumtsev A, Tualle-Brouri R, Laurat J, Grangier P (2006) Science 312:83. https://doi.org/10.1126/science.1122858
103. Neergaard-Nielsen JS, Nielsen BM, Hettich C, Mølmer K, Polzik ES (2006) Phys Rev Lett 97:083604. https://doi.org/10.1103/PhysRevLett.97.083604
104. Lund AP, Jeong H, Ralph TC, Kim MS (2004) Phys Rev A 70:020101. https://doi.org/10.1103/PhysRevA.70.020101
105. Ourjoumtsev A, Jeong H, Tualle-Brouri R, Grangier P (2007) Nature 448:784. https://doi.org/10.1038/nature06054
106. Gilchrist A, Nemoto K, Munro WJ, Ralph TC, Glancy S, Braunstein SL, Milburn GJ (2004) J Opt B: Quantum Semiclassical Opt 6:S828. https://doi.org/10.1088/1464-4266/6/8/032
107. Hacker B, Welte S, Daiss S, Shaukat A, Ritter S, Li L, Rempe G (2019) Nat Photonics 13:110. https://doi.org/10.1038/s41566-018-0339-5
108. Sun F-X, Fang Y, He Q, Liu Y (2023) Sci Bull 68:1366. https://doi.org/10.1016/j.scib.2023.06.006
109. Jeong H, Kim MS (2002) Phys Rev A 65:042305. https://doi.org/10.1103/PhysRevA.65.042305
110. Lund AP, Ralph TC, Haselgrove HL (2008) Phys Rev Lett 100:030503. https://doi.org/10.1103/PhysRevLett.100.030503
111. Su D, Dhand I, Ralph TC (2022) Phys Rev A 106:042614. https://doi.org/10.1103/PhysRevA.106.042614
112. Braunstein SL, Kimble HJ (1998) Phys Rev Lett 80:869. https://doi.org/10.1103/PhysRevLett.80.869

113. van Enk SJ, Hirota O (2001) Phys Rev A 64:022313. https://doi.org/10.1103/PhysRevA.64.022313
114. Lee N, Benichi H, Takeno Y, Takeda S, Webb J, Huntington E, Furusawa A (2011) Science 332:330. https://doi.org/10.1126/science.1201034
115. van Loock P (2011) Laser & Photonics Rev 5:167. https://doi.org/10.1002/lpor.201000005
116. Nielsen MA, Chuang IL (2010) Introduction to quantum mechanics. In: Quantum Computation and Quantum Information: 10th Anniversary Edition. Cambridge University Press, Cambridge, UK (Chap. 2), pp 60–119
117. Einstein A, Podolsky B, Rosen N (1935) Phys Rev 47:777. https://doi.org/10.1103/PhysRev.47.777
118. Horodecki R, Horodecki P, Horodecki M, Horodecki K (2009) Rev Mod Phys 81:865. https://doi.org/10.1103/RevModPhys.81.865
119. Reid MD, Drummond PD, Bowen WP, Cavalcanti EG, Lam PK, Bachor HA, Andersen UL, Leuchs G (2009) Rev Mod Phys 81:1727. https://doi.org/10.1103/RevModPhys.81.1727
120. Shor P (1994) In: Proceedings 35th annual symposium on foundations of computer science, pp 124–134. https://doi.org/10.1109/SFCS.1994.365700
121. Grover LK (1996) In: Proceedings of the twenty-eighth annual ACM symposium on Theory of Computing, series and number STOC '96. Association for Computing Machinery, address New York, NY, USA, pp 212–219. https://doi.org/10.1145/237814.237866
122. Shor PW (1999) SIAM Rev 41:303. https://doi.org/10.1137/S0036144598347011
123. Nielsen MA, Chuang IL (2010) The quantum Fourier transform and its applications. In: Quantum computation and quantum information: 10th Anniversary Edition. Cambridge University Press, Cambridge, UK (Chap. 5), pp 216–247
124. Nielsen MA, Chuang IL (2010) Quantum search algorithms. In: Quantum computation and quantum information: 10th Anniversary Edition. Cambridge University Press, Cambridge, UK (Chap. 6), pp 248–276
125. Bennett CH, Brassard G (2014) Theoretical computer science series theoretical aspects of quantum cryptography – celebrating 30 years of BB84 560(7). https://doi.org/10.1016/j.tcs.2014.05.025
126. Ekert AK (1991) Phys Rev Lett 67:661. https://doi.org/10.1103/PhysRevLett.67.661
127. Nielsen MA, Chuang IL (2010) Quantum information theory. In: Quantum computation and quantum information: 10th Anniversary Edition. Cambridge University Press, Cambridge, UK (Chap. 12), pp 528–607
128. Gurvits L (2004) J Comput Syst Sci Ser Special Issue on STOC 2003(69):448. https://doi.org/10.1016/j.jcss.2004.06.003
129. Nielsen MA, Chuang IL (2010) Entropy and information. In: Quantum computation and quantum information: 10th Anniversary Edition. Cambridge University Press, Cambridge, UK (Chap. 11), pp 500–527
130. Plenio MB, Virmani S (2007) Q Inf & Comput 7:1
131. Amico L, Fazio R, Osterloh A, Vedral V (2008) Rev Mod Phys 80:517. https://doi.org/10.1103/RevModPhys.80.517
132. Peres A (1996) Phys Rev Lett 77:1413. https://doi.org/10.1103/PhysRevLett.77.1413
133. Horodecki M, Horodecki P, Horodecki R (1996) Phys Lett A 223:1. https://doi.org/10.1016/S0375-9601(96)00706-2
134. Brunner N, Cavalcanti D, Pironio S, Scarani V, Wehner S (2014) Rev Mod Phys 86:419. https://doi.org/10.1103/RevModPhys.86.419
135. Born M (1926) Z Phys 37:863. https://doi.org/10.1007/BF01397477
136. Pitowsky I (1986) J Math Phys 27:1556. https://doi.org/10.1063/1.527066
137. Cirel'son BS (1980) Lett Math Phys 4:93. https://doi.org/10.1007/BF00417500
138. Popescu S, Rohrlich D (1994) Found Phys 24:379. https://doi.org/10.1007/BF02058098
139. Capasso V, Fortunato D, Selleri F (1973) Int J Theor Phys 7:319. https://doi.org/10.1007/BF00669912
140. Gisin N (1991) Phys Lett A 154:201. https://doi.org/10.1016/0375-9601(91)90805-I

141. Home D, Selleri F (1991) La Rivista del Nuovo Cimento (1978–1999) 14(1). https://doi.org/10.1007/BF02811227
142. Werner RF (1989) Phys Rev A 40:4277. https://doi.org/10.1103/PhysRevA.40.4277
143. Barrett J (2002) Phys Rev A 65:042302. https://doi.org/10.1103/PhysRevA.65.042302
144. Vogel W, Welsch D-G (2006) In: Quantum optics. Wiley-VCH Verlag GmbH & Co. KGaA, Weinheim, Germany (Chap. 2), pp 15–72

Chapter 3
Non-classical States of Light After Strong-Laser Field Processes in Atoms

> Using light that in turn controls electrons that in turn create light.
>
> – Paul Corkum, at *the International Conference of Electronic, Atomic and Photonics Collisions (ICPEAC) XXXIII*, (2023).

The invention of the laser by Maiman in [1] led to significant advancements in science and technology that profoundly influenced the latter half of the 20th century and beyond, revolutionizing numerous fields [2]. The widespread adoption of lasers became feasible due to the availability of a wide range of laser sources, each differing in intensity, frequency, beam spatial distribution, pulse duration and other properties. Consequently, laser technology has been effectively tailored to meet the specific requirements of various scientific communities.

In fact, laser light and technologies have advanced to the point that, nowadays, it is possible to generate laser sources with intensity high enough ($I \sim 10^{11}$–10^{16} W/cm^2)[1] to compete with atomic forces, enabling highly nonlinear optical processes [3]. These advancements, coupled with their remarkable applications, have firmly consolidated strong-field physics and attosecond science as well-established fields in Physics [4], with their primary focus being the study of the interaction between high-intensity laser fields and matter. Indeed, the origins of strong-field physics date back shortly after the invention of the laser, with the seminal work by

---

[1] It is nowadays possible to go beyond this laser intensity, reaching values in the order of $I \sim 10^{20}$–10^{23} W/cm^2, although the processes initiated by these sources are beyond the scope of this Thesis.

J. Rivera-Dean, *Non-classical States of Light*, Springer Theses,
https://doi.org/10.1007/978-3-031-73769-5_3

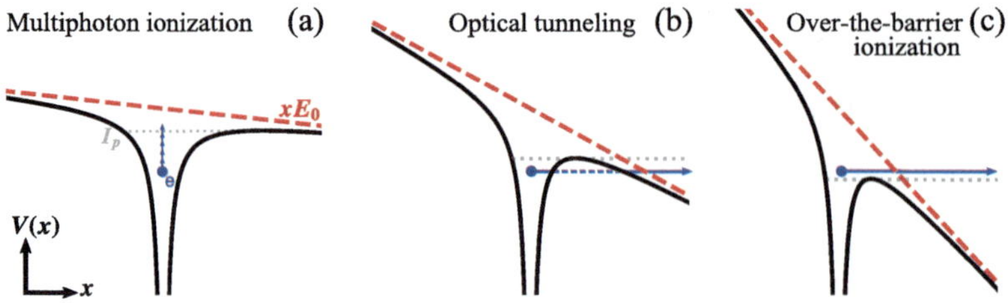

Fig. 3.1 **Schematic representation of the different strong-field ionization regimes**. In **a** the multiphoton ionization regime is presented, where the field slightly perturbs the atomic potential (represented with the black curve), while the electron (represented with the blue dot) absorbs many photons (represented by the blue arrows) to surpass the ionization potential (represented by the gray dashed line). As the field intensity increases while the driving frequency decreases, we enter into the optical tunneling domain shown in (**b**). In this case, the electric field strongly perturbs the atomic potential, forming a barrier through which the electron can tunnel out. When the intensity of the field surpasses a certain threshold, we enter into the over-the-barrier ionization regime shown in (**c**), in which the electron is directly found in the continuum

L. V. Keldysh [5]. In this groundbreaking contribution, Keldysh laid the foundations for the field, and his ideas continue to influence developments in the community to this day [6].

In his theoretical work, Keldysh analyzed from a primarily academic research perspective, what occurs in the regime where the applied electromagnetic field can no longer be considered a perturbative quantity anymore but rather is on par with the atomic potential. In this context, Keldysh proposed several mechanisms by which the electrons can become ionized, paving the way for one of the most intriguing phenomena in strong-field physics: above-threshold ionization (ATI). In ATI, an electron ionizes by absorbing more energy than required by the ionization threshold of the corresponding system [7]. The involved ionization mechanisms include multiphoton ionization, where the electrons absorb multiple photons in order to get ionized [8, 9] (see Fig. 3.1a); and optical tunneling, where the oscillating field distorts the atomic potential, forming a potential barrier through which electrons can tunnel out [10] (see Fig. 3.1b). This leads to the emission of high-energy and high-momentum photoelectrons.[2]

In addition to the intriguing dynamics resulting from the interaction between highly intense fields and matter, strong-field physics has yielded extremely valuable applications relevant to various scientific disciplines. Many of these applications rely on the elastic or inelastic recollision of photoelectrons with the parent system after ionization, driven back by the applied electric field. Elastic collisions can be harnessed for laser-induced electron diffraction (LIED) [11–15] and holography [16, 17], where the rescattered electrons acquire structural information about the parent system. On the other hand, inelastic recollisions can lead to the high-harmonic generation process (HHG), where high-energy electrons recombine with their parent

[2] By *photoelectrons*, we mean electrons that have been emitted from a material upon the interaction with an electromagnetic field.

system, resulting in the emission of high-frequency coherent radiation [18–20]. This radiation is typically emitted as a periodic series of ultrashort bursts, which can combine to form a comb of harmonics of the driving laser, spanning dozens, hundreds or even thousands of harmonic orders. Due to its natural energy and time scales, which correspond to those of typical nuclei and electron dynamics, HHG has found diverse applications, including high-resolution spectroscopy [21–23], molecular tomography [24], the generation of extreme-ultraviolet (XUV) radiation sources [25, 26], the production of attosecond pulses [27, 28], non-linear XUV optics [29–36] and attosecond science [37, 38].

In this chapter, based on our works in Refs. [39–46], we commence with a review of the fundamentals of intense laser-matter interactions from a semiclassical perspective. This analysis provides the necessary foundation for a detailed exploration of above-threshold ionization and high-harmonic generation processes. Following this, we shift our focus to describing these phenomena within a quantum optical framework. We begin with a survey of the current state of the field in this direction, and subsequently introduce the formalism for studying the aforementioned processes in the quantum optical regime. Specifically, we emphasize the investigation of non-classical properties of light resulting from these processes.

3.1 Semiclassical Analysis of Strongly Driven Laser-Atom Interactions

The seminal work by L. V. Keldysh was expanded upon by many authors [6, 47], notably including the contributions of F. H. M. Faisal [48] and H. R. Reiss [49], whose analyses led to what is nowadays known as Keldysh-Faisal-Reiss (KFR) theory. This theory represents one of the earliest instances of the so-called Strong-Field Approximation (SFA). Over time, the SFA has undergone substantial evolution and, nowadays, it encompasses a range of techniques, each with its own characteristics, used to study the interaction between strong laser fields and matter. A comprehensive historical overview and evolution of the SFA are available in several references, including Refs. [47, 50]. In this section, our objective is to provide a practical and comprehensive presentation by deriving the main equations of the semiclassical theory. We follow the pioneering work of M. Lewenstein and collaborators [51], outlined as well in Ref. [47], with the aim of describing ATI and HHG processes from a semiclassical perspective within the tunneling regime, illustrated in Fig. 3.1b.

The applicability of the SFA, as considered in this text, is contingent upon the fulfillment of specific conditions concerning the applied electromagnetic field in relation to the properties of the atomic system [5, 47, 49, 51–54]. Firstly, the frequency of the applied electromagnetic field,[3] denoted as $\omega_{\mathbf{k}_L}$, should be significantly smaller than the ionization potential I_p of the atomic species, i.e., $\hbar\omega_{\mathbf{k}_L} \ll I_p$. This condition

[3] When working with pulsed sources consisting of a continuous set of frequencies, $\omega_{\mathbf{k}_L}$ refers to the pulse's central frequency.

ensures that ionization from the parent system necessitates a substantial number of photons for the process to occur.

Secondly, to satisfy tunneling conditions, it is imperative that the electric field can sufficiently distort the atomic potential, creating a barrier through which electrons can tunnel out. This distortion primarily depends on the strength of the electric field. The Keldysh parameter [5, 50] provides a rough estimation of the boundaries between the tunneling and multiphoton regimes. It is defined as

$$\gamma = \sqrt{\frac{I_p}{2U_p}} \tag{3.1}$$

where $U_p = (\mathrm{e}^2 E_0^2)/(4m_\mathrm{e}\omega_{\mathbf{k}_L}^2)$ represents the ponderomotive energy, with e and m_e denoting the electron's charge and mass, respectively. The ponderomotive energy characterizes the average kinetic energy of a free electron driven by an oscillatory electric field with a peak amplitude E_0. Consequently, the Keldysh parameter portrays a competition between the electron's binding energy to the corresponding atom and its energy during the free oscillation within the electric field. Therefore, when $\gamma \gg 1$, the field only slightly perturbs the atomic potential, leading to multiphoton processes. However, for the SFA to be applicable we must operate in the $\gamma \lesssim 1$ regime. In this regime, the force induced by the external field becomes comparable to that of the atomic potential, creating conditions conducive to tunneling events.[4] For typical atomic systems, these conditions are met when working with intensities of $I \sim 10^{14}$–10^{15} W/cm^2 and wavelengths belonging to the infrared (IR) regime (λ ~700–5000 nm) [47, 51]. Furthermore, if the field strength exceeds a certain critical value [55], the peak of the potential barrier may exceed the energy level of the electron, resulting in over-the-barrier ionization processes, as pictured in Fig. 3.1c.

Lastly, once the electron, initially in the atomic ground state $|\mathrm{g}\rangle$, reaches the continuum through tunneling, it rapidly drives away from the parent ion. Thus, the effects of the atomic effective potential on the electron are expected to be small, and are treated perturbatively.

Under these conditions, the *standard* version of the SFA makes the following assumptions [47, 51]:

1. the contribution of all bound states, except for the atomic ground state $|\mathrm{g}\rangle$, can be neglected;
2. the depletion of the ground state can considered to be negligible, as ionization processes are weak;
3. the electronic continuum states are selected from the basis of scattering states which satisfy $\hat{H}_{\mathrm{at}} |\mathbf{v}\rangle = \frac{1}{2} m_\mathrm{e} \mathbf{v}^2 |\mathbf{v}\rangle$, where $\hat{H}_{\mathrm{at}}$ represents the atomic Hamiltonian.

[4] It is important to note that the Keldysh parameter provides a rough estimation and does not definitively determine whether tunneling or multiphoton processes are taking place.

Beyond this *standard* version of the SFA, applicable for *non-relativistic* laser intensities,[5] typically $I < 10^{16}$ W/cm^2, several methods have been proposed to extend the versatility of the SFA. These methods encompass various considerations, including Coulomb effects[6] such as the Perelomov-Popov-Terent'ev (PPT) method [57], the Coulomb-corrected SFA methods (CCSFA) [50, 58–60] and the Coulomb quantum-orbit SFA methods (CQSFA) [61–63]. Others incorporate spin considerations to account for relativistic terms as they appear in the Dirac's equation [64, 65], while some methods combine both aspects [66]. Furthermore, it has been recently possible to emulate strong-field processes in analog quantum simulators [67–69].

3.1.1 Dynamics of a Bound Electron Interacting with an Intense Electromagnetic Field

When describing the interaction between a strong-laser field and an atomic system, it is customary to employ the single-active-electron (SAE) approximation. According to this approximation, it is assumed that each electron interacts independently with the electromagnetic field, while all other electrons collectively create an effective potential that combines with the potential generated by the atomic core. Thus, under this assumption and the dipole approximation,[7] the Hamiltonian characterizing the interaction between a classical field and the atom can be expressed, in the length gauge, as

$$\hat{H}(t) = \frac{\hat{\mathbf{p}}^2}{2m_\mathrm{e}} + \mathrm{e}\mathbf{E}_{\mathrm{cl}}(t)\cdot\hat{\mathbf{r}} + \hat{V}_{\mathrm{at}}(\hat{\mathbf{r}}), \tag{3.2}$$

where $\mathbf{E}_{\mathrm{cl}}(t)$ represents the applied classical electric field[8] and $\hat{V}_{\mathrm{at}}(\hat{\mathbf{r}})$ denotes the effective atomic potential. Defining the atomic Hamiltonian as $\hat{H}_{\mathrm{at}} = \hat{\mathbf{p}}^2/(2m_\mathrm{e}) + \hat{V}_{\mathrm{at}}(\hat{\mathbf{r}})$, the Schrödinger equation governing the electron's dynamics is given by

$$\begin{aligned} i\hbar\frac{\partial\,|\psi(t)\rangle}{\partial t} &= \left[\frac{\hat{\mathbf{p}}^2}{2m_\mathrm{e}} + \mathrm{e}\mathbf{E}_{\mathrm{cl}}(t)\cdot\hat{\mathbf{r}} + \hat{V}_{\mathrm{at}}(\hat{\mathbf{r}})\right]|\psi(t)\rangle \\ &= \left[\hat{H}_{\mathrm{at}} + \mathrm{e}\mathbf{E}_{\mathrm{cl}}(t)\cdot\hat{\mathbf{r}}\right]|\psi(t)\rangle\,, \end{aligned} \tag{3.3}$$

---

[5] By non-relativistic laser intensities, we mean laser-matter interactions where the photoelectrons that are released do not acquire relativistic velocities.

[6] See Chap. 1 in Ref. [56] for a nice summary about other Coulomb-corrected SFA methods.

[7] Typical atomic sizes are on the order of 1 Å, and even though the electron undergoes some excursions in strong-field processes, they are much smaller than the applied wavelengths, which are in the hundreds and thousands of nanometers range. Therefore, in this regime, we expect the dipole approximation to hold.

[8] Hereupon, we denote classical quantities with the subscript "cl" particularly in cases where they could be confused with operators such as the electric field operator or the vector potential operator.

under the initial condition $|\psi(t_0)\rangle = |\text{g}\rangle$, i.e., the electron being initially located in the ground state energy level.

In accordance with the first of the assumptions presented earlier, the SFA proposes the following ansatz for solving the aforementioned differential equation

$$|\psi(t)\rangle = a(t)|\text{g}\rangle + \int d\mathbf{v}\, b(\mathbf{v}, t)\, |\mathbf{v}\rangle\,, \tag{3.4}$$

where other bound states aside from the ground state are not included. We use $a(t)$ and $b(\mathbf{v}, t)$ to represent the probability amplitudes of finding the electron in the ground state $|\text{g}\rangle$ and in a scattering state $|\mathbf{v}\rangle$ at time t, respectively. By introducing this ansatz into Eq. (3.3) and projecting it with respect to $|\text{g}\rangle$ and $|\mathbf{v}\rangle$, we obtain the following set of differential equations

$$i\hbar\frac{\partial a(t)}{\partial t} = -I_p a(t) + \mathsf{e}\int d\mathbf{v}\, \mathbf{E}_{\text{cl}}(t)\cdot\langle \text{g}|\hat{\mathbf{r}}|\mathbf{v}\rangle b(\mathbf{v}, t), \tag{3.5}$$

$$i\hbar\frac{\partial b(\mathbf{v}, t)}{\partial t} = \frac{1}{2}m_{\mathsf{e}}\mathbf{v}^2 b(\mathbf{v}, t) + \mathsf{e}\mathbf{E}_{\text{cl}}(t)\cdot\langle \mathbf{v}|\hat{\mathbf{r}}|\text{g}\rangle a(t) + \mathsf{e}\int d\mathbf{v}'\, \mathbf{E}_{\text{cl}}(t)\cdot\langle \mathbf{v}|\hat{\mathbf{r}}|\mathbf{v}'\rangle b(\mathbf{v}', t), \tag{3.6}$$

describing the evolution of these two probability amplitudes. Let us analyze each of these equations separately. In Eq. (3.5), the first term on the right-hand side represents the ground state energy, while the second one introduces depletion effects due to ionization processes mediated by the classical electric field $\mathbf{E}_{\text{cl}}(t)$. Therefore, in accordance with the second of the SFA assumptions presented earlier, we consider the depletion of the ground state to be negligible. Consequently, we approximate this equation as follows

$$i\hbar\frac{\partial a(t)}{\partial t} \approx -I_p a(t), \tag{3.7}$$

which is solved straightforwardly by $a(t) = e^{iI_p(t-t_0)/\hbar}a(t_0)$.

Concerning Eq. (3.6), the first term on the right-hand side corresponds to the scattering energy of the photoelectron, while the second and the third ones represent processes in which the electron transitions either from the ground state $|\text{g}\rangle$ or another scattering state $|\mathbf{v}'\rangle$, respectively, to the scattering state $|\mathbf{v}\rangle$. As for the latter, we decompose the continuum-continuum matrix element $\langle \mathbf{v}|\hat{\mathbf{r}}|\mathbf{v}'\rangle$ into its most singular part and other additional terms that encompass rescattering effects [47, 51, 53, 54]. In other words, we expand these terms as

$$\mathsf{e}\left\langle \mathbf{v}|\hat{\mathbf{r}}|\mathbf{v}'\right\rangle = i\hbar\mathsf{e}\delta(\mathbf{v}-\mathbf{v}') + \hbar\mathbf{g}(\mathbf{v}, \mathbf{v}'), \tag{3.8}$$

where $\mathbf{g}(\mathbf{v}, \mathbf{v}')$ represents rescattering effects, which we treat pertubatively [54]. Likewise, we expand $b(\mathbf{v}, t)$ up to first-order perturbation theory

$$b(\mathbf{v}, t) \approx b^{(0)}(\mathbf{v}, t) + b^{(1)}(\mathbf{v}, t), \tag{3.9}$$

with $b^{(n)}(\mathbf{v}, t)$ denoting the nth-order pertubation term. Therefore, by introducing Eqs. (3.8) and (3.9) into Eq. (3.6), we arrive at the following set of equations for zeroth and first-order perturbation theory

$$i\hbar\frac{\partial b^{(0)}(\mathbf{v},t)}{\partial t} = \frac{1}{2}m_e\mathbf{v}^2 b^{(0)}(\mathbf{v},t) + \mathbf{E}_{\mathrm{cl}}(t)\cdot\mathbf{d}(\mathbf{v})a(t) + i\hbar e\mathbf{E}_{\mathrm{cl}}(t)\cdot\nabla_{\mathbf{v}} b^{(0)}(\mathbf{v},t), \tag{3.10}$$

$$\begin{aligned} i\hbar\frac{\partial b^{(1)}(\mathbf{v},t)}{\partial t} &= \frac{1}{2}m_e\mathbf{v}^2 b^{(1)}(\mathbf{v},t) + i\hbar e\mathbf{E}_{\mathrm{cl}}(t)\cdot\nabla_{\mathbf{v}} b^{(1)}(\mathbf{v},t) \\ &+ \hbar\int \mathrm{d}\mathbf{v}'\, \mathbf{E}_{\mathrm{cl}}(t)\cdot\mathbf{g}(\mathbf{v},\mathbf{v}')b^{(0)}(\mathbf{v}',t), \end{aligned} \tag{3.11}$$

where we denote $\mathbf{d}(\mathbf{v}) \equiv e\langle\mathbf{v}|\hat{\mathbf{r}}|g\rangle$.

Both Eqs. (3.10) and (3.11) share a similar mathematical form: they are first-order differential equations with well-defined homogeneous and inhomogeneous parts. Consequently, their solutions can be expressed as the sum of the solution to the homogeneous part and a particular solution to the inhomogeneous part. Specifically, for the zeroth-order component, we obtain for the homogeneous part

$$i\hbar\frac{\partial b^{(0)}_{\mathrm{hom}}(\mathbf{v},t)}{\partial t} = \frac{1}{2}m_e\mathbf{v}^2 b^{(0)}_{\mathrm{hom}}(\mathbf{v},t) + i\hbar e\mathbf{E}_{\mathrm{cl}}(t)\cdot\nabla_{\mathbf{v}} b^{(0)}_{\mathrm{hom}}(\mathbf{v},t), \tag{3.12}$$

which resembles the evolution of a free particle with charge e, subjected to an electromagnetic field $\mathbf{E}_{\mathrm{cl}}(t)$ as demonstrated in Eq. (2.29). Hence, by following the procedure outlined in Sect. 2.2.4, the solution to this equation can be represented as

$$b^{(0)}_{\mathrm{hom}}(\mathbf{p},t) = \exp\Big[-\frac{i}{2m_e\hbar}\int_{t_0}^{t}\mathrm{d}\tau\Big[\mathbf{p}+e\mathbf{A}(\tau)\Big]^2\Big]b^{(0)}_{\mathrm{hom}}(\mathbf{p},t_0), \tag{3.13}$$

where it is important to recall that $\mathbf{p} = m_e\mathbf{v} - e\mathbf{A}_{\mathrm{cl}}(t)$ represents the canonical momentum, and $\mathbf{A}_{\mathrm{cl}}(t)$ is the vector potential defined by $\mathbf{E}_{\mathrm{cl}}(t) = -\partial\mathbf{A}_{\mathrm{cl}}(t)/\partial t$. With this result in mind, a solution to Eq. (3.10) can be expressed as

$$\begin{aligned} b^{(0)}(\mathbf{p},t) &= b^{(0)}_{\mathrm{hom}}(\mathbf{p},t) \\ &\quad -\frac{i}{\hbar}\int_{t_0}^{t}\mathrm{d}t_1\exp\Big[-\frac{i}{2m_e\hbar}\int_{t_1}^{t}\mathrm{d}\tau\Big[\mathbf{p}+e\mathbf{A}(\tau)\Big]^2\Big]\mathbf{E}_{\mathrm{cl}}(t_1)\cdot\mathbf{d}\big(\mathbf{p}+e\mathbf{A}_{\mathrm{cl}}(t_1)\big)a(t_1) \\ &= -\frac{i}{\hbar}\int_{t_0}^{t}\mathrm{d}t_1 e^{-iS(\mathbf{p},t,t_1)/\hbar}\mathbf{E}_{\mathrm{cl}}(t_1)\cdot\mathbf{d}\big(\mathbf{p}+e\mathbf{A}_{\mathrm{cl}}(t_1)\big), \end{aligned} \tag{3.14}$$

where in transitioning from the first to the second equality, we introduced the solution to Eq. (3.7) and considered the initial conditions $a(t_0) = 1$ and $b(\mathbf{v}, t_0) = 0$. Additionally, in Eq. (3.14), we have defined

$$S(\mathbf{p}, t, t_1) \equiv \frac{1}{2m_e} \int_{t_1}^{t} d\tau \Big[\mathbf{p} + e\mathbf{A}(\tau)\Big]^2 - I_p(t_1 - t_0), \tag{3.15}$$

which is commonly referred to as the semiclassical action due to its association with Feynmann's path integral approach and the concept of quantum orbits [70]. We further elaborate on this concept in forthcoming subsections.

From a physical perspective, Eq. (3.14) describes a process in which the electron transitions from the ground state $|g\rangle$ to a continuum state $|\mathbf{p} + e\mathbf{A}_{cl}(t)\rangle$ at time t_1. Subsequently, from t_1 to t (with $t \geq t_1$), it propagates in the continuum under the influence of the laser field. This process is further extended by the inclusion of the first-order perturbation component, which introduces additional scattering effects. To be precise, by following a procedure akin to the one previously described, we can express a solution to Eq. (3.9) as

$$\begin{aligned} b^{(1)}(\mathbf{p}, t) = -\frac{1}{\hbar} \int d\mathbf{p}' \int_{t_0}^{t} dt_2 \int_{t_0}^{t_2} dt_1 \exp\Big[-\frac{i}{2m_e\hbar} \int_{t_2}^{t} d\tau \Big[\mathbf{p} + e\mathbf{A}(\tau)\Big]^2\Big] \\ \times \mathbf{E}_{cl}(t_2) \cdot \mathbf{g}\big(\mathbf{p} + e\mathbf{A}_{cl}(t_2), \mathbf{p}' + e\mathbf{A}_{cl}(t_2)\big) \\ \times e^{-iS(\mathbf{p},t_2,t_1)/\hbar} \mathbf{E}_{cl}(t_1) \cdot \mathbf{d}\big(\mathbf{p}' + e\mathbf{A}_{cl}(t_1)\big), \end{aligned} \tag{3.16}$$

where in addition to the processes described earlier, the electron undergoes scattering with the nucleus at time t_2, transitioning from a state with kinetic momentum $m_e\mathbf{v}' = \mathbf{p}' + e\mathbf{A}_{cl}(t_2)$ to another state with kinetic momentum $m_e\mathbf{v} = \mathbf{p} + e\mathbf{A}_{cl}(t_2)$. Subsequently, it propagates under the influence of the electric field until the final time t. Therefore, by combining these results in Eq. (3.4), we obtain

$$|\psi(t)\rangle = a(t)|g\rangle + \int d\mathbf{p}\, b^{(0)}(\mathbf{p}, t)\, |\mathbf{p} + e\mathbf{A}_{cl}(t)\rangle + \int d\mathbf{p}\, b^{(1)}(\mathbf{p}, t)\, |\mathbf{p} + e\mathbf{A}_{cl}(t)\rangle\,, \tag{3.17}$$

which constitutes the primary focus of attention in the forthcoming subsections, where we describe the processes of above-theshold ionization and high-harmonic generation, pictorially depicted in Fig. 3.2. On top of these, one could also find other processes where the emitted photoelectron scatters with the parent ion, resulting in the ionization of a second electron. These processes are referred to as non-sequential double ionization (NSDI) [71–76], and lead to a sudden change in the energy distribution of the emitted electron pairs as a function of the intensity, typically referred to as *knee* structure. This sudden change marks the transition from sequential to non-sequential ionization processes [72, 73].

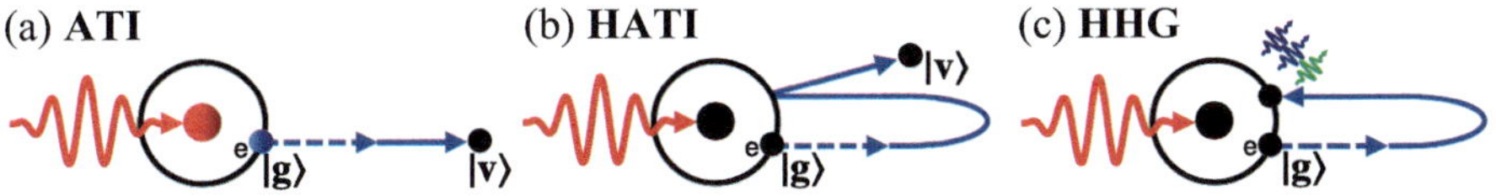

Fig. 3.2 Schematic representation of different strong-field processes. In **a**, the above-threshold ionization process is depicted, where a strong input IR field (represented by the red curve) ionizes an electron (represented by the blue circle) initially in the atomic ground state $|\mathrm{g}\rangle$, releasing it from its parent nucleus (represented by the red circle). The generated photoelectron then propagates, driven by the applied laser field, and is later found in the scattering state $|\mathbf{v}\rangle$. Due to the oscillating nature of the field, the electron can return to the parent ion and collide elastically with it in a process called high-order ATI (HATI), shown in (**b**), resulting in the state $|\mathbf{v}\rangle$. Alternatively, the electron could recombine with the parent system in a process called high-harmonic generation, shown in (**c**). In this case, upon the electron's return to the ground state, the energy acquired during its propagation in the continuum is released in the form of high-harmonic radiation (represented by the purple, blue, and green curves)

3.1.1.1 Above-Threshold Ionization

As previously mentioned, above-threshold ionization processes involve the ionization of an electron with more energy than required by the corresponding atomic potential. This remarkable phenomenon was first observed within the multiphoton ionization regime by P. Agostini and collaborators [7]. In their experiment, they observed the ionization of xenon atoms involving seven photons, one more than required to get ionized according to the input source's frequency. These experimental results are presented in Fig. 3.3a, where the first peak of the circular-dotted curve presents the absorption of six photons, while the secondary, lower peak corresponds to the absorption of seven photons. In 1983, just four years later after the previous experiment, advancements in laser technology enabled the generation of laser sources with increased power, thus allowing the exploration of optical tunneling processes. This was demonstrated by Chin et al. in Ref. [10], where they observed the generation of Xe^{+}, Xe^{2+} and Xe^{3+} ions using laser wavelengths of $\lambda_{\mathbf{k}_L} = 9.55\ \mu$m and $\lambda_{\mathbf{k}_L} = 10.55\ \mu$m. These results could not be adequately explained within the perturbative framework of multiphoton ionization, according to which the absorption of hundreds of photons was required. Instead, they aligned with the predictions of L. V. Keldysh regarding tunneling ionization.

Since then, non-perturbative ATI processes have been extensively studied and reported in numerous works [73, 77–79], with the ATI spectra being the most characteristic observable [80–82]. The ATI spectrum represents the number of electron counts as a function of their kinetic energy, indicating the probability of finding an electron with a given kinetic energy. One of its primary features when dealing with non-perturbative ATI, is the presence of a plateau of peaks, separated by a distance of $\hbar\omega_{\mathbf{k}_L}$ along the x axis, where $\omega_{\mathbf{k}_L}$ is the central wavelength of the source. This plateau extends to very high values of kinetic energies, typically on the order of $10U_p$. An example of this is shown for the case of Xe in Fig. 3.3b, originally reported in Ref. [77]. In this experiment, Xe atoms were excited with a laser source emitting

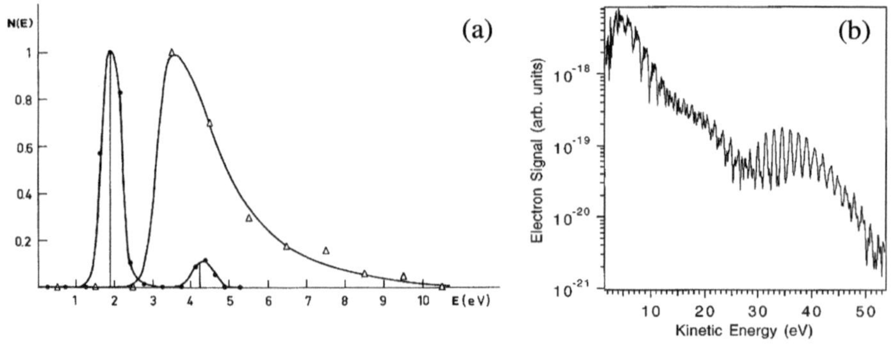

Fig. 3.3 Experimental above-threshold ionization spectra obtained under different ionization regimes. In **a**, the energy spectra of emitted photoelectrons after multiphoton ionization processes taking place in xenon atoms are shown. Circular markers represent the case of excitation with a laser field of $I = 8 \times 10^{12}$ W/cm^2 and wavelength $\lambda_{\mathbf{k}_L} \approx 531$ nm, while triangular markers represent excitation with a laser field of $I = 4 \times 10^{13}$ W/cm^2 and wavelength $\lambda_{\mathbf{k}_L} \approx 1057$ nm. In **b**, the energy spectra of emitted photoelectrons are shown under regimes where tunneling ionization dominates in the same media. Here, the peak intensity of the employed field is $I = 1.5 \times 10^{14}$ W/cm^2, and its associated wavelength $\lambda_{\mathbf{k}_L} = 800$ nm. In this case, the laser pulse duration is 120 fs. These two figures have been adapted from Ref. [7] (reprinted figure with permission of P. Agostini et al., *Phys. Rev. Lett.* **42** 17, pp. 1127-1130 (1979) Copyright (2023) by the American Physical Society) and Ref. [77] (reprinted figure with permission of P. Hansch, M. A. Walker and L. D. Van Woerkom, *Phys. Rev. A* **55** 4, pp. R2535-R2538 (1997) Copyright (2023) by the American Physical Society), respectively

pulses with a peak intensity of $I = 1.5 \times 10^{14}$ W/cm^2, a duration $\Delta t \approx 120$ fs (45 optical cycles) and central wavelength of $\lambda_{\mathbf{k}_L} = 800$ nm.

From a theoretical perspective, the ATI spectra can be computed within the SFA framework by projecting the quantum state in Eq. (3.17) onto a scattering state $|\mathbf{p} + \mathrm{e}\mathbf{A}_{\text{cl}}(t)\rangle$ [47, 80–82], leading to

$$M(\mathbf{p}, t) = \langle \mathbf{p} + \mathrm{e}\mathbf{A}_{\text{cl}}(t)|\psi(t)\rangle = b^{(0)}(\mathbf{p}, t) + b^{(1)}(\mathbf{p}, t), \tag{3.18}$$

and taking the square of the norm of this quantity:

$$\begin{aligned} M_{\text{tot}}(\mathbf{p}) &= \lim_{t\to\infty} |b^{(0)}(\mathbf{p}, t) + b^{(1)}(\mathbf{p}, t)|^2 \\ &= \lim_{t\to\infty} |b^{(0)}(\mathbf{p}, t)|^2 + \lim_{t\to\infty} |b^{(1)}(\mathbf{p}, \tau)|^2 + \lim_{t\to\infty} \left\{\left[b^{(0)}(\mathbf{p}, t)\right]^* b^{(1)}(\mathbf{p}, t)\right\} + \text{c.c.}, \end{aligned} \tag{3.19}$$

where the limit $t \to \infty$ has been taken, as in the case of finite and isolated pulses, the canonical and kinetic momenta asymptotically coincide.

In Eq. (3.19), the first and second terms of the last equality are typically referred to as the *direct ATI* and *rescattering ATI* contributions, respectively, with the latter leading to the high-order ATI (HATI) mechanism. On the one hand, the direct ATI term corresponds to processes in which the electron ionizes and is directly measured, as pictorially represented in Fig. 3.2a. On the other hand, the rescattering ATI

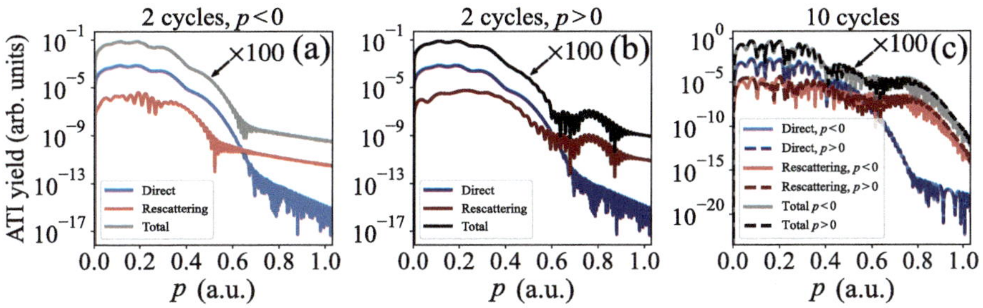

Fig. 3.4 Numerically computed ATI spectra. Here, we consider the ATI spectra obtained after the interaction of xenon atoms with a linearly polarized laser pulse with sinusoidal squared envelope, peak intensity $I = 8 \times 10^{13}$ W/cm^2 and central wavelength $\lambda_{\mathbf{k}_L} = 800$ nm. In panels **a** and **b**, the laser-pulse duration was set to $\Delta t \approx 5$ fs (2 optical cycles), and respectively consider photoelectrons propagating in the backward ($p < 0$) and forward ($p > 0$) direction, shown with light and dark colors, respectively. In **c**, a longer pulse of duration $\Delta t \approx 27$ fs (10 optical cycles) was considered instead. In these plots, blue curves show the direct ATI contribution, the red curves show the rescattering contributions leading to HATI, while the black curves show a combination of both. This figure has been computed using the software QPROP [84]

term is related to processes in which the electron scatters with the parent ion, resulting in a change in its kinetic energy, as shown in Fig. 3.2b. More specifically, the direct ATI contribution becomes more important for photoelectron kinetic energies approximately lower than $2U_p$. In contrast, in the high-energy part of the spectrum, roughly between $2U_p$ and $10U_p$, the rescattering contribution dominates [83], while more complicated electron dynamics lead to extra plateaus in the generated ATI spectra [70].

The behaviors outlined in the previous paragraph are illustrated in Fig. 3.4, which displays the theoretical 1D ATI spectra[9] for a Xe atom excited by a sinusoidal squared pulse of peak intensity 8×10^{13} W/cm^2, central wavelength $\lambda_{\mathbf{k}_L} = 800$ nm and of two different durations: in (a) and (b) we have $\Delta t \approx 5$ fs (2 optical cycles), while in (c), $\Delta t \approx 27$ fs (10 optical cycles). As seen in the figure, using long laser pulses consisting of many optical cycles leads to better-resolved ATI peaks, and a more symmetric contribution of electrons that propagate either in the forward ($p > 0$) or backward ($p < 0$) directions. However, in both scenarios we distinctly observe direct and rescattering features. The former defines an initial plateau that persists until $p \approx 0.4$ a.u., after which a second plateau emerges attributed to rescattering effects.

The more or less symmetric behavior of the ATI spectra for backward and forward electrons depending on the pulse duration, can be better understood when considering a quantum-orbit picture of ATI processes [70, 83]. As discussed in Sect. 2.2.4, the integrals defining the various contributions of ATI can be viewed as an interference of all the possible paths, i.e. the quantum orbits, the electron can take to reach

[9] This figure was generated using the QPROP package [84] software, a 3D-TDSE solver. However, the analysis here is done for a single dimension, defined by the laser field polarization. This is the direction along which electrons propagate either in the *backward* or *forward* direction.

a kinetic momentum $\mathbf{p}$ at time t. Due to the presence of the I_p factor in the semiclassical action presented in Eq. (3.15), which satisfies $\omega_{\mathbf{k}_L} \ll I_p$ by construction of the problem, the integrand becomes a highly oscillatory function, causing most of these trajectories to interfere destructively. Consequently, the integrals in Eqs. (3.14) and (3.16) become computationally challenging. In this context, the saddle-point approximation becomes an extremely valuable tool, simplifying the computation of these integrals while providing an intuitive conceptual picture of the underlying physical process.

The saddle-point approximation is based on the fact that, given an integral of the general form[10]

$$I = \int_C \mathrm{d}z\, f(z) e^{i\lambda g(z)} \quad z \in \mathbb{C}, \tag{3.20}$$

where C denotes the integration contour[11] and the integrand is an analytic function in z, one can modify the contour's shape as desired while keeping its endpoints unchanged, as well as its result. In the case where $g(z) \equiv \mathrm{Re}[g(z)] + i\mathrm{Im}[g(z)]$ hugely affects both the amplitude and phase of $e^{i\lambda g(z)}$, or for large values of $\lambda \in \mathbb{R}$, it is possible to seek contour deformations that can effectively freeze the oscillations of this function. This transformation also allows the absolute value of $e^{ig(z)}$ to pass through a maximum spot and then decay afterwards, significantly simplifying the evaluation of Eq. (3.20). This is indeed possible because, according to the Cauchy-Riemann conditions, if we express $z = x + iy$, then

$$\frac{\partial \mathrm{Re}[g(z)]}{\partial x} = \frac{\partial \mathrm{Im}[g(z)]}{\partial y}, \qquad \frac{\partial \mathrm{Re}[g(z)]}{\partial y} = -\frac{\partial \mathrm{Im}[g(z)]}{\partial x}, \tag{3.21}$$

which leads to the following properties:

1. if $z_0 = x_0 + iy_0$ is a critical point where $\partial g(z)/\partial z|_{z_0} = 0$, then it follows that $\partial \mathrm{Re}[g(z)]/\partial x = \partial \mathrm{Re}[g(z)]/\partial y = 0$ and $\partial \mathrm{Im}[g(z)]/\partial x = \partial \mathrm{Im}[g(z)]/\partial y = 0$, meaning that both the real and imaginary components of the function $g(z)$ are stationary[12];
2. the gradients of the real and imaginary components of $\mathrm{Re}[g(z)]$ and $\mathrm{Im}[g(z)]$ are orthogonal, i.e. $\nabla \mathrm{Re}[g(z)] \cdot \nabla \mathrm{Im}[g(z)] = 0$, implying that if we choose, for instance, the contour along the direction of steepest descent of $\mathrm{Im}[g(z)]$, then $\mathrm{Re}[g(z)]$ would remain constant.

---

[10] We refer the reader to Refs. [85, 86] and references therein for an introduction to the saddle-point approximation within a strong-field background, as well as to Chaps. 1 and 2 of Ref. [56].

[11] We assume that the integration contour ends somewhere where the contributions to the integral are negligible.

[12] This can be easily verified by considering that we can express $\partial \mathrm{Re}[g(z)]/\partial z = (1/2)\{\partial \mathrm{Re}[g(z)]/\partial x - i\partial \mathrm{Re}[g(z)]/\partial y\}$ using the chain rule, and similarly to $\partial \mathrm{Im}[g(z)]/\partial z = \partial \mathrm{Im}[g(z)]/\partial y + i\partial \mathrm{Im}[g(z)]/\partial x$. Thus, at the critical point, both imaginary and real parts must independently be zero.

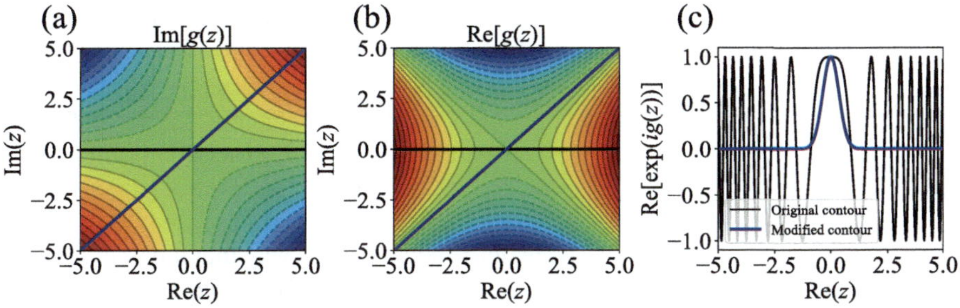

Fig. 3.5 Illustration of the saddle-point approximation applied to a complex-valued function. In this example, we consider the function $e^{\lambda g(z)}$, where $g(z) = z^2$, which we aim to integrate over the interval $(-\infty, \infty)$. In **a** and **b**, we provide a contour plot displaying the real and imaginary components of $g(z)$ as functions of the real and imaginary parts of z. Initially, the integration is represented by the horizontal black curve, resulting to a highly oscillating integral, as depicted by the black curve in panel (**c**). When employing the saddle-point approximation, we choose an alternative integration contour shown in blue in panels (**a**) and (**b**), along which the integrand exhibits a behavior as shown by the blue curve in panel (**c**)

An example of a simple function $g(z) = z^2$ is shown in Fig. 3.5, where in panels (a) and (b) the imaginary and real components of $g(z)$ are shown as a function of the real and imaginary parts of z. The black line in these plots represents the original integration contour, along which $g(z)$ exhibits pronounced oscillations, as shown in panel (c). However, if we instead choose an integration contour that follows the steepest descent direction of $\text{Im}[g(z)]$ and passes through the critical point[13] $z_0 = 0$, freezing $\text{Re}[g(z)]$, as indicated by the blue curve in Fig. 3.5, we obtain a Gaussian integral that is easier to handle. This Gaussian becomes more peaked as λ increases. In general, for any analytic function $g(z)$, if we expand it in a Taylor series around z_0 as $g(z) \approx g(z_0) + (1/2)[\partial^2 g(z)/\partial z^2|_{z_0}](z - z_0)^2$ and change the integration contour by introducing $dz = d(z - z_0) = e^{i\theta}d\mathsf{z}$, where θ determines the new direction in the complex plane, we can approximate Eq. (3.20) as

$$I \approx f(z_0)e^{i\varphi}e^{i\lambda g(z_0)} \int_C d\mathsf{z}\, e^{-\lambda[\partial g(z)/\partial z|_{z_0}]\mathsf{z}^2/2} = f(z_0)e^{i\varphi}e^{i\lambda g(z_0)} \sqrt{\frac{2\pi}{\lambda|\frac{\partial^2 g(z)}{\partial z^2}|_{z_0}|}}, \tag{3.22}$$

where we have assumed that $f(z)$ is a slowly varying function, causing the integrand to approach zero at infinity.

[13] In fact, following the Cauchy-Riemann conditions, it can be easily shown that this point can only be a saddle-point for $\text{Re}[g(z)]$ and $\text{Im}[g(z)]$ separately.

Thus, applying this technique to our strong-field analysis, we can approximate the time-dependent integral in Eq. (3.14) as

$$b^{(0)}(\mathbf{p},t) \approx \sum_{t_{\text{ion}}} \sqrt{\frac{i2\pi\hbar}{\left|\frac{\partial^2 S(\mathbf{p},t,t_1)}{\partial t_1{}^2}\right|_{t_1=t_{\text{ion}}}}}\mathbf{E}_{\text{cl}}(t_{\text{ion}})\cdot\mathbf{d}\big(\mathbf{p}+\mathsf{e}\mathbf{A}_{\text{cl}}(t_{\text{ion}})\big)e^{-iS(\mathbf{p},t,t_{\text{ion}})/\hbar}, \tag{3.23}$$

where in this case the contour has been deformed in such a way that it passes through all the saddle points $\{t_{\text{ion}}\}$ of $S(\mathbf{p},t,t_1)$. One of the most important aspects of these saddle points is their clear physical interpretation, that can be assessed by explicitly writing their saddle-point equation

$$\left.\frac{\partial S(\mathbf{p},t,t_1)}{\partial t_1}\right|_{t_1=t_{\text{ion}}} = 0 \Rightarrow \frac{1}{2m_{\mathsf{e}}}\big[\mathbf{p}+\mathbf{A}_{\text{cl}}(t_{\text{ion}})\big]^2 + I_p = 0. \tag{3.24}$$

This equation describes an ionization event in which the electron acquires the enough kinetic energy to surpass the ionization potential of the corresponding atomic system. Therefore, t_{ion} corresponds with the time at which the electron undergoes ionization from its parent atom. However, as observed from Eq. (3.24), all the terms on the left-hand side are positive. The only way for this equation to be satisfied is for t_{ion} to be a complex number with non-vanishing imaginary part. Although this might initially appear unusual, the presence of complex times is expected in tunneling processes, where the electron passes through the barrier formed by the combined effects of the electric field and the atomic potential. In this context, the imaginary part of the complex time represents the time spent by the electron within the barrier, i.e., the time at which the electron *enters* in the barrier [85]. This introduces a decay in the amplitude described by Eq. (3.23), quantifying the likelihood of such tunneling events. On the other side, the real part of t_{ion} corresponds to the time at which the electron emerges into the continuum, i.e. *exits* the barrier [85], contributing to Eq. (3.23) with a phase resembling the Feynmann's path trajectory formulation. Thus, within this formalism, the set $\{t_{\text{ion}}\}$ characterizes the most probable trajectories the electron can follow, and therefore define the concept of quantum orbits in direct ATI events.

In Fig. 3.6 we present the solutions to Eq. (3.24) for a linearly polarized laser field with a sinusoidal squared envelope, a central wavelength $\lambda_{\mathbf{k}_L} = 800$ nm, peak intensity $I = 1 \times 10^{14}$ W/cm^2 and a pulse duration of $\Delta t \approx 21$ fs (8 optical cycles). As shown in panel (a), electrons ionizing with $p = 0$ always appear at the maximum and minimum values of the field (blue circles). When $p \neq 0$, the ionization time shifts either towards the left or right side of these critical points, depending on whether $p < 0$ (red downward triangle) or $p > 0$ (green upward triangle), respectively. The probability of electron ionization is higher in those regions where the electric field is stronger, as illustrated in panel (b), where the imaginary part of the ionization times is smaller. This leads to a less depleted probability amplitude when compared with those regions where the imaginary part becomes larger, i.e., when the field is

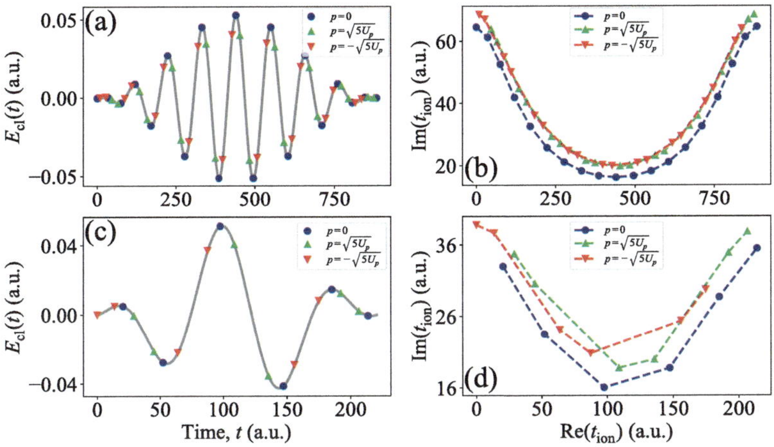

Fig. 3.6 Ionization times obtained from the saddle-point equation for ATI. We considered the excitation of hydrogen atoms subjected to a linearly polarized laser pulse with sinusoidal-squared envelope, peak intensity $I = 1 \times 10^{14}$ W/cm^2, central wavelength $\lambda_{\mathbf{k}_L} = 800$ nm and two different pulse durations. In panels **a** and **b**, the pulse duration was set to $\Delta t \approx 21$ fs (8 optical cycles), while in panels **c** and **d** it was $\Delta t \approx 5$ fs (2 optical cycles). In panels **a** and **c**, the real part of the ionization times is presented, overlaid within the electric field for three different photoelectron momenta. In panels **b** and **d** the imaginary part of the ionization time is shown as a function of their corresponding real parts

weaker. Furthermore, for larger momentum values, $\text{Im}(t_{\text{ion}})$ increases, making direct ionization processes with these momenta less likely.

However, it is worth noting that the curves for $p > 0$ and $p < 0$ exhibit symmetry, arising from the symmetric nature of the considered long pulse. For shorter pulses, around $\Delta t \approx 5$ fs (2 optical cycles), the carrier-envelope phase (CEP), i.e. the phase difference between the field's envelope and the carrier wave, plays a crucial role in determining the pulse's symmetry. Specifically, it leads to enhanced contributions of either positive or negative momentum values [80, 87]. This is demonstrated in Fig. 3.6c and d, where the same laser parameters are used, but with a reduced pulse duration of $\Delta t \approx 5$ fs. In this scenario, the absence of time-reversal symmetry within the driving pulse causes ionization times happening at the maximum and minimum values to be non-equivalent. Consequently, this results in a difference in the probability of finding electrons with $p > 0$ compared to those with $p < 0$.

3.1.1.2 High-Harmonic Generation

In contrast to HATI processes, in high-harmonic generation the rescattered photoelectron recombines with the ground state of the parent ion, releasing the kinetic energy it has acquired (plus the ionization potential of the atom) in the form of high-order harmonic radiation. Consequently, one of the most characteristic observables

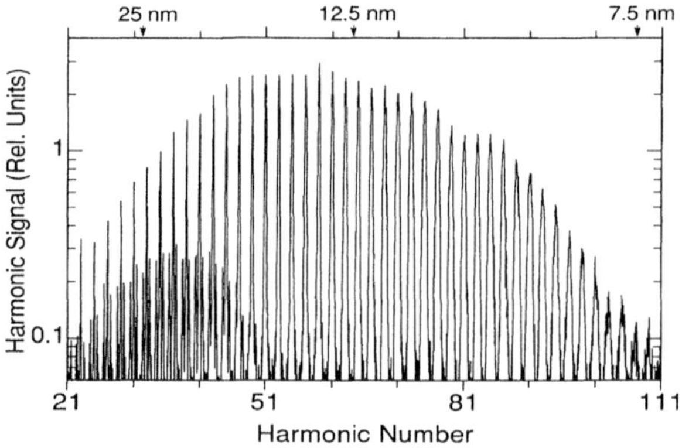

Fig. 3.7 Experimental high-harmonic generation (HHG) spectrum. In this experiment, neon gas was excited with a 800 nm titanium-sapphire laser with a peak intensity $I = 1.3 \times 10^{15}$ W/cm^2 and a pulse duration of $\Delta t = 125$ fs. The displayed HHG spectrum exhibits two distinct features: an initial increase at small harmonic orders attributed to the grating response, and a subsequent decrease at higher energies due to the harmonic cutoff, which extends up the 111th harmonic order. This figure has been adapted from Ref. [88] (reprinted figure with permission of J. J. Mackin, J. D. Kmetec and C. L. Gordon, *Phys. Rev. Lett.* **70** 6, pp. 766-769 (1993) Copyright (2023) by the American Physical Society)

in HHG is the HHG spectra, which measures the intensity of the emitted radiation as a function of its frequency. Similar to ATI, the main distinction between the perturbative (multiphoton) and non-perturbative (tunneling) pictures on the HHG spectrum, lies in the presence of a plateau in the measured spectrum [19, 20], which can extend into the XUV [88–90] and even the soft-X ray regime [91–93]. This plateau structure is also characterized by other remarkable features: specially when considering linearly polarized driving fields exciting the atomic system, it exhibits well-defined harmonic peaks located at odd harmonic orders, as depicted in Fig. 3.7, followed by a sharp cutoff at around $\hbar\omega_{\mathbf{k}} \approx 3U_p + I_p$ [27].

From a theoretical standpoint, an atom's response to an electromagnetic field is proportional to the induced dipole, which can be computed from a quantum mechanical framework as [85]

$$\langle \mathbf{d}(t) \rangle = \mathsf{e} \big\langle \psi(t) | \hat{\mathbf{r}} | \psi(t) \big\rangle, \tag{3.25}$$

with $|\psi(t)\rangle$ given in Eq. (3.17). Introducing the latter into Eq. (3.25) while neglecting continuum-continuum contributions[14] which are usually negligible [51], we arrive to

$$\langle \mathbf{d}(t) \rangle \approx \mathsf{e} \int \mathrm{d}\mathbf{p}\, b^{(0)}(\mathbf{p}, t) a^*(t) \mathbf{d}^*\big(\mathbf{p} + \mathsf{e}\mathbf{A}_{\mathrm{cl}}(t)\big) + \text{c.c.}, \tag{3.26}$$

[14] To be specific, we neglect terms of the form $\langle \mathbf{v}' | \hat{\mathbf{r}} | \mathbf{v} \rangle$.

where we also considered that typical ground states have a well-defined parity, opposite to that of $\hat{\mathbf{r}}$, which implies $\langle g|\hat{\mathbf{r}}|g\rangle = 0$. By incorporating the expression of $b^{(0)}(\mathbf{p}, t)$ from Eq. (3.14) and that found for $a(t)$, we can expand Eq. (3.26) as follows

$$\langle \mathbf{d}(t)\rangle = -\frac{i\mathrm{e}}{\hbar}\int \mathrm{d}\mathbf{p} \int_{t_0}^{t} \mathrm{d}t_1 e^{-iI_p(t-t_0)/\hbar}\mathbf{d}\big(\mathbf{p}+\mathrm{e}\mathbf{A}_{\mathrm{cl}}(t)\big)e^{-iS(\mathbf{p},t,t_1)/\hbar} \times \mathbf{E}_{\mathrm{cl}}(t_1)\cdot \mathbf{d}^*\big(\mathbf{p}+\mathrm{e}\mathbf{A}_{\mathrm{cl}}(t_1)\big). \tag{3.27}$$

From the time-dependent dipole moment in Eq. (3.27), the HHG spectrum along the polarization mode μ can be computed[15] by performing the Fourier transform of $\boldsymbol{\epsilon}_{\mathbf{k},\mu}\cdot\langle \mathbf{d}(t)\rangle$, which we shall refer to as $\langle d_\mu(\omega_{\mathbf{k}})\rangle$ [85]. Therefore, the electromagnetic field mode with frequency $\omega_{\mathbf{k}}$ appears after HHG processes with an intensity given by

$$\begin{aligned} I(\omega_{\mathbf{k}}) &\propto \omega_{\mathbf{k}}^4 |d_\mu(\omega_{\mathbf{k}})|^2 \\ &\propto \omega_{\mathbf{k}}^4 \bigg| \int_{t_0}^{t} \mathrm{d}t_2 \int \mathrm{d}\mathbf{p} \int_{t_0}^{t_2} \mathrm{d}t_1 \boldsymbol{\epsilon}_{\mathbf{k},\mu}\cdot \mathbf{d}\big(\mathbf{p}+\mathrm{e}\mathbf{A}_{\mathrm{cl}}(t_2)\big)e^{-i\tilde{S}(\mathbf{p},t_2,t_1)/\hbar} \\ &\qquad \times \mathbf{E}_{\mathrm{cl}}(t_1)\cdot \mathbf{d}^*\big(\mathbf{p}+\mathrm{e}\mathbf{A}_{\mathrm{cl}}(t_1)\big)\bigg|^2, \end{aligned} \tag{3.28}$$

where the action $\tilde{S}(\mathbf{p}, t_2, t_1)$ is given by

$$\tilde{S}(\mathbf{p}, t_2, t_1) = \frac{1}{2m_\mathrm{e}}\int_{t_1}^{t_2} \mathrm{d}\tau\big[\mathbf{p}+\mathrm{e}\mathbf{A}_{\mathrm{cl}}(\tau)\big]^2 + I_p(t_2-t_1) - \hbar\omega_{\mathbf{k}}t_2. \tag{3.29}$$

Similarly to what happens in the ATI spectra, the presence of $I_p t_1$ and $\omega_{\mathbf{k}} t_1$ for the high-order harmonics causes the integrands appearing in Eq. (3.28) to oscillate significantly, making the five-dimensional integral a numerically challenging task. Instead, one can rely on the saddle-point approximation, which allows us to approximate $d_\mu(\omega_{\mathbf{k}})$ as

$$d_\mu(\omega_{\mathbf{k}}) \approx \sum_{t_{\mathrm{re}}}\sum_{\mathbf{p}_s}\sum_{t_{\mathrm{ion}}} G(\mathbf{p}_s, t_{\mathrm{re}}, t_{\mathrm{ion}})\boldsymbol{\epsilon}_{\mathbf{k},\mu}\cdot \mathbf{d}\big(\mathbf{p}_s+\mathrm{e}\mathbf{A}_{\mathrm{cl}}(t_{\mathrm{re}})\big)e^{-i\tilde{S}(\mathbf{p},t_{\mathrm{re}},t_{\mathrm{ion}})/\hbar} \times \mathbf{E}_{\mathrm{cl}}(t_{\mathrm{ion}})\cdot \mathbf{d}^*\big(\mathbf{p}+\mathrm{e}\mathbf{A}_{\mathrm{cl}}(t_{\mathrm{ion}})\big), \tag{3.30}$$

[15] To be more correct, the spectrum of the scattered radiation should be described by the dipole correlation function, as shown in Ref. [94]. Calculations performed with the dipole moment instead of their correlation at different times is, though, appropriate for reproducing experimental results with many atoms, i.e. independent electrons, that effectively *see* the same field. In this case, the different dipole correlation functions do actually become uncorrelated.

with $G(\mathbf{p}_s, t_{\text{re}}, t_{\text{ion}})$ representing the prefactors of Eq. (3.22), which involve second-order derivatives of the action $\tilde{S}(\mathbf{p}, t_2, t_1)$ with respect to the different integration variables.

In Eq. (3.30), the saddle-point approximation has been applied simultaneously to each of the three integrals. Consequently, we get three different sets of saddle points, denoted as $\{t_{\text{re}}\}$ for the t_2 integration variable, $\{\mathbf{p}_s\}$ for $\mathbf{p}$, and $\{t_{\text{ion}}\}$ for t_1, which are obtained from the following set of saddle-point equations [47, 51, 81, 85]

$$\frac{\partial \tilde{S}(\mathbf{p}, t, t_1)}{\partial t_1}\bigg|_{\substack{t_2=t_{\text{re}},\mathbf{p}=\mathbf{p}_s\\ t_1=t_{\text{ion}}}} = 0 \Rightarrow \frac{1}{2m_{\text{e}}}\big[\mathbf{p}_s + \text{e}\mathbf{A}_{\text{cl}}(t_{\text{ion}})\big]^2 = I_p, \tag{3.31}$$

$$\nabla_{\mathbf{p}}\tilde{S}(\mathbf{p}, t, t_1)\bigg|_{\substack{t_2=t_{\text{re}},\mathbf{p}=\mathbf{p}_s\\ t_1=t_{\text{ion}}}} = 0 \Rightarrow \int_{t_{\text{ion}}}^{t_{\text{re}}} \mathrm{d}\tau\big[\mathbf{p}_s + \text{e}\mathbf{A}_{\text{cl}}(\tau)\big] = 0, \tag{3.32}$$

$$\frac{\partial \tilde{S}(\mathbf{p}, t, t_1)}{\partial t_1}\bigg|_{\substack{t_2=t_{\text{re}},\mathbf{p}=\mathbf{p}_s\\ t_1=t_{\text{ion}}}} = 0 \Rightarrow \frac{1}{2m_{\text{e}}}\big[\mathbf{p}_s + \text{e}\mathbf{A}_{\text{cl}}(t_{\text{re}})\big]^2 + I_p = \hbar\omega_{\mathbf{k}}. \tag{3.33}$$

These three equations define the set of quantum orbits the electron performs when it undergoes a HHG process. As with ATI processes, Eq. (3.31) describes a tunneling event happening at time t_{ion}, i.e. the ionization time, while Eqs (3.32) and (3.33) describe the additional dynamics the electron undergoes. Specifically, Eq. (3.32) imposes that the total displacement performed by the electron between times t_{ion} and t_{re} must be equal to zero, while Eq. (3.33) is an energy balance equation. In this equation, the kinetic energy acquired by the electron, plus the ionization potential energy, must be equal to the frequency of the emitted radiation $\omega_{\mathbf{k}}$. Hence, this equation describes the energy conditions required for the electron to return to the ground state and, for this reason, we refer to t_{re} as the recombination time.

This set of equations explicitly introduces what is typically referred to as the *three-step* model or the *simple man's* model [27, 71, 95], which is pictorially summarized in Fig. 3.2c. According to this model, the electron undergoes three distinct steps during a non-perturbative HHG process:

1. ionization from the parent ion via a tunneling event through the barrier formed by the atomic potential and the laser electric field;
2. propagation in the continuum driven by the laser field;
3. return to the parent ion, where it recombines, emitting high-harmonic radiation.

In this picture, it becomes clear that the cutoff frequency $\omega_{\mathbf{k}_c}$ is determined by the maximum kinetic energy the electron can attain while freely oscillating under the influence of the field. It can be shown that this energy is given by [27]

$$\hbar\omega_{\mathbf{k}_c} \approx 3.17U_p + I_p. \tag{3.34}$$

Each of the aforementioned steps is respectively described by Eqs. (3.31), (3.32) and (3.33), whose solutions lead to complex-valued saddle points [85, 86, 96, 97].

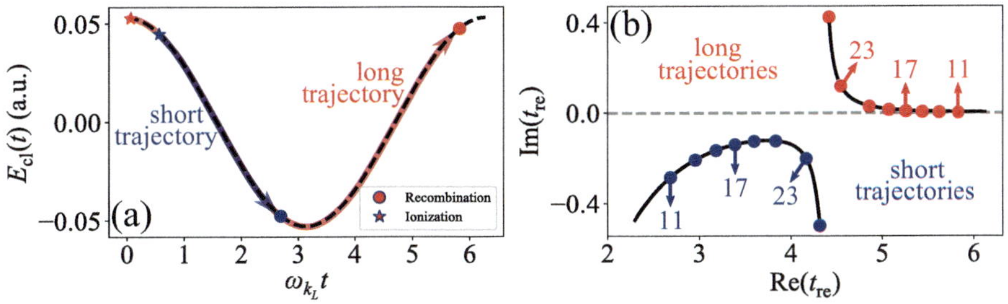

Fig. 3.8 Short and long trajectories derived from HHG saddle-point equations. We considered the excitation of hydrogen atoms by a monochromatic laser field with peak intensity of $I = 1 \times 10^{14}$ W/cm^2 and a wavelength $\lambda_{\mathbf{k}_L} = 800$ nm. In **a**, a schematic representation of the short trajectories (in blue) and long trajectories (in red) obtained for the 11th harmonic mode is presented overlaid within the electric field. Star markers correspond to the ionization time, while circular markers correspond to the recombination time. In **b**, we illustrate the imaginary part of the recombination time as a function of their real part. In this case, the harmonic cutoff occurs at approximately the 21st harmonic mode

Specifically, when solving these equations for a specific value of $\mathbf{p}_s$, one typically encounters two kind of trajectories, commonly referred to as short and long trajectories [98, 99], which differ in their duration, as shown in Fig. 3.8a when considering $\omega_{\mathbf{k}} = 11\omega_{\mathbf{k}_L}$. As the frequency of the generated radiation reaches the cutoff, these two trajectories tend to merge, as illustrated in panel (b), leading to a Stokes transition [100, 101], where the contribution of the short trajectory starts to grow exponentially and must be discarded [86, 97]. Finally, when considering more than a single cycle of the field, as presented in Fig. 3.8, additional saddle points corresponding to higher-order returns may arise. However, from an experimental perspective, it has been shown that the contribution of these higher-order return trajectories becomes negligible due to phase-matching conditions [70].

When introducing these saddle points into Eq. (3.22), and taking its absolute value, HHG spectra like the ones shown in Fig. 3.9a and b are obtained. These spectra were computed for a hydrogen atom with ionization potential $I_p = 0.5$ a.u., excited by a linearly polarized laser field with a central wavelength of $\lambda_{\mathbf{k}_L} = 800$ nm and a duration of approximately $\Delta t \approx 27$ fs (10 optical cycles). For each curve in panel (a) and each point along the x axis in panel (b), the HHG spectrum has been computed for different field strengths. From these plots, it becomes evident that the SFA suffices for reproducing the main characteristics of the HHG emission: the presence of a plateau extending up the cutoff frequency (represented by the dashed curves), followed by a sharp cutoff. However, when compared to experimental results such as the one shown in Fig. 3.7, it becomes apparent that in the central region there is a lack of harmonic peaks in the theoretical results, specially for large field strengths. This significant difference arises due to the interference between distinct electron trajectories contributing to harmonic radiation [102, 103], some of which lead to destructive interferences. Effects such as propagation [70] or the consideration of an intensity-dependent phase due to the laser's beam width [104], result in destructive interference

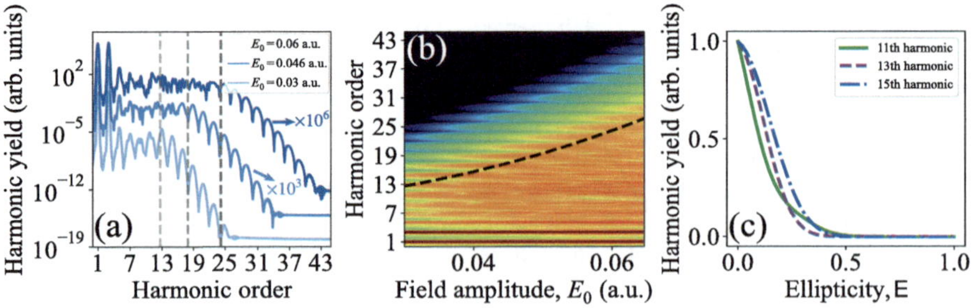

Fig. 3.9 Numerically computed HHG spectra. In panels **a** and **b**, we consider the interaction of hydrogen atoms with a linearly polarized laser field with sinusoidal-squared envelope, central wavelength $\lambda_{\mathbf{k}_L} = 800$ nm and $\Delta t \approx 27$ fs of duration (10 optical cycles). We present the harmonic spectra as a function of the harmonic order for various electric field strengths. In **c**, we maintain the same laser parameters but we instead consider a generally elliptically polarized field with varying ellipticity E. In this case, we depict the intensity of three harmonic orders as a function of E. For representational purposes, each harmonic intensity has been normalized to its maximum value. This figure has been generated using the saddle-point approximation, using the `RB-SFA` Mathematica package [105]

of some trajectories within the plateau region, leading to well-resolved theoretical harmonic spectra that are in better agreement with experimental observations.

Although our discussion thus far has mainly focused on linearly polarized laser fields, the quantum-orbit picture of HHG processes provides a comprehensive understanding of what could occur when considering fields with generic elliptical polarization. In this context, it is common to introduce the ellipticity E, defined as the ratio between the absolute values of the field amplitudes along the two polarization axes, satisfying[16] $0 \leq \mathsf{E} \leq 1$, where the lower bound is satisfied for a linearly polarized field while the upper bound for a circularly polarized field. If we consider a field with $\mathsf{E} > 0$, it may result in the generated photoelectron missing the parent ion when driven back by the field, leading to a lack of recombination events compared to the case with $\mathsf{E} = 0$ [71]. In fact, the larger E becomes, the lower the conversion efficiency of the harmonics, as shown in Fig. 3.9c, where the normalized intensity of different harmonic orders is depicted as a function of E. This effect has been extensively studied theoretically [106, 107] and experimentally [108–113], as well as in more complex setups in which two distinct fields with different ellipticities are combined to produce bright circularly polarized harmonics [114–117].

[16] In this definition, we consider that the field with the lowest amplitude (in absolute value) is allocated in the numerator, such that $0 \leq \mathsf{E} \leq 1$.

3.2 Quantum Optics Meets Strong-Field Physics: A State of the Art

As discussed thus far, the study of atomic systems under the influence of strong laser fields has predominantly relied on classical [6] or semiclassical models [118]. While early (and not so early) attempts to incorporate quantum electrodynamical approaches aimed primarily at validating semiclassical models, or proposing novel mechanisms for the generation of harmonic radiation, a quantum optical analysis was not the main focus. The landscape began to shift with the emergence of experiments dedicated to exploring quantum optical signatures within the generated field following HHG processes. This transition in research emphasis prompted a new direction within the strong-field community. Notably, concepts such as non-classical behavior of light, entanglement between light and matter, as well as among the diverse light modes, and photon statistics gained prominence in the study of strong-field physics.

Consequently, quantum optical-based approaches in strong-field physics can be broadly categorized into three main domains: early (and not so early) work, first observations of quantum optical signatures in the field, and the study of non-classical light in strong-field physics.

3.2.1 *Early (and Not So Early) Work*

One of the earliest quantum electrodynamical calculations conducted in the context of strong-field physics was done by Guo and Åberg in Ref. [119]. In this study, they addressed multiphoton ionization processes by solving Dirac's equation in the presence of a generic elliptically polarized field. Their results were found to be in good agreement with those obtained when classical fields were considered within the non-relativistic limit. Building upon their initial work, these same authors expanded their investigation by proposing a scattering theory to elucidate multiphoton ionization processes [120]. This study exhibited favorable concordance with alternative scattering-theoretical methods, demonstrating their equivalence particularly in the large-photon-number limit [121]. Furthermore, D. S. Guo and G. W. F. Drake extended the aforementioned analysis to encompass multiphoton ionization processes induced by circularly polarized fields [122].

In a similar but parallel direction, the work conducted by P. Lu and collaborators culminated in the development of a non-perturbative quantum electrodynamics theory to elucidate strong-laser matter interactions [123]. This comprehensive framework served as the foundation for their investigations into HHG and ATI, as well as their interrelation [124]. In these seminal studies, the authors derived mathematical expressions for the transition probabilities governing ATI and HHG processes while considering the initial state of light as a Fock state characterized by a large photon number. Surprisingly, despite the initial state having an electric field mean value of zero, the HHG spectra exhibited striking similarities to those predicted by semiclassical theories, faithfully reproducing the anticipated cutoff laws. Building upon this foundation, the researchers extended their analytical capabilities to explore other strong-field phenomena, including HATI [125] and NSDI [126].

While these studies successfully linked their observations to typical strong-field dynamics, others proposed new mechanisms. One such example is the work conducted by Gao et al. in Ref. [127]. Their quantum electrodynamical extension to HHG processes revealed that transitions between continuum-continuum states could significantly impact the resulting HHG spectra. However, these results have been the subject of intense debate and scrutiny within the scientific community [128, 129]. Another noteworthy example comes from the research efforts of A. Bogatskaya, E. Volkova and A. Popov, who introduced a quantum optical perturbative approach for studying spontaneous transitions in atomic systems excited by strong-fields. Their aim was to evaluate semiclassical theories and determine their range of validity [130]. Subsequently, these authors applied their approach to describe HHG spectra, revealing the presence of both odd and even harmonics, the latter emerging during the *bremsstrahlung* process where radiation is emitted due to the deceleration of the active electron within the parent's ion potential [131]. In a related vein, D. B. Milošević developed a quantum theory of photon emission in a single-step process during ATI. This theory predicted probabilities of this process which were orders of magnitude higher than those associated with HHG processes [132].

The utilization of quantum optical-based calculations to investigate HHG processes have proven invaluable in deriving accurate analytical formulas as well. In this regard, the work of Sundaram and Milonni in Ref. [94] represents a significant milestone in the analytical calculation of HHG spectra. Specifically, they demonstrated the fact that these calculations should rely on the computation of dipole correlation functions, rather than the expectation value of the induced dipole.[17] Furthermore, they derived the correct prefactor $\omega_{\mathbf{k}}^4$ for the intensity of the harmonic spectra—a practice that was not commonplace within the scientific community at that time.

The work by B. Sundaram and P. W. Milonni also demonstrated, from a semiclassical frame, that when two-level systems are subjected to a sufficiently intense field, they can also give rise to the generation of harmonic radiation. This discovery, besides highlighting that harmonic generation is not exclusive to atomic systems, paved the way for simplified quantum-optical HHG analyses in which continuum states for the target are not a prerequisite. Along these lines, S. Varrò and colleagues made significant contributions. Specifically, in Ref. [133], they employed a quantum treatment for the driving field while handling other field modes classically. This approach proved valuable in understanding the distinct roles played by each term in the light-matter interaction Hamiltonian in the context of HHG processes. This analysis was further elaborated upon in Ref. [134], where they also reversed the treatment—employing a classical driving field and quantized harmonics. This alternative perspective allowed them to successfully replicate typical HHG spectra.

[17] As previously noted in another footnote, in the many-atom regime, the dipole correlation functions effectively become decorrelated, and the results are analogous to using the mean value of the induced dipole.

3.2.2 *First Observations of Quantum Optical Signatures in the Field*

Much of the experimental groundwork concerning the quantum optical analysis of strong-field physics processes was conducted by the research group led by P. Tzallas in Crete, Greece. Their theoretical insights and experimental investigations have laid the foundation for many of the results presented in this Thesis.

The earliest experimental indications of the quantum optical nature of high-harmonic generation were realized in Ref. [135]. In this study, I. K. Kominis, G. Kolliopoulos, D. Charalambidis, and P. Tzallas provided a quantum optical interpretation of the recombination process in HHG. They demonstrated that photons emitted during short and long trajectories could be combined using Hadamard-like gates, creating an interference pattern that was experimentally resolved.

However, the major breakthrough that indeed revolutionized the field occurred in Refs. [136, 137], where the authors successfully predicted [136] and observed [137] the existence of correlations between the generated harmonics and the input driving field. To elaborate, the theoretical analysis developed in Ref. [136] involved an extension of Lewenstein SFA model [51], with the main distinction being the treatment of Volkov states from a quantum optical perspective, the driving field treated quantum mechanically. Although the XUV modes were not explicitly considered, this work theoretically demonstrated that the characteristics of the HHG process leave an imprint on the IR photon statistics.

These theoretical findings received experimental support in Ref. [137], which introduced a scheme for measuring correlations between the driving infrared field and the generated harmonics, subsequently referred to as the Quantum Spectrometer. This experimental setup is illustrated in Fig. 3.10a and features a modified version of typical HHG generation setups. An intense driving field is directed into a chamber containing an atomic gas (in their case Xe) for harmonic generation. The key distinction with usual setups is that, following the interaction, a grating separates the generated harmonics from the input driving field. The harmonics are then detected by a photomultiplier PMT, while the IR after the interaction by a photodiode PD1. Another detector, PD2, serves to eliminate energy fluctuations from the incoming laser field in the final results. Consequently, when energy fluctuations remain below a certain threshold (in the experiments the threshold was set to 1%), they observed that the reduction in the signal recorded by PD1, denoted as S_{PD1} and which corresponds to an enhancement of $S_{\text{PD2}} - S_{\text{PD1}}$, exhibits the same nonlinear behavior with respect the input intensity as the recorded harmonic signal, i.e. S_{PMT}. This is depicted in panel (b) with the red and blue data, respectively. Therefore, the more harmonic radiation is generated (larger S_{PMT}), the more radiation is absorbed from the input driving field (lower S_{PD1} or greater $S_{\text{PD2}} - S_{\text{PD1}}$), suggesting the presence of a correlation between both signals.

To highlight the existence of these correlations, the two signals were recorded simultaneously and plotted against each other, resulting in the correlation map shown in Fig. 3.10c. These points correspond to various strong-field processes occuring in

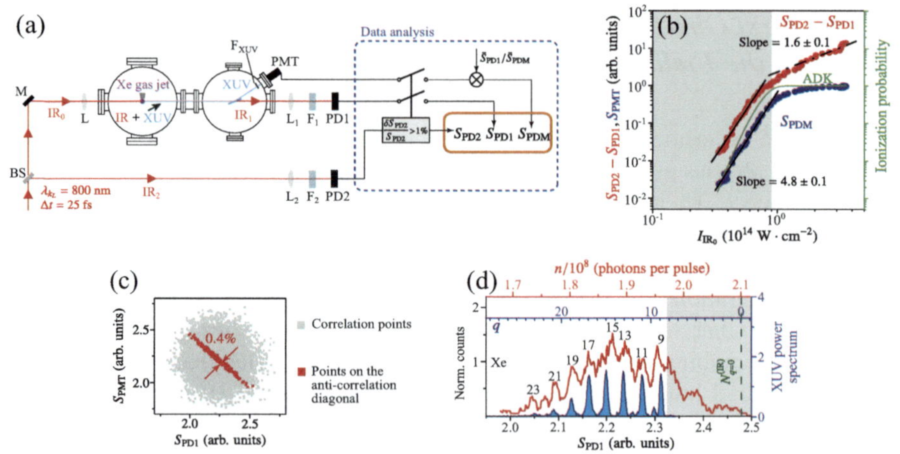

Fig. 3.10 Experimental setup demonstrating the correlation between the driving field and the generated harmonics in HHG processes. In **a**, we present the experimental configuration, where BS symbols represent beam splitters, L denotes focusing lenses, F signifies filters, PD stands for photodetectors and PMT indicates photomultipliers. The output signals from this setup correspond to S_{PD1}, S_{PD2} and S_{PDM}. These signals are plotted in panel (**b**) as a function of the laser intensity. In **c**, S_{PDM} data points are depicted as a function of the corresponding values measured for S_{PD1}, with the red points highlighting the so-called anticorrelation diagonal. Finally, in **d**, points lying on the anticorrelation diagonal are displayed when HHG is driven on Xe atoms. The measured high-harmonics (after filtering orders $q < 9$) are shown in blue. This figure has been adapted from Ref. [137]

the chamber, including HHG. However, due to the dependency found in panel (b), only the points lying along the diagonal (red points in panel (c)), also referred to as the *anti-correlation* diagonal, can account for HHG mechanisms. Consequently, by considering only the S_{PD1} points along this diagonal, the red curve in panel (d) was obtained, where peaks in the measured IR field are correlated with those in the generated harmonic signal (in blue). Remarkably, although this experiment was conducted in atomic systems, similar findings were also reported in semiconductor materials [138].

3.2.3 Non-Classical Light in Strong-Field Physics

The research conducted by the group of P. Tzallas raised a significant question concerning the nature of light after strong-field processes, transitioning it from a purely academic inquiry to a practical consideration.

This question was initially addressed from a theoretical perspective by the group of M. Lewenstein in Ref. [139] and the group of I. Kaminer and O. Cohen in Ref. [140]. In these two works, it was found that the state of the electron andthe field becomes

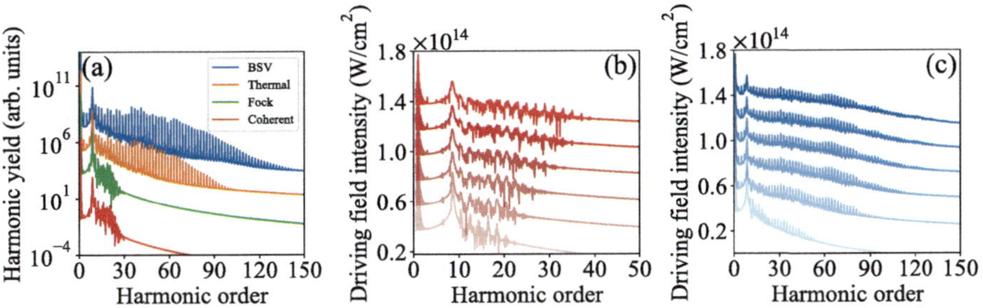

Fig. 3.11 HHG spectra driven by light with different photon statistics. In **a**, HHG processes are driven by a driving field with a central wavelength $\lambda_{\mathbf{k}_L} = 800$ nm and a peak intensity $I = 1 \times 10^{14}$ W/cm^2, each exhibiting different photon statistics. In this panel, the spectra have been vertically shifted for clarity. In panels **b** and **c**, particular attention is given to light fields with Poissonian statistics, represented by pure coherent states, and sub-Poissonian statistics, achieved using bright-squeezed vacuum (BSV) states, respectively. In these two cases, the spectra are computed using different driving field intensities, shown along the y axis. This figure has been adapted from Ref. [143]. Reproduced with permission from Springer Nature and the authors

entangled following strong-field processes. Furthermore, it was suggested that HHG could potentially lead to the generation of massively entangled states between the field modes themselves. While the work in Ref. [139] considered an extension of Lewenstein SFA model under a quantum optical perspective, the study in Ref. [140] accounted for intermediate bound states and highlighted their relevance in the harmonic generation spectrum from a single-electron perspective. Although, these intermediate states contribute incoherently in the emission from many-atoms, within the single-atom regime the authors of Ref. [140] showed that these may give rise to subtle signatures of sub-Poissonian behaviors. This observation was subsequently analyzed by the group of S. Varrò within their two-level system framework in Refs. [141, 142].

In this context, the groups of I. Kaminer and O. Cohen have been exceptionally productive in analyzing the influence of non-Poissonian photon statistics [143, 144] and non-classical correlations in matter systems [145] on HHG processes. In Ref. [143], A. Gorlach and colleagues developed a quantum optical description of HHG processes based on the generalized positive P representation, which elegantly allows for the incorporation of various photon statistics. Their work specifically focused on cases involving coherent, Fock states, thermal and bright squeezed vacuum (BSV) light sources, yielding the results outlined in Fig. 3.11a. For the first two cases, they found that both types of light sources (red and green curves) yielded identical HHG spectra, in agreement with the analysis in Ref. [124]. However, they obtained surprising results for the other two sources (blue and orange curves). The use of these sources lead to extended cutoffs, with a different intensity scaling compared to coherent states of light, as exemplified in panels (b) and (c) for the cases of coherent and BSV sources, respectively. Moreover, using these photon statistics can also result in better-resolved harmonic peaks, a feature related to the super-Poissonian nature of the considered sources. This methodology was further refined by M. Even-Tzur and collaborators in Ref. [144], where they developed a quantum optical version of Lewenstein SFA model, termed qSFA. In addition to the expected semiclassical action, they found that for optical states with non-vanishing electric field amplitudes, the photon statistics introduced an extra phase term that exerts an effective force on the electron, influencing its dynamics. Moreover, they recently showed that the use of intense squeezed states as drivers can lead to high-harmonic squeezed states [146]. In a similar direction, and using the same methodology, Y. Fang and coworkers studied in Ref. [147] the consequences of using BSV light for driving ATI processes, resulting in the presence of asymmetric photoelectron distributions.

On a different study, A. Pizzi and colleagues conducted a study of HHG in many-atom systems [145]. They demonstrated that the presence of non-classical correlations in the initial state of the system could lead to non-trivial geometries of the Wigner function of the generated harmonic radiation, even though the photon statistics appeared to be classical. They also revealed that these non-classical correlations in the driven many-atom system resulted in the presence of non-trivial correlations between the harmonic modes.

In a parallel direction, the group of M. Lewenstein, in collaboration with that of P. Tzallas, expanded upon the initial experiments conducted by the Crete group, demonstrating the generation of high photon-number coherent state superpositions [39, 41, 44]. These superpositions could exhibit potential entanglement features between different harmonic modes [42, 148], and were used to create larger superpositions between two coherent states, as well as triple coherent state superpositions [40]. Furthermore, the high-photon nature of the generated coherent states leveraged to induce nonlinear optics interactions in Beta Barium Borate (BBO) crystals [46]. The approach presented in these works was also extended to different media such as diatomic molecules [149] or solid state systems[18] [151, 152], and was further applied in atomic systems to study ATI processes [41, 44], revealing that they could potentially lead to light-matter entanglement features [43]. A review covering all these observations can be found in Ref. [45]. Moreover, by employing the generalized positive P representation model developed by the groups of I. Kaminer and O. Cohen, P. Stammer recently demonstrated that HHG can be driven by incoherent light, subsequently leading to incoherent radiation [153].

In addition to these works that have explored the quantum optical nature of light after strong-field processes, other studies have also delved into the presence of entanglement in different strong-field-based contexts. In this regard, M. J. J. Vrakking demonstrated that, through the combination of an ionizing strong-field and a delayed XUV pulse, one could generate entangled states between the parent ion and the resulting photoelectron [154]. Remarkably, this concept was experimentally validated in Ref. [155]. Along similar lines, M. Ruperti and colleagues investigated these entangled states in a distinct context by examining the presence of Bell nonlocality [156]. Finally, A. S. Maxwell, L. B. Madsen and M. Lewenstein demonstrated the generation of entangled states among electrons with different orbital angular momentum

[18] In this regard, I. A. Gonoskov and collaborators elaborated a discussion that shows similar results to those developed by the group of Lewenstein [150]. However, this is further discussed in Chap. 4.

after NSDI processes. They further developed entanglement witnesses that could facilitate the identification of these entangled states [157].

3.3 Quantum Optical Description of Above-Threshold Ionization and High-Harmonic Generation Processes in Atoms

In the upcoming sections of this chapter, we engage in a more extensive exploration of the studies conducted by the groups of M. Lewenstein and P. Tzallas. We commence with a thorough investigation of ATI processes, which result in the generation of intricate coherent state superpositions and the establishment of light-matter entanglement. Following this, we delve into HHG, where we present the *conditioning on HHG* operation and analyze its associated consequences.

The analysis presented within this section utilizes Lewenstein SFA model to describe the light-matter interaction. Furthermore, for the quantum optical description of the electromagnetic field, we consider a discrete set of modes positioned at harmonic orders of the laser's central frequency $\omega_{\mathbf{k}_L}$, that is the set $\{\omega_{\mathbf{k}_q} = q\omega_{\mathbf{k}_L} : q = 1, 2, \ldots, q_c\}$, with $q_c\omega_{\mathbf{k}_L}$ the cutoff frequency. It is essential to acknowledge that this choice introduces an inherent limitation to our analysis, as accurately describing electromagnetic fields of finite duration typically requires encompassing the entire continuum spectrum. However, in this analysis, we introduce a dimensionless function $0 \leq f(t) \leq 1$ to account for the laser field envelope. Consequently, the electric field operator we use hereupon is given by

$$\hat{\mathbf{E}}(t) = -if(t)\sum_{q=1}^{q_c} \mathbf{g}(\omega_{\mathbf{k}_q})\Big[\hat{a}^\dagger_{\mathbf{k}_q} - \hat{a}_{\mathbf{k}_q}\Big], \tag{3.35}$$

where we define $\mathbf{g}(\omega_{\mathbf{k}_q}) \equiv \boldsymbol{\epsilon}_{\mathbf{k}_q}\sqrt{\hbar\omega_{\mathbf{k}_q}/(2\varepsilon_0 V)}$. To extend our calculations to the continuous limit, it is technically necessary to consider the limit $V \to \infty$. However, since we intentionally constrain ourselves to a discrete set of modes, we compute the quantization volume V with the following considerations: for electromagnetic fields with a field strength $E_0 = 0.053$ a.u. (corresponding to an intensity of $I = 10^{14}$ W/cm^2) and frequency $\omega_{\mathbf{k}_L} = 0.057$ a.u. (corresponding to a wavelength of $\lambda_{\mathbf{k}_L} = 800$ nm), the amplitude of the associated coherent states $|\alpha_{\mathbf{k}_L}|$ is in the order of 10^6. Thus, by evaluating the mean value of Eq. (3.35) corresponding to this frequency mode, we get an estimate of $V \sim 10^{14}$ a.u for the quantization volume. This estimation ensures that $\mathbf{g}(\omega_{\mathbf{k}_q})$ behaves as a perturbative quantity at the single-atom level.

Furthermore, it is essential to note that in Eq. (3.35), we have omitted the polarization dependence. This choice is made under the assumption of linear polarization for the sake of simplicity. Nevertheless, it is important to emphasize that the expressions derived in the subsequent sections can be readily extended to accommodate arbitrary elliptical polarizations.

3.3.1 Above-Threshold Ionization

In this section, based on our work in Ref. [43], we explore above-threshold ionization from a quantum optical perspective, following a discussion similar to that presented in Sect. 3.1.1. However, for the majority of this analysis, we emphasize direct ionization processes, excluding rescattering effects of the emitted photoelectron with the parent ion. This limitation confines our analysis to photoelectron energies below, at most, $\mathcal{E}(\mathbf{p}) = 3U_p$.

Under the single-active-electron and dipole approximations, the Hamiltonian that describes the interaction between an atom and a quantized field can be expressed in the length gauge as

$$\hat{H}(t) = \hat{H}_{\text{at}} + \mathsf{e}\hat{\mathbf{E}}(t) \cdot \hat{\mathbf{r}} + \hat{H}_{\text{field}}, \tag{3.36}$$

which is analogous to the expression in Eq. (2.147), but where we have included in the effective atomic potential the polarization effects appearing in the aforementioned equation. It is important to note that, within our approximation of a discrete set of modes, the field's Hamiltonian is represented by $\hat{H}_{\text{field}} = \sum_{q=1}^{q_c} \hbar\omega_{\mathbf{k}_q} \hat{a}^{\dagger}_{\mathbf{k}_q} \hat{a}_{\mathbf{k}_q}$. In this context, the initial state of the combined electron-light system is given by

$$|\Psi(t = t_0)\rangle = |\text{g}\rangle \otimes |\alpha_{\mathbf{k}_L}\rangle \bigotimes_{q=2}^{q_c} |0_{\mathbf{k}_q}\rangle, \tag{3.37}$$

with the electron initially in the ground state of the atom, the infrared driving mode in a coherent state with amplitude $\alpha_{\mathbf{k}_L}$, while all the other modes are in a vacuum state.[19]

From Eq. (3.36), it follows that the time-dependent Schrödinger equation describing the dynamics of the entire system is given by

$$i\hbar \frac{\partial |\Psi(t)\rangle}{\partial t} = \left[\hat{H}_{\text{at}} + \mathsf{e}\hat{\mathbf{E}}(t) \cdot \hat{\mathbf{r}} + \hat{H}_{\text{field}}\right] \tag{3.38}$$

and to solve it using the SFA, we first perform some unitary transformations, which are pictorially described in Fig. 3.12:

1. First, we move to the interaction picture with respect to the free-field Hamiltonian $\hat{H}_{\text{field}}$, i.e. $\hat{U}_{\text{field}}(t) = e^{-i\hat{H}_{\text{field}}t/\hbar}$, so that the transformed Schrödinger equation reads

$$i\hbar \frac{\partial |\bar{\Psi}(t)\rangle}{\partial t} = \left[\hat{H}_{\text{at}} + \mathsf{e}\hat{\mathbf{E}}(t) \cdot \hat{\mathbf{r}}\right] |\bar{\Psi}(t)\rangle \quad \text{with} \quad |\Psi\rangle = \hat{U}_{\text{field}}(t)|\bar{\Psi}(t)\rangle, \tag{3.39}$$

and the electric field operator acquires an extra time-dependence, resulting in

---

[19] Hereupon, we refer to the mode $q = 1$ as L, such that $\alpha_{\mathbf{k}_{q=1}} \equiv \alpha_{\mathbf{k}_L}$.

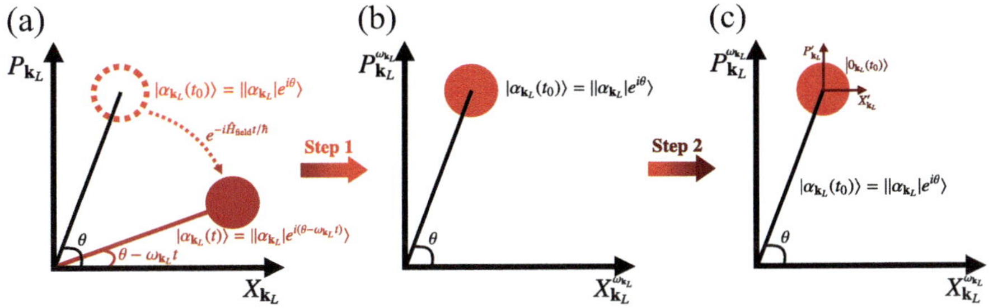

Fig. 3.12 Schematic representation illustrating unitary transformations on the quantum optical modes. In **a**, the initial frame of reference is presented, with the initial coherent state depicted by the red dashed circle. When this coherent state evolves freely, i.e. according to $\hat{H}_{\text{field}}$, it begins to rotate around the origin. In **b**, we operate in the interaction picture with respect to $\hat{H}_{\text{field}}$, and the frame of reference oscillates with frequency $\omega_{\mathbf{k}_L}$ (denoted here with the $\omega_{\mathbf{k}_L}$ superscript). Finally, in **c**, we shift our frame of reference so that its origin coincides with the location of the coherent state. In this new frame, the initial quantum optical state is represented as a vacuum state $|0_{\mathbf{k}_L}\rangle$

$$\hat{\mathbf{E}}(t) = -if(t)\sum_{q=1}^{q_c}\mathbf{g}(\omega_{\mathbf{k}_q})\Big[\hat{a}^{\dagger}_{\mathbf{k}_q}e^{i\omega_{\mathbf{k}_q}t} - \hat{a}_{\mathbf{k}_q}e^{-i\omega_{\mathbf{k}_q}t}\Big]. \tag{3.40}$$

2. Second, we work in the displaced frame with respect to the input IR field by introducing the displacement operation $\hat{D}_{\mathbf{k}_L}(\alpha_{\mathbf{k}_L})$, such that the transformed equation reads

$$i\hbar\frac{\partial|\tilde{\Psi}(t)\rangle}{\partial t} = \Big[\hat{H}_{\text{at}} + \mathsf{e}\mathbf{E}_{\text{cl}}(t)\cdot\hat{\mathbf{r}} + \mathsf{e}\hat{\mathbf{E}}(t)\cdot\hat{\mathbf{r}}\Big]|\tilde{\Psi}(t)\rangle \quad \text{with} \quad |\bar{\Psi}(t)\rangle = \hat{D}_{\mathbf{k}_L}(\alpha_{\mathbf{k}_L})|\tilde{\Psi}(t)\rangle, \tag{3.41}$$

where the electric field operator splits into a *classical* term describing the mean value of the field

$$\begin{aligned}\mathbf{E}_{\text{cl}}(t) &= \text{tr}[\hat{\mathbf{E}}(t)|\alpha_{\mathbf{k}_L}\rangle\langle\alpha_{\mathbf{k}_L}|\bigotimes_{q=2}^{q_c}|0_{\mathbf{k}_q}\rangle\langle 0_{\mathbf{k}_q}|]\\ &= -if(t)\mathbf{g}(\omega_{\mathbf{k}_L})\Big[\alpha^*_{\mathbf{k}_L}e^{i\omega_{\mathbf{k}_L}t} - \alpha_{\mathbf{k}_L}e^{-i\omega_{\mathbf{k}_L}t}\Big],\end{aligned} \tag{3.42}$$

and another term $\hat{\mathbf{E}}(t)$ identical to the one in Eq. (3.40) which accounts for the quantum fluctuations.[20] Within this frame of reference, the initial state of the system shifts to

[20] It is worth noting that the same effects should be observed for the electric polarization operator appearing in Eq. (2.147). Thus, we keep the classical term which effectively contributes to the unperturbed atomic potential, while we neglect quantum optical fluctuation terms since they are second order in $\mathbf{g}(\omega_{\mathbf{k}_q})$.

$$|\tilde{\Psi}(t_0)\rangle = |\mathrm{g}\rangle \bigotimes_{q=1}^{q_c} |0_{\mathbf{k}_q}\rangle \equiv |\mathrm{g}\rangle \otimes |\bar{0}\rangle \tag{3.43}$$

where $|\bar{0}\rangle$ denotes the vacuum state in all modes.

In the spirit of the Lewenstein SFA model [51], we solve the Schrödinger equation in Eq. (3.41) by considering the following ansatz

$$|\tilde{\Psi}(t)\rangle = a(t)|\mathrm{g}\rangle \otimes |\Phi_{\mathrm{g}}(t)\rangle + \int \mathrm{d}\mathbf{v}\, b(\mathbf{v}, t)\, |\mathbf{v}\rangle \otimes |\Phi(\mathbf{v}, t)\rangle\,, \tag{3.44}$$

which is a generalization of the SFA ansatz in Eq. (3.4) to the quantum optical domain. In this expression $|\Phi_{\mathrm{g}}(t)\rangle$ and $|\Phi(\mathbf{v}, t)\rangle$ represent the quantum optical state when the electron is found in states $|\mathrm{g}\rangle$ and $|\mathbf{v}\rangle$, respectively, with $a(t)$ and $b(\mathbf{v}, t)$ the associated probability amplitudes.

Analogously to the analysis presented in the semiclassical description in Sect. 3.1.1, we now project Eq. (3.41), evaluated with the ansatz in Eq. (3.44), with respect to the ground state $|\mathrm{g}\rangle$ and a continuum state $|\mathbf{v}\rangle$. This results in a set of coupled differential equations describing the behavior of $a(t)|\Phi_{\mathrm{g}}(t)\rangle$ and $b(\mathbf{v}, t)\,|\Phi(\mathbf{v}, t)\rangle$, respectively. Specifically, when projecting with respect to $|\mathbf{v}\rangle$, we obtain

$$\begin{aligned} i\hbar\frac{\partial}{\partial t}\Big(b(\mathbf{v}, t)\,|\Phi(\mathbf{v}, t)\rangle\Big) &= \frac{1}{2}m_{\mathrm{e}}\mathbf{v}^2 b(\mathbf{v}, t)\,|\Phi(\mathbf{v}, t)\rangle + \mathrm{e}\Big(E_{\mathrm{cl}}(t) + \hat{\mathbf{E}}(t)\Big)\cdot \mathbf{d}(\mathbf{v})a(t)|\Phi_{\mathrm{g}}(t)\rangle \\ &\quad + i\hbar\mathrm{e}\Big(E_{\mathrm{cl}}(t) + \hat{\mathbf{E}}(t)\Big)\cdot \nabla_{\mathbf{v}}\big(b(\mathbf{v}, t)\,|\Phi(\mathbf{v}, t)\rangle\big), \end{aligned} \tag{3.45}$$

where, as mentioned earlier, we have restricted our analysis to direct ATI processes, consequently neglecting the $\mathbf{g}(\mathbf{v}, \mathbf{v}')$ terms. Similar to the semiclassical analysis, we obtain a differential equation with well-defined homogeneous and inhomogeneous parts. Its solution can be expressed as a linear combination between the solution to the homogeneous equation and a particular solution to the inhomogeneous. For the homogeneous component, we can represent it as follows[21]

$$\begin{aligned} i\hbar\frac{\partial b_{\mathrm{h}}(\mathbf{v}, t)}{\partial t}|\Phi_{\mathrm{h}}(\mathbf{v}, t)\rangle + i\hbar b_{\mathrm{h}}(\mathbf{v}, t)\frac{\partial|\Phi_{\mathrm{h}}(\mathbf{v}, t)\rangle}{\partial t} \\ = \frac{1}{2}m_{\mathrm{e}}\mathbf{v}^2 b_{\mathrm{h}}(\mathbf{v}, t)|\Phi_{\mathrm{h}}(\mathbf{v}, t)\rangle \\ + i\hbar\mathrm{e}\mathbf{E}_{\mathrm{cl}}(t)\cdot\Big[\big(\nabla_{\mathbf{v}} b_{\mathrm{h}}(\mathbf{v}, t)\big)|\Phi_{\mathrm{h}}(\mathbf{v}, t)\rangle + b_{\mathrm{h}}(\mathbf{v}, t)\big(\nabla_{\mathbf{v}}|\Phi_{\mathrm{h}}(\mathbf{v}, t)\rangle\big)\Big] \\ + i\hbar\mathrm{e}\hat{\mathbf{E}}(t)\cdot\Big[\big(\nabla_{\mathbf{v}} b_{\mathrm{h}}(\mathbf{v}, t)\big)|\Phi_{\mathrm{h}}(\mathbf{v}, t)\rangle + b_{\mathrm{h}}(\mathbf{v}, t)\big(\nabla_{\mathbf{v}}|\Phi_{\mathrm{h}}(\mathbf{v}, t)\rangle\big)\Big]. \end{aligned} \tag{3.46}$$

On the right-hand side of this equation, we have the sum of three different terms. The first term represents the energy of the scattered electron. The second describes

[21] Here, we refer to the homogeneous solution with the subindex "h" instead of "hom" as done in Sect. 3.1.1 for brevity.

the influence of the field's average value on the electron's trajectory. The third, defines the quantum optical fluctuations around this mean value. Regarding the third term, we can distinguish different contributions to these quantum optical fluctuations. The term $(\nabla_{\mathbf{v}} b_{\mathrm{h}}(\mathbf{v},t))|\Phi_{\mathrm{h}}(\mathbf{v},t)\rangle$ accounts for the back-action of the electron's trajectory on the quantum optical state of the field, while $b_{\mathrm{h}}(\mathbf{v},t)(\nabla_{\mathbf{v}}|\Phi_{\mathrm{h}}(\mathbf{v},t)\rangle)$ represents the effect of this quantum optical fluctuations on the electron's trajectory. In the following, we adopt a weak-coupling-like approach, assuming that the electron is not significantly affected by the quantum optical fluctuations. This approximation aligns with more comprehensive analysis of the electronic trajectories in quantum fields [144] specifically when considering coherent states of light as drivers. Consequently, we omit the last term we have just described, allowing us to approximate Eq. (3.46) as follows

$$
\begin{aligned}
i\hbar\frac{\partial b_{\mathrm{h}}(\mathbf{v},t)}{\partial t}|\Phi_{\mathrm{h}}(\mathbf{v},t)\rangle + i\hbar b_{\mathrm{h}}(\mathbf{v},t)\frac{\partial|\Phi_{\mathrm{h}}(\mathbf{v},t)\rangle}{\partial t}& \\
\approx \frac{1}{2}m_{\mathrm{e}}\mathbf{v}^2 b_{\mathrm{h}}(\mathbf{v},t)|\Phi_{\mathrm{h}}(\mathbf{v},t)\rangle& \\
+ i\hbar\mathrm{e}\mathbf{E}_{\mathrm{cl}}(t)\cdot\big[\big(\nabla_{\mathbf{v}} b_{\mathrm{h}}(\mathbf{v},t)\big)|\Phi_{\mathrm{h}}(\mathbf{v},t)\rangle + b_{\mathrm{h}}(\mathbf{v},t)\big(\nabla_{\mathbf{v}}|\Phi_{\mathrm{h}}(\mathbf{v},t)\rangle\big)\big]& \\
+ i\hbar\mathrm{e}\hat{\mathbf{E}}(t)\cdot\big(\nabla_{\mathbf{v}} b_{\mathrm{h}}(\mathbf{v},t)\big)|\Phi_{\mathrm{h}}(\mathbf{v},t)\rangle,&
\end{aligned}
\tag{3.47}
$$

and further rewrite it as

$$
\begin{aligned}
&\left[i\hbar\frac{\partial b_{\mathrm{h}}(\mathbf{v},t)}{\partial t} - i\hbar\mathrm{e}\mathbf{E}_{\mathrm{cl}}(t)\cdot\big(\nabla_{\mathbf{v}} b_{\mathrm{h}}(\mathbf{v},t)\big) - \frac{1}{2}m_{\mathrm{e}}\mathbf{v}^2 b_{\mathrm{h}}(\mathbf{v},t)\right]|\Phi_{\mathrm{h}}(\mathbf{v},t)\rangle \\
&\quad+\left[i\hbar b_{\mathrm{h}}(\mathbf{v},t)\frac{\partial|\Phi_{\mathrm{h}}(\mathbf{v},t)\rangle}{\partial t} - i\hbar\mathrm{e}\mathbf{E}_{\mathrm{cl}}(t)\cdot\big(\nabla_{\mathbf{v}}|\Phi_{\mathrm{h}}(\mathbf{v},t)\rangle\big) b_{\mathrm{h}}(\mathbf{v},t)\right. \\
&\quad\left.- i\hbar\mathrm{e}\mathbf{E}_{\mathrm{cl}}(t)\cdot\big(\nabla_{\mathbf{v}} b_{\mathrm{h}}(\mathbf{v},t)\big)|\Phi_{\mathrm{h}}(\mathbf{v},t)\rangle\right] = 0.
\end{aligned}
\tag{3.48}
$$

To solve this equation, our first step is to find a function $b(\mathbf{v},t)$ which can nullify the first bracket in this expression. In other words, we seek a solution to the following differential equation

$$
i\hbar\frac{\partial b_{\mathrm{h}}(\mathbf{v},t)}{\partial t} - i\hbar\mathrm{e}\mathbf{E}_{\mathrm{cl}}(t)\cdot\big(\nabla_{\mathbf{v}} b_{\mathrm{h}}(\mathbf{v},t)\big) - \frac{1}{2}m_{\mathrm{e}}\mathbf{v}^2 b_{\mathrm{h}}(\mathbf{v},t) = 0,
\tag{3.49}
$$

which actually describes the motion of an electron influenced by an electromagnetic field $\mathbf{E}_{\mathrm{cl}}(t)$, as illustrated in Eqs. (2.29) and (3.12). The solution to this equation was provided in Eq. (3.13), which necessitates the introduction of the canonical momentum $\mathbf{p} = m_{\mathrm{e}}\mathbf{v} - \mathrm{e}\mathbf{A}_{\mathrm{cl}}(t)$. When we incorporate this solution into Eq. (3.48), we are left with only the second bracket term

$$
\begin{aligned}
i\hbar b_{\mathrm{h}}(\mathbf{v},t)\frac{\partial|\Phi_{\mathrm{h}}(\mathbf{v},t)\rangle}{\partial t} &- i\hbar \mathrm{e}\mathbf{E}_{\mathrm{cl}}(t)\cdot\big(\nabla_{\mathbf{v}}|\Phi_{\mathrm{h}}(\mathbf{v},t)\rangle\big)b_{\mathrm{h}}(\mathbf{v},t)\\
&- i\hbar \mathrm{e}\mathbf{E}_{\mathrm{cl}}(t)\cdot\big(\nabla_{\mathbf{v}}b_{\mathrm{h}}(\mathbf{v},t)\big)|\Phi_{\mathrm{h}}(\mathbf{v},t)\rangle = 0.
\end{aligned}
\tag{3.50}
$$

Before we proceed further, let us calculate the gradient of $b_{\mathrm{h}}(\mathbf{v},t)$ with respect to $\mathbf{v}$, as it is necessary for evaluating the differential equation above

$$
\nabla_{\mathbf{v}}b_{\mathrm{h}}(\mathbf{v},t) = -\frac{i}{\hbar m_{\mathrm{e}}}\left[\int_{t_0}^{t}\mathrm{d}\tau\Big(\mathbf{p}+\mathrm{e}\mathbf{A}_{\mathrm{cl}}(\tau)\Big)\right]b_{\mathrm{h}}(\mathbf{v},t) \equiv -\frac{i}{\hbar}\Delta\mathbf{r}(\mathbf{p},t,t_0)b_{\mathrm{h}}(\mathbf{v},t),
\tag{3.51}
$$

where $\Delta\mathbf{r}(\mathbf{p},t,t_0)$ denotes the electronic displacement in the continuum between times t_0 and t. Introducing this expression into Eq. (3.50), we can factor out the $b_{\mathrm{h}}(\mathbf{v},t)$ function and rewrite the differential equation as follows

$$
i\hbar\frac{\partial|\Phi_{\mathrm{h}}(\mathbf{v},t)\rangle}{\partial t} - i\hbar\mathrm{e}\mathbf{E}_{\mathrm{cl}}(t)\cdot\big(\nabla_{\mathbf{v}}|\Phi_{\mathrm{h}}(\mathbf{v},t)\rangle\big) = \mathrm{e}\hat{\mathbf{E}}(t)\cdot\Delta\mathbf{r}(\mathbf{p},t,t_0)|\Phi_{\mathrm{h}}(\mathbf{v},t)\rangle,
\tag{3.52}
$$

where the left-hand side term corresponds to the total derivative of $|\Phi_{\mathrm{h}}(\mathbf{v},t)\rangle$ with respect to time,

$$
i\hbar\frac{\mathrm{d}|\Phi_{\mathrm{h}}(\mathbf{v},t)\rangle}{\mathrm{d}t} = \mathrm{e}\hat{\mathbf{E}}(t)\cdot\Delta\mathbf{r}(\mathbf{p},t,t_0)|\Phi_{\mathrm{h}}(\mathbf{p},t)\rangle.
\tag{3.53}
$$

The differential equation we have just obtained bears a resemblance to the one presented in Eq. (2.85), which describes the generation of coherent states from an oscillating charge current. In this case, since we are working in the length gauge, the coupling occurs between the displacement carried out by the electron in the continuum and the electric field operator. By following the same procedure as done to solve Eq. (2.85), we arrive at

$$
|\Phi_{\mathrm{h}}(\mathbf{p},t)\rangle = \hat{\mathcal{D}}\big(\boldsymbol{\delta}(\mathbf{p},t,t_0)\big)|\Phi_{\mathrm{h}}(\mathbf{p},t_0)\rangle,
\tag{3.54}
$$

with $\hat{\mathcal{D}}(\boldsymbol{\delta}(\mathbf{p},t,t_0)) \equiv \prod_{q=1}^{q_c} e^{i\varphi_q(\mathbf{p},t)}\hat{D}_q(\delta_q(\mathbf{p},t,t_0))$ denoting the multimode displacement operator, where $\hat{D}_q(\cdot)$ is given in Eq. (2.77). In this expression, $\delta_q(\mathbf{p},t,t_0)$ corresponds to the Fourier transform of the electronic displacement, while the additional phase factor $\varphi_q(\mathbf{p},t,t_0)$ results from applying the BCH formula in our analysis (see Sect. 2.3.3.2). These quantities are respectively given by

$$
\delta_q(\mathbf{p},t,t_0) = -\frac{\mathrm{e}}{\hbar}\sqrt{\frac{\hbar\omega_{\mathbf{k}_q}}{2\varepsilon_0 V}}\int_{t_0}^{t}\mathrm{d}\tau\Big(\boldsymbol{\epsilon}_{\mathbf{k}_q}\cdot\Delta\mathbf{r}(\mathbf{p},\tau,t_0)\Big)e^{i\omega_{\mathbf{k}_q}\tau},
\tag{3.55}
$$

$$\varphi_q(\mathbf{p}, t, t_0) = \frac{\mathrm{e}^2}{\hbar^2} \frac{\hbar\omega_{\mathbf{k}_q}}{2\varepsilon_0 V} \int_{t_0}^{t} \mathrm{d}t_2 \int_{t_0}^{t_2} \mathrm{d}t_1 \Big(\boldsymbol{\epsilon}_{\mathbf{k}_q} \cdot \Delta\mathbf{r}(\mathbf{p}, t_2, t_0)\Big)\Big(\boldsymbol{\epsilon}_{\mathbf{k}_q} \cdot \Delta\mathbf{r}(\mathbf{p}, t_1, t_0)\Big) \times \sin(\omega_{\mathbf{k}_q}(t_2 - t_1)). \tag{3.56}$$

Using this solution to the homogeneous equation, under the weak-coupling approximation and accounting for the initial conditions, we obtain for Eq. (3.45)

$$b(\mathbf{p}, t)|\Phi(\mathbf{p}, t)\rangle = -\frac{i}{\hbar} \int_{t_0}^{t} \mathrm{d}t_1 e^{-\frac{i}{2\hbar m\mathrm{e}} \int_{t_0}^{t} \mathrm{d}\tau [\mathbf{p} + \mathrm{e}\mathbf{A}_{\mathrm{cl}}(\tau)]^2} \hat{\mathcal{D}}\big(\boldsymbol{\delta}(\mathbf{p}, t, t_1)\big) \times \big(\mathbf{E}_{\mathrm{cl}}(t_1) + \hat{\mathbf{E}}(t_1)\big) \cdot \mathbf{d}\big(\mathbf{p} + \mathrm{e}\mathbf{A}_{\mathrm{cl}}(t_1)\big) a(t_1)|\Phi_{\mathrm{g}}(t_1)\rangle. \tag{3.57}$$

Regarding the projection of Eq. (3.41) with respect to the ground state $|\mathrm{g}\rangle$, we obtain

$$i\hbar\frac{\partial}{\partial t}\big(a(t)|\Phi_{\mathrm{g}}(t)\rangle\big) = -I_p a(t)|\Phi_{\mathrm{g}}(t)\rangle + \int \mathrm{d}\mathbf{v} b(\mathbf{v}, t)\big(\mathbf{E}_{\mathrm{cl}}(t) + \hat{\mathbf{E}}(t)\big) \cdot \mathbf{d}(\mathbf{v}) \, |\Phi(\mathbf{v}, t)\rangle \,, \tag{3.58}$$

and in line with the Lewenstein SFA model, we assume the depletion of the ground state to be negligible. Thus, we can approximate the previous differential equation as

$$i\hbar\frac{\partial}{\partial t}\big(a(t)|\Phi_{\mathrm{g}}(t)\rangle\big) \approx -I_p a(t)|\Phi_{\mathrm{g}}(t)\rangle, \tag{3.59}$$

whose solution can be found straightforwardly and, together with the initial conditions, leads to

$$a(t)|\Phi_{\mathrm{g}}(t)\rangle = e^{iI_p(t-t_0)/\hbar}|\bar{0}\rangle. \tag{3.60}$$

Combining Eqs. (3.57) and (3.60), we finally get for the quantum optical version of the SFA ansatz

$$|\tilde{\Psi}(t)\rangle = e^{iI_p(t-t_0)/\hbar}|\mathrm{g}\rangle \otimes |\bar{0}\rangle - \frac{i}{\hbar}\int \mathrm{d}\mathbf{p} \int_{t_0}^{t} \mathrm{d}t_1 e^{-iS(\mathbf{p},t,t_1)/\hbar} \hat{\mathcal{D}}\big(\boldsymbol{\delta}(\mathbf{p}, t, t_1)\big) \times \big(\mathbf{E}_{\mathrm{cl}}(t_1) + \hat{\mathbf{E}}(t_1)\big) \cdot \mathbf{d}\big(\mathbf{p} + \mathrm{e}\mathbf{A}_{\mathrm{cl}}(t_1)\big) \, |\mathbf{p} + \mathrm{e}\mathbf{A}_{\mathrm{cl}}(t)\rangle \otimes |\bar{0}\rangle, \tag{3.61}$$

which has two contributions: the first one describes a situation where the electron is completely unaffected by the field, and consequently the quantum optical state does not experience any change; the second one describes an ionization process, where the electron reaches the continuum at time t_1 and propagates driven by the field. In this second scenario, the electronic propagation leads to a displacement $\boldsymbol{\delta}(\mathbf{p}, t, t_1)$ on the electromagnetic field modes.

Finally, we return to the undisplaced frame of reference, that is, the one defined in Eq. (3.39). In this frame, we get

$$\begin{aligned}|\bar{\Psi}(t)\rangle &= e^{iI_p(t-t_0)/\hbar}\hat{D}_{\mathbf{k}_L}(\alpha_{\mathbf{k}_L})|\mathrm{g}\rangle \otimes |\bar{0}\rangle \\ &\quad - \frac{i}{\hbar}\int \mathrm{d}\mathbf{p}\int_{t_0}^{t} \mathrm{d}t_1 e^{-iS(\mathbf{p},t,t_1)/\hbar}\hat{\mathcal{D}}\big(\boldsymbol{\delta}(\mathbf{p},t,t_1)\big)e^{i\mathrm{Im}\{\alpha_{\mathbf{k}_L}\delta^*_{\mathbf{k}_L}(\mathbf{p},t,t_1)\}} \\ &\quad \times \mathbf{d}\big(\mathbf{p}+\mathrm{e}\mathbf{A}_{\mathrm{cl}}(t_1)\big)\cdot\Bigg[\mathbf{E}_{\mathrm{cl}}(t_1)\,|\mathbf{p}+\mathrm{e}\mathbf{A}_{\mathrm{cl}}(t)\rangle \otimes |\alpha_{\mathbf{k}_L}\rangle \bigotimes_{q=2}^{q_c}|0_{\mathbf{k}_q}\rangle \\ &\quad + \sum_{q=1}^{q_c}\mathbf{g}(\omega_{\mathbf{k}_q})e^{i\omega_{\mathbf{k}_q}t}\hat{D}_{\mathbf{k}_L}(\alpha_{\mathbf{k}_L})\hat{a}^\dagger_{\mathbf{k}_q}\,|\mathbf{p}+\mathrm{e}\mathbf{A}_{\mathrm{cl}}(t)\rangle \otimes |\bar{0}\rangle\Bigg],\end{aligned} \tag{3.62}$$

where we have moved $\hat{D}(\alpha_{\mathbf{k}_L})$ to the end of the expression, taking into account the commutation relation properties of the displacement operator. Given that we are working with strong-laser fields of field strengths on the order of 10^7 V/cm, or even larger, we expect the third line of Eq. (3.62) to be much larger than the fourth line. Therefore we approximate the previous state as follows

$$\begin{aligned}|\bar{\Psi}(t)\rangle &\simeq e^{iI_p(t-t_0)/\hbar}\hat{D}_{\mathbf{k}_L}(\alpha_{\mathbf{k}_L})|\mathrm{g}\rangle \otimes |\bar{0}\rangle \\ &\quad - \frac{i}{\hbar}\hat{D}_{\mathbf{k}_L}(\alpha_{\mathbf{k}_L})\int \mathrm{d}\mathbf{p}\int_{t_0}^{t} \mathrm{d}t_1 e^{-iS(\mathbf{p},t,t_1)/\hbar}\hat{\mathcal{D}}\big(\boldsymbol{\delta}(\mathbf{p},t,t_1)\big) \\ &\qquad \times \mathbf{d}\big(\mathbf{p}+\mathrm{e}\mathbf{A}_{\mathrm{cl}}(t_1)\big)\cdot\mathbf{E}_{\mathrm{cl}}(t_1)\,|\mathbf{p}+\mathrm{e}\mathbf{A}_{\mathrm{cl}}(t)\rangle \otimes |\bar{0}\rangle,\end{aligned} \tag{3.63}$$

where we have moved the displacement operator to the front of the expression. It is worth noting that the third and fourth lines in Eq. (3.62) represent the behavior of the mean value of the field and its quantum fluctuations, respectively. Thus, even though we are neglecting the latter, we still observe non-classical features in the resulting state through the coupling between the electron and the field. Specifically, from the second line in Eq. (3.63), we can see that the resulting displacement $\boldsymbol{\delta}(\mathbf{p}, t, t_1)$ varies depending on the ionization conditions. This results in a light-matter entangled state where the electronic degrees of freedom couple with the electromagnetic field ones.

In the following, we explore different non-classical properties of the generated state. More specifically, we focus on characterizing the amount of entanglement between light and matter, and explore the Wigner function representation of the IR mode. To achieve this, we adopt a 1D model for hydrogen atoms, representing the momentum $\mathbf{p}$ using a single coordinate p aligned along the laser's polarization direction.

3.3.1.1 Dependence of the Light's Displacement on the Electronic Motion

As mentioned earlier, the displacement $\boldsymbol{\delta}(p, t, t_1)$ corresponds to the back-action of the electronic motion on the electromagnetic field. This quantity is of fundamental interest in the analysis we perform here because it ultimately determines the non-classical features of the state in Eq. (3.63), particularly in situations where we assume the electron to be in a continuum state. In this section, our goal is to study the relationship between this quantity and the characteristics of the electronic motion.

Let us consider one of the most standard scenarios in the analysis of strong-field physics: a hydrogen atom excited by a linearly polarized laser field with a sinusoidal squared envelope having $\lambda_{\mathbf{k}_L} = 800$ nm ($\omega_{\mathbf{k}_L} = 0.057$ a.u.), $I = 10^{14}$ W/cm^2 ($E_0 = 0.053$ a.u.) and a duration of $\Delta t \approx 13.3$ fs (5 optical cycles). Under these conditions, as observed in Fig. 3.13, we find that the absolute value of $\delta_q(p, t, t_1)$ is on the order of 10^{-4}. This suggests that in this regime, the electronic propagation in the continuum has a minimal effect on the initial coherent state of the field. It is important to note that these observations are within the direct ionization regime, since we consider $\mathcal{E}(p) < 2.5U_p$. However, even at this level, some interesting features can be discerned.

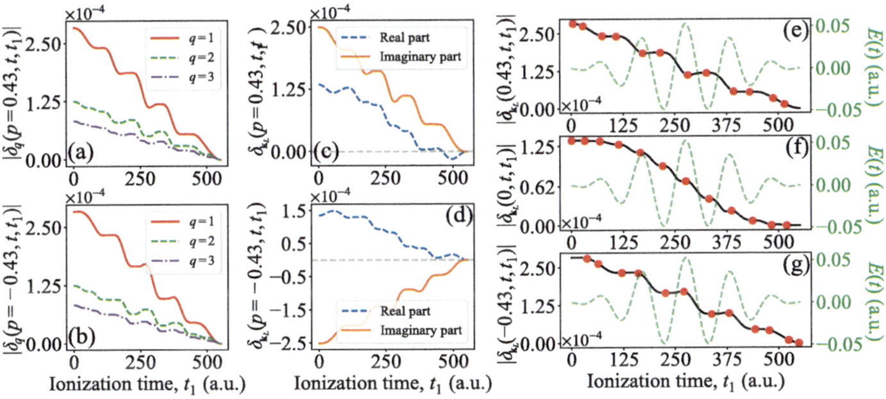

Fig. 3.13 **Dependence of the ATI-induced optical displacement as a function of the ionization time.** These plots are generated considering a hydrogen atom interacting with a linearly polarized field featuring a sinusoidal-squared envelope with peak intensity $I = 1 \times 10^{14}$ W/cm^2, central wavelength $\lambda_{\mathbf{k}_L} = 800$ nm, and pulse duration $\Delta t \approx 13.3$ fs (5 optical cycles). In panels **a** and **b**, the norm of the ATI-induced quantum optical displacement $\delta_q(p, t, t_1)$ is depicted as a function of ionization time for electrons propagating in the forward ($p > 0$) and backward ($p < 0$) directions, respectively. Various harmonic orders are considered. In panels **c** and **d**, emphasis is placed on the fundamental mode, illustrating the real (dashed blue curve) and imaginary (orange solid curve) parts of $\delta_{\mathbf{k}_L}(p, t, t_1)$. In panels **e–g**, the norm of this displacement is presented for three different values of momenta, respectively, with the real part of the ATI ionization time indicated by the red dots. The ionization time has been computed using the saddle-point approximation. Furthermore, the electric field used in the calculations is represented in these last three panels with the green dashed curve

Firstly, we observe that the electronic motion actually gives rise to coherent state displacements in the fundamental mode as well as on other harmonic modes, as seen in Fig. 3.13a and b. Specifically, this displacement decreases with increasing harmonic orders. This phenomenon aligns with recent findings by D. B. Milošević in Ref. [132], where a more detailed study of this process is conducted. Additionally, it is worth noting from panels (c) and (d) that the magnitude of the obtained displacements remains the same for different ionization times, regardless of the electronic propagation direction. However, the propagation direction becomes crucial in determining the phase of $\delta_q(p, t, t_1)$, the latter being in general a complex number. Specifically, the imaginary part of $\delta_q(p, t, t_1)$ differs in sign depending on whether the electron is propagating along the forward (panel (c)) or backward (panel (d)) direction. Furthermore, it is evident that the earlier the electronic ionization occurs, the larger the contribution $|\delta_q(p, t, t_1)|$ becomes. This is an expected behavior since the earlier ionization takes place, the more time the electron spends propagating in the continuum, and consequently contributing to $\delta_q(p, t, t_1)$.

Focusing on the fundamental mode, the ionization conditions of the electron, that is the ionization time and the corresponding kinetic energy, also play an important role in the behavior of $\delta_{\mathbf{k}_L}(p, t, t_1)$. This is illustrated in Fig. 3.13e–g, such that the absolute value of this quantity is plotted with respect to the ionization times. The red points in these plots, correspond to the real part of the ionization time obtained from the saddle-point equations. Three different values of p are considered: $p = 0.43$ a.u., $p = 0.00$ a.u. and $p = -0.43$ a.u., respectively.

In these plots we observe that, depending on the outgoing electron's momentum, the value of $|\delta_{\mathbf{k}_L}(p, t, t_1)|$ differs. Specifically, for $p = 0.00$ a.u., we observe that, among each ionization time, there is a step in $|\delta_{\mathbf{k}_L}(p, t, t_1)|$ which is not observed at all ionization times for the other two non-vanishing values of momentum. Moreover, for these two momenta, we find that two consecutive ionization times lead to approximately the same value of $|\delta_{\mathbf{k}_L}(p, t, t_1)|$.

In order to delve in more detail on this matter, let us consider the two ionization times located at the right and left of $t_1 = 300$ a.u. in Fig. 3.13e, though the same discussion is valid for panel (g). Here, we observe that if the electron ionizes close to the maximum of the field, it yields almost the same contribution to $|\delta_{\mathbf{k}_L}(p, t, t_1)|$ as if it ionizes close to the field's minimum. It is worth noting that in the first case, the photoelectron's momentum and the field are pointing in the same direction, and therefore the field exerts a decelerating force on the generated electron, due to the negative charge of the electron. Conversely, in the second case, these two vectors are pointing in opposite directions, and the electron experiences an accelerating force, leading to an increased value of $|\delta_{\mathbf{k}_L}(p, t, t_1)|$ when compared with later ionization times. This behavior contrasts with what happens for $p = 0.00$ a.u. in panel (f), where ionization values placed at maximum or minimum values of the field all lead to an increase of $|\delta_{\mathbf{k}_L}(p, t, t_1)|$. In this case, the field always exerts an accelerating force on the electron, which points towards opposite directions depending on the specific ionization time.

3.3.1.2 Dependence of the Light's Displacement on the Laser Field Parameters

With the parameters used in the previous subsection, the displacement $|\delta_{\mathbf{k}_L}(p, t, t_1)|$ in the quantum optical state becomes negligible. This quantity could naturally be enhanced, for instance, by considering many-atom ATI processes. However, in this section we focus on single-atom dynamics and investigate the dependence of $\delta_{\mathbf{k}_L}(p, t, t_1)$ on various laser parameters that can be experimentally controlled, such as laser intensity and frequency.

Given that $\delta_{\mathbf{k}_L}(p, t, t_1)$ depends on the electronic propagation, it is expected that the longer the electron remains in the continuum, the higher its acquired kinetic energy, leading to a larger $|\delta_{\mathbf{k}_L}(p, t, t_1)|$. Thus, we anticipate this quantity to be proportional to some power of the ponderomotive energy U_p. To determine the exact power, we examine the dependence of $|\delta_{\mathbf{k}_L}(p, t, t_1)|$ as a function of the field strength and frequency in Fig. 3.14. Here, we set $t_1 = T/2$, with T representing the total duration of the laser pulse, corresponding to a field maximum where the probability of electron ionization is at its peak. Furthermore, we restrict ourselves to peak strengths satisfying $E_0 < 0.147$ a.u., ensuring that over-the-barrier ionization events are less likely than tunneling ones when considering hydrogenic atoms [55]. We also use $\omega_{\mathbf{k}_L} \in [0.008, 0.04]$ a.u., corresponding to $\lambda_{\mathbf{k}_L}$ in the range 1–5 μm, making tunneling events more probable than multiphoton ionization processes.

In Fig. 3.14a, we present the behavior of $|\delta_{\mathbf{k}_L}(p, t, t_1)|$ as a function of the laser's frequency for three different peak field strengths while setting $p = \sqrt{2U_p}$. Here, we observe that, for a fixed value of the field peak strength, the generated displacement is inversely proportional to $\omega_{\mathbf{k}_L}^2$. This contrasts with the behavior of electron displacement, which is inversely proportional to $\omega_{\mathbf{k}_L}$, a distinction arising from the relationship between the field's and electron's displacement. Specifically, the former

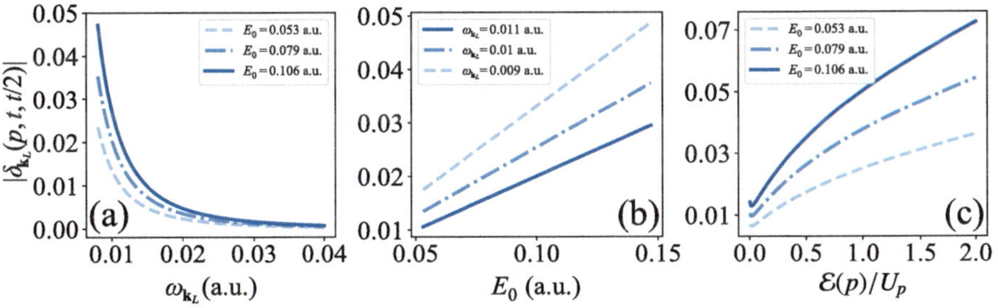

Fig. 3.14 Dependence of the ATI-induced optical displacement on laser parameters. In panel **a**, we demonstrate the variation of $|\delta_{\mathbf{k}_L}(p, t, T/2)|$ with the laser field's central frequency for three different electric field strengths. Conversely, in **b**, this quantity is displayed as a function of the laser field's central frequency for three different field strengths. In **c**, the laser's central wavelength is fixed to $\lambda_{\mathbf{k}_L} \approx 5\ \mu$m (corresponding to $\omega_{\mathbf{k}_L} = 0.009$ a.u.), and $|\delta_{\mathbf{k}_L}(p, t, T/2)|$ is presented as a function of the photoelectron's kinetic energy for three different field strengths. In this last panel, we have restricted to $p > 0$, although similar results are obtained for $p < 0$

is defined as the Fourier transform of the latter, with the second presenting an oscillatory motion with frequency matching that of the field, yielding the additional factor of $\omega_{\mathbf{k}_L}^{-1}$ after integration.

In Fig. 3.14b, we illustrate the behavior of $|\delta_{\mathbf{k}_L}(p, t, t_1)|$ as a function of the field's peak amplitude E_0 for three different frequencies within the mid-IR (MIR) regime ($\lambda_{\mathbf{k}_L} \sim 4$–$5\ \mu$m). This behavior displays a linear dependence on E_0. Consequently, by combining this and the behavior depicted in the previous paragraph, we can conclude that $|\delta_{\mathbf{k}_L}(p, t, t_1)| \propto \sqrt{U_p}/\omega_{\mathbf{k}_L}$. Thus, as the electron acquires greater kinetic energy in the continuum, the value of $|\delta_{\mathbf{k}_L}(p, t, t_1)|$ becomes larger. Specifically, when considering $\omega_{\mathbf{k}_L} \approx 0.01$ a.u., we obtain $|\delta_{\mathbf{k}_L}(p, t, t_1)| \approx 10^{-2}$, a value that suggests the presence of resolvable non-classical features in the generated state.

Beyond the laser parameters, we have also observed that the field's displacement is influenced by the final kinetic energy of the photoelectron. In Fig. 3.14c, we further emphasize this point by demonstrating the dependence of $|\delta_{\mathbf{k}_L}(p, t, t_1)|$ on the electron's kinetic energy for three different field peak strengths. Here, $\omega_{\mathbf{k}_L} = 0.009$ a.u. ($\lambda_{\mathbf{k}_L} \approx 5\ \mu$m), and we set $p \geq 0$, although similar features are observed for $p < 0$. This plot reveals a linear relationship between the field's displacement and the electron's momentum, specifically, with the square root of the photoelectron energy $\mathcal{E}(p)$. However, it is worth noting that for $p = 0$ a.u., we get non-zero contributions to the displacement, which are originated by the electron's oscillation driven by the field.

In subsequent sections, with the objective of achieving $|\delta_{\mathbf{k}_L}(p, t, t_1)|$ on the order of 10^{-2}, we predominantly employ MIR laser sources unless otherwise specified. It is worth noting that within the strong-field community, this range of parameters has already been experimentally implemented in the analysis of ATI [158, 159]. These studies have unveiled a spikelike low-energy structure in the low energy region of ATI spectra, which is not accounted for by the SFA. These effects have been attributed to disruptions in the electron's momentum due to the Coulomb potential shortly after ionization [160]. Although this effect may impact the final field displacement, we anticipate this perturbation to be relatively small, given its association with low-momentum values where $|\delta_{\mathbf{k}_L}(p, t, t_1)| < 10^{-2}$, as depicted in Fig. 3.13. Consequently, in the upcoming sections, we continue to apply the Lewenstein SFA model.

3.3.1.3 Conditioning Onto Electrons with a Fixed Kinetic Energy

The state presented in Eq. (3.63) represents the total state of the joint system after the interaction with the applied laser field.[22] Since our focus is on analyzing ATI, we require the electron to be in the continuum. We enforce this constraint through the following projective operation

[22] HHG processes can be further introduced in this expression by considering depletion effects on the ground state population in Eq. (3.58). However, a discussion about HHG is provided in Sect. 3.3.2 using a different approach that allows the inclusion of many-atom effects.

$$\hat{P}_{\text{ATI}} = \int \mathrm{d}p \, |p + \mathrm{e}A_{\text{cl}}(t)\rangle\langle p + \mathrm{e}A_{\text{cl}}(t)| \,, \tag{3.64}$$

such that the conditioning to an ATI state reads

$$\begin{aligned} |\bar{\Psi}_{\text{ATI}}(t)\rangle &= \hat{P}_{\text{ATI}}|\bar{\Psi}(t)\rangle \\ &= \hat{D}_{\mathbf{k}_L}(\alpha_{\mathbf{k}_L})\bigg[\int \mathrm{d}p \int_{t_0}^{t} \mathrm{d}t_1 M(p, t_1)\hat{\mathcal{D}}\big(\delta(p, t, t_1)\big)\, |p + \mathrm{e}A_{\text{cl}}(t)\rangle \otimes |\bar{0}\rangle\bigg], \end{aligned} \tag{3.65}$$

where $M(p, t_1)$ represents the integrand of the probability amplitude of finding an electron with momentum p, that is

$$M(p, t_1) = e^{-iS(p,t,t_1)/\hbar} E_{\text{cl}}(t_1) d\big(p + \mathrm{e}A_{\text{cl}}(t)\big). \tag{3.66}$$

In the forthcoming subsections, one of the primary objectives is to explore the entanglement between light and matter of Eq. (3.65). However, a significant challenge in this case stems from the presence of continuum states for both the field and the electron, as well as the presence of a double integral. Therefore, in the following sections, we simplify the scenario by focusing on electrons with specific kinetic energy values. Consequently, we introduce the projector

$$\hat{P}(p) = |-p\rangle\langle -p| + |p\rangle\langle p| \,, \tag{3.67}$$

which constraints ourselves to the subspace of electrons with kinetic energy $\mathcal{E}(p)$. Each term in this expression distinguishes between electrons propagating in the backward or forward direction. When applying this operator onto the state in Eq. (3.65), we obtain

$$\begin{aligned} |\bar{\Psi}_{\text{ATI}}(p, t)\rangle &= \hat{P}(p)|\bar{\Psi}_{\text{ATI}}(t)\rangle \\ &= \frac{1}{\sqrt{\mathcal{N}}}\Big[\, |p\rangle \otimes |\Phi(p, t)\rangle + |-p\rangle \otimes |\Phi(-p, t)\rangle \Big], \end{aligned} \tag{3.68}$$

where $\mathcal{N}$ is a normalization constant and where we have defined

$$|\Phi(p, t)\rangle = \hat{D}_{\mathbf{k}_L}(\alpha_{\mathbf{k}_L})\bigg[\int_{t_0}^{t} \mathrm{d}t_1 M(\pm p, t_1)\hat{\mathcal{D}}\big(\delta(\pm p, t, t_1)\big)|\bar{0}\rangle\bigg]. \tag{3.69}$$

3.3.1.4 Wigner Function Characterization

The quantum optical state in Eq. (3.69), which is obtained for a fixed value of momentum p, is expressed as a summation over all possible ionization times t_1 of coherent states, whose amplitudes depend on t_1. In essence, this state is defined as a superposition of coherent states, with their complex amplitudes varying with the ionization time and the electron's final momentum. As discussed in Sect. 2.3.3.4, general states of this form exhibit strong non-classical features under specific conditions, particularly the presence of negative regions in their Wigner function representation. Therefore, this section aims to investigate the conditions under which the states outlined in Eq. (3.69) exhibit these features, specifically when considering the IR field mode.

To compute the Wigner function, we utilize Eq. (2.76), where we perform the substitution of $\hat{\rho}$ by $|\Phi(p,t)\rangle\langle\Phi(p,t)|$ in Eq. (3.69). By combining these two expressions, we obtain

$$\begin{aligned} W(\beta) = \int_0^t \mathrm{d}t_1 \int_{t_0}^t & \mathrm{d}t_2 M^*(p,t_1) M(p,t_2) C_{q\geq 2}(p,t,t_1,t_2) e^{i(\varphi_{\mathbf{k}_L}(p,t_2)-\varphi_{\mathbf{k}_L}(p,t_1))} \\ & \times \langle 0_{\mathbf{k}_L}| \hat{D}^\dagger_{\mathbf{k}_L}\big(\delta_{\mathbf{k}_L}(p,t,t_1)\big) \hat{D}^\dagger_{\mathbf{k}_L}(\alpha_{\mathbf{k}_L}) \hat{D}_{\mathbf{k}_L}(\beta) \\ & \times \hat{\Pi} \hat{D}_{\mathbf{k}_L}(-\beta) \hat{D}_{\mathbf{k}_L}(\alpha_{\mathbf{k}_L}) \hat{D}_{\mathbf{k}_L}\big(\delta_{\mathbf{k}_L}(p,t,t_2)\big) |0_{\mathbf{k}_L}\rangle, \end{aligned} \tag{3.70}$$

where we defined

$$C_{q\geq 2}(p,t,t_1,t_2) = \prod_{q\geq 2}^{q_c} \langle \delta_q(p,t,t_1)|\, \delta_q(p,t,t_2)\rangle e^{i(\varphi_q(p,t_2)-\varphi_q(p,t_1))}. \tag{3.71}$$

As discussed in Section 2.3.2.3, the Wigner function remains unchanged when displacement operations are performed. Therefore, we can combine the displacement operators for β and $\alpha_{\mathbf{k}_L}$ in the expression above and represent the resulting Wigner functions in terms of the variable $\tilde{\beta}_{\mathbf{k}_L} \equiv \beta - \alpha_{\mathbf{k}_L}$ without affecting the outcomes beyond an overall displacement. Consequently, we arrive to

$$\begin{aligned} W(\tilde{\beta}_{\mathbf{k}_L}) &= \int_0^t \mathrm{d}t_1 \int_{t_0}^t \mathrm{d}t_2 M^*(p,t_1) M(p,t_2) C_{q\geq 2}(p,t,t_1,t_2) e^{i(\varphi_{\mathbf{k}_L}(p,t_2)-\varphi_{\mathbf{k}_L}(p,t_1))} \\ & \quad \times \langle 0_{\mathbf{k}_L}| \hat{D}^\dagger_{\mathbf{k}_L}\big(\delta_{\mathbf{k}_L}(p,t,t_1)\big) \hat{D}_{\mathbf{k}_L}(\tilde{\beta}_{\mathbf{k}_L}) \hat{\Pi} \hat{D}_{\mathbf{k}_L}(-\tilde{\beta}_{\mathbf{k}_L}) \hat{D}_{\mathbf{k}_L}\big(\delta_{\mathbf{k}_L}(p,t,t_2)\big)|0_{\mathbf{k}_L}\rangle, \\ &= \int_0^t \mathrm{d}t_1 \int_{t_0}^t \mathrm{d}t_2 M^*(p,t_1) M(p,t_2) C_{q\geq 2}(p,t,t_1,t_2) \\ & \qquad \times e^{i(\varphi_{\mathbf{k}_L}(p,t_2)-\varphi_{\mathbf{k}_L}(p,t_1))} w(\tilde{\beta}_{\mathbf{k}_L},\delta_1,\delta_2), \end{aligned} \tag{3.72}$$

where in transitioning from the first to the second equality, we have condensed the matrix element of the displacement and parity operators in $w(\tilde{\beta}_{\mathbf{k}_L}, \delta_1, \delta_2)$, with δ_i a shorthand notation for $\delta_{\mathbf{k}_L}(p, t, t_i)$. By utilizing the property of displacement operators presented in Eq. (2.103), we can express $w(\tilde{\beta}_{\mathbf{k}_L}, \delta_1, \delta_2)$ as

$$w(\tilde{\beta}_{\mathbf{k}_L}, \delta_1, \delta_2) = e^{\frac{1}{2}[\tilde{\beta}^*_{\mathbf{k}_L}(\delta_2-\delta_1)-\tilde{\beta}_{\mathbf{k}_L}(\delta_2-\delta_1)^*]}\langle 0_{\mathbf{k}_L}|\hat{D}_{\mathbf{k}_L}(\tilde{\beta}_{\mathbf{k}_L}-\delta_1)\hat{\Pi}\hat{D}_{\mathbf{k}_L}(-\tilde{\beta}_{\mathbf{k}_L}+\delta_2)|0_{\mathbf{k}_L}\rangle|0\rangle, \tag{3.73}$$

and taking into account the properties of the parity operator

$$\hat{\Pi}\hat{\Pi} = \mathbb{1}, \quad \hat{\Pi}\hat{D}(\alpha)\hat{\Pi} = \hat{D}(-\alpha) \quad \hat{\Pi}|0\rangle = |0\rangle, \tag{3.74}$$

we can rewrite Eq. (3.73) as

$$\begin{aligned} &w(\tilde{\beta}_{\mathbf{k}_L}, \delta_1, \delta_2) \\ &= e^{\frac{1}{2}[\tilde{\beta}^*_{\mathbf{k}_L}(\delta_2-\delta_1)-\tilde{\beta}_{\mathbf{k}_L}(\delta_2-\delta_1)^*]}\langle 0_{\mathbf{k}_L}|\hat{D}_{\mathbf{k}_L}(\tilde{\beta}_{\mathbf{k}_L}-\delta_1)\hat{\Pi}\hat{D}_{\mathbf{k}_L}(\tilde{\beta}_{\mathbf{k}_L}-\delta_2)|0_{\mathbf{k}_L}\rangle \\ &= e^{\frac{1}{2}[\tilde{\beta}^*_{\mathbf{k}_L}(\delta_2-\delta_1)-\tilde{\beta}_{\mathbf{k}_L}(\delta_2-\delta_1)^*]}e^{\frac{1}{2}[(\tilde{\beta}_{\mathbf{k}_L}-\delta_1)(\tilde{\beta}_{\mathbf{k}_L}-\delta_2)^*-(\tilde{\beta}_{\mathbf{k}_L}-\delta_1)^*(\tilde{\beta}_{\mathbf{k}_L}-\delta_2)]} \\ &\times \langle 0_{\mathbf{k}_L}|2\tilde{\beta}_{\mathbf{k}_L}-\delta_1-\delta_2\rangle \\ &= e^{\frac{1}{2}[\tilde{\beta}^*_{\mathbf{k}_L}(\delta_2-\delta_1)-\tilde{\beta}_{\mathbf{k}_L}(\delta_2-\delta_1)^*]}e^{\frac{1}{2}[(\tilde{\beta}_{\mathbf{k}_L}-\delta_1)(\tilde{\beta}_{\mathbf{k}_L}-\delta_2)^*-(\tilde{\beta}_{\mathbf{k}_L}-\delta_1)^*(\tilde{\beta}_{\mathbf{k}_L}-\delta_2)]}e^{-\frac{1}{2}|2\tilde{\beta}_{\mathbf{k}_L}-\delta_1-\delta_2|^2}. \end{aligned} \tag{3.75}$$

Then, by introducing Eq. (3.75) into Eq. (3.72), we finally arrive to

$$\begin{aligned} W(\tilde{\beta}_{\mathbf{k}_L}) = &\int_0^t \mathrm{d}t_1 \int_{t_0}^t \mathrm{d}t_2 M^*(p, t_1)M(p, t_2)C_{q\geq 2}(p, t, t_1, t_2)e^{i(\varphi_{\mathbf{k}_L}(p,t_2)-\varphi_{\mathbf{k}_L}(p,t_1))} \\ &\times e^{\frac{1}{2}[\tilde{\beta}^*_{\mathbf{k}_L}(\delta_2-\delta_1)-\tilde{\beta}_{\mathbf{k}_L}(\delta_2-\delta_1)^*]}e^{\frac{1}{2}[(\tilde{\beta}_{\mathbf{k}_L}-\delta_1)(\tilde{\beta}_{\mathbf{k}_L}-\delta_2)^*-(\tilde{\beta}_{\mathbf{k}_L}-\delta_1)^*(\tilde{\beta}_{\mathbf{k}_L}-\delta_2)]} \\ &\times e^{-\frac{1}{2}|2\tilde{\beta}_{\mathbf{k}_L}-\delta_1-\delta_2|^2}, \end{aligned} \tag{3.76}$$

where, in computing this double integral, we utilize two different approaches based on the laser parameter regime in use. When working with NIR frequencies, we employ fully numerical approaches. However, considering that for typical laser field strengths used in strong-field physics $M(p, t)$ becomes a highly oscillating function, we can apply the saddle-point approximation to compute this quantity. This becomes particularly useful when dealing with MIR laser sources, where longer pulse durations[23] make the fully numerical approach more time-consuming. For comprehensive explanations of these two methods, we refer the reader to Appendix 8.1.

In Fig. 3.15a and b, we focus on the latter case, when considering $\lambda_{\mathbf{k}_L} \approx 5\ \mu$m and field intensity $I = 4 \times 10^{14}$ W/cm^2. Specifically, we display the Wigner functions for the IR state when considering electrons with $\mathcal{E}(p) = 2.2U_p$ propagating

[23] In the numerical analysis, we fix the laser pulse to have a total of 5 optical cycles, which for MIR frequencies leads to longer pulses when compared to NIR ones.

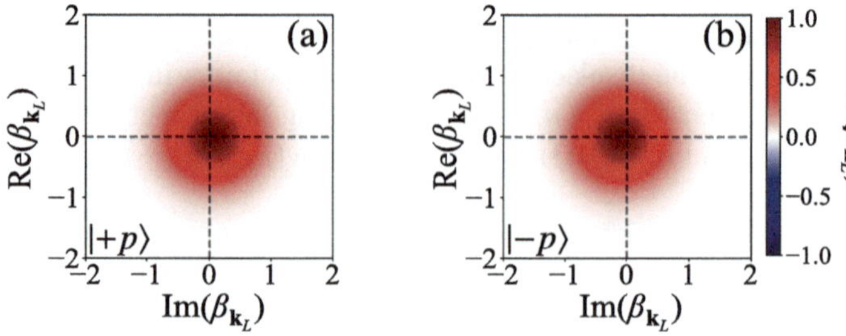

Fig. 3.15 Wigner function for the fundamental mode for single-electrons propagating in opposite directions. We considered a hydrogen atom interacting with a laser field featuring a sinusoidal-squared envelope, central wavelength $\lambda_{\mathbf{k}_L} \approx 5\ \mu$m, peak intensity $I = 4 \times 10^{14}$ W/cm^2 (equivalently $E_0 = 0.106 <$ a.u.) and a duration of 5 optical cycles. In **a**, the Wigner function of the fundamental mode when the electron is found propagating along the forward direction with kinetic energy $\mathcal{E}(p) = 2.2U_p$. Conversely, in **b**, we depict the Wigner function when the electron is observed propagating in the opposite direction with the same kinetic energy. In this case, the Wigner function has been computed using a saddle-point analysis (see Appendix 8.1.1). Furthermore, their values have been normalized to the maximum value

in the forward and backward directions, respectively. The obtained Wigner function exhibit a Gaussian-like behavior, devoid of non-classical signatures such as the presence of negative regions. However, a significant distinction arises between the resulting distributions for electrons propagating forward and backward, concerning their position in phase-space. Specifically, for $p > 0$ the Wigner distribution slightly shifts towards positive values of the x axis, as evident in panel (a). Conversely, for $p < 0$ we observe the opposite behavior, as illustrated in panel (b).

While this analysis has primarily focused on single-electron dynamics, as previously mentioned, we expect the observation of non-classical features in the Wigner function within a multi-atom regime. This expectation arises from the fact that, in more realistic scenarios, the total optical displacements appearing in the final quantum optical state stem from the coherent contribution of multiple electrons undergoing the ATI process. These contributions would accumulate, resulting in small overlaps among the various coherent states in the superposition presented in Eq. (3.69). However, characterizing this many-body effect becomes computationally demanding, particularly in typical experimental implementations where the number of atoms participating in these processes can easily exceed 10^6. Therefore, with the aim of conducting a proof-of-principle analysis to support our hypothesis, we consider a scenario in which all atoms collectively participate in the process as a single entity. We account for many-body effects in a phenomenological manner by multiplying the generated displacement $\delta_{\mathbf{k}_L}(p, t, t_1)$ by a factor N_{at} representing the number of atoms.[24] Hence, we describe the quantum optical state as follows

[24] Here, we assume that each atom undergoing ATI contributes with a single electron.

$$|\Phi^{(N)}(p,t)\rangle = \hat{D}_{\mathbf{k}_L}(\alpha_{\mathbf{k}_L})\left[\int_{t_0}^{t} \mathrm{d}t_1 M(\pm p, t_1)\hat{\mathcal{D}}\big(N_{\text{at}}\delta(\pm p, t, t_1)\big)|\bar{0}\rangle\right], \tag{3.77}$$

corresponding to the scenario where all the photoelectrons get ionized collectively, and propagate along the same direction.

The results obtained from this phenomenological approach, where $\lambda_{\mathbf{k}_L} = 800$ nm and therefore the fully numerical approach has been used for their computation, are presented in Fig. 3.16. In this figure, the upper and lower rows correspond to scenarios with $N_{\text{at}} \approx 10^4$ and $N_{\text{at}} \approx 2 \times 10^4$, respectively. Each column represents different values of p: $p = 0.00$ a.u. for the first column, $p = 0.43$ a.u. for the second, and $p = -0.43$ a.u. for the third. In contrast to Fig. 3.15, negative values are observed in the Wigner function for all cases. These negativities result from the structure of our state, which is expressed as a linear combination of various coherent states, each with distinct amplitudes dependent on the ionization time and the electron momentum. Additionally, a factor of N_{at} multiplies this displacement, enhancing the amplitude differences between the coherent states in the superposition. For the scenario with $N_{\text{at}} \approx 10^4$, shown in Fig. 3.16a–c, there are relatively few coherent states within the superposition for which $\langle N_{\text{at}}\delta_{\mathbf{k}_L}(p,t,t_i)|N_{\text{at}}\delta_{\mathbf{k}_L}(p,t,t_j)\rangle \to 0$ for $t_i \neq t_j$. Consequently, the corresponding Wigner representation can be approximated

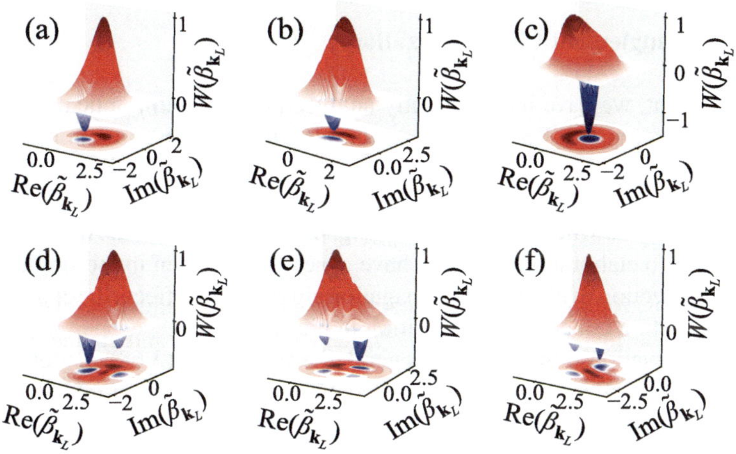

Fig. 3.16 Wigner function for the fundamental mode in the phenomenological many-electron scenario. We considered a hydrogen atom interacting with a laser field featuring a sinusoidal-squared envelope, central wavelength $\lambda_{\mathbf{k}_L} = 800$ nm, peak intensity $I = 1 \times 10^{14}$ W/cm^2 and a duration of 5 optical cycles. In the first row (panels (**a**)–(**c**)) we used $N_{\text{at}} \approx 10^4$ while in the second row (panels (**d**)–(**f**)) $N_{\text{at}} \approx 2 \times 10^4$. Each of the columns represents different electron momenta. In the first column (panels (**a**) and (**d**)), second column (panels (**b**) and (**e**)) and third column (panels (**c**) and (**f**)), we set $p = 0.00$ a.u., $p = 0.43$ a.u. and $p = -0.43$ a.u., respectively. In this case, the Wigner functions were computed using the fully-numerical approach presented in Appendix 8.1.2. Furthermore, their values have been normalized to the maximum value

by an unbalanced superposition of two coherent states with similar, but yet different, amplitudes (see Fig. 2.12d). As N_{at} increases, the real and imaginary components of $\delta_{\mathbf{k}_L}(p, t, t_1)$ cover a broader range of values, resulting in a greater number of states in the superposition for which $\langle N_{\text{at}}\delta_{\mathbf{k}_L}(p, t, t_i)|N_{\text{at}}\delta_{\mathbf{k}_L}(p, t, t_j)\rangle \to 0$. This results in a more complicated structure of the obtained distributions, as observed in Fig. 3.16d–f.

Another important aspect to emphasize is the variation observed when considering different values of p. On one hand, for $N_{\text{at}} \approx 10^4$ (first row), these variations manifest in the shape of the Wigner functions, particularly in the depth of the minima and the overall orientation of the distribution. These distinctions arise from the different relative phases between the probability amplitudes that weight the coherent states within the superposition, as well as the coherent states' phases. On the other hand, for $N_{\text{at}} \approx 2 \times 10^4$ (second row), apart from the structural differences among the obtained Wigner functions, their position in phase-space also differs. Specifically, we observe that for positive values of momentum (Fig. 3.16e), the maximum peak of the distribution is located towards positive values of $\text{Im}(\tilde{\beta}_{\mathbf{k}_L})$, while for negative values (Fig. 3.16f), we find the opposite behavior. It is worth noting that these observations align with those presented in Fig. 3.13c and d. Moreover, for $p = 0.00$ a.u. (Fig. 3.16d) the maximum peak is centered around zero. Finally, we anticipate these differences between positive and negative momenta to strongly depend on the chosen CEP phase. Specifically, if the CEP phase differs in π from the one used here, we expect the behaviors of $p > 0$ and $p < 0$ to be reverted.

3.3.1.5 Entanglement Characterization

Up to this point, we have independently analyzed the quantum optical components of the state described in Eq. (3.68) by examining their Wigner functions. However, from a more general perspective, the state in this equation exhibits another non-classical feature: entanglement. The non-zero impact of entanglement has been indirectly demonstrated when observing the Wigner distributions of the quantum optical components. To elaborate further, we have observed that even in the single-electron regime, the direction of electronic propagation significantly affects the characteristics of this distribution concerning its position in phase-space.

These observations imply that, in general, the state in Eq. (3.68) cannot be written as a product state and is, therefore, entangled. Furthermore, this state exhibits a *hybrid entangled* structure [161], since we effectively[25] have a finite-dimensional Hilbert space, spanned by $\{|p\rangle, |-p\rangle\}$, which is coupled to an infinite-dimensional Hilbert space, spanned for instance by the (over-complete) coherent state basis or the Fock basis.

[25] It is worth noting that in our analysis, we consider the use of linearly polarized fields, which justifies the reasonable approximation of conducting a 1D analysis for the electron along the field's polarization axis, as mentioned previously. Hence, due to this specific arrangement, we can assume that only two momentum components propagate in the forward and backward directions with respect to the polarization direction.

The analysis of hybrid entangled states must be, in general, thoroughly investigated depending on the state's specific structure [161–163]. For the case of pure states, quantifying the amount of entanglement can be achieved using the entanglement entropy [161] (refer to Sect. 2.3.4.1 for a definition). To do so, instead of working with the states $|\tilde{\Phi}(\pm p, t)\rangle \equiv |\Phi(\pm p, t)\rangle / \mathcal{N}_{\pm}$, where $\mathcal{N}_{\pm}$ is the normalization constant of the state, which generally have a non-vanishing overlap, we choose to work with the orthonormal set $\{|u\rangle, |v\rangle\}$. This change of basis allows us to effectively treat the quantum optical states as belonging to a Hilbert space of dimension 2. This new set of states are defined as follows

$$|u\rangle \equiv \frac{1}{2\mu}\Big[|\tilde{\Phi}(p,t)\rangle + e^{-i\Theta}|\tilde{\Phi}(-p,t)\rangle\Big], \quad |v\rangle \equiv \frac{1}{2\nu}\Big[|\tilde{\Phi}(p,t)\rangle - e^{-i\Theta}|\tilde{\Phi}(-p,t)\rangle\Big], \tag{3.78}$$

where the parameters appearing in these states are given by

$$\mu = \sqrt{\frac{1 + |\langle\tilde{\Phi}(p,t)|\tilde{\Phi}(-p,t)\rangle|}{2}}, \qquad \nu = \sqrt{1-\mu^2}, \tag{3.79}$$
$$\langle\tilde{\Phi}(p,t)|\tilde{\Phi}(-p,t)\rangle = e^{i\Theta}|\langle\tilde{\Phi}(p,t)|\tilde{\Phi}(-p,t)\rangle|.$$

In this new basis, Eq. (3.68) can be rewritten as

$$\begin{aligned}
&|\tilde{\Psi}_{\text{ATI}}(t)\rangle \\
&= \frac{1}{\sqrt{\mathcal{N}}}\Big[\mu\,|u\rangle \otimes \Big(\sqrt{\mathcal{N}_+}\,|p\rangle + e^{i\theta}\sqrt{\mathcal{N}_-}\,|-p\rangle\Big) + \nu\,|u\rangle \otimes \Big(\sqrt{\mathcal{N}_+}\,|p\rangle - e^{i\theta}\sqrt{\mathcal{N}_-}\,|-p\rangle\Big)\Big],
\end{aligned} \tag{3.80}$$

and after tracing out the $\{|u\rangle, |v\rangle\}$ modes, an operation needed for computing the entanglement entropy, we arrive to the following reduced density matrix

$$\begin{aligned}
\hat{\rho} = {}& \frac{\mathcal{N}_+}{\mathcal{N}}\,|p\rangle\langle p| + e^{-i\theta}\big(\mu^2 - \nu^2\big)\frac{\sqrt{\mathcal{N}_+\mathcal{N}_-}}{\mathcal{N}}\,|p\rangle\langle -p| \\
&+ \frac{\mathcal{N}_-}{\mathcal{N}}\,|-p\rangle\langle -p| + e^{i\theta}\big(\mu^2 - \nu^2\big)\frac{\sqrt{\mathcal{N}_+\mathcal{N}_-}}{\mathcal{N}}\,|-p\rangle\langle p|,
\end{aligned} \tag{3.81}$$

whose associated Schmidt matrix is given by

$$\mathsf{S} = \frac{1}{\mathcal{N}}\begin{pmatrix} \mathcal{N}_+ & e^{-i\theta}\big(\mu^2-\nu^2\big)\sqrt{\mathcal{N}_+\mathcal{N}_-} \\ e^{i\theta}\big(\mu^2-\nu^2\big)\sqrt{\mathcal{N}_+\mathcal{N}_-} & \mathcal{N}_- \end{pmatrix}, \tag{3.82}$$

and that has eigenvalues

$$\lambda_{\pm} = \frac{1}{2}\left[1 \pm \sqrt{1 - 4\big(1 - |\langle\tilde{\Phi}(p,t)|\tilde{\Phi}(-p,t)\rangle|^2\big)\frac{\mathcal{N}_+\mathcal{N}_-}{\mathcal{N}}}\right]. \tag{3.83}$$

In terms of these eigenvalues, we can express the entanglement entropy as

$$\mathcal{S}(\hat{\rho}) = -\lambda_+^2 \log_2 \lambda_+^2 - \lambda_-^2 \log_2 \lambda_-^2, \tag{3.84}$$

which is shown in Fig. 3.17a for a laser field with amplitude $E_0 = 0.106$ a.u. and three different frequencies within the MIR regime. From this figure, we observe that the entanglement entropy is small in the single electron regime, although non-vanishing. Furthermore, it tends towards zero as $\mathcal{E}(p) \to 0$, and increases as $\mathcal{E}(p)$ grows larger. For decreasing frequencies, the degree of entanglement also becomes larger, which is a direct consequence of the behavior observed for the displacement $\delta_{\mathbf{k}_L}(p, t, t_1)$ as a function of the frequency, presented in Fig. 3.14a. It is important to note that even though the amount of entanglement increases with the photoelectron energy, extending this analysis beyond the range of values considered would necessitate the inclusion of HATI terms. However, this falls outside the scope of the formulas derived within this section.

In Fig. 3.17b and c, we present the behavior of the absolute values of the overlap between the states $|\tilde{\Phi}(p,t)\rangle$ and $|\tilde{\Phi}(-p,t)\rangle$, as well as that of the normalization constant $\mathcal{N}_+$, respectively, as functions of the photoelectron energy. It is essential to note that these two quantities are critical in determining the amount of entanglement, as evident from Eqs. (3.83) and (3.84). Specifically, when $|\langle\tilde{\Phi}(p,t)|\tilde{\Phi}(-p,t)\rangle| \to 1$, it results in $\lambda_+ \to 1$ ($\lambda_- \to 0$), indicating separable states. The same outcome can be achieved if $\mathcal{N}_\pm \to 1$ and $\mathcal{N}_\mp \to 0$. However, as observed in these figures, we are closer to the first scenario: the overlap between the states plays a crucial role in

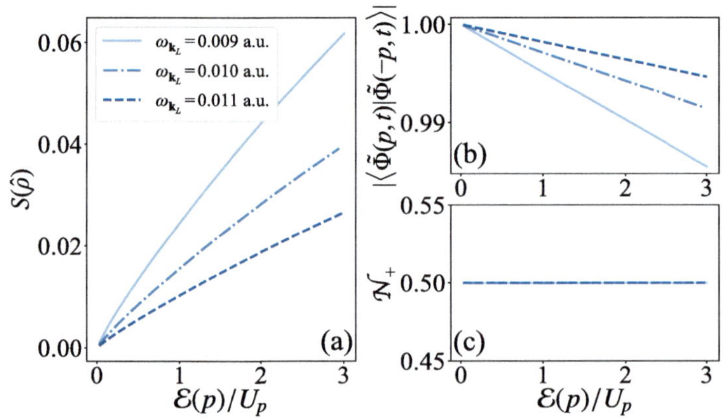

Fig. 3.17 **Behavior of entanglement entropy as a function of the photoelectron energy**. In **a**, we depict the behavior of $\mathcal{S}(\hat{\rho})$, with $\hat{\rho}$ defined in Eq. (3.81), as a function of the photoelectron energy for various central frequencies of the driving laser source. The field intensity of the linearly polarized sinusoidal-squared laser field has been fixed to $I = 4 \times 10^{14}$ W/cm^2, while its duration to 5 optical cycles. In **b**, we display the overlap between the quantum optical states present in the entangled superposition given in Eq. (3.68). In **c**, we show their respective normalization constants as a function of the photoelectron energy

determining the degree of entanglement. Conversely, $\mathcal{N}_+$ remains constant for all possible values of the photoelectron energy.[26] This feature is somewhat expected as we are using multicycle pulses. In this case, the probability of obtaining photoelectrons propagating in the forward and backward directions, ultimately determined by $\mathcal{N}_\pm$, is the same. Thus, we anticipate these results to be significantly influenced by the duration of the laser pulse and, for short enough pulses, by the CEP as well.

3.3.2 High-Harmonic Generation

In this section, based on our works in Refs. [39, 41, 42, 44, 45], we explore high-harmonic generation from a quantum optical perspective. To be more precise, we extend our analysis from ATI to HHG utilizing the differential equation previously derived for ATI, particularly the one presented in Eq. (3.41). We express this equation as follows

$$i\hbar\frac{\partial|\tilde{\Psi}(t)\rangle}{\partial t} = \Big[\hat{H}_{\text{sc}}(t) + \mathsf{e}\hat{\mathbf{E}}(t)\cdot\hat{\mathbf{r}}\Big]|\tilde{\Psi}(t)\rangle, \tag{3.85}$$

where $\hat{H}_{\text{sc}}(t) \equiv \hat{H}_{\text{at}} + \mathsf{e}\mathbf{E}_{\text{cl}}(t)\cdot\hat{\mathbf{r}}$ represents the semiclassical Hamiltonian describing the electron dynamics under the influence of a classical field $\mathbf{E}_{\text{cl}}(t)$. To investigate HHG processes, we introduce an additional transformation on top of the ones already performed to arrive to Eq. (3.85). This transformation involves working in the interaction picture with respect to $\hat{H}_{\text{sc}}(t)$. Consequently, we obtain

$$i\hbar\frac{\partial|\Psi'(t)\rangle}{\partial t} = \mathsf{e}\hat{\mathbf{E}}(t)\cdot\hat{\mathbf{r}}(t)|\Psi'(t)\rangle \quad \text{with} \quad |\tilde{\Psi}(t)\rangle = \hat{U}_{\text{sc}}(t)|\Psi'(t)\rangle, \tag{3.86}$$

where the position operator becomes time-dependent, i.e., $\hat{\mathbf{r}} = \hat{U}^\dagger_{\text{sc}}(t)\hat{\mathbf{r}}\hat{U}_{\text{sc}}(t)$, with $\hat{U}_{\text{sc}}(t)$ representing the time-evolution operator describing the semiclassical dynamics, given by

$$\hat{U}_{\text{sc}}(t) = \exp\Big[-\frac{i}{\hbar}\hat{\mathcal{T}}\int_{t_0}^{t} \mathrm{d}\tau\,\hat{H}_{\text{sc}}(\tau)\Big]|\Psi'(t)\rangle. \tag{3.87}$$

This transformation enables us to include all the semiclassical dynamics described in Sect. 3.1.1.2 within the time-dependent dipole operator $\mathsf{e}\hat{\mathbf{r}}(t)$. It is important to note that this transformation does not affect the initial state of the system, as represented by Eq. (3.43).

As mentioned in previous sections, there are several hallmarks about the HHG process. A crucial one, important when solving Eq. (3.85), is that it is a parametric process, meaning that the initial and final quantum optical states are the same.

[26] It is worth noting that the behavior of $\mathcal{N}_-$ can be obtained as $1 - \mathcal{N}_+$.

Consequently, to study the quantum optical state after these processes we project this equation with respect to $|\text{g}\rangle$, resulting in

$$i\hbar\frac{\partial|\Psi'(t)\rangle}{\partial t} = \mathrm{e}\hat{\mathbf{E}}(t)\cdot\langle\text{g}|\hat{\mathbf{r}}(t)|\Psi'(t)\rangle, \tag{3.88}$$

and, at this point, we introduce the SFA version of the identity, that is

$$\mathbb{1} = |\text{g}\rangle\langle\text{g}| + \int \mathrm{d}\mathbf{v}\, |\mathbf{v}\rangle\langle\mathbf{v}|, \tag{3.89}$$

where all intermediate bound states have been neglected in virtue of the Lewenstein SFA model. By doing this, we arrive to

$$i\hbar\frac{\partial|\tilde{\Phi}_{\text{g}}(t)\rangle}{\partial t} = \mathrm{e}\hat{\mathbf{E}}(t)\cdot\langle\hat{\mathbf{r}}(t)\rangle|\tilde{\Phi}_{\text{g}}(t)\rangle + \mathrm{e}\hat{\mathbf{E}}(t)\cdot\int \mathrm{d}\mathbf{v}\langle\text{g}|\hat{\mathbf{r}}(t)\,|\mathbf{v}\rangle\,|\tilde{\Phi}(\mathbf{v},t)\rangle, \tag{3.90}$$

where we have defined $|\tilde{\Phi}_{\text{g}}(t)\rangle \equiv \langle\text{g}|\Psi'(t)\rangle$ and $|\tilde{\Phi}(\mathbf{v},t)\rangle \equiv \langle\mathbf{v}|\Psi'(t)\rangle$. As customary in these processes, we obtain a differential equation with well-defined homogeneous and inhomogeneous parts. Following a methodology similar to that presented in previous sections, we can express the solution to this differential equation as

$$|\tilde{\Phi}_{\text{g}}(t)\rangle = \hat{\mathcal{D}}\big(\boldsymbol{\chi}(t,t_0)\big)|\tilde{\Phi}_{\text{g}}(t_0)\rangle - \frac{i}{\hbar}\int_{t_0}^{t}\mathrm{d}t_1\int \mathrm{d}\mathbf{v}\,\hat{\mathcal{D}}\big(\boldsymbol{\chi}(t,t_1)\big)\hat{\mathbf{E}}(t_1)\cdot\langle\text{g}|\hat{\mathbf{r}}(t)\,|\mathbf{v}\rangle\,|\tilde{\Phi}(\mathbf{v},t_1)\rangle. \tag{3.91}$$

The first term in this expression corresponds to a multimode displacement of $\boldsymbol{\chi}(t,t_0)$ on the quantum optical state of the field due to HHG processes, while the second term represents quantum optical fluctuations arising from the back-action of the ionized photoelectrons. By invoking the SFA assumptions, we consider the latter to be small in comparison to the ground-state amplitude, an approximation we expect to be valid in the regime of small ground state depletion. Consequently, we approximate Eq. (3.91) as follows

$$|\tilde{\Phi}_{\text{g}}(t)\rangle \approx \hat{\mathcal{D}}\big(\boldsymbol{\chi}(t,t_0)\big)|\bar{0}\rangle, \tag{3.92}$$

where we have explicitly incorporated the initial conditions. As in earlier expressions, $\hat{\mathcal{D}}(\boldsymbol{\chi}(t,t_0)) \equiv \prod_{q=1}^{q_c} e^{i\tilde{\varphi}_q(t)}\hat{D}_q(\chi_q(t,t_0))$ denotes the multimode displacement operator, where in this case the amplitude of the displacements, $\chi_q(t,t_0)$, and the phases, $\tilde{\varphi}_q(t)$, are determined as follows

$$\chi_q(t,t_0) = -\frac{\mathrm{e}}{\hbar}\sqrt{\frac{\hbar\omega_{\mathbf{k}_q}}{2\varepsilon_0 V}}\int_{t_0}^{t}\mathrm{d}\tau\,\Big(\boldsymbol{\epsilon}_{\mathbf{k}_q}\cdot\langle\hat{\mathbf{r}}(t)\rangle\Big)e^{i\omega_{\mathbf{k}_q}t}, \tag{3.93}$$

$$\tilde{\varphi}_q(t) = \frac{\mathrm{e}^2}{\hbar^2}\frac{\hbar\omega_{\mathbf{k}_q}}{2\varepsilon_0 V}\int_{t_0}^{t} \mathrm{d}t_1 \int_{t_0}^{t_1} \mathrm{d}t_2 \Big(\boldsymbol{\epsilon}_{\mathbf{k}_q}\cdot\langle\hat{\mathbf{r}}(t_1)\rangle\Big)\Big(\boldsymbol{\epsilon}_{\mathbf{k}_q}\cdot\langle\hat{\mathbf{r}}(t_2)\rangle\Big)\sin(\omega_{\mathbf{k}_q}(t_1-t_2)). \tag{3.94}$$

Reversing the transformations that have been applied thus far, we finally arrive to

$$|\bar{\Psi}(t)\rangle = \hat{D}_{\mathbf{k}_L}(\alpha_{\mathbf{k}_L})\hat{U}_{\text{sc}}(t,t_0)\Big[|\text{g}\rangle\otimes|\tilde{\Phi}_\text{g}(t)\rangle + |\mathbf{v}\rangle\otimes|\tilde{\Phi}(\mathbf{v},t)\rangle\Big], \tag{3.95}$$

and since in HHG processes the electron ends up recombining with the ground state of the system, we project the equation above onto $|\text{g}\rangle$, resulting in

$$|\bar{\Phi}_\text{g}(t)\rangle = \langle \text{g}|\bar{\Psi}(t)\rangle \approx \hat{D}_{\mathbf{k}_L}(\alpha_{\mathbf{k}_L})\hat{\mathcal{D}}\big(\boldsymbol{\chi}(t,t_0)\big)|\bar{0}\rangle, \tag{3.96}$$

where we have incorporated the SFA assumptions: $\langle \text{g}|U_{\text{sc}}(t,t_0)|\text{g}\rangle \approx 1$, i.e. the ground state depletion to be negligible, which in turn implies $\langle \text{g}|U_{\text{sc}}(t,t_0)\,|\mathbf{v}\rangle \approx 0$.

Hence, we conclude that the dominant contribution to the quantum optical state after events where the electron ends up in the ground state, is a product state composed of displaced coherent states. However, it is crucial to emphasize that this state has been derived under the assumption of small depletion of the ground state amplitude or, in other words, by disregarding the contribution of dipole moment correlations [42, 94, 148]. It can be demonstrated that when taking these correlations into consideration, they result in both squeezing in the field modes and entanglement [164].

In the context of ATI, we observed that the generated displacement resulted from the electronic oscillation in the continuum. For the case of HHG, the obtained displacement is contingent upon the effective oscillation that the electron undergoes when transitioning cyclically to and from the ground state. This oscillation-like behavior leads to the generation of harmonic radiation, consistent with the semiclassical predictions, and the resulting displacement encapsulates the dynamics experienced by the electron. This can be better understood by examining the energy of the emitted radiation, which can be calculated as follows

$$\langle \hat{H}_{\text{field}}\rangle = \lim_{t\to\infty}\sum_{q=1}^{q_c}\hbar\omega_{\mathbf{k}_q}|\chi_q(t,t_0)|^2 = \sum_{q=1}^{q_c}\frac{\mathrm{e}^2\omega_{\mathbf{k}_q}^2}{2\varepsilon_0 V}\left|\int_{-\infty}^{\infty}\mathrm{d}\tau\,\Big(\boldsymbol{\epsilon}_{\mathbf{k}_q}\cdot\langle\hat{\mathbf{r}}(t)\rangle\Big)e^{i\omega_{\mathbf{k}_q}t}\right|^2, \tag{3.97}$$

where $\mathrm{e}\langle\hat{\mathbf{r}}(t)\rangle$ is given in Eq. (3.27). We observe that the energy of the generated radiation is proportional to the absolute value of the Fourier transform of the time-dependent dipole moment, akin to the outcome obtained within the semiclassical framework (see Eq. (3.28)). Therefore, applying the saddle-point approximation to the integrals in Eq. (3.97) results in the same saddle-point equations as described in Eqs. (3.31)–(3.33), thereby yielding a description of the electron dynamics identical to the semiclassical picture. Yet, this expression differs from that in Eq. (3.28) in that the energy of the emitted radiation scales as $\omega_{\mathbf{k}_q}^2$ instead of $\omega_{\mathbf{k}_q}^4$. This discrepancy

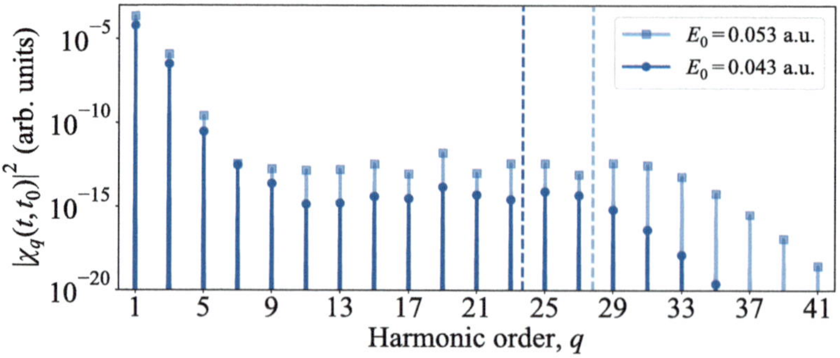

Fig. 3.18 Amplitude of the HHG-induced displacements as a function of the harmonic order. We consider a helium atom interacting with a monochromatic laser field with $\lambda_{\mathbf{k}_L} = 800$ nm and two different field strengths. The resulting spectrum is represented with the circular and square markers for peak intensities $I = 1 \times 10^{14}$ W/cm^2 and $I = 6.6 \times 10^{13}$ W/cm^2, respectively. The spectra were computed using the saddle-point approximation via the RB-SFA Mathematica package [105]

can be reconciled by considering a continuous range of frequencies, wherein $\sum_q \to V/((2\pi)^3 c^3) \int \mathrm{d}\Omega\, \mathrm{d}\omega\, \omega^2$, with $\mathrm{d}\Omega$ representing an infinitesimal solid angle element.

In Fig. 3.18, we present the mean-photon number $|\chi_q(t, t_0)|^2$ for different harmonic modes. This analysis is conducted for helium ($I_p \approx 0.9$ a.u.) interacting with a monochromatic field of $\omega_{\mathbf{k}_L} = 0.057$ a.u. ($\lambda_{\mathbf{k}_L} = 800$ nm) and varying field strengths. As observed, the generated spectra exhibit the expected features: a plateau of frequencies consisting of odd harmonics and a cutoff frequency governed by Eq. (3.34). The cutoff frequency is indicated in the figure with the vertical dashed lines, and it increases with the input field strength.

Finally, we note that the generated shift in the single-atom regime is very small, as the coupling factor $\mathbf{g}(\omega_{\mathbf{k}_q})$ is generally a perturbative quantity. However, we can extend this solution to a more realistic scenario where N_{at} atoms participate in the dynamics independently but in a phase-matched way, each contributing with a single electron. In this scenario, each atom provides a displacement $\boldsymbol{\chi}(t, t_0)$ to Eq. (3.96), such that the final state reads

$$|\bar{\Phi}_{\text{g}}(t)\rangle = \left[\bigotimes_{i=1}^{N_{\text{at}}} \langle \text{g}|\right] |\bar{\Psi}(t)\rangle \approx \hat{D}_{\mathbf{k}_L}(\alpha_{\mathbf{k}_L}) \hat{\mathcal{D}}\big(N_{\text{at}} \boldsymbol{\chi}(t, t_0)\big) |\bar{0}\rangle. \tag{3.98}$$

3.3.2.1 The *Conditioning on HHG* Operation From a Theoretical and Experimental Perspective

The state presented in Eq. (3.96) lacks non-classical features in the sense of Sect. 2.3.2, as it is formulated as a product state consisting of different optical modes, each represented as a coherent state. Consequently, for each field mode, we obtain

Poissonian statistics and positive (Gaussian) Wigner functions. To generate non-classical states of light from these, additional operations on the generated state are necessary. In this regard, we demonstrate that the conditioning on harmonic generation events, as performed in Refs. [137, 138] (see Fig. 3.10), allows us to seemingly create coherent state superpositions. In this section, we explore these postselection operations, which we shall refer to as *conditioning on HHG* operation, both from a theoretical and experimental perspective.

A crucial observation towards the conditioning on HHG operation is that, as demonstrated in Ref. [137], there exists a correlation between the excitation of the fundamental and the harmonic modes (see Fig. 3.10d). This correlation arises from the fact that the generated displacements originate from the same dipole moment, as shown in the first term of Eq. (3.90). Consequently, the wavepacket mode that is actually excited during the HHG processes can be described by a creation operator of the form $\hat{b}^{\dagger}(t) \propto \sum_{q=1}^{q_c} \hat{a}^{\dagger}_{\mathbf{k}_q} e^{i\omega_{\mathbf{k}_q} t}$, resulting in the global coherent state presented in Eq. (3.96). However, only the effective modes described by this wavepacket mode are the ones that become excited. Let us denote these excitations as $\{|\tilde{n}\rangle\}$, which account for the generated correlations. In this sense, HHG corresponds to those cases where these wavepacket modes get excited, i.e., whenever $\tilde{n} \neq 0$. This insight allows us to introduce a POVM set taking into consideration events in which an HHG excitation has occurred or not [42, 148]

$$\hat{M}_{\text{HHG}} = \{\hat{P}_{\tilde{0}}, \hat{P}_{\tilde{n}\neq 0}\}, \tag{3.99}$$

where $\hat{P}_{\tilde{0}} \equiv |\tilde{0}\rangle\langle\tilde{0}|$ represents the scenario in which excitations have occurred, indicating the absence of harmonic radiation emission. Conversely, $\hat{P}_{\tilde{n}\neq 0} \equiv \sum_{\tilde{n}=1}^{\infty} |\tilde{n}\rangle\langle\tilde{n}|$ accounts for situations where harmonic radiation has been generated. It is worth noting that these operators span a complete set, and therefore it holds that $P_{\tilde{0}} + \hat{P}_{\tilde{n}\neq 0} = \sum_{\tilde{n}=0}^{\infty} |\tilde{n}\rangle\langle\tilde{n}| = \mathbb{1}$, allowing us to express $\hat{P}_{\tilde{n}\neq 0} = \mathbb{1} - \hat{P}_{\tilde{0}}$. Thus, bearing in mind that the case of zero excitations corresponds to the quantum optical state prior to the interaction, i.e. $|\tilde{0}\rangle \equiv \hat{D}_{\mathbf{k}_L}(\alpha_{\mathbf{k}_L}) \bigotimes_{q=1}^{q_c} |0_{\mathbf{k}_q}\rangle \equiv \hat{D}_{\mathbf{k}_L}(\alpha_{\mathbf{k}_L})|\bar{0}\rangle$, the final state of the system conditioned on the generation of harmonic radiation can be expressed, up to a normalization factor, as follows

$$\begin{aligned} |\bar{\Phi}_{\text{HHG}}\rangle &= P_{\tilde{n}\neq 0}|\bar{\Phi}_{\text{g}}(t)\rangle \\ &= \hat{D}_{\mathbf{k}_L}(\alpha_{\mathbf{k}_L})\Big[\,|\chi(t,t_0)\rangle - \xi(t,t_0)|\bar{0}\rangle\Big] \\ &= \hat{D}_{\mathbf{k}_L}(\alpha_{\mathbf{k}_L})\Bigg[\bigotimes_{q=1}^{q_c}|\chi_q(t,t_0)\rangle - \xi(t,t_0)\bigotimes_{q=1}^{q_c}|0_{\mathbf{k}_q}\rangle\Bigg], \end{aligned} \tag{3.100}$$

where $\xi(t, t_0)$ is the overlap between the initial and the final state after the light-matter interaction within the interaction picture, that is

$$\xi(t, t_0) = \prod_{q=1}^{q_c} \langle 0_{\mathbf{k}_q} | \chi_q(t, t_0) \rangle. \tag{3.101}$$

In general terms, the state right after the conditioning operation corresponds to a *massively* entangled state involving all the field modes that become excited during the HHG process. We emphasize the word *massive* because the number of modes encompassed by the entangled state can vary significantly, depending on the harmonic generation conditions, ranging from dozens to hundreds, or even thousands of harmonic orders. As mentioned earlier, the specific conditions under which HHG occurs, such as the laser field parameters and the atomic species, critically determine the properties of the generated radiation, including its intensity and the cutoff frequency. Therefore, by adjusting these parameters, it becomes possible to control the number of field modes that become entangled after the conditioning on HHG operation, as well as the degree of entanglement. More precisely, the degree of entanglement in the state is contingent on the parameter $\xi(t, t_0)$, as demonstrated in Ref. [42]. This becomes evident when setting $\xi(t, t_0) \to 0$, which results in a separable state. According to the definition of $\xi(t, t_0)$ provided in Eq. (3.101), this condition is met when $\chi_q(t, t_0) \gg 1$, that is, when the initial and final states before and after the interaction with the atomic media become completely distinguishable. This parameter can be regulated by controlling the gas density in the interaction region, as we illustrate later in this section.

Furthermore, the state described by Eq. (3.100) can be utilized for the generation of coherent state superpositions with a high-photon number. To accomplish this, one must project the harmonic modes onto the final coherent state in which they are found, that is, with respect to $\bigotimes_{q=2}^{q_c} |\chi_q(t, t_0)\rangle$. In simpler terms, measuring the intensity of the harmonic modes is necessary. This step yields

$$\begin{aligned} |\bar{\Phi}_{\text{IR}}(t, t_0)\rangle &= \hat{D}_{\mathbf{k}_L}(\alpha_{\mathbf{k}_L}) \Big[|\chi_{\mathbf{k}_L}(t, t_0)\rangle - \xi_{\text{IR}}(t, t_0) |\xi_{\text{HH}}(t, t_0)|^2 |0_{\mathbf{k}_L}\rangle \Big] \\ &\equiv \hat{D}_{\mathbf{k}_L}(\alpha_{\mathbf{k}_L}) \Big[|\chi_{\mathbf{k}_L}(t, t_0)\rangle - \tilde{\xi}(t, t_0) |0_{\mathbf{k}_L}\rangle \Big], \end{aligned} \tag{3.102}$$

where we denote $\xi_{\text{IR}}(t, t_0) \equiv \langle 0_{\mathbf{k}_L} | \chi_{\mathbf{k}_L}(t, t_0)\rangle$ and $\xi_{\text{HH}}(t, t_0) \equiv \prod_{q=2}^{q_c} \langle 0_{\mathbf{k}_q} | \chi_q(t, t_0)\rangle$. For the sake of simplicity, we represent $\tilde{\xi}(t, t_0)$ as the product of the former with the squared norm of the latter,[27] in other words, $\tilde{\xi}(t, t_0) \equiv \xi_{\text{IR}}(t, t_0) |\xi_{\text{HH}}(t, t_0)|^2$. This state generally represents an unbalanced coherent state superposition where the

[27] In many cases, the conditioning on HHG operation is approximated by $\hat{P}_{\tilde{0}} \approx \mathbb{1} - |\alpha_{\mathbf{k}_L}\rangle\langle\alpha_{\mathbf{k}_L}|$ applied to the single-mode state $|\alpha_{\mathbf{k}_L} + \chi_{\mathbf{k}_L}\rangle$, i.e., without considering the presence of harmonic modes. This approximation, valid for small intensities of the measured harmonic radiation, simplifies calculations specifically in those situations where we are only interested in the properties of the fundamental mode.

probability amplitudes depend on the magnitude of the coherent state displacements, as described by $\tilde{\xi}(t, t_0)$. Depending on the values of this parameter,[28] we can identify three regimes:

- In the case where $\tilde{\xi} \to 0$, the second term in Eq. (3.102) becomes negligible, and Eq. (3.102) simplifies to

$$|\bar{\Phi}_{\text{IR}}(t, t_0)\rangle \approx \hat{D}_{\mathbf{k}_L}(\alpha_{\mathbf{k}_L})|\chi_{\mathbf{k}_L}(t, t_0)\rangle, \tag{3.103}$$

  that is, a coherent state.
- In the case where $0 < \tilde{\xi} < 1$, we generally get an unbalanced coherent state superposition where one of the superposition terms dominates over the other.
- The scenario where $\tilde{\xi} \to 1$, i.e., when $\chi_{\mathbf{k}_L}(t, t_0) \ll 1$, can be readily analyzed by expanding the coherent state in Eq. (3.102) in the Fock basis as follows

$$\begin{aligned} |\bar{\Phi}_{\text{IR}}(t, t_0)\rangle &= \hat{D}_{\mathbf{k}_L}(\alpha_{\mathbf{k}_L})\left[|0_{\mathbf{k}_L}\rangle + \sum_{n=1}^{\infty} \frac{\chi^n_{\mathbf{k}_L}(t, t_0)}{\sqrt{n!}}|n_{\mathbf{k}_L}\rangle - \tilde{\xi}|0_{\mathbf{k}_L}\rangle\right] \\ &= \hat{D}_{\mathbf{k}_L}(\alpha_{\mathbf{k}_L})\left[(1 - \tilde{\xi})|0_{\mathbf{k}_L}\rangle + \sum_{n=1}^{\infty} \frac{\chi^n_{\mathbf{k}_L}(t, t_0)}{\sqrt{n!}}|n_{\mathbf{k}_L}\rangle\right], \end{aligned} \tag{3.104}$$

  and within the considered limit, we can express the state above as follows

$$|\bar{\Phi}_{\text{IR}}(t, t_0)\rangle \approx \hat{D}_{\mathbf{k}_L}(\alpha_{\mathbf{k}_L})|1_{\mathbf{k}_L}\rangle + O\big(\chi^2_{\mathbf{k}_L}(t, t_0)\big), \tag{3.105}$$

  where the dominant term corresponds to a displaced Fock state.

From a practical perspective, implementing the conditioning operation and subsequently measuring the generated state, is performed similarly to what was illustrated in Fig. 3.10, albeit with some differences aimed at probing the properties of the resulting state. The experimental scheme is depicted in Fig. 3.19, where, like in Fig. 3.10, the intense laser field is initially split into two branches, denoted as IR_t and IR_r. The former is directed toward the atomic gas jet, while the latter serves as a reference. After the HHG interaction region within the Quantum Spectrometer section (highlighted with the blue chart), the generated harmonics and the fundamental are separated, with the former being detected by the PD_{HH} photodetector. Meanwhile, the fundamental mode is further divided by the BS_3 beam splitter, which reflects 10% of the intensity. The reflected beam is then measured by the PD_{out} photodetector, while the transmitted beam is attenuated and directed into a homodyne detection setup. Here, it interferes with the IR_r beam. The trajectory of this second beam, since its origin, is different to the one of IR_t: first, it undergoes a controlled phase shift φ, achieved by altering the optical path length, and it undergoes a division where a small portion of the generated field (IR_0) is detected by the PD_0. This is used to

[28] In the following, we represent $\tilde{\xi} \equiv \tilde{\xi}(t, t_0)$.

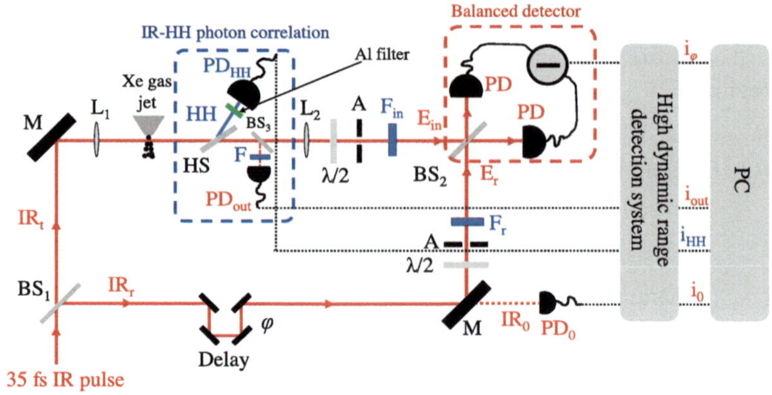

Fig. 3.19 **Experimental setup for the conditioning on HHG operation**. In this setup, the various components are labeled as follows: BS symbols represent beam splitters (BS_1 with an 80:20 transmission-reflection ratio, while BS_2 a 50:50 ratio and BS_3 a 90:10 ratio, approximately), IR symbols denote the beams with subscripts indicating whether they are reflected or transmitted, M depicts mirrors, L signifies focusing lenses, HS represents harmonic separators that reflect the harmonic modes and allow the infrared (IR) to pass through, PD stands for detectors, Al denotes an aluminum filter, A represents an aperture, F denotes neutral density filters, φ is the controlled phase shift added to the local oscillator, and "i" indicates the measured photocurrent values. The blue box highlights the Quantum Spectrometer layout as depicted in Fig. 3.10a, while the red box represents a homodyne detection system

measure the shot-to-shot energy fluctuations of the driving field. The other portion, is used as a local oscillator in the homodyne detection scheme.

While the outcomes recorded by PD_{HH}, PD_{out} and PD_0 are used for executing the conditioning operation, similar to the case in Fig. 3.10, the homodyne detection system is employed to reconstruct the Wigner function of the resulting state, as outlined in Fig. 2.6. More precisely, the Wigner function is computed for the points lying along the anticorrelation diagonal of Fig. 3.10c. By doing so, we obtain the experimental Wigner function presented in Fig. 3.20c, denoted as $W_{\text{exp}}(\tilde{\beta}_{\mathbf{k}_L})$. This experimental Wigner function, presents an asymmetric vulcano-like shape with a minimum value of $\min_{\tilde{\beta}_{\mathbf{k}_L}} W_{\text{exp}}(\tilde{\beta}_{\mathbf{k}_L}) = -0.050 \pm 0.002$. This asymmetricity is due to the unbalanced nature of the superposition, with the maximum value of the Wigner distribution centered around $|\chi_{\mathbf{k}_L}(t, t_0)| \equiv |\chi_{\mathbf{k}_L}|$, in this case estimated to be 1.2 approximately. To validate the experimental implementation of the homodyne detection scheme, we show, in panel (a), the Wigner function of the final state of the field in the absence of interactions. Consequently, no conditioning operation is performed. In this case, there is no laser-matter interaction, and the coherent state remains unchanged, thus yielding to a Gaussian-like Wigner function.

Conversely, in Fig. 3.20b and d the theoretical Wigner function, denoted as $W_{\text{th}}(\tilde{\beta}_{\mathbf{k}_L})$, is shown for the two cases just outlined. Specifically, for the second case corresponding to that of Eq. (3.102), the Wigner function is given as follows

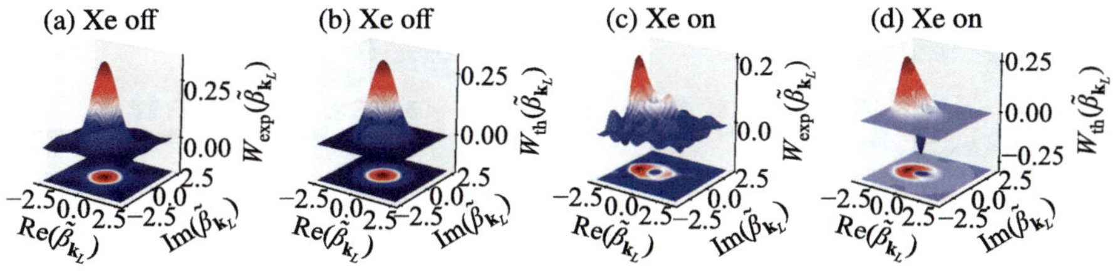

Fig. 3.20 Experimental and theoretical Wigner functions after the conditioning on HHG operation. The experimental Wigner functions are denoted as $W_{\exp}(\tilde{\beta}_{\mathbf{k}_L})$ (panels (**a**) and (**c**)) while the theoretical Wigner functions are labeled as $W_{\mathrm{th}}(\tilde{\beta}_{\mathbf{k}_L})$ (panels (**b**) and (**d**)). In panels **a** and **b**, no conditioning on HHG operation was considered, corresponding to the case where the atomic gas jet, composed of xenon atoms, does not introduce atoms in the interaction region. In contrast, in panels **c** and **d**, this operation has been implemented, resulting in $|\chi_{\mathbf{k}_L}| \approx 1.2$

$$
\begin{aligned}
W(\beta_{\mathbf{k}_L}) &= \frac{2}{\mathcal{N}_{\mathrm{IR}}\pi}\Big[\langle\bar{\Phi}_{\mathrm{IR}}(t,t_0)|\hat{D}_{\mathbf{k}_L}(\alpha_{\mathbf{k}_L})\Big]\hat{D}_{\mathbf{k}_L}(\tilde{\beta}_{\mathbf{k}_L})\hat{\Pi}\hat{D}_{\mathbf{k}_L}(\tilde{\beta}_{\mathbf{k}_L})\Big[\hat{D}_{\mathbf{k}_L}(-\alpha_{\mathbf{k}_L})|\bar{\Phi}_{\mathrm{IR}}(t,t_0)\rangle\Big] \\
&= \frac{2}{\mathcal{N}_{\mathrm{IR}}\pi}\Bigg[e^{-2|\tilde{\beta}_{\mathbf{k}_L}-\chi_{\mathbf{k}_L}(t,t_0)|^2} + e^{-|\chi_{\mathbf{k}_L}(t,t_0)|^2}e^{-2|\tilde{\beta}_{\mathbf{k}_L}|^2} \\
&\qquad - \Big(e^{2\tilde{\beta}_{\mathbf{k}_L}\chi^*_{\mathbf{k}_L}(t,t_0)} + e^{2\tilde{\beta}^*_{\mathbf{k}_L}\chi_{\mathbf{k}_L}(t,t_0)}\Big)e^{-|\chi_{\mathbf{k}_L}|^2}e^{-2|\tilde{\beta}_{\mathbf{k}_L}|^2}\Bigg],
\end{aligned}
\tag{3.106}
$$

which for the case of $|\chi_{\mathbf{k}_L}| = 1.2$ considered in panel (d), this theoretical calculation closely resembles the overall shape of the measured Wigner function, although it yields a minimum value of $\min_{\tilde{\beta}_{\mathbf{k}_L}}[W_{\mathrm{th}}(\tilde{\beta}_{\mathbf{k}_L})] = -0.188$. This result is in contrast with the value of -0.050 ± 0.002 obtained for the experimental case. It is worth noting that the value of $|\chi_{\mathbf{k}_L}| = 1.2$ was chosen such that the maximum peaks of both the experimental result and the theoretical prediction coincide.

One of the main advantages about the generated superpositions is their tunability, which allows for the adjustment of their shape by controlling various experimental parameters. For instance, it is possible to tune the value of $|\chi_{\mathbf{k}_L}|$, thus enabling the generated coherent state superposition to span a range from a displaced Fock state to the larger unbalanced coherent state superposition in Eq. (3.102). This tunability is illustrated in Fig. 3.21, where the experimentally measured Wigner functions (in panels (a) and (c)) are shown alongside their corresponding theoretical predictions (in panels (b) and (d)).

From a practical perspective, the tuning of the generated coherent state superposition was achieved by varying the number of atoms N_{at} in the interaction region, which was controlled experimentally by adjusting the delay between the pulse arrival at the gas jet and the opening of the Xe gas nozzle. Using this approach, we get $|\chi_{\mathbf{k}_L}| \approx 0.5$ for Fig. 3.21a and b, while $|\chi_{\mathbf{k}_L}| \approx 0.1$ for panels (c) and (d). Although the theoretical predictions closely resemble the overall shape of the measured Wigner function, as observed in Fig. 3.20 there are still discrepancies in the minimum values of the Wigner function between them. Specifically, in panels (a) and (c) we observe $\min_{\tilde{\beta}_{\mathbf{k}_L}}[W_{\exp}(\tilde{\beta}_{\mathbf{k}_L})] = -0.050 \pm 0.002$ and $\min_{\tilde{\beta}_{\mathbf{k}_L}}[W_{\exp}(\tilde{\beta}_{\mathbf{k}_L})] = -0.052 \pm 0.002$,

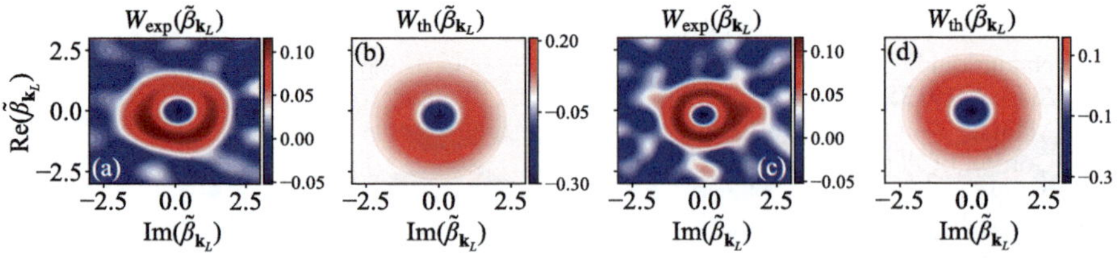

Fig. 3.21 Experimental and theoretical Wigner functions after the conditioning on HHG operation for varying atomic densities. The experimental Wigner functions in panels **a** and **c** have been obtained by adjusting the atomic density in the interaction region, as described in the text. Similarly, the same effect has been replicated in the theoretical plots in **b** and **d** by varying N_{at} in Eq. (3.98). The resulting values of $|\chi_{\mathbf{k}_L}|$ for these cases are $|\chi_{\mathbf{k}_L}| \approx 0.5$ for **a** and **b**, while $|\chi_{\mathbf{k}_L}| \approx 0.1$ for **c** and **d**

respectively, while for their theoretical predictions we find $\min_{\tilde{\beta}_{\mathbf{k}_L}}[W_{\text{th}}(\tilde{\beta}_{\mathbf{k}_L})] = -0.292$ and $\min_{\tilde{\beta}_{\mathbf{k}_L}}[W_{\text{th}}(\tilde{\beta}_{\mathbf{k}_L})] = -0.317$.

3.3.2.2 About the Reduced Negativities in the Measured Wigner Functions

In the previous subsection, we defined the conditioning on HHG operation from a theoretical perspective and presented how it can be conducted experimentally. As a result of this approach, high-photon number coherent state superpositions were generated and their experimentally reconstructed Wigner functions, alongside their theoretical predictions, were presented. Nevertheless, discrepancies were observed between them, most notably concerning their negative regions. On one hand, the reduction in the absolute value of the measured Wigner function negativities could be attributed to noise contributions from the measurement devices. On the other hand, it might also arise, in part, due to effective interactions with a generic environment. These could encompass decoherence effects and the neglect of correlations with other systems, such as between the different field modes.[29]

The main aim of this subsection is to investigate how decoherence accounts for the diminished values of the Wigner function's minimum. To address this, we consider a simple noise mode where the state in Eq. (3.102) interacts with a Gaussian environment, the effect of which can be modeled as that of an ideal beam splitter. In one mode, we input an ancillary vacuum state, representing an ideal environment, denoted as $|0_{\text{anc}}\rangle$, while in the other, we have the state under consideration [165]. According to this model, the state after the interaction can be expressed as

$$\hat{\rho} = \text{tr}_{\text{anc}}\Big[\hat{B}(\theta)|\bar{\Phi}_{\text{IR}}(t, t_0)\rangle\langle\bar{\Phi}_{\text{IR}}(t, t_0)| \otimes |0_{\text{anc}}\rangle\langle 0_{\text{anc}}|\,\hat{B}^{\dagger}(\theta)\Big], \tag{3.107}$$

[29] This aspect is emphasized in the following chapter when analyzing the interaction with diatomic molecules.

where the trace is taken over the ancillary modes. The unitary transformation describing the beam splitter is given by $\hat{B}(\theta) = \exp[\theta(\hat{a}_{\mathbf{k}_L}\hat{a}^\dagger_{\text{anc}} - \hat{a}^\dagger_{\mathbf{k}_L}\hat{a}_{\text{anc}})]$, which for the case of acting on two arbitrary coherent states $|\alpha\rangle$ and $|\beta\rangle$, we have $\hat{B}(\theta)\,|\alpha\rangle \otimes |\beta\rangle = |\alpha\cos(\theta) + \beta\sin(\theta)\rangle \otimes |-\alpha\sin(\theta) + \beta\cos(\theta)\rangle$. In the operator $\hat{B}(\theta)$, $\hat{a}_{\text{anc}}$ ($\hat{a}^\dagger_{\text{anc}}$) denotes the annihilation (creation) operator acting on the ancillary mode, while the angle θ is related to the transmission efficiency η_t as $\eta_t = cos^2(\theta)$.

After incorporating this Gaussian noise, we obtain for the resulting density matrix

$$
\begin{aligned}
\hat{\rho} = \frac{1}{\mathcal{N}}\Big[&|(\alpha_{\mathbf{k}_L} + \chi_{\mathbf{k}_L})\cos(\theta)\rangle\langle(\alpha_{\mathbf{k}_L} + \chi_{\mathbf{k}_L})\cos(\theta)| + \xi^2|\alpha_{\mathbf{k}_L}\cos(\theta)\rangle\langle\alpha_{\mathbf{k}_L}\cos(\theta)| \\
&- \xi e^{i\text{Im}(\alpha_{\mathbf{k}_L}\chi_{\mathbf{k}_L})}\langle\alpha_{\mathbf{k}_L}\sin(\theta)|(\alpha_{\mathbf{k}_L} + \chi_{\mathbf{k}_L})\sin(\theta)\rangle|(\alpha_{\mathbf{k}_L} + \chi_{\mathbf{k}_L})\cos(\theta)\rangle\langle\alpha_{\mathbf{k}_L}\cos(\theta)| \\
&- \xi e^{-i\text{Im}(\alpha_{\mathbf{k}_L}\chi_{\mathbf{k}_L})}\langle(\alpha_{\mathbf{k}_L} + \chi_{\mathbf{k}_L})\sin(\theta)|\alpha_{\mathbf{k}_L}\sin(\theta)\rangle|\alpha_{\mathbf{k}_L}\cos(\theta)\rangle\langle(\alpha_{\mathbf{k}_L} + \chi_{\mathbf{k}_L})\cos(\theta)|\Big],
\end{aligned}
\tag{3.108}
$$

where $\mathcal{N}$ is a normalization constant, and we have written $\chi_{\mathbf{k}_L} \equiv \chi_{\mathbf{k}_L}(t, t_0)$ for brevity. We observe that in the case of perfect transmission, i.e. $\eta_t = 1$ or equivalently $\theta = 0$, this last expression recovers the pure state in Eq. (3.102). In contrast, in the case where all the amplitude has been transferred to the ancillary modes, i.e. $\eta_t = 0$ or equivalently $\theta = \pi/2$, it leads to a vacuum state.

In Fig. 3.22, we present the Wigner functions obtained from Eq. (3.108). Specifically, we considered two different values of $\chi_{\mathbf{k}_L}$: $|\chi_{\mathbf{k}_L}| = 1.2$ (upper row) and

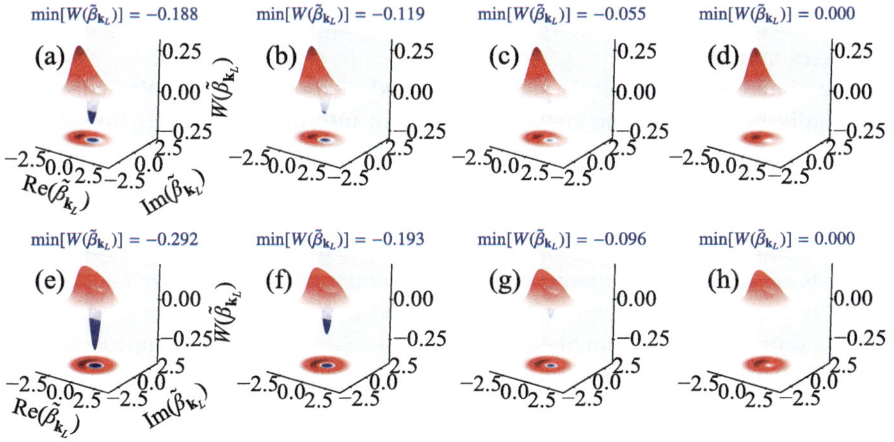

Fig. 3.22 Decoherence effects on the generated Wigner functions. The theoretically computed Wigner functions in this figure are the result of the interaction of the conditioned on HHG states, as described in Eq. (3.102), with a Gaussian environment. In the first row (panels **(a)**–**(d)**), we set $|\chi_{\mathbf{k}_L}| = 1.2$, while in the second row (panels **(e)**–**(h)**), we set $|\chi_{\mathbf{k}_L}| = 0.5$. In each column, a different value of the effective transmission efficiency η_t is considered: $\eta_t = 1.0$ for the first column, $\eta_t = 0.83$ for the second column, $\eta_t = 0.67$ for the third column and $\eta_t = 0.5$ for the fourth column. The minimum value of the Wigner function is written in blue at the top of each panel

$|\chi_{\mathbf{k}_L}| = 0.5$ (lower row), along with four different values of η_t, in particular $\eta_t \in \{1.00, 0.83, 0.67, 0.5\}$ from the leftmost plot to the rightmost one.

As observed in this figure, for the considered values of η_t, decreasing values of the transmission efficiency maintain the same vulcano-shape structure of the Wigner function as obtained for the perfect transmission case, while reducing the Wigner function negativity. More specifically, this reduction follows a linear trend with the transmission efficiency, although the rate of reduction is smaller for increasing[30] $|\chi_{\mathbf{k}_L}|$. Therefore, the primary takeaway from this noise analysis is that the presence of entanglement with additional degrees of freedom, in this case the ancillary modes, could potentially explain the reduced Wigner negativities, along other factors related to experimental noise.

3.3.2.3 Applications of the Obtained Coherent State Superpositions I: Generating Coherent State Superpositions with Three Different Elements

In the following subsections, we explore different applications of the generated coherent state superpositions, which can also be considered as potential approaches for further benchmarking the experimental techniques. In this specific subsection we study how these states can be used in conjunction with other linear optical elements to design coherent state superpositions composed of three states, as well as larger coherent state superpositions than the initial ones in Sect. 3.3.2.4. These two first subsections are based on our work in Ref. [40]. Additionally, in Sect. 3.3.2.5 which is based on our work in Ref. [46], we investigate their potential for driving nonlinear processes in matter.

The generation of coherent states composed of more than two elements could potentially be useful in the context of quantum information science. Specifically, they have shown utility in the violation of Bell inequalities within the continuous variable regime [166], as well as in quantum teleportation protocols [167]. In this subsection, we explore two methods that allow for the generation of high-photon number coherent state superpositions consisting of three different elements. These methods are based on the conditioning on HHG operation presented in Sect. 3.3.2.1. These two approaches are visually summarized in Fig. 3.23, and are based on the combination of the Quantum Spectrometer (QS) method with linear optical elements (panel (a)), or with other QS approaches that involve additional conditioning on HHG operations (panel (b)).

Linear-optics-based scheme. First, we place our focus on the scheme presented in Fig. 3.23a, where an initial coherent state of amplitude $\sqrt{2}\alpha_{\mathbf{k}_L}$ is split into two different paths by a 50:50 beam splitter. This results in the product state $|\alpha_{\mathbf{k}_L}\rangle \otimes |\alpha_{\mathbf{k}_L}\rangle$, with each mode representing different spatial modes. These modes propagate

[30] Specifically, in the limit where $|\chi_{\mathbf{k}_L}|$ is large enough for $\tilde{\xi} \approx 0$, the reduction rate mentioned here is zero.

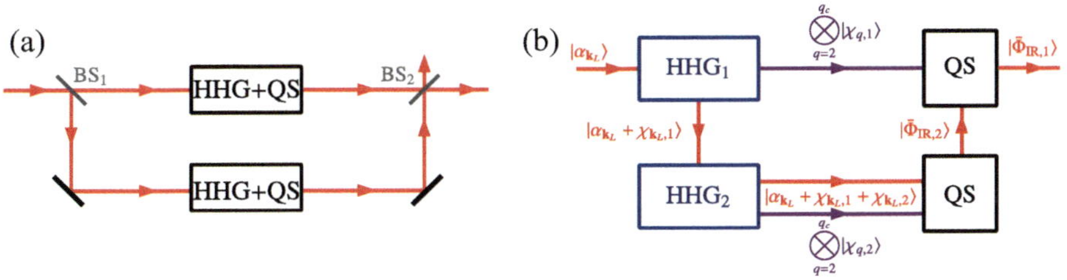

Fig. 3.23 Schemes for generating triple coherent state superpositions based on the conditioning on HHG operation. In **a**, a setup utilizing linear optical elements mixing the conditioned on HHG state generated independently in each branch of the interferometer. In **b**, a setup based on combining sequential conditioning on HHG approaches, where the HHG boxes represent a harmonic generation media, while the QS boxes implement the Quantum Spectrometer approach detailed in Fig. 3.10

independently and reach two independent atomic gas jets, where the HHG process takes place. By performing the conditioning on HHG operation separately for each mode, we obtain for the joint quantum optical state

$$|\Phi_{\mathrm{IR},1}\rangle = \frac{1}{\sqrt{\mathcal{N}}}\Big[|\alpha_{\mathbf{k}_L} + \chi_{\mathbf{k}_L}\rangle - \tilde{\xi}\,|\alpha_{\mathbf{k}_L}\rangle\Big] \otimes \Big[|\alpha_{\mathbf{k}_L} + \chi_{\mathbf{k}_L}\rangle - \tilde{\xi}\,|\alpha_{\mathbf{k}_L}\rangle\Big], \tag{3.109}$$

where we have imposed that both atomic gas jets are subjected to the same conditions, resulting in identical HHG processes in both optical paths and, consequently, equal values for the generated shifts $\chi_{\mathbf{k}_L}$. Furthermore, we have assumed that $\alpha_{\mathbf{k}_L}, \chi_{\mathbf{k}_L} \in \mathbb{R}$, as this is something that can be controlled by adjusting the laser source's CEP, which we assume to be phase-locked. It is important to note that if the applied laser field is sufficiently long, the magnitude of $\chi_{\mathbf{k}_L}$ remains constant regardless of the chosen CEP, which only determines its phase. However, for shorter pulses where the CEP strongly influences the pulse shape, the magnitude of $\chi_{\mathbf{k}_L}$ may vary depending on the specific CEP value.

Finally, we carefully choose the optical paths for the two optical modes to ensure that, upon reaching the entrance of the second 50:50 beam splitter (BS$_2$), both modes have acquired the same overall phase. Under these conditions, we can express the final quantum optical state after the BS$_2$ as follows

$$\begin{aligned} |\bar{\Phi}_{\mathrm{IR},2}\rangle = \frac{1}{\sqrt{\mathcal{N}}}\Big[&|0_{\mathbf{k}_L}\rangle \otimes \Big(\Big|\sqrt{2}(\alpha_{\mathbf{k}_L} + \chi_{\mathbf{k}_L})\Big\rangle + \bar{\xi}^2\,\Big|\sqrt{2}\alpha_{\mathbf{k}_L}\Big\rangle\Big) \\ &- \bar{\xi}\Big(\Big|\tfrac{\chi_{\mathbf{k}_L}}{\sqrt{2}}\Big\rangle + \Big|-\tfrac{\chi_{\mathbf{k}_L}}{\sqrt{2}}\Big\rangle\Big) \otimes \Big|\sqrt{2}\alpha_{\mathbf{k}_L} + \tfrac{\chi_{\mathbf{k}_L}}{\sqrt{2}}\Big\rangle\Big], \end{aligned} \tag{3.110}$$

which leads to an entangled state between the two optical modes. This entangled state enables us to readily obtain unbalanced high-photon number coherent state superpositions composed of three distinct optical modes. To illustrate this, let us examine the following POVM set acting on the leftmost optical mode

$$\big\{|0_{\mathbf{k}_L}\rangle\langle 0_{\mathbf{k}_L}|, \mathbb{1} - |0_{\mathbf{k}_L}\rangle\langle 0_{\mathbf{k}_L}|\big\}, \tag{3.111}$$

such that the first POVM element corresponds to a situation where destructive interference occurs in the first mode, resulting in no photons detected at one of the BS_2 output modes. The second element, on the other hand, identifies with the case where IR radiation is indeed detected. In essence, this POVM set serves as a heralding operation triggered by the detection of radiation at one of the beam splitter's outputs. Applying each of these projective operations to the state in Eq. (3.110) yields, up to a normalization constant, the following state at the other mode

$$\begin{aligned}|\phi_0\rangle &= \big(|0_{\mathbf{k}_L}\rangle\langle 0_{\mathbf{k}_L}|\big)|\bar{\Phi}_{\text{IR},2}\rangle \\ &= \frac{1}{\sqrt{\mathcal{N}}}\Big[\Big|\sqrt{2}(\alpha_{\mathbf{k}_L}+\chi_{\mathbf{k}_L})\Big\rangle + \bar{\xi}^2\Big|\sqrt{2}\alpha_{\mathbf{k}_L}\Big\rangle - 2\bar{\xi}\,\langle 0_{\mathbf{k}_L}|\tfrac{\xi\chi_{\mathbf{k}_L}}{2}\rangle\Big|\sqrt{2}\alpha_{\mathbf{k}_L}+\tfrac{\chi_{\mathbf{k}_L}}{\sqrt{2}}\Big\rangle\Big],\end{aligned} \tag{3.112}$$

when considering the projection onto a vacuum state, while for the other projector we obtain

$$|\phi_1\rangle = \big(\mathbb{1} - |0_{\mathbf{k}_L}\rangle\langle 0_{\mathbf{k}_L}|\big)|\bar{\Phi}_{\text{IR},2}\rangle = \frac{1}{\sqrt{\mathcal{N}}}\Big|\sqrt{2}\alpha_{\mathbf{k}_L}+\tfrac{\chi_{\mathbf{k}_L}}{\sqrt{2}}\Big\rangle. \tag{3.113}$$

Thus, we observe that while the first operation provides us with the desired coherent state superposition, the second leads to a shifted coherent state. Furthermore, it is important to remark that the shape of the generated coherent state superposition, specifically $|\phi_0\rangle$, is strongly influenced by the amplitude of the HHG-induced shift $|\chi_{\mathbf{k}_L}|$. Similar to what was found after the standard conditioning on HHG operation presented in Sect. 3.3.2.1, in the regime where $|\chi_{\mathbf{k}_L}| \ll 1$, the state in Eq. (3.112) can be well approximated by a displaced Fock state. On the other hand, in the opposite regime, when $|\chi_{\mathbf{k}_L}| \gg 1$, the probability amplitudes $\bar{\xi}$ and $\langle 0_{\mathbf{k}_L}|\chi_{\mathbf{k}_L}/\sqrt{2}\rangle$ tend to zero, and we obtain a coherent state of amplitude $\sqrt{2}(\alpha_{\mathbf{k}_L}+\chi_{\mathbf{k}_L})$.

These different regimes are presented in Fig. 3.24a–f, where the Wigner function of $|\phi_0\rangle$ in Eq. (3.112) is shown for different values of $|\chi_{\mathbf{k}_L}|$. As described earlier, for the smallest value of $|\chi_{\mathbf{k}_L}|$, we recover the Wigner function of a displaced Fock state (see panel (a)). As this quantity becomes larger, the resulting Wigner function tends to the Gaussian shape expected for coherent states (see panel (f)). In between, we get sets of triple coherent state superpositions where the probability amplitude varies along with $|\chi_{\mathbf{k}_L}|$. Specifically, for $|\chi_{\mathbf{k}_L}| \in [0.2, 0.8)$, the displaced vacuum state $\hat{D}_{\mathbf{k}_L}(\sqrt{2}\alpha_{\mathbf{k}_L})|0_{\mathbf{k}_L}\rangle$ in Eq. (3.112) dominates the superposition, represented by the leftmost peak in panels (b) to (f), while for the range $|\chi_{\mathbf{k}_L}| \in [0.85, \infty)$, the first coherent state in the superposition becomes more important, leading to the rightmost peak in the aforementioned plots. In between, the contribution of the last term in the

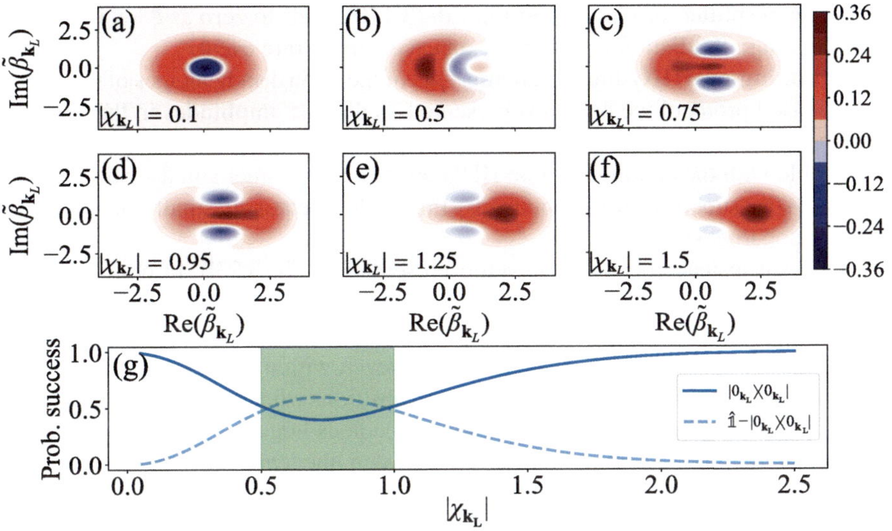

Fig. 3.24 Results for the linear-optics-based scheme for triple coherent state superpositions. In panels **a**–**f**, we present the Wigner functions obtained when implementing the linear-optics-based scheme for generating triple coherent state superpositions. Different values of $|\chi_{\mathbf{k}_L}|$ have been considered, ranging from 0.1 up to 1.5 in increasing order from **a** to **f**, indicated at the bottom of each panel. In **g**, the success probability of performing each of the measurements indicated in the POVM set in Eq. (3.111) is shown as a function of $|\chi_{\mathbf{k}_L}|$. The green dashed region represents the range of values for which the three coherent states in the superposition provide significant contribution to the state, while being substantially different

superposition becomes significant, playing a fundamental role in the observation of Wigner negativities due to their negative probability amplitude.[31]

In order for the heralding measurement presented in Eq. (3.111) to be useful, it is important for it to have a non-negligible probability of success, specifically for the $|0_{\mathbf{k}_L}\rangle\langle 0_{\mathbf{k}_L}|$ operation. This probability is computed as

$$P_{\text{success}}(\chi_{\mathbf{k}_L}) = \mathrm{tr}\big[\big(|0_{\mathbf{k}_L}\rangle\langle 0_{\mathbf{k}_L}| \otimes \hat{\mathbb{1}}\big)|\bar{\Phi}_{\text{IR},2}\rangle\langle\bar{\Phi}_{\text{IR},2}|\big], \tag{3.114}$$

and is shown in Fig. 3.24g as a function of $|\chi_{\mathbf{k}_L}|$ with the solid blue curve, while the dashed curve presents the probability of success for the other projective operator. In this plot, the green shaded region highlights the range in which the three coherent states in the superposition contribute significantly. While within this range, the probability of success goes through a minimum, the lowest value for this probability is $\min_{\chi_{\mathbf{k}_L}}[P_{\text{success}}(\chi_{\mathbf{k}_L})] = 0.4$ which is, nonetheless, far from negligible. For the limiting cases when $|\chi_{\mathbf{k}_L}| \ll 1$ and $|\chi_{\mathbf{k}_L}| \gg 1$, the probability in Eq. (3.114) tends to one. In these two cases, either almost no radiation goes through the heralded mode

[31] It can be shown that, by switching off the contribution of this last term, no negative regions are observed for the resulting Wigner functions.

of the beam splitter, or the last term in Eq. (3.110) tends to zero as $\bar{\xi} \to 0$, making the heralding of radiation in this mode an exceedingly rare event.

Conditioning-on-HHG-based scheme. One of the main disadvantages of the linear-optics-based protocol we have just presented, is that the amplitudes in the coherent state superposition differ by a factor of $\chi_{\mathbf{k}_L}/\sqrt{2}$. Considering that the values of $\chi_{\mathbf{k}_L}$ achievable with the conditioning on HHG operation are rather small—otherwise no coherent state superposition is obtained—this leads to relatively small superpositions of three coherent states.

Here, we instead focus on the setup presented in Fig. 3.23b. In contrast to the linear-optics-based scheme, the one considered here cannot be extended to other coherent state superposition generation methods, as it fundamentally depends on the conditioning on the HHG mechanism. However, within this context, this specific setup proves useful for the generation of superpositions where the difference between the corresponding coherent state amplitudes becomes larger.

One of the fundamental differences between the two schemes presented in Fig. 3.23 is that in the one depicted in panel (b), the HHG process is consistently excited by the same optical mode. Therefore, if we consider the initial coherent state of the applied field as $|\alpha_{\mathbf{k}_L}\rangle$, after the two different harmonic generation media denoted in the figure as HHG_1 and HHG_2, the resulting quantum optical state is given by

$$|\bar{\Phi}_{\text{g}}\rangle = |\alpha_{\mathbf{k}_L} + \chi_{\mathbf{k}_L,1} + \chi_{\mathbf{k}_L,2}\rangle \bigotimes_{q=2}^{q_c} \big[|\chi_{q,1}\rangle \otimes |\chi_{q,2}\rangle\big], \tag{3.115}$$

accordingly to Eq. (3.96). In this expression, $\chi_{q,1}$ and $\chi_{q,2}$ represent the coherent state amplitudes of the harmonic modes after interaction with the HHG_1 and HHG_2 regions, defined on spatially different optical modes. For all harmonic modes q, these two coherent state amplitudes are generally different, as HHG_1 was excited first with a coherent state of amplitude $\alpha_{\mathbf{k}_L}$ whereas HHG_2, which comes after, was excited with a coherent state of amplitude $\alpha_{\mathbf{k}_L} + \chi_{\mathbf{k}_L,1}$. Furthermore, the generated harmonic modes are uncorrelated among themselves but correlated to the fundamental mode through the operator $\hat{b}_j^\dagger(t) \propto \hat{a}_{\mathbf{k}_L}^\dagger e^{i\omega_{\mathbf{k}_L} t} + \sum_{q=2}^{q_c} \hat{a}_{\mathbf{k}_q,j}^\dagger e^{i\omega_{\mathbf{k}_q} t}$, with $j = 1, 2$, which defines the collective excitations allowing for the conditioning on HHG operation, as outlined in Sect. 3.3.2.1. Thus, the main implication this has is that, if we now apply the conditioning on HHG operation using the harmonic modes $\{\chi_{q,2}\}_{q=2}^{q_c}$, the projector acting on the state is given by

$$\hat{P}_{\tilde{0},2} = \hat{D}_{\mathbf{k}_L}(\alpha_{\mathbf{k}_L} + \chi_{\mathbf{k}_L,1})|0_{\mathbf{k}_L}\rangle\langle 0_{\mathbf{k}_L}| \bigotimes_{q=2}^{q_c} \big[|\chi_{q,1}\rangle\langle\chi_{q,1}| \otimes |0_{q,2}\rangle\langle 0_{q,2}|\big]\hat{D}_{\mathbf{k}_L}^\dagger(\alpha_{\mathbf{k}_L} + \chi_{\mathbf{k}_L,1}), \tag{3.116}$$

and according to Eq. (3.102), after measuring the high-harmonics obtained from the second excitation process, we arrive at the following separable state

$$|\bar{\Phi}_{\text{IR},2}\rangle = \frac{1}{\sqrt{\mathcal{N}_2}}\Big[|\alpha_{\mathbf{k}_L} + \chi_{\mathbf{k}_L,1} + \chi_{\mathbf{k}_L,2}\rangle - \bar{\xi}_2|\alpha_{\mathbf{k}_L} + \chi_{\mathbf{k}_L,1}\rangle\Big]\bigotimes_{q=2}^{q_c}|\chi_{q,1}\rangle, \tag{3.117}$$

where $\mathcal{N}_2$ is a normalization constant and

$$\bar{\xi}_2 \equiv \langle\alpha_{\mathbf{k}_L} + \chi_{\mathbf{k}_L,1}|\alpha_{\mathbf{k}_L} + \chi_{\mathbf{k}_L,1} + \chi_{\mathbf{k}_L,2}\rangle\prod_{q=2}^{q_c}|\langle 0_{q,2}|\chi_{q,2}\rangle|^2. \tag{3.118}$$

Using the harmonics $\{\chi_{q,1}\}_{q=2}^{q_c}$ generated in HHG$_1$, it is possible to implement a second conditioning on HHG operation governed by the projector

$$\hat{P}_{\tilde{0},1} = \hat{D}_{\mathbf{k}_L}(\alpha_{\mathbf{k}_L})|0_{\mathbf{k}_L}\rangle\langle 0_{\mathbf{k}_L}|\bigotimes_{q=2}^{q_c}|0_{q,2}\rangle\langle 0_{q,2}|\hat{D}^{\dagger}_{\mathbf{k}_L}(\alpha_{\mathbf{k}_L}), \tag{3.119}$$

which, applied on Eq. (3.117) leads to

$$|\bar{\Phi}_{\text{IR},2}\rangle = \frac{1}{\sqrt{\mathcal{N}_1}}\Big[|\alpha_{\mathbf{k}_L} + \chi_{\mathbf{k}_L,1} + \chi_{\mathbf{k}_L,2}\rangle - \bar{\xi}_2|\alpha_{\mathbf{k}_L} + \chi_{\mathbf{k}_L,1}\rangle - \bar{\xi}_1|\alpha_{\mathbf{k}_L}\rangle\Big], \tag{3.120}$$

where $\mathcal{N}_1$ is the new normalization constant of the state after projection, and $\bar{\xi}_1 \equiv [\langle\alpha_{\mathbf{k}_L}|\alpha_{\mathbf{k}_L} + \chi_{\mathbf{k}_L,1} + \chi_{\mathbf{k}_L,2}\rangle - \bar{\xi}_2\langle\alpha_{\mathbf{k}_L}|\alpha_{\mathbf{k}_L} + \chi_{\mathbf{k}_L,1}\rangle]\prod_{q=2}^{q_c}|\langle 0_{q,1}|\chi_{q,1}\rangle|^2$.

The state found after all these operations is indeed a coherent state superposition, where the difference in amplitude between each of the coherent states in the superposition is $\chi_{\mathbf{k}_L,1}$ and $\chi_{\mathbf{k}_L,2}$. Assuming that the IR field mode is not significantly affected by the interaction in HHG$_1$, and therefore by $\chi_{\mathbf{k}_L,1}$, one can assume that $\chi_{\mathbf{k}_L,1}$ and $\chi_{\mathbf{k}_L,2}$ are approximately the same, i.e., $\chi_{\mathbf{k}_L,1} \approx \chi_{\mathbf{k}_L,2} \equiv \chi_{\mathbf{k}_L}$. Within this assumption, we observe that the amplitude difference between the coherent states in the superposition is $\chi_{\mathbf{k}_L}$, which is a factor $\sqrt{2}$ larger than that obtained by the linear-optics-based scheme depicted earlier.

However, one of the main drawbacks of this approach is that the $\bar{\xi}_1$ might be very small for reduced values of $\chi_{\mathbf{k}_L}$, effectively turning Eq. (3.120) into a similar coherent state superposition to that of Eq. (3.102). This aspect can be circumvented by varying the mean photon number of the generated superposition state between Eqs. (3.116) and Eq. (3.119), that is, over Eq. (3.117). Specifically, we consider the introduction of a displacement operator $\hat{D}_{\mathbf{k}_L}(-3\chi_{\mathbf{k}_L})$, which can be achieved using beam splitters and coherent states [168]. Applying this operator onto Eq. (3.117) results in

$$\begin{aligned}|\tilde{\Phi}_{\text{IR},2}\rangle &= \hat{D}_{\mathbf{k}_L}(-3\chi_{\mathbf{k}_L})|\bar{\Phi}_{\text{IR},2}\rangle\\ &= \frac{1}{\sqrt{\mathcal{N}_2}}\Big[|\alpha_{\mathbf{k}_L} - \chi_{\mathbf{k}_L}\rangle - \bar{\xi}_2|\alpha_{\mathbf{k}_L} - 2\chi_{\mathbf{k}_L}\rangle\Big]\bigotimes_{q=2}^{q_c}|\chi_{q,1}\rangle,\end{aligned} \tag{3.121}$$

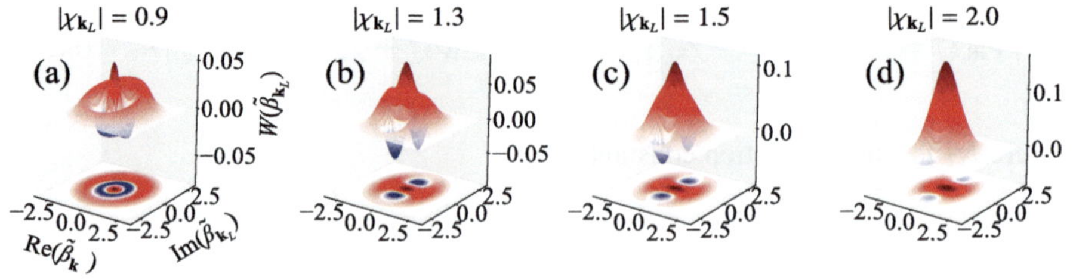

Fig. 3.25 Wigner functions obtained for the conditioning-on-HHG-based scheme. From **a** to **d**, the resulting Wigner functions are shown for different values of $|\chi_{\mathbf{k}_L}|$. More specifically, from left to right, we have $|\chi_{\mathbf{k}_L}| \in \{0.9, 1.3, 1.5, 2.0\}$

and applying then the $\hat{P}_{\tilde{0},1}$ operator in Eq. (3.120), we finally get

$$|\tilde{\Phi}_{\mathrm{IR},2}\rangle = \frac{1}{\sqrt{\tilde{\mathcal{N}}_1}}\Big[-\bar{\xi}'_1|\alpha_{\mathbf{k}_L}\rangle + |\alpha_{\mathbf{k}_L} - \chi_{\mathbf{k}_L}\rangle - \bar{\xi}_2|\alpha_{\mathbf{k}_L} - 2\chi_{\mathbf{k}_L}\rangle\Big], \tag{3.122}$$

where $\bar{\xi}'_1 \equiv \bar{\xi}_2(1 - \langle\alpha_{\mathbf{k}_L}|\alpha_{\mathbf{k}_L} - 2\chi_{\mathbf{k}_L}\rangle)\prod_{q=2}^{q_c}|\langle 0_{q,1}|\chi_{q,1}\rangle|^2$. Thus, the addition of this extra operation allows us to obtain a more balanced state since, as $\chi_{\mathbf{k}_L}$ grows larger, $\bar{\xi}'_1 \to \bar{\xi}_2$.

The Wigner function for the state in Eq. (3.122) is presented in Fig. 3.25 for different values of $|\chi_{\mathbf{k}_L}|$. As can be observed, for the smallest value of this parameter, shown in panel (a), the resulting Wigner function is not entirely symmetric around the origin due to the fact that $\bar{\xi}'_1 < \bar{\xi}_2$. Notably, for increasing values of $|\chi_{\mathbf{k}_L}|$ this asymmetry diminishes as $\bar{\xi}'_1 \to \bar{\xi}_2$. Thus, we observe that this effect occurs earlier than the vanishing of the first and last terms in Eq. (3.122) due to the evanescence of $\bar{\xi}'_1$ and $\bar{\xi}_2$. Moreover, it is remarkable that for $|\chi_{\mathbf{k}_L}| = 2.0$, small negative regions in the resulting Wigner functions are still present. Beyond this value, the resulting Wigner functions tend toward a Gaussian-like shape.

3.3.2.4 Applications of the Obtained Coherent State Superpositions II: Enlarging the Coherent State Superpositions

The methods we have described in the previous subsection allow us to obtain coherent state superpositions composed of three different coherent states. In this subsection, we draw inspiration from the approaches presented in Refs. [169, 170], where the combination of two optical cat states (see Sect. 2.3.3.4) in a beam splitter, along with the use of heralding operations, enables the construction of enlarged versions of the input cat states.

On a similar basis, we utilize the state in Eq. (3.122) as the input for a beam splitter with transmittance $T = \cos^2(\theta)$, in combination with the state $|(\alpha_{\mathbf{k}_L} - \chi_{\mathbf{k}_L})\tan(\theta)\rangle$. This leads to

$$
\begin{aligned}
|\tilde{\Phi}\rangle &= \hat{B}(\pi/4)|\bar{\Phi}_{\mathrm{IR},1}\rangle|\alpha_{\mathbf{k}_L} + \chi_{\mathbf{k}_L}\rangle \\
&= \frac{1}{\sqrt{\tilde{\mathcal{N}}}}\Big[-\tilde{\xi}_1' \,\big|\tilde{\alpha}_{\mathbf{k}_L} + \chi_{\mathbf{k}_L}\cos(\theta)\big\rangle \otimes \big|-\chi_{\mathbf{k}_L}\sin(\theta)\big\rangle + |\tilde{\alpha}_{\mathbf{k}_L}\rangle \otimes |0_{\mathbf{k}_L}\rangle \\
&\quad - \tilde{\xi}_2 \,\big|\tilde{\alpha}_{\mathbf{k}_L} - \chi_{\mathbf{k}_L}\cos(\theta)\big\rangle \otimes \big|\chi_{\mathbf{k}_L}\sin(\theta)\big\rangle \Big],
\end{aligned}
\tag{3.123}
$$

where we have introduced the notation $\tilde{\alpha}_{\mathbf{k}_L} \equiv (\alpha_{\mathbf{k}_L} - \chi_{\mathbf{k}_L})/\cos(\theta)$. With this state, we now apply a heralding operation to the reflected (rightmost) mode, which is spanned by the POVM set

$$
\left\{\mathbb{1}\otimes|0_{\mathbf{k}_L}\rangle\langle 0_{\mathbf{k}_L}|,\, \mathbb{1}\otimes\sum_{n=1}^{\infty}|(2n)_{\mathbf{k}_L}\rangle\langle(2n)_{\mathbf{k}_L}|,\, \mathbb{1}\otimes\sum_{n=1}^{\infty}|(2n+1)_{\mathbf{k}_L}\rangle\langle(2n+1)_{\mathbf{k}_L}|\right\},
\tag{3.124}
$$

where the leftmost element considers events where no radiation is measured in the second mode, the middle one heralds events where an even number of photons are found, while the rightmost one conditions on events where an odd number of photons are detected. As expected, the sum of these three operators yields the identity when expressed in the Fock basis.

Applying the even-heralding and odd-heralding measurements on the state presented in Eq. (3.123), yields the following states in the transmitted mode of the beam splitter

$$
|\tilde{\Phi}_{\text{even}}\rangle = \frac{1}{\sqrt{\mathcal{N}_{\text{even}}}}\big[-\tilde{\xi}_1'|\alpha_{\mathbf{k}_L} + \chi_{\mathbf{k}_L}\cos(\theta)\rangle - \tilde{\xi}_2|\alpha_{\mathbf{k}_L} - \chi_{\mathbf{k}_L}\cos(\theta)\rangle\big],
\tag{3.125}
$$

$$
|\tilde{\Phi}_{\text{odd}}\rangle = \frac{1}{\sqrt{\mathcal{N}_{\text{even}}}}\big[\tilde{\xi}_1'|\alpha_{\mathbf{k}_L} + \chi_{\mathbf{k}_L}\cos(\theta)\rangle - \tilde{\xi}_2|\alpha_{\mathbf{k}_L} - \chi_{\mathbf{k}_L}\cos(\theta)\rangle\big],
\tag{3.126}
$$

which correspond to even- and odd-like cat states, as discussed in Sect. 2.3.3.4. The probability of successfully obtaining these states is given by

$$
P_{\text{even}}\big(\theta, \chi_{\mathbf{k}_L}\big) = \frac{1}{\tilde{\mathcal{N}}}e^{-|\chi_{\mathbf{k}_L}\sin(\theta)|^2}\Big[-1 + \cosh(\chi_{\mathbf{k}_L}\sin(\theta))\Big],
\tag{3.127}
$$

$$
P_{\text{even}}\big(\theta, \chi_{\mathbf{k}_L}\big) = \frac{1}{\tilde{\mathcal{N}}}e^{-|\chi_{\mathbf{k}_L}\sin(\theta)|^2}\sinh(\chi_{\mathbf{k}_L}\sin(\theta)).
\tag{3.128}
$$

As evident from Eqs. (3.125) and (3.126), the difference in amplitude between the two coherent states in the superposition varies as a function of $2\chi_{\mathbf{k}_L}\cos(\theta)$. This feature enables us to achieve enhancements larger than those of Refs. [169, 170], where the use of 50:50 beam splitters was the limiting factor. In this scenario, the primary bottleneck is the success probability of performing the desired projective operation, as described in Eqs. (3.127) and (3.128).

In Fig. 3.26, we present this probability as a function of $|\tilde{\chi}_{\mathbf{k}_L}| \equiv |\chi_{\mathbf{k}_L}\cos(\theta)|$ for three beam splitters with different transmissivity-reflectivity ratios: 75:25 for the solid

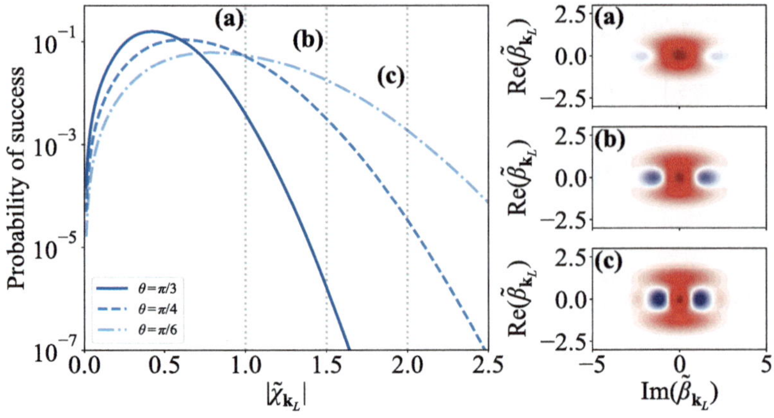

Fig. 3.26 Probability of success in performing the even-photon number heralding measurement. The probability of success is depicted as a function of $|\tilde{\chi}_{\mathbf{k}_L}|$ for different values of the transmissivity of the beam splitter under consideration. In particular, we utilize a transmission-reflection ratio of 75:25 ($\theta = \pi/3$) for the solid curve, 50:50 ($\theta = \pi/4$) for the dashed curve and 25:75 ($\theta = \pi/3$) for the dashed-dotted curve. The vertical dotted lines represent the values of $|\tilde{\chi}_{\mathbf{k}_L}|$ for which the Wigner functions of these states are represented in the rightmost column

curve, 50:50 for the dashed curve and 25:75 for the dashed-dotted curve. The vertical dotted lines indicate the generated coherent state superpositions whose Wigner functions are displayed in the right panels. In this plot, we observe that increasing values of $|\tilde{\chi}_{\mathbf{k}_L}|$ lead to an almost exponential decrease in the probability of success for all three considered ratios. Of particular interest is the 25:75 beam splitter configuration (where $\theta = \pi/6$), as it results in a difference in amplitude between the coherent states within the superposition of $\sqrt{3}\chi_{\mathbf{k}_L}$ and exhibits greater robustness as $|\tilde{\chi}_{\mathbf{k}_L}|$ increases. Notably, for $|\tilde{\chi}_{\mathbf{k}_L}| \geq 1.5$, it clearly outperforms the other two cases, yielding a success probability of 1.8×10^{-3} for $|\tilde{\chi}_{\mathbf{k}_L}| = 2.0$, a value of significant interest for quantum information science applications [171].

In contrast, in Fig. 3.27, we present the results for the case in which an odd number of photons is obtained from the heralding measurement. As evident from the right panels, the generated state corresponds to an odd-like cat state. Although for values of $|\tilde{\chi}_{\mathbf{k}_L}| \leq 1$ the success probability behaves slightly differently when compared to the results in Fig. 3.26, for values larger than that threshold, it exhibits nearly the same scaling behavior. Similar to the previous scenario, the 25:75 configuration (where $\theta = \pi/6$) leads to the maximum values of the success probability, reaching a value of 3.4×10^{-3} for $|\tilde{\chi}_{\mathbf{k}_L}| = 2.0$.

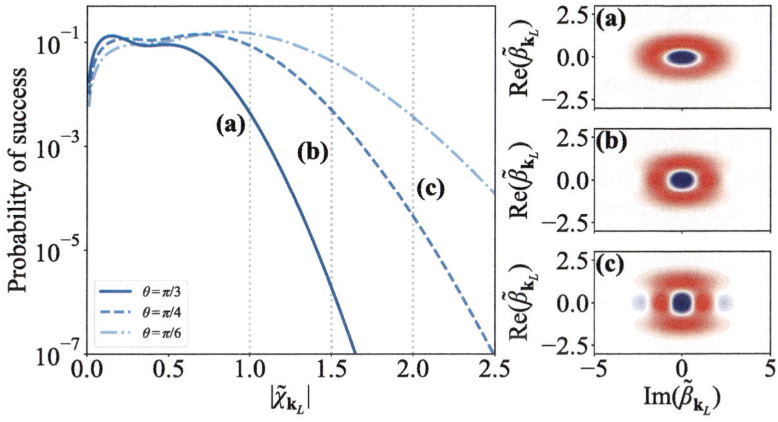

Fig. 3.27 Probability of success in performing the odd-photon number heralding measurement. The probability of success is depicted as a function of $|\bar{\chi}_{\mathbf{k}_L}|$ for different values of the transmissivity of the beam splitter under consideration. In particular, we utilize a transmission-reflection ratio of 75:25 ($\theta = \pi/3$) for the solid curve, 50:50 ($\theta = \pi/4$) for the dashed curve and 25:75 ($\theta = \pi/3$) for the dashed-dotted curve. The vertical dotted lines represent the values of $|\bar{\chi}_{\mathbf{k}_L}|$ for which the Wigner functions of these states are represented in the rightmost column

3.3.2.5 Applications of the Obtained Coherent State Superpositions III: Nonlinear Optical Processes Driven with Coherent State Superpositions

One of the most remarkable features of the resulting coherent state superpositions after the conditioning on HHG operation is their high-photon number. In conventional HHG experiments in atomic systems, the driving laser source has peak intensities on the order of $I \sim 10^{13}$–10^{14} W/cm^2. Consequently, the magnitude of $|\alpha_{\mathbf{k}_L}|$ in Eq. (3.102) falls within the range of 10^6–10^7. This indicates that the approach described in Sect. 3.3.2.1 provides an effective means of generating coherent state superpositions suitable for inducing nonlinear optical phenomena in matter. In this subsection, we demonstrate through a combination of the Quantum Spectrometer approach with an interferometric setup, that the intensity of the generated states is sufficiently high to drive nonlinear processes in a BBO crystal. These processes lead to signatures in the measured intensity of the second-order generated harmonic regarding the driving field's non-classical nature.

Before delving into the methods, it is important to note that the optimal theoretical framework for describing interferometric setups from a quantum mechanical perspective is the Heisenberg picture. For this one, transformations are applied to the operators while the states remain unchanged. However, in our case, this presents a challenge because, considering finite-duration pulses, the electric field operator defined in Eq. (3.40) does not accurately account for delays induced in the envelope. Specifically, when we introduce a phase shift $\omega_{\mathbf{k}_L}\tau$ into the field, we obtain the following when using the equation mentioned above

$$\hat{\mathbf{E}}_{\mathbf{k}_L}(t) \longrightarrow \hat{\mathbf{E}}_{\mathbf{k}_L}(t+\tau) = -if(t)\mathbf{g}(\omega_{\mathbf{k}_L})\big[\hat{a}^\dagger_{\mathbf{k}_L} e^{i\omega_{\mathbf{k}_L}(t+\tau)} - \hat{a}_{\mathbf{k}_L} e^{-i\omega_{\mathbf{k}_L}(t+\tau)}\big], \quad (3.129)$$

which accounts for the phase shift but not for the delay in the envelope. Thus, the correct approach to address this situation is to consider the multimode nature of pulsed fields. Nonetheless, this poses a significant numerical challenge due to the exceedingly large Hilbert space dimensions we would have to deal with. As an alternative, we maintain the single-mode nature of the applied field but include all the time-dependences in the coherent state amplitude, encompassing the pulse envelope. In other words, we define

$$|\tilde{\alpha}_{\mathbf{k}_L}(t)\rangle = |f(t)\tilde{\alpha}_{\mathbf{k}_L} e^{-i\omega_{\mathbf{k}_L} t}\rangle, \quad (3.130)$$

and work with time-independent operators. Thus, adding a phase shift $\omega_{\mathbf{k}_L}\tau$ transforms the state as

$$|\tilde{\alpha}_{\mathbf{k}_L}(t)\rangle \longrightarrow |\tilde{\alpha}_{\mathbf{k}_L}(t+\tau)\rangle = |f(t+\tau)\tilde{\alpha}_{\mathbf{k}_L} e^{-i\omega_{\mathbf{k}_L}(t+\tau)}\rangle, \quad (3.131)$$

for which the expected value of the electric field operator reads

$$\begin{aligned} \mathbf{E}_{\text{cl}}(t) &= \langle\tilde{\alpha}_{\mathbf{k}_L}(t+\tau)|\hat{\mathbf{E}}_{\mathbf{k}_L}|\tilde{\alpha}_{\mathbf{k}_L}(t+\tau)\rangle \\ &= -i\mathbf{g}(\omega_{\mathbf{k}_L})\big[f^*(t+\tau)\tilde{\alpha}^*_{\mathbf{k}_L} e^{i\omega_{\mathbf{k}_L}(t+\tau)} - f(t+\tau)\tilde{\alpha}_{\mathbf{k}_L} e^{-i\omega_{\mathbf{k}_L}(t+\tau)}\big], \end{aligned} \quad (3.132)$$

properly accounting for the pulse delay.

Bearing this in mind, the setup considered for this purpose is presented in Fig. 3.28. In this setup, a linearly polarized IR pulse with a peak intensity $I \approx 10^{14}$ W/cm^2 and duration of $\Delta t \approx 25$ fs (9 optical cycles) is focused on an atomic gas jet composed of Argon atoms, where the HHG process takes place (represented by the olive green box). As a consequence of this interaction, harmonic photons with orders $q \leq 33$ are

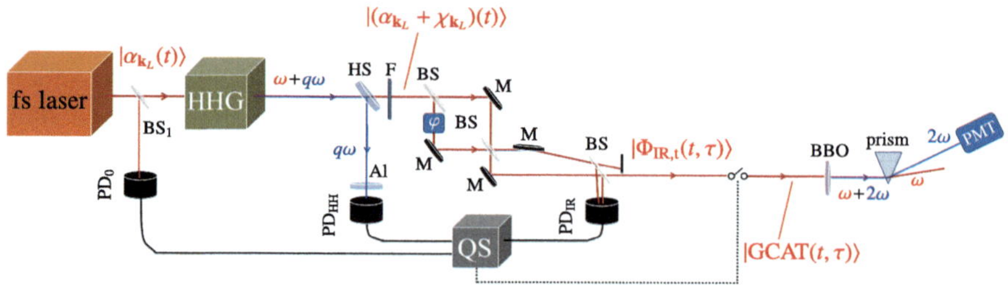

Fig. 3.28 Experimental setup used for driving nonlinear interactions in a crystal with coherent state superpositions. The experimental layout employs the same notation as presented in Fig. 3.19. In this case, the IR beam after the HHG interaction (output of the olive box), enters a Mach-Zender interferometer where, in one of the arms, the field acquires a controllable delay $\varphi = \omega_{\mathbf{k}_L}\tau$. The transmitted mode obtained at the output of the interferometer serves both for performing the conditioning on HHG operation, and drive nonlinear optical processes in the BBO crystal

generated, and the input driving field mode, characterized by $|\tilde{\alpha}_{\mathbf{k}_L}(t)\rangle$, gets shifted to $|(\tilde{\alpha}_{\mathbf{k}_L} + \tilde{\chi}_{\mathbf{k}_L})(t)\rangle$. After the interaction, the IR field passes through a filter F that attenuates the state to $|(\alpha_{\mathbf{k}_L} + \chi_{\mathbf{k}_L})(t)\rangle$, where $\alpha_{\mathbf{k}_L} = \eta\tilde{\alpha}_{\mathbf{k}_L}$ and $\chi_{\mathbf{k}_L} = \eta\tilde{\chi}_{\mathbf{k}_L}$ are the attenuated coherent state amplitudes with η the attenuation factor. This attenuation factor is chosen such that the mean photon number of the generated coherent state at this point is $\langle\hat{n}_{\mathbf{k}_L}\rangle \approx 1.5 \times 10^2$. Then, it enters a Mach-Zender interferometer after encountering a 50:50 beam splitter that initially divides the state into two optical modes, each receiving an equal share of amplitude. Thus, after this first beam splitter, the quantum optical state of the system reads

$$|\Phi_{\text{IR}}\rangle = \left|\frac{1}{\sqrt{2}} f(t)\big(\alpha_{\mathbf{k}_L} + \chi_{\mathbf{k}_L}\big)e^{-i\omega_{\mathbf{k}_L}t}\right\rangle \otimes \left|\frac{1}{\sqrt{2}} f(t)\big(\alpha_{\mathbf{k}_L} + \chi_{\mathbf{k}_L}\big)e^{-i\omega_{\mathbf{k}_L}t}\right\rangle. \quad (3.133)$$

Then, a relative time-delay of τ between the two modes is induced which, in agreement with Eq. (3.131), leads to

$$\begin{aligned} |\Phi_{\text{IR}}\rangle = &\left|\frac{1}{\sqrt{2}} f(t+\tau/2)\big(\alpha_{\mathbf{k}_L} + \chi_{\mathbf{k}_L}\big)e^{-i\omega_{\mathbf{k}_L}(t+\tau/2)}\right\rangle \\ &\otimes \left|\frac{1}{\sqrt{2}} f(t-\tau/2)\big(\alpha_{\mathbf{k}_L} + \chi_{\mathbf{k}_L}\big)e^{-i\omega_{\mathbf{k}_L}(t-\tau/2)}\right\rangle, \end{aligned} \quad (3.134)$$

and then both modes enter in another 50:50 beam splitter that mixes them. The transmitted mode is then characterized by

$$|\Phi_{\text{IR,t}}(t,\tau)\rangle = \left|\frac{1}{2}\big(\alpha_{\mathbf{k}_L} + \bar{\chi}_{\mathbf{k}_L}\big)\big(f_+ e^{-i\omega(t+\tau/2)} + f_- e^{-i\omega(t-\tau/2)}\big)\right\rangle, \quad (3.135)$$

where $f_\pm \equiv f(t \pm \tau/2)$. It is worth noting that, up to this point, we have focused on the evolution of the IR field mode, leaving the harmonics untouched until they are measured by the PD_{HH} photodetector. Therefore, the conditioning on HHG operation results in the following state, up to a normalization factor

$$|\text{GCAT}(t,\tau)\rangle = |\Phi_{\text{IR},t}(t,\tau)\rangle - \bar{\xi}(t,\tau)|\alpha_{\mathbf{k}_L}(t)\rangle, \quad (3.136)$$

with $\bar{\xi}(t,\tau) \equiv \langle\alpha_{\mathbf{k}_L}(t)|\Phi_{\text{IR},t}(t,\tau)\rangle \prod_{q=2}^{q_c} |\langle 0_{\mathbf{k}_L}|\chi_q\rangle|^2$.

This state is the one interacting with the nonlinear crystal and generating the second harmonic radiation, which is later measured by the photomultiplier PMT in Fig. 3.28. The signal generated by this second harmonic radiation is proportional to the square of the driving field's intensity, and therefore a function of the induced delay.[32] Specifically, we have

---

[32] In Appendix 8.2, we present a proof-of-principle numerical analysis where the interaction with the BBO medium is numerically accounted for, and the intensity of the second harmonic explicitly computed.

$$S_{2\omega} = \eta_{2\omega} \int \mathrm{d}\tau \langle \hat{a}^{\dagger}_{\mathbf{k}_L} \hat{a}_{\mathbf{k}_L} \hat{a}^{\dagger}_{\mathbf{k}_L} \hat{a}_{\mathbf{k}_L} \rangle (t, \tau), \tag{3.137}$$

where $\eta_{2\omega}$ is a constant characterizing the efficiency of the second harmonic generation process. In this expression, the operator for which the expected value is calculated can be expressed as $\langle (\hat{a}^{\dagger}_{\mathbf{k}_L})^2 (\hat{a}_{\mathbf{k}_L})^2 \rangle + \langle \hat{a}^{\dagger}_{\mathbf{k}_L} \hat{a}_{\mathbf{k}_L} \rangle$, utilizing the commutation properties of creation and annihilation operators. From this expression, it can be observed that the measured signal scales as $\langle \hat{n}_{\mathbf{k}_L} \rangle^2 + \langle \hat{n}_{\mathbf{k}_L} \rangle$. Thus, given that in the presented scenario we are working with high-photon number IR states, we neglect the second term and approximate the $S_{2\omega}$ signal as

$$S_{2\omega} \approx \eta_{2\omega} \int \mathrm{d}\tau \langle (\hat{a}^{\dagger}_{\mathbf{k}_L})^2 (\hat{a}_{\mathbf{k}_L})^2 \rangle (t, \tau). \tag{3.138}$$

In Fig. 3.29, the results of numerically evaluating Eq. (3.138) for different quantum optical states inducing the nonlinear interaction are presented. In the case of coherent states shown in panel (a), the obtained signal exhibits oscillations at a frequency 2ω, which are embedded in a Gaussian-like envelope. This envelope decays to a constant value as the delay between the pulses increases. However, these observations change significantly when, instead of considering a coherent state, we use the $|\mathrm{GCAT}(t, \tau)\rangle$ in Eq. (3.136). The results for this case are presented in panels (b) and (c), where beating patterns in the envelope are observed. Specifically, three beats are noticeable, with a prominent one occurring in the range -20 fs $< \tau <$ 20 fs and two smaller ones at $\tau \approx \pm 45$ fs. Moreover, as $|\chi_{\mathbf{k}_L}|$ increases, the modulation depth of the beat located between -20 fs $< \tau <$ 20 fs decreases, as does the value of $S_{2\omega}$ when evaluated at $\tau = 0$. To highlight the differences between the results in Fig. 3.29 and provide better comparison, the red lines represent the cycle-averaged $S_{2\omega}$ signal.

Before delving into the details of the experimental results, it is worth noting that the observed differences between the leftmost panel and the others in Fig. 3.29,

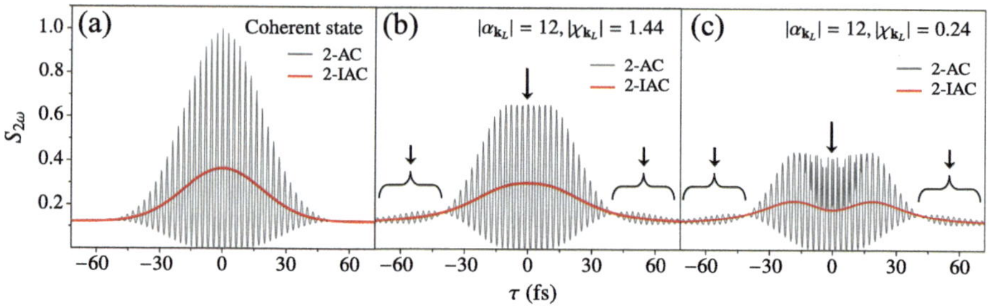

Fig. 3.29 **Theoretical calculations of the second harmonic generated signal.** In **a**, the $S_{2\omega}$ signal (black curve) is computed for a coherent state of amplitude $|\alpha_{\mathbf{k},\mu}|^2 = 12$, while in panels **b** and **c** for the conditioned on HHG state in Eq. (3.136). In these two latter plots, the values $|\chi_{\mathbf{k}_L}| = 1.44$ and $|\chi_{\mathbf{k}_L}| = 0.24$ have been used. The red curves in each of the panels represent the average value of $S_{2\omega}$ over the optical cycles. The arrows in the figure highlight the differences between panel (**a**) and panels (**b**) and (**c**)

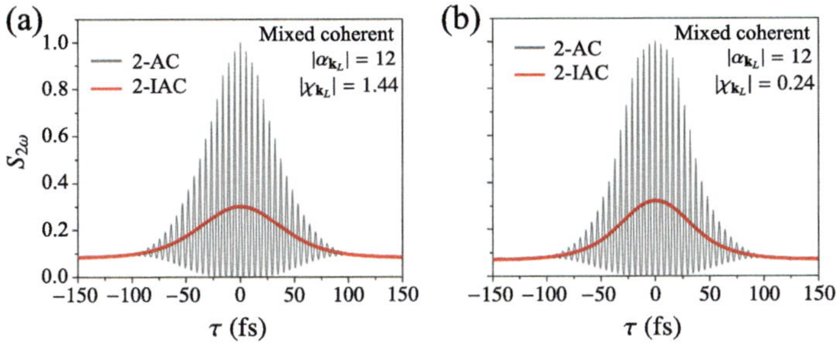

Fig. 3.30 Theoretical calculations of the second harmonic generation signal using mixed coherent states. The $S_{2\omega}$ is computed for the density matrix described in Eq. (3.139), which corresponds to the density matrix of Eq. (3.136) with the coherences set to zero. In panel **a**, we set $|\chi_{\mathbf{k}_L}| = 1.44$, while in panel **b**, $|\chi_{\mathbf{k}_L}| = 0.24$. In both cases, we set $|\alpha_{\mathbf{k}_L}| = 12$. As before, the red curves represent the average $S_{2\omega}$ signal averaged over the optical cycles

can be attributed to the non-classical effects arising from the quantum interference between the two coherent states in Eq. (3.136). To verify this, we numerically evaluate Eq. (3.138) with the classical mixture

$$\hat{\rho}(t,\tau) = |\Phi_{\text{IR,t}}(t,\tau)\rangle\langle\Phi_{\text{IR,t}}(t,\tau)| + |\bar{\chi}(t,\tau)|^2 |\alpha_{\mathbf{k}_l}\rangle\langle\alpha_{\mathbf{k}_L}|, \tag{3.139}$$

for the two values of $|\chi_{\mathbf{k}_L}|$ considered in Fig. 3.29. This classical mixture corresponds to the density matrix of Eq. (3.136), but with vanishing values for the coherences of the matrix which are the ones introducing quantum interference effects. When considering this classical mixture, as shown in Fig. 3.30, we observe that the lack of these coherence terms leads to the absence of extra beatings in the signal as well as of the reduced amplitude of the central beating. These features become more evident when comparing the red curves of Fig. 3.29. Therefore, we can confidently conclude that the characteristics of the $S_{2\omega}$ signal when driven with the state in Eq. (3.136) indeed arise from the quantum interference between the states in the superposition. While the contribution of such mixtures and decoherence effects cannot be entirely disregarded in experimental measurements, we exclude them from our theoretical analysis. The rationale behind this consideration lies in the fact that the primary features of Fig. 3.29 are governed by the quantum interference between the coherent states in the superposition.

From an experimental standpoint, the signal $S_{2\omega}$ was recorded for the maximum and minimum values of $|\chi_{\mathbf{k}_L}|$ achievable within the experimental capabilities, which correspond to those used in the theoretical analysis. This was practically achieved by varying the atomic density in the HHG region, following a similar procedure used to obtain the experimental Wigner functions in Fig. 3.21. In Fig. 3.31 the experimental results are depicted, with the red curve representing the cycle-averaged data computed from the experimental measurements. Panels (a) and (c) show the $S_{2\omega}$ signal measured

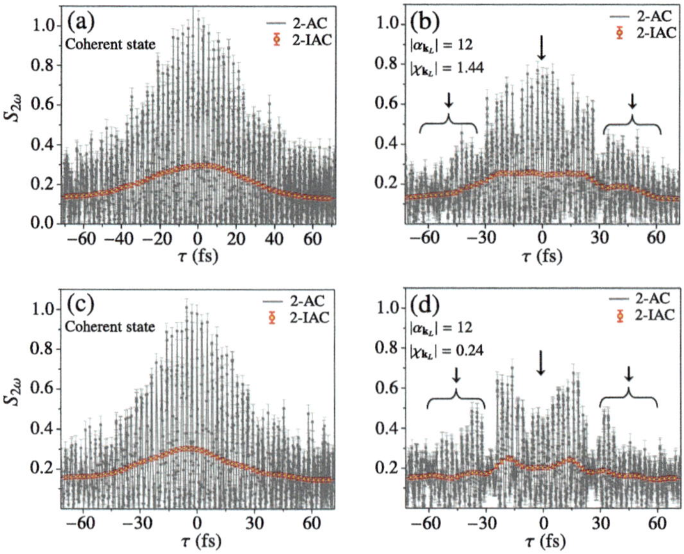

Fig. 3.31 Experimental measurements of the second harmonic generated signal. In panels **a** and **c**, the experimentally measured $S_{2\omega}$ signal is presented for the case where no conditioning on HHG operation is applied. Conversely, in panels **b** and **d**, the results are shown for those cases where such operation has indeed been applied. In these plots, it is estimated that, prior to the interaction with the BBO crystal, $|\alpha_{\mathbf{k}_L}| \approx 12$. The estimates for $|\chi_{\mathbf{k}_L}|$ in panels **b** and **d** are $|\chi_{\mathbf{k}_L}| \approx 1.44$ and $|\chi_{\mathbf{k}_L}| \approx 0.24$, respectively. In these plots, the red curve with yellow markers represents the measured $S_{2\omega}$ signal after averaging over the optical cycles. These data have been measured under identical experimental conditions

without implementing the conditioning on HHG operation, resulting in a Gaussian-like envelope, consistent with the theoretical expectations. In contrast, panels (b) and (d) display the results obtained when the conditioning on HHG operation was applied in the setup.

Consistent with the theoretical predictions, the measured signals exhibit three beatings located within the time window -20 fs $< \tau < 20$ fs and at $\tau \approx \pm 45$ fs, as indicated by the black arrows in the figure. Furthermore, the cycle-averaged signal (red curves) exhibit features that align with the theoretical expectations. For the highest value of $|\chi_{\mathbf{k}_L}|$ in Fig. 3.31b, it shows a plateau-like structure within the time window -20 fs $< \tau < 20$ fs. For the smallest value of $|\chi_{\mathbf{k}_L}|$ in Fig. 3.31d, the presence of two maxima around $\tau \approx \pm 20$ fs is observed.

Finally, although less evident from Fig. 3.31, the experimental measures also reveal that the value of $S_{2\omega}$, when evaluated at $\tau = 0$, is smaller when driving the BBO interaction with Eq. (3.136) than with a coherent state. To quantitatively highlight this feature, these values are plotted as a function of $|\chi_{\mathbf{k}_L}|$ in Fig. 3.32 using red circular markers. Additionally, the modulation depth $\mathrm{M} \equiv 2(S_{2\omega}^{\max} - S_{2\omega}^{\min})/(S_{2\omega}^{\max} + S_{2\omega}^{\min})$ is presented with the blue triangular markers. This modulation depth is computed from the red curves shown in Fig. 3.31. The experimental values are compared with

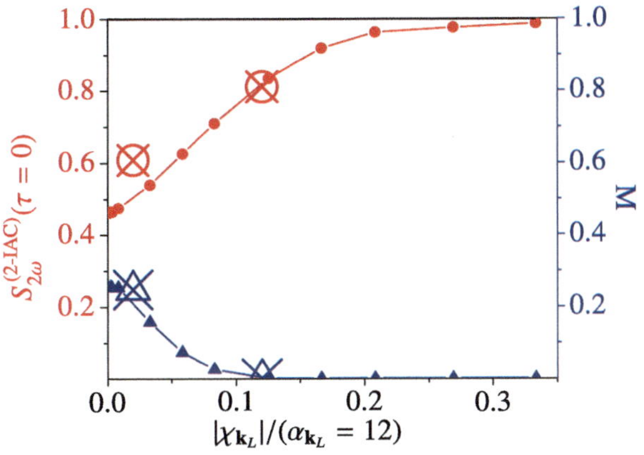

Fig. 3.32 Experimental values for the second harmonic generation signal at zero delay and its modulation depth. The red curve with circular markers represents the theoretically calculated $S_{2\omega}$ signal at zero delay ($\tau = 0$) as a function of $|\chi_{\mathbf{k}_L}|$. The large red circular markers with a cross inside correspond to the experimentally measured values. In blue, the modulation depth M, as defined in the text, is shown instead. The triangular dotted curve corresponds to the theoretical value while the larger markers with a cross inside correspond to the experimentally measured values. The data is fitted to the theoretical curves to determine the values of $|\chi_{\mathbf{k}_L}|$ used in the theoretical plots

the theoretical expectations represented by the solid curves. These results not only demonstrate good agreement between the experiment and theory, but also provide further justification for the choices of $|\chi_{\mathbf{k}_L}|$ in the theoretical analysis.

3.4 Discussion and Outlook

In this chapter, we have presented a quantum optical analysis of strong-field-driven processes in atomic systems. We have observed that these processes can produce states with seemingly non-classical features either directly, or through the use of conditioning operations. Our particular focus has been on two of the most characteristic phenomena in strong-field physics: above-threshold ionization and high-harmonic generation.

For ATI, we conducted a single-electron analysis through which we observed that electron dynamics have a significant impact on the quantum optical state. This impact is manifested in terms of displacements applied to the initial coherent state of the field. These displacements are contingent on the instant of time at which the electron ionizes and its final kinetic momentum. Thus, conditioning the joint electron-light state on the measurement of electrons with specific momentum values leads to the generation of coherent state superpositions. While the presence of non-classical behavior, in terms of Wigner function negativities, is not observed at the single-atom level, considering multiple atoms participating in the process may potentially reveal such behaviors.

This possibility was demonstrated through a phenomenological analysis. On the other hand, when restricting our focus to electrons with a fixed value of the kinetic energy but propagating in opposite directions, we observed a non-negligible degree of entanglement already at the single-atom level. This is particularly pronounced when working with mid-infrared frequencies and sufficiently high intensities to induce tunneling ionization processes in atoms.

In this regard, one could contemplate the following possibilities to further expand our work:

- **Inclusion of rescattering effects**. The analysis performed in this chapter primarily focused on direct ATI electrons. However, it was observed that higher photoelectron energies might result in increased entanglement values. To surpass the $\mathcal{E}(p) \approx 3U_p$ limit considered in our analysis, it is essential to incorporate high-order ATI processes. One approach to address this is by extending our analysis to account for first-order perturbation theory terms, $\mathbf{g}(\mathbf{v}, \mathbf{v}')$, which encompass rescattering effects [44, 47].
- **Refined multi-atom description**. While phenomenological many-atomic analysis was utilized to glimpse the effects of multiple atoms on the fundamental mode's Wigner function, this approach falls short when characterizing entanglement. Developing more accurate models in this context becomes crucial, specially for realistic scenarios where it is practically impossible to isolate single-atom responses. Ideas in this direction could benefit from the work of A. Pizzi and colleagues [145].
- **Continuous variable entanglement**. The entanglement characterization between light and matter in our analysis focused on instances where electrons have fixed kinetic energy, effectively treating the electronic degrees of freedom as two-dimensional. However, electron energies span a continuum spectrum, necessitating continuous variable entanglement characterization methods [172–174]. Thus, an extension of our work in this regard would be of practical interest, even when constraining to a fixed value of kinetic energy, in order to properly account for the experimental errors associated with the accuracy of the employed detectors.
- **Beyond the weak-coupling approximation**. One of the approximations we employed in our analysis was that quantum optical fluctuations' impact on electron dynamics could be considered negligible. While this assumption aligns with findings from the theoretical analysis by M. Even-Tzur and coworkers [144], in this work they also showed that the use of non-classical driving fields could affect electron dynamics in non-trivial ways. Additionally, in Ref. [147], non-symmetric ATI spectra were observed due to these effects. Therefore, we anticipate this deviation from the weak-coupling approximation, particularly when non-classical drivers are used, to significantly affect the various non-classical measures used throughout this chapter, specifically the entanglement between field and electron. Therefore, it is advisable to extend our discussions to more generic scenarios that do not rely on the weak-coupling assumption while also accommodating the use of non-classical driving fields. A potential direction in this regard, as recently proposed in Ref. [66], could involve a coherent-spin-state-based formalism.

For HHG, we conducted an analysis analogous to that of ATI, up to a unitary transformation. This analysis revealed that, assuming the depletion of the ground state probability amplitude to be negligible, the final quantum optical state, resulting from a parametric process in which the electron begins and ends the dynamics in the ground state of the system, can be expressed as a product of coherent states. Each of the coherent states in this product corresponds to a displaced version of its initial state, with the displacement determined by the spectral amplitude of the HHG emission. While this state exhibits classical features (see Sect. 2.3.3.2), applying conditioning operations which restrict to HHG processes could lead to the generation of coherent state superpositions.

What distinguishes these states from those in the existing literature [175–179], is their natural high-photon number. However, they do have a limitation: the difference in amplitude between the coherent states in the superposition cannot be made arbitrarily large. Nevertheless, their high-photon number property makes them ideal candidates for driving nonlinear optical process in matter, a subject that has been explored from both a theoretical and experimental perspective in the context of second harmonic generation in a BBO crystal. Furthermore, we have proposed various schemes in which these coherent state superpositions are initially injected, enabling an increase of one extra coherent state in the superposition. These states could then be used to create enlarged and more balanced versions of the input superpositions.

In this context, potential follow-ups to our work could include the following:

- **Beyond the weak-depletion assumption**. As mentioned earlier, the analysis presented in this chapter was based on the weak-depletion of the ground state amplitude. However, recently, in Ref. [164], we have shown that going beyond this approximation leads to the generation of squeezed states on the harmonic modes, particularly on the fundamental driving field (see Fig. 3.33). The generated squeezing is highly tunable, as the squeezing direction strongly depends on the characteristics of the HHG process, as for instance with the carrier-envelope phase of the employed laser source. In a broader context, since the squeezing operation involves the different optical modes in the HHG process, this naturally leads to the presence of entangled states between the field modes. Future work in this direction will focus on characterizing the amount of entanglement in this context, along with further investigations into the generated squeezing.
- **Driving nonlinear optics with coherent state superpositions**. As we have seen, the generation of high-photon number coherent state superpositions allows for inducing nonlinear optical processes in materials such as BBO. Therefore, it is pertinent to gain a deeper understanding of the properties of the generated radiation and its potential applications, along with those of the proposed scheme, in quantum information science. This exploration could encompass areas such as quantum metrology [180] or quantum sensing [181].

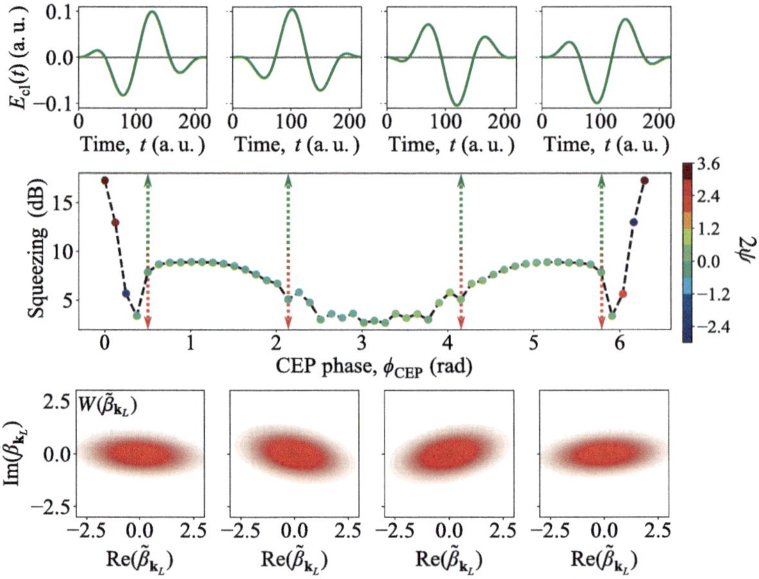

Fig. 3.33 Beyond the weak-depletion assumption: squeezing effects following HHG processes. In the middle panel, the amount of squeezing in the fundamenal mode is shown as a function of the CEP of the employed laser source. A total of $N_{\mathrm{at}} = 5 \times 10^{13}$ atoms are considered to independently contribute to the HHG process. The arrows in this plot indicate the corresponding electric field values (uppermost row) and the resulting Wigner functions (lowermost row) associated with these squeezing values. The color code in the middle panel represents the phase of the squeezing parameter ψ

- **Connections with practical applications in quantum information science**. While we have identified different aspects about the potential interest of the generated state for quantum information science, practical applications in this direction remain largely unexplored. In Chap. 6, following insights from the works developed in Chap. 5, we will further elaborate on this matter.

References

1. Maiman TH (1960) Phys Rev Lett 4:564. https://doi.org/10.1103/PhysRevLett.4.564
2. Slusher RE (1999) Rev Modern Phys 71:S471. https://doi.org/10.1103/RevModPhys.71.S471
3. Protopapas M, Keitel CH, Knight PL (1997) Rep Progr Phys 60:389. https://doi.org/10.1088/0034-4885/60/4/001
4. Nobel Prize Outreach AB 2024 (2023) The Nobel Prize in Physics 2023. https://www.nobelprize.org/prizes/physics/2023/summary/. Accessed 4 Oct 2023
5. Keldysh (1965) Sov Phys JETP 20:1307. http://www.jetp.ac.ru/cgi-bin/e/index/e/20/5/p1307?a=list
6. DiMauro L, Frolov M, Ishikawa KL, Ivanov M (2014) J Phys B: Atomic, Mol Opt Phys 47:200301. https://doi.org/10.1088/0953-4075/47/20/200301

7. Agostini P, Fabre F, Mainfray G, Petite G, Rahman NK (1979) Phys Rev Lett 42:1127. https://doi.org/10.1103/PhysRevLett.42.1127
8. Voronov GS. Delone NB (1965) JETP Lett 1:66. http://www.jetpletters.ac.ru/ps/1590/article_24398.shtml
9. Voronov GS, Delone NB (1966) Sov Phys JETP 23:54. http://www.jetp.ac.ru/cgi-bin/e/index/r/50/1/p78?a=list
10. Chin SL, Farkas G, Yergeau F (1983) J Phys B: Atomic Mol Phys 16:L223. https://doi.org/10.1088/0022-3700/16/8/003
11. Spanner M, Smirnova O, Corkum PB, Ivanov MY (2004) J Phys B: Atomic, Mol Opt Phys 37:L243. https://doi.org/10.1088/0953-4075/37/12/L02
12. Yurchenko SN, Patchkovskii S, Litvinyuk IV, Corkum PB, Yudin GL (2004) Phys Rev Lett 93:223003. https://doi.org/10.1103/PhysRevLett.93.223003
13. Blaga CI, Xu J, DiChiara AD, Sistrunk E, Zhang K, Agostini P, Miller TA, DiMauro LF, Lin CD (2012) Nature 483:194. https://doi.org/10.1038/nature10820
14. Pullen MG, Wolter B, Le A-T, Baudisch M, Hemmer M, Senftleben A, Schröter CD, Ullrich J, Moshammer R, Lin CD, Biegert J (2015) Nat Commun 6:7262. https://doi.org/10.1038/ncomms8262
15. Wolter B, Pullen MG, Le A-T, Baudisch M, Doblhoff-Dier K, Senftleben A, Hemmer M, Schröter CD, Ullrich J, Pfeifer T, Moshammer R, Gräfe S, Vendrell O, Lin CD, Biegert J (2016) Science 354:308. https://doi.org/10.1126/science.aah3429
16. Huismans Y, Rouzée A, Gijsbertsen A, Jungmann JH, Smolkowska AS, Logman PSWM, Lépine F, Cauchy C, Zamith S, Marchenko T, Bakker JM, Berden G, Redlich B, van der Meer AFG, Muller HG, Vermin W, Schafer KJ, Spanner M, Ivanov MY, Smirnova O, Bauer D, Popruzhenko SV, Vrakking MJJ (2011) Science 331:61. https://doi.org/10.1126/science.1198450
17. Figueira de Morisson Faria C, Maxwell AS (2020) Rep Progr Phys 83:034401. https://doi.org/10.1088/1361-6633/ab5c91
18. Burnett NH, Baldis HA, Richardson MC, Enright GD (1977) Appl Phys Lett 31:172. https://doi.org/10.1063/1.89628
19. McPherson A, Gibson G, Jara H, Johann U, Luk TS, McIntyre IA, Boyer K, Rhodes CK (1987) JOSA B 4:595. https://doi.org/10.1364/JOSAB.4.000595
20. Ferray M, L'Huillier A, Li XF, Lompre LA, Mainfray G, Manus C (1988) J Phys B: Atomic, Mol Opt Phys 21: L31. https://doi.org/10.1088/0953-4075/21/3/001
21. Gohle C, Udem T, Herrmann M, Rauschenberger J, Holzwarth R, Schuessler HA, Krausz F, Hänsch TW (2005) Nature 436:234. https://doi.org/10.1038/nature03851
22. Silva REF, Jiménez-Galán A, Amorim B, Smirnova O, Ivanov M (2019) Nat Photon 13:849. https://doi.org/10.1038/s41566-019-0516-1
23. Alcalà J, Bhattacharya U, Biegert J, Ciappina M, Elu U, Graß T, Grochowski PT, Lewenstein M, Palau A, Sidiropoulos TPH, Steinle T, Tyulnev I (2022) Proc Natl Acad Sci 119:e2207766119. https://doi.org/10.1073/pnas.2207766119
24. Itatani J, Levesque J, Zeidler D, Niikura H, Pépin H, Kieffer JC, Corkum PB, Villeneuve DM (2004) Nature 432:867. https://doi.org/10.1038/nature03183
25. Drescher M, Hentschel M, Kienberger R, Tempea G, Spielmann C, Reider GA, Corkum PB, Krausz F (2001) Science 291:1923. https://doi.org/10.1126/science.1058561
26. Silva F, Teichmann SM, Cousin SL, Hemmer M, Biegert J (2015) Nat Commun 6:6611. https://doi.org/10.1038/ncomms7611
27. Krause JL, Schafer KJ, Kulander KC (1992) Phys Rev Lett 68:3535. https://doi.org/10.1103/PhysRevLett.68.3535
28. Popmintchev T, Chen MC, Popmintchev D, Arpin P, Brown S, Ališauskas S, Andriukaitis G, Balčiunas T, Mücke OD, Pugzlys A, Baltuška A, Shim B, Schrauth SE, Gaeta A, Hernández-García C, Plaja L, Becker A, Jaron-Becker A, Murnane MM, Kapteyn HC (2012) Science 336:1287. https://doi.org/10.1126/science.1218497
29. Kobayashi Y, Sekikawa T, Nabekawa Y, Watanabe S (1998) Opt Lett 23:64. https://doi.org/10.1364/OL.23.000064

30. Midorikawa K, Nabekawa Y, Suda A (2008) Progr Quant Electron 32:43. https://doi.org/10.1016/j.pquantelec.2008.04.001
31. Tsatrafyllis N, Bergues B, Schröder H, Veisz L, Skantzakis E, Gray D, Bodi B, Kuhn S, Tsakiris GD, Charalambidis D, Tzallas P (2016) Sci Rep 6:21556. https://doi.org/10.1038/srep21556
32. Chatziathanasiou S, Kahaly S, Skantzakis E, Sansone G, Lopez-Martens R, Haessler S, Varju K, Tsakiris GD, Charalambidis D, Tzallas P (2017) Photonics 4:26. https://doi.org/10.3390/photonics4020026
33. Bergues B, Rivas DE, Weidman M, Muschet AA, Helml W, Guggenmos A, Pervak V, Kleineberg U, Marcus G, Kienberger R, Charalambidis D, Tzallas P, Schröder H, Krausz F, Veisz L (2018) Optica 5:237. https://doi.org/10.1364/OPTICA.5.000237
34. Nayak A, Orfanos I, Makos I, Dumergue M, Kühn S, Skantzakis E, Bodi B, Varju K, Kalpouzos C, Banks HIB, Emmanouilidou A, Charalambidis D, Tzallas P (2018) Phys Rev A 98:023426. https://doi.org/10.1103/PhysRevA.98.023426
35. Senfftleben B, Kretschmar M, Hoffmann A, Sauppe M, Tümmler J, Will I, Nagy T, Vrakking MJJ, Rupp D, Schütte B (2020) J Phys: Photon 2:034001. https://doi.org/10.1088/2515-7647/ab87fd
36. Orfanos I, Makos I, Liontos I, Skantzakis E, Major B, Nayak A, Dumergue M, Kühn S, Kahaly S, Varju K, Sansone G, Witzel B, Kalpouzos C, Nikolopoulos LAA, Tzallas P, Charalambidis D (2020) J Phys: Photon 2:042003. https://doi.org/10.1088/2515-7647/aba172
37. Corkum PB, Krausz F (2007) Nat Phys 3:381. https://doi.org/10.1038/nphys620
38. Krausz F, Ivanov M (2009) Rev Modern Phys 81:163. https://doi.org/10.1103/RevModPhys.81.163
39. Lewenstein M, Ciappina MF, Pisanty E, Rivera-Dean J, Stammer P, Lamprou T, Tzallas P (2021) Nat Phys 17:1104. https://doi.org/10.1038/s41567-021-01317-w
40. bibitemch3riveraspsdeanspsnewsps2021 Rivera-Dean J, Stammer P, Pisanty E, Lamprou T, Tzallas P, Lewenstein M, Ciappina MF (2021) J Comput Electron 20:2111. https://doi.org/10.1007/s10825-021-01789-2
41. Rivera-Dean J, Lamprou T, Pisanty E, Stammer P, Ordóñez AP, Maxwell AS, Ciappina MF, Lewenstein M, Tzallas P (2022) Phys Rev A 105:033714. https://doi.org/10.1103/PhysRevA.105.033714
42. Stammer P, Rivera-Dean J, Lamprou T, Pisanty E, Ciappina MF, Tzallas P, Lewenstein M (2022) Phys Rev Lett 128:123603. https://doi.org/10.1103/PhysRevLett.128.123603
43. Rivera-Dean J, Stammer P, Maxwell AS, Lamprou T, Tzallas P, Lewenstein M, Ciappina MF (2022) Phys Rev A 106:063705. https://doi.org/10.1103/PhysRevA.106.063705
44. Stammer P, Rivera-Dean J, Maxwell A, Lamprou T, Ordóñez A, Ciappina MF, Tzallas P, Lewenstein M (2023) PRX Quant 4:010201. https://doi.org/10.1103/PRXQuantum.4.010201
45. Bhattacharya U, Lamprou T, Maxwell AS, Ordóñez A, Pisanty E, Rivera-Dean J, Stammer P, Ciappina MF, Lewenstein M, Tzallas P (2023) Rep Progr Phys 86:094401. https://doi.org/10.1088/1361-6633/acea31
46. Lamprou T, Rivera-Dean J, Stammer P, Lewenstein M, Tzallas P (2023) Nonlinear optics using intense optical Schrödinger "cat" states. https://doi.org/10.48550/arXiv.2306.14480, arXiv:2306.14480 [physics, physics:quant-ph]
47. Amini K, Biegert J, Calegari F, Chacón A, Ciappina MF, Dauphin A, Efimov DK, Figueira de Morisson Faria C, Giergiel K, Gniewek P, Landsman AS, Lesiuk M, Mandrysz M, Maxwell AS, Moszyński R, Ortmann L, Pérez-Hernández JA, Picón A, Pisanty E, Prauzner-Bechcicki J, Sacha K, Suárez N, Zaïr A, Zakrzewski J, Lewenstein M (2019) Rep Progr Phys 82:116001. https://doi.org/10.1088/1361-6633/ab2bb1
48. Faisal FHM, Kamiński JZ (1997) Phys Rev A 56:748. https://doi.org/10.1103/PhysRevA.56.748
49. Reiss HR (1980) Phys Rev A 22:1786. https://doi.org/10.1103/PhysRevA.22.1786
50. Popruzhenko SV (2014) J Phys B: Atomic, Mol Opt Phys 47:204001. https://doi.org/10.1088/0953-4075/47/20/204001

51. Lewenstein M, Balcou P, Ivanov MY, L'Huillier A, Corkum PB (1994) Phys Rev A 49:2117. https://doi.org/10.1103/PhysRevA.49.2117
52. Faisal FHM (1973) J Phys B: Atomic Mol Phys 6:L89. https://doi.org/10.1088/0022-3700/6/4/011
53. Grochmalicki J, Kuklinski JR, Lewenstein M (1986) J Phys B: Atomic Mol Phys 19:3649. https://doi.org/10.1088/0022-3700/19/22/010
54. Lewenstein M, Kulander KC, Schafer KJ, Bucksbaum PH (1995) Phys Rev A 51:1495. https://doi.org/10.1103/PhysRevA.51.1495
55. Mulser P, Bauer D (2010) High power laser-matter interaction. In: Mulser P, Bauer D (eds) Springer tracts in modern physics. Springer, Berlin, Heidelberg, pp 267–330. https://doi.org/10.1007/978-3-540-46065-7_7
56. Pisanty Alatorre E (2016) Electron dynamics in complex time and complex space. Doctoral thesis, school Imperial College, London. https://core.ac.uk/download/pdf/77020798.pdf
57. Perelomov A, Popov V, Terent'ev M (1965) Sov Phys JETP 24:207. http://www.jetp.ac.ru/cgi-bin/e/index/e/24/1/p207?a=list
58. Popruzhenko SV, Paulus GG, Bauer D (2008) Phys Rev A 77:053409. https://doi.org/10.1103/PhysRevA.77.053409
59. Popruzhenko S, Bauer D (2008) J Modern Opt 55:2573. https://doi.org/10.1080/09500340802161881
60. Popruzhenko SB, Mur VD, Popov VS, Bauer D (2009) J Exp Theoret Phys 108:947. https://doi.org/10.1134/S1063776109060053
61. Lai X-Y, Poli C, Schomerus H, Figueira de Morisson Faria C (2015) Phys Rev A 92:043407. https://doi.org/10.1103/PhysRevA.92.043407
62. Maxwell AS, Al-Jawahiry A, Das T, Figueira de Morisson Faria C (2017) Phys Rev A 96:023420. https://doi.org/10.1103/PhysRevA.96.023420
63. Maxwell AS, Al-Jawahiry A, Lai XY, Figueira de Morisson Faria C (2018) J Phys B: Atomic, Mol Opt Phys 51:044004. https://doi.org/10.1088/1361-6455/aa9e81
64. Reiss HR (1990) JOSA B 7:574. https://doi.org/10.1364/JOSAB.7.000574
65. Reiss HR (1990) Phys Rev A 42:1476. https://doi.org/10.1103/PhysRevA.42.1476
66. Maxwell AS, Madsen LB (2024) Phys Rev A 110:033108. https://doi.org/10.1103/PhysRevA.110.033108
67. Sala S, Förster J, Saenz A (2017) Phys Rev A 95:011403. https://doi.org/10.1103/PhysRevA.95.011403
68. Senaratne R, Rajagopal SV, Shimasaki T, Dotti PE, Fujiwara KM, Singh K, Geiger ZA, Weld DM (2018) Nat Commun 9:2065. https://doi.org/10.1038/s41467-018-04556-3
69. Argüello-Luengo J, Rivera-Dean J, Stammer P, Maxwell AS, Weld DM, Ciappina MF, Lewenstein M (2024) PRX Quant 5:010328. https://doi.org/10.1103/PRXQuantum.5.010328
70. Salières P, Carré B, Le Déroff L, Grasbon F, Paulus GG, Walther H, Kopold R, Becker W, Milošević DB, Sanpera A, Lewenstein M (2001) Science 292:902. https://doi.org/10.1126/science.108836
71. Corkum PB (1993) Phys Rev Lett 71:1994. https://doi.org/10.1103/PhysRevLett.71.1994
72. l'Huillier A, Lompre LA, Mainfray G, Manus C (1983) Phys Rev A 27:2503. https://doi.org/10.1103/PhysRevA.27.2503
73. Walker B, Sheehy B, DiMauro LF, Agostini P, Schafer KJ, Kulander KC (1994) Phys Rev Lett 73:1227. https://doi.org/10.1103/PhysRevLett.73.1227
74. Feuerstein B, Moshammer R, Fischer D, Dorn A, Schröter CD, Deipenwisch J, Crespo Lopez-Urrutia JR, Höhr C, Neumayer P, Ullrich J, Rottke H, Trump C, Wittmann M, Korn G, Sandner W (2001) Phys Rev Lett 87:043003. https://doi.org/10.1103/PhysRevLett.87.043003
75. Figueira de Morisson Faria C, Liu X (2011) J Modern Opt 58:1076. https://doi.org/10.1080/09500340.2010.543958
76. Becker W, Liu X, Ho PJ, Eberly JH (2012) Rev Modern Phys 84:1011. https://doi.org/10.1103/RevModPhys.84.1011
77. Hänsch P, Walker MA, Van Woerkom LD (1997) Phys Rev A 55:R2535. https://doi.org/10.1103/PhysRevA.55.R2535

78. Paulus GG, Nicklich W, Xu H, Lambropoulos P, Walther H (1994) Phys Rev Lett 72:2851. https://doi.org/10.1103/PhysRevLett.72.2851
79. Walker B, Sheehy B, Kulander KC, DiMauro LF (1996) Phys Rev Lett 77:5031. https://doi.org/10.1103/PhysRevLett.77.5031
80. Milošević DB, Paulus GG, Bauer D, Becker W (2006) J Phys B: Atomic, Mol Opt Phys 39:R203. https://iopscience.iop.org/article/10.1088/0953-4075/39/14/R01/meta
81. Lewenstein M, L'Huillier A (2009) Strong field laser physics. In: Brabec T (ed) Springer in optical sciences. Springer, New York, NY, pp 147–183. https://doi.org/10.1007/978-0-387-34755-4_7
82. Agostini P, DiMauro LF (2012) Advances in atomic, molecular, and optical physics, vol 61. In: Berman P, Arimondo E, Lin C (eds) Academic Press, pp 117–158. https://doi.org/10.1016/B978-0-12-396482-3.00003-X
83. Paulus GG, Grasbon F, Dreischuh A, Walther H, Kopold R, Becker W (2000) Phys Rev Lett 84:3791. https://doi.org/10.1103/PhysRevLett.84.3791
84. Bauer D, Koval P (2006) Comput Phys Commun 174:396. https://doi.org/10.1016/j.cpc.2005.11.001
85. Smirnova O, Ivanov M (2014) Multielectron high harmonic generation: simple man on a complex plane. In: Attosecond and XUV physics. Wiley, chapter 7, pp 201–256. https://doi.org/10.1002/9783527677689.ch7
86. Nayak A, Dumergue M, Kühn S, Mondal S, Csizmadia T, Harshitha NG, Füle M, Upadhyay Kahaly M, Farkas B, Major B, Szaszkó-Bogár V, Földi P, Majorosi S, Tsatrafyllis N, Skantzakis E, Neoričić L, Shirozhan M, Vampa G, Varjú K, Tzallas P, Sansone G, Charalambidis D, Kahaly S (2019) Saddle point approaches in strong field physics and generation of attosecond pulses. Phys Rep 833:1. https://doi.org/10.1016/j.physrep.2019.10.002
87. Paulus GG, Grasbon F, Walther H, Villoresi P, Nisoli M, Stagira S, Priori E, De Silvestri S (2001) Nature 414:182. https://doi.org/10.1038/35102520
88. Macklin JJ, Kmetec JD, Gordon CL (1993) Phys Rev Lett 70:766. https://doi.org/10.1103/PhysRevLett.70.766
89. Perry MD, Mourou G (1994) Science 264:917. https://doi.org/10.1126/science.264.5161.917
90. Preston SG, Sanpera A, Zepf M, Blyth WJ, Smith CG, Wark JS, Key MH, Burnett K, Nakai M, Neely D, Offenberger AA (1996) Phys Rev A 53:R31. https://doi.org/10.1103/PhysRevA.53.R31
91. L'Huillier A, Balcou P (1993) Phys Rev Lett 70:774. https://doi.org/10.1103/PhysRevLett.70.774
92. Chang Z, Rundquist A, Wang H, Murnane MM, Kapteyn HC (1997) Phys Rev Lett 79:2967. https://doi.org/10.1103/PhysRevLett.79.2967
93. Spielmann C, Burnett NH, Sartania S, Koppitsch R, Schnürer M, Kan C, Lenzner M, Wobrauschek P, Krausz F (1997) Science 278:661. https://doi.org/10.1126/science.278.5338.661
94. Sundaram B, Milonni PW (1990) Phys Rev A 41:6571. https://doi.org/10.1103/PhysRevA.41.6571
95. Kulander KC, Schafer KJ, Krause JL (1993) Super-intense laser-atom physics. In: Piraux B, L'Huillier A, Rzążewski K (eds) NATO ASI. Springer US, Boston, MA, pp 95–110. https://doi.org/10.1007/978-1-4615-7963-2_10
96. Kopold R, Becker W, MiloÅžević DB (2002) J Modern Opt 49:1987. https://doi.org/10.1080/09500340210140731
97. Pisanty E, Ciappina MF, Lewenstein M (2020) J Phys: Photon 2:034013. https://doi.org/10.1088/2515-7647/ab8f1e
98. Lewenstein M, Salières P, L'Huillier A (1995) Phys Rev A 52:4747. https://doi.org/10.1103/PhysRevA.52.4747
99. Zaïr A, Holler M, Guandalini A, Schapper F, Biegert J, Gallmann L, Keller U, Wyatt AS, Monmayrant A, Walmsley IA, Cormier E, Auguste T, Caumes JP, Salières P (2008) Phys Rev Lett 100:143902. https://doi.org/10.1103/PhysRevLett.100.143902

100. Figueira de Morisson Faria C, Schomerus H, Becker W (2002) Phys Rev A 66:043413. https://doi.org/10.1103/PhysRevA.66.043413
101. Milošević DB, Becker W (2002) Phys Rev A 66:063417. https://doi.org/10.1103/PhysRevA.66.063417
102. Schafer KJ, Kulander KC (1997) Phys Rev Lett 78:638. https://doi.org/10.1103/PhysRevLett.78.638
103. Salières P, Antoine P, de Bohan A, Lewenstein M (1998) Phys Rev Lett 81:5544. https://doi.org/10.1103/PhysRevLett.81.5544
104. Kim J-H, Shin HJ, Lee DG, Nam CH (2000) Phys Rev A 62:055402. https://doi.org/10.1103/PhysRevA.62.055402
105. Pisanty E (2020) RB-SFA: high harmonic generation in the strong field approximation via mathematica, GitHub. https://doi.org/10.5281/zenodo.592519, https://github.com/episanty/RB-SFA. Accessed 24 Oct 2023
106. Corkum PB, Burnett NH, Ivanov MY (1994) Opt Lett 19:1870. https://doi.org/10.1364/OL.19.001870
107. Antoine P, L'Huillier A, Lewenstein M, Salières P, Carrú B (1996) Phys Rev A 53:1725. https://doi.org/10.1103/PhysRevA.53.1725
108. Budil KS, Salières P, L'Huillier A, Ditmire T, Perry MD (1993) Phys Rev A 48:R3437. https://doi.org/10.1103/PhysRevA.48.R3437
109. Burnett NH, Kan C, Corkum PB (1995) Phys Rev A 51:R3418. https://doi.org/10.1103/PhysRevA.51.R3418
110. Weihe FA, Dutta SK, Korn G, Du D, Bucksbaum PH, Shkolnikov PL (1995) Phys Rev A 51:R3433. https://doi.org/10.1103/PhysRevA.51.R3433
111. Weihe FA, Bucksbaum PH (1996) JOSA B 13:157. https://doi.org/10.1364/JOSAB.13.000157
112. Antoine P, Carré B, L'Huillier A, Lewenstein M (1997) Phys Rev A 55:1314. https://doi.org/10.1103/PhysRevA.55.1314
113. Schulze D, Dörr M, Sommerer G, Ludwig J, Nickles PV, Schlegel T, Sandner W, Drescher M, Kleineberg U, Heinzmann U (1998) Phys Rev A 57:3003. https://doi.org/10.1103/PhysRevA.57.3003
114. Milošević DB, Becker W, Kopold R (2000) Phys Rev A 61:063403. https://doi.org/10.1103/PhysRevA.61.063403
115. Fleischer A, Kfir O, Diskin T, Sidorenko P, Cohen O (2014) Nat Photon 8:543. https://doi.org/10.1038/nphoton.2014.108
116. Pisanty E, Sukiasyan S, Ivanov M (2014) Phys Rev A 90:043829. https://doi.org/10.1103/PhysRevA.90.043829
117. Kfir O, Grychtol P, Turgut E, Knut R, Zusin D, Popmintchev D, Popmintchev T, Nembach H, Shaw JM, Fleischer A, Kapteyn H, Murnane M, Cohen O (2015) Nat Photon 9:99. https://doi.org/10.1038/nphoton.2014.293
118. Amico L, Fazio R, Osterloh A, Vedral V (2008) Rev Modern Phys 80:517. https://doi.org/10.1103/RevModPhys.80.517
119. Guo DS, Åberg T (1988) J Phys A: Math General 21:4577. https://doi.org/10.1088/0305-4470/21/24/013
120. Guo D-S, Åberg T, Crasemann B (1989) Phys Rev A 40:4997. https://doi.org/10.1103/PhysRevA.40.4997
121. Åberg T, Guo D-S, Ruscheinski J, Crasemann B (1991) Phys Rev A 44:3169. https://doi.org/10.1103/PhysRevA.44.3169
122. Guo D-S, Drake GWF (1992) Phys Rev A 45:6622. https://doi.org/10.1103/PhysRevA.45.6622
123. Gao L, Li X, Fu P, Freeman RR, Guo D-S (2000) Phys Rev A 61:063407. https://doi.org/10.1103/PhysRevA.61.063407
124. Fu P, Wang B, Li X, Gao L (2001) Phys Rev A 64:063401. https://doi.org/10.1103/PhysRevA.64.063401

125. Wang B, Gao L, Li X, Guo D-S, Fu P (2007) Phys Rev A 75:063419. https://doi.org/10.1103/PhysRevA.75.063419
126. Wang B, Guo Y, Chen J, Yan Z-C, Fu P (2012) Phys Rev A 85:023402. https://doi.org/10.1103/PhysRevA.85.023402
127. Gao J, Shen F, Eden JG (1998) Phys Rev Lett 81:1833. https://doi.org/10.1103/PhysRevLett.81.1833
128. Chen J, Chen SG, Liu J (2000) Phys Rev Lett 84:4252. https://doi.org/10.1103/PhysRevLett.84.4252
129. Eden JG, Gao J, Shen F (2000) Phys Rev Lett 84:4253. https://doi.org/10.1103/PhysRevLett.84.4253
130. Bogatskaya A, Volkova E, Popov A (2016) Europhys Lett 116:14003. https://doi.org/10.1209/0295-5075/116/14003
131. Bogatskaya AV, Volkova EA, Popov AM (2017) J Exp Theoret Phys 125:587. https://doi.org/10.1134/S1063776117090114
132. Milošević DB (2023) Phys Rev A 108:033110. https://doi.org/10.1103/PhysRevA.108.033110
133. Gombkötő A, Varró S, Mati P, Földi P (2020) Phys Rev A 101:013418. https://doi.org/10.1103/PhysRevA.101.013418
134. Földi P, Magashegyi I, Gombkötő A, Varró S (2021) Photonics 8:263. https://doi.org/10.3390/photonics8070263
135. Kominis IK, Kolliopoulos G, Charalambidis D, Tzallas P (2014) Phys Rev A 89:063827. https://doi.org/10.1103/PhysRevA.89.063827
136. Gonoskov IA, Tsatrafyllis N, Kominis IK, Tzallas P (2016) Sci Rep 6:32821. https://doi.org/10.1038/srep32821
137. Tsatrafyllis N, Kominis IK, Gonoskov IA, Tzallas P (2017) Nat Commun 8:15170. https://doi.org/10.1038/ncomms15170
138. Tsatrafyllis N, Kühn S, Dumergue M, Foldi P, Kahaly S, Cormier E, Gonoskov I, Kiss B, Varju K, Varro S, Tzallas P (2019) Phys Rev Lett 122:193602. https://doi.org/10.1103/PhysRevLett.122.193602
139. Rivera-Dean J (2019), Quantum-optical analysis of high-order harmonic generation. Master thesis, school Universitat Politècnica de Catalunya. https://upcommons.upc.edu/handle/2117/168580
140. Gorlach A, Neufeld O, Rivera N, Cohen O, Kaminer I (2020) Nat Commun 11:4598. https://doi.org/10.1038/s41467-020-18218-w
141. Gombkötő A, Földi P, Varró S (2021) Phys Rev A 104:033703. https://doi.org/10.1103/PhysRevA.104.033703
142. Varró S (2021) Photonics 8:269. https://doi.org/10.3390/photonics8070269
143. Gorlach A, Tzur ME, Birk M, Krúger M, Rivera N, Cohen O, Kaminer I (2023) Nat Phys 19:1689. https://doi.org/10.1038/s41567-023-02127-y
144. Even Tzur M, Birk M, Gorlach A, Krúger M, Kaminer I, Cohen O (2023) Nat Photon 17:501. https://doi.org/10.1038/s41566-023-01209-w
145. Pizzi A, Gorlach A, Rivera N, Nunnenkamp A, Kaminer I (2023) Nat Phys 19:551. https://doi.org/10.1038/s41567-022-01910-7
146. Tzur ME, Birk M, Gorlach A, Kaminer I, Krúger M, Cohen O (2024) Phys Rev Res 6:033079. https://doi.org/10.1103/PhysRevResearch.6.033079
147. Fang Y, Sun F-X, He Q, Liu Y (2023) Phys Rev Lett 130:253201. https://doi.org/10.1103/PhysRevLett.130.253201
148. Stammer P (2022) Phys Rev A 106:L050402. https://doi.org/10.1103/PhysRevA.106.L050402
149. Rivera-Dean J, Stammer P, Maxwell AS, Lamprou T, Pisanty E, Tzallas P, Lewenstein M, Ciappina MF (2024) Phys Rev A 109:033706. https://doi.org/10.1103/PhysRevA.109.033706
150. Gonoskov I, Sondenheimer R, Hünecke C, Kartashov D, Peschel U, Gräfe S (2024) Phys Rev B 109:125110. https://doi.org/10.1103/PhysRevB.109.125110

151. Rivera-Dean J, Stammer P, Maxwell AS, Lamprou T, Ordóñez AF, Pisanty E, Tzallas P, Lewenstein M, Ciappina MF (2023) Entanglement and non-classical states of light in a strong-laser driven solid-state system. https://doi.org/10.48550/arXiv.2211.00033, arXiv:2211.00033 [physics, physics:quant-ph]
152. Rivera-Dean J, Stammer P, Maxwell AS, Lamprou T, Ordóñez AF, Pisanty E, Tzallas P, Lewenstein M, Ciappina MF (2024)Nonclassical states of light after high-harmonic generation in semiconductors: a Bloch-based perspective. https://doi.org/10.1103/PhysRevB.109.035203
153. Stammer P (2024) Phys Rev Res 6:L032033. https://doi.org/10.1103/PhysRevResearch.6.L032033
154. Vrakking MJ (2021) Phys Rev Lett 126:113203. https://doi.org/10.1103/PhysRevLett.126.113203
155. Koll L-M, Maikowski L, Drescher L, Witting T, Vrakking MJ (2022) Phys Rev Lett 128:043201. https://doi.org/10.1103/PhysRevLett.128.043201
156. Ruberti M, Averbukh V, Mintert F (2024) Phys Rev X 14:041402. https://doi.org/10.1103/PhysRevX.14.041042
157. Maxwell AS, Madsen LB, Lewenstein M (2022) Nat Commun 13:4706. https://doi.org/10.1038/s41467-022-32128-z
158. Quan W, Lin Z, Wu M, Kang H, Liu H, Liu X, Chen J, Liu J, He XT, Chen SG, Xiong H, Guo L, Xu H, Fu Y, Cheng Y, Xu Z (2009) Phys Rev Lett 103:093001. https://doi.org/10.1103/PhysRevLett.103.093001
159. Blaga CI, Catoire F, Colosimo P, Paulus GG, Muller HG, Agostini P, DiMauro LF (2009) Nat Phys 5:335. https://doi.org/10.1038/nphys1228
160. Liu C, Hatsagortsyan KZ (2010) Phys Rev Lett 105:113003. https://doi.org/10.1103/PhysRevLett.105.113003
161. van Loock P (2011) Laser Photon Rev 5:167. https://doi.org/10.1002/lpor.201000005
162. Kreis K, van Loock P (2012) Phys Rev A 85:032307. https://doi.org/10.1103/PhysRevA.85.032307
163. Massé G, Coudreau T, Keller A, Milman P (2020) Phys Rev A 102:062406. https://doi.org/10.1103/PhysRevA.102.062406
164. Stammer P, Rivera-Dean J, Maxwell AS, Lamprou T, Argüello-Luengo J, Tzallas P, Ciappina MF, Lewenstein M (2024) Entanglement and squeezing of the optical field modes in high harmonic generation. https://doi.org/10.1103/PhysRevLett.132.143603
165. Leonhardt U (1993) Phys Rev A 48:3265. https://doi.org/10.1103/PhysRevA.48.3265
166. Wenger J, Hafezi M, Grosshans F, Tualle-Brouri R, Grangier P (2003) Phys Rev A 67:012105. https://doi.org/10.1103/PhysRevA.67.012105
167. van Enk SJ (2003), https://doi.org/10.1103/PhysRevLett.91.017902 Phys Rev Lett 91:017902
168. Paris MGA (1996) Phys Lett A 217:78. https://doi.org/10.1016/0375-9601(96)00339-8
169. Laghaout A, Neergaard-Nielsen JS, Rigas I, Kragh C, Tipsmark A, Andersen UL (2013) Phys Rev A 87:043826. https://doi.org/10.1103/PhysRevA.87.043826
170. Sychev DV, Ulanov AE, Pushkina AA, Richards MW, Fedorov IA, Lvovsky AI (2017) Nat Photon 11:379. https://doi.org/10.1038/nphoton.2017.57
171. Gilchrist A, Nemoto K, Munro WJ, Ralph TC, Glancy S, Braunstein SL, Milburn GJ (2004) J Opt B: Quant Semiclassical Opt 6:S828. https://doi.org/10.1088/1464-4266/6/8/032
172. van Loock P, Furusawa A (2003) Phys Rev A 67:052315. https://doi.org/10.1103/PhysRevA.67.052315
173. Gärttner M, Haas T, Noll J (2023) Phys Rev Lett 131:150201. https://doi.org/10.1103/PhysRevLett.131.150201
174. Gärttner M, Haas T, Noll J (2023) Phys Rev A 108:042410. https://doi.org/10.1103/PhysRevA.108.042410
175. Zavatta A, Viciani S, Bellini M (2004) Science 306:660. https://doi.org/10.1126/science.1103190
176. Ourjoumtsev A, Tualle-Brouri R, Laurat J, Grangier P (2006) Science 312:83. https://doi.org/10.1126/science.1122858

177. Neergaard-Nielsen JS, Nielsen BM, Hettich C, Mølmer K, Polzik ES (2006) Phys Rev Lett 97:083604. https://doi.org/10.1103/PhysRevLett.97.083604
178. Ourjoumtsev A, Jeong H, Tualle-Brouri R, Grangier P (2007) Nature 448:784. https://doi.org/10.1038/nature06054
179. Hacker B, Welte S, Daiss S, Shaukat A, Ritter S, Li L, Rempe G (2019) Nat Photon 13:110. https://doi.org/10.1038/s41566-018-0339-5
180. Giovannetti V, Lloyd S, Maccone L (2011) Nat Photon 5:222. https://doi.org/10.1038/nphoton.2011.35
181. Pirandola S, Bardhan BR, Gehring T, Weedbrook C, Lloyd S (2018) Nat Photon 12:724. https://doi.org/10.1038/s41566-018-0301-6

Chapter 4
Non-classical States of Light After High-Harmonic Generation in Molecular and Solid Systems

My whole life, all I ever did was fight... But I can't give up either. I can't fight my own nature. That's the paradox, John... You see?

— Dutch van der Linde in *Red Dead Redemption* (2010)

The analysis of high-harmonic generation processes to more complicated samples than atoms, such as molecular or solid-state systems, has proven to yield essential insights into the studied materials. This is due to their intricate internal structure, which introduces additional electron dynamics that influence the characteristics of the generated radiation. These alterations can subsequently be employed to extract important information about the material's inner structure, facilitating further characterization and revealing novel properties that may lead to advancements in technology [1].

The interest in HHG processes occurring in molecular targets primarily arises from the additional degrees of freedom they offer. Molecules inherently consist of multiple atoms, with each atom interacting with the others to form an overall structure minimizing the total energy of the molecule. The characteristics of this internal structure dictate how molecules interact with a driving field, as well as their orientation with respect to the field's polarization. In the context of HHG (driven by fields with $I \sim 10^{14}$ W/cm^2), this orientation ultimately determines the properties of the emitted radiation. These modifications are of utmost importance for various applications in strong-field physics involving molecular systems. An example of this is ultrafast imaging of chiral molecules[1] [2, 3], where the specific polarization properties of the driving field [4, 5] allow discrimination between molecules with

[1] A molecule is said to be chiral if it cannot be superimposed with its mirrored image.

J. Rivera-Dean, *Non-classical States of Light*, Springer Theses,
https://doi.org/10.1007/978-3-031-73769-5_4

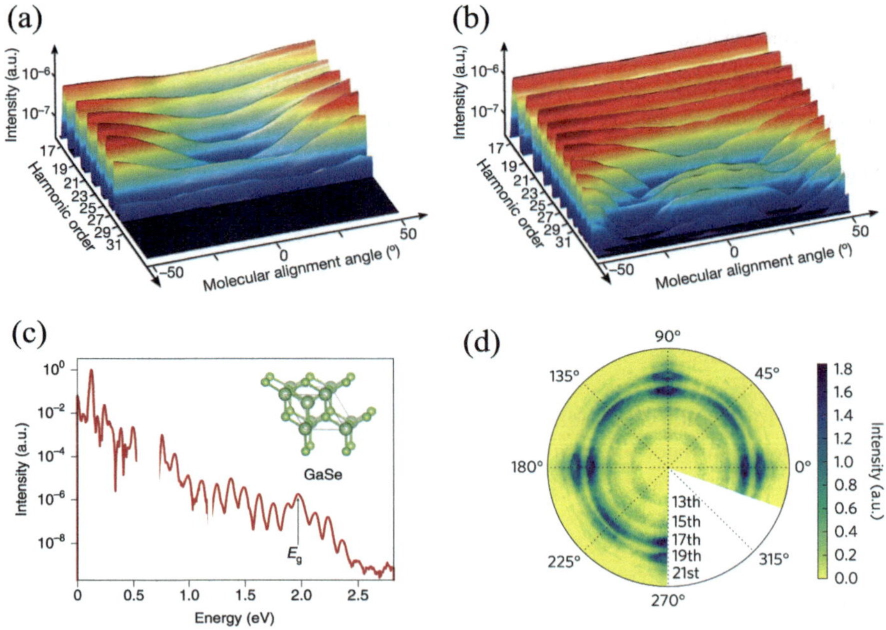

Fig. 4.1 Experimentally measured HHG spectra from molecular and solid-state samples. In panels **a** and **b**, we display the HHG spectra obtained from exciting a CO_2 molecule with a linearly polarized laser field at a wavelength $\lambda_{\mathbf{k}_L} = 800$ nm and two peak intensities ($I \approx 1.1 \times 10^{14}$ W/cm^2 in **a** and $I \approx 1.8 \times 10^{14}$ W/cm^2 in **b**), plotted as a function of the molecular alignment with respect to the laser's polarization. In **c**, the measured spectrum results from the interaction of a GaSe crystal with a driving laser field at $\lambda_{\mathbf{k}_L} \approx 10$ μm and peak intensity $I \approx 6.5 \times 10^{12}$ W/cm^2. Panel **d** depicts the harmonic spectrum's dependence on the solid orientation for a MgO crystal, determined by the angle it forms with respect to the field's polarization, using a driving laser with $\lambda_{\mathbf{k}_L} \approx 1.3$ μm and peak intensity $I \approx 1.3 \times 10^{13}$ W/cm^2. Panels **a** and **b** have been obtained from Ref. [14], panel **c** from Ref. [20] (an adapted version from Ref. [21]), and panel **d** from Ref. [22]. Reproduced with permission from Springer Nature

opposite chiralities based on the characteristics of the emitted radiation [4–7]. Furthermore, the measured HHG spectrum also encodes valuable information about the electronic orbital structure. This provides a means to extract molecular parameters with sub-Angstrom spatial and attosecond temporal resolutions [8–12], as well as structural [13] and electron-dynamical [14–16] information of simple molecules. In this regard, HHG spectroscopy has also played a vital role for extending these studies to more complex targets [17–19].

Solid-state systems introduce an additional level of complexity when compared to atoms and molecules. In solids, atoms are densely packed and their electrons are subject to numerous interactions, including significant electron-electron correlation effects. The arrangement of atoms within the solid critically determines the properties of the entire system, specially electron dynamics when exposed to an electric field. However, unlike atoms and molecules, solid-state systems have an intensity limit on the fields that can be applied to them. This limit is determined by the material's damage-threshold: beyond this value, the sample can be damaged, either by the driving field or the products of the interaction, rendering it unusable [20, 23, 24]. This restriction leads to the use of intensities on the order of $10^{11} - 10^{12}$ W/cm^2, unlike the typical $10^{14} - 10^{15}$ W/cm^2 common in strong-field studies with atomic and molecular systems. Nevertheless, despite this natural upper limit on the intensity, it is still possible to observe HHG processes in solid-state systems [25, 26]. In fact, these processes provide valuable information about the solid-state system. For example, they can be used to accurately reconstruct the material's energy dispersion relations [27–29], or to probe the potential to which electrons are subjected within the solid in real space [22, 30, 31]. Moreover, HHG has demonstrated significant potential towards quantum technology applications, particularly in the study of quantum materials [1]. These materials exhibit exotic and unique quantum properties, often associated with strong electron-electron correlations. In this regard, high-harmonic generation has been shown to probe phase transitions with femtosecond temporal precision [32–38], as well as to detect topological order [39–45]. Furthermore, it has recently been employed to distinguish between different phases of specific superconductor materials [46], which are key platforms in the development of quantum information science applications [47].

This chapter, which is based on our works in Refs. [48–50], is divided in two main sections that explore from a quantum optical perspective, how the inherent structure of molecular and solid-state systems influences the non-classical properties of the states after the interaction. Each section begins with a concise review of a suitable quantum mechanical characterization of these systems. Then, we delve into a quantum optical description of the strong laser-matter interaction, and evaluate the non-classical features of the state after the interaction by examining various quantum optical observables and quantum information measures. We emphasize the impact of additional electron dynamics, absent in atomic systems, on these non-classical indicators.

4.1 High-Harmonic Generation Processes in Hydrogen Molecular Ions

In this section, we perform a quantum optical analysis of HHG processes in H_2^+, the simplest molecular system composed by two atomic centers and an electron delocalized between them. The presence of these two nuclei in the molecule already leads to some important differences on the electron dynamics followed by HHG processes, when compared to its atomic counterpart. For instance in this case the active electron can ionize from one center and either recombine with the same one or with the opposite. This leads to interference features already observable on the HHG spectrum [8, 51–55]. Moreover, the dissociate photoionization mechanisms for these molecules have been extensively studied in the literature, specially when considering H_2 and D_2, which ultimately leads to H_2^+. This has been done either through direct ionization using a laser field at different frequencies (see e.g. Ref. [56]), or by experiments involving two time-delayed laser pulses at different frequencies. This second type of experiments further allow for studying the electronic charge distribution within the molecule (see e.g. Ref. [57] and references therein).

4.1.1 A Linear Combination of Atomic Orbitals Picture of H_2^+ Molecules

A significant advantage of studying H_2^+ molecules is the ability to characterize their lowest electronic energy eigenstates as simple linear combinations of atomic orbitals. This method is referred to in the literature as the Linear Combination of Atomic Orbitals (LCAO) technique [58, 59]. However, before delving into its details, it is necessary to introduce some approximations that allow us to focus exclusively on the electronic degrees of freedom.

In general, the Hamiltonian characterizing an H_2^+ molecule, composed of two centers which we shall refer to as L, for the left center, and R for the right center, is given by

$$\hat{H}_{\text{mol}} = \frac{\hat{\mathbf{p}}_{\text{e}}}{2m_{\text{e}}} + \frac{\hat{\mathbf{p}}_L}{2m_L} + \frac{\hat{\mathbf{p}}_R}{2m_R} + \hat{V}\big(\hat{\mathbf{r}}_{\text{e}}, \hat{\mathbf{r}}_L, \hat{\mathbf{r}}_R\big), \tag{4.1}$$

where $\{\hat{\mathbf{p}}_{\text{e}}, \hat{\mathbf{p}}_L, \hat{\mathbf{p}}_R\}$ are the momentum operators for the electron and the two protons, respectively, while $\{\hat{\mathbf{r}}_{\text{e}}, \hat{\mathbf{r}}_L, \hat{\mathbf{r}}_R\}$ are their position operators, and $\{m_{\text{e}}, m_L, m_R\}$ are their masses.[2] Consequently, the first three terms represent the kinetic energies of the three particles that are involved in the structure of H_2^+ while the $\hat{V}(\hat{\mathbf{r}}_{\text{e}}, \hat{\mathbf{r}}_L, \hat{\mathbf{r}}_R)$ term comprises the different interactions involved among them. It specifically includes the electron-nucleus attraction and the repulsion between the molecular centers.

From this Hamiltonian, one can observe that the dynamics of this system involve two distinct time scales associated with electronic and nuclear motions. Specifically, the mass ratio between a proton and an electron is approximately $m_L/m_{\text{e}} \approx 1836.15$. Given that protons are more massive than electrons, it is expected that the nuclear dynamics at short time scales are less pronounced than that resulting from the electronic part. More specifically, it has been found that the vibrational nuclear dynamics in H_2^+ caused by the repulsion between the two centers of the molecule, occurs on the order of 10 fs [60, 61]. In contrast, when considering HHG processes driven by laser fields at 800 nm, the electron ionization and recombination processes happen on the order of 1 fs. Due to these disparate time scales, we make the assumption that, during the electron dynamics, the nucleus can be considered fixed in space. This approximation is commonly referred to as the Born-Oppenheimer approximation [59, 62],

---

[2] It is worth noting that, for the case of H_2^+, $m_L = m_R$.

which implies that the Hamiltonian effectively describing the electronic behavior within the molecule can be written as

$$\hat{H}_{\text{mol}} \approx \frac{\hat{\mathbf{p}}}{2m} + \hat{V}_{\text{nuc}}(\hat{\mathbf{r}}) + V(R), \tag{4.2}$$

where we have omitted the e subscript in the definition of the position and momentum operators compared to Eq. (4.1). In this expression, $\hat{V}_{\text{nuc}}(\hat{\mathbf{r}})$ represents the potential that attracts the electron to the molecular nucleus, while $V(R)$ signifies the repulsion between the molecular centers. Under the Born-Oppenheimer approximation, this latter term remains constant as we assume that R, which will be referred to as the *bond length*, is held fixed.

Finding the ground and excited states of the Hamiltonian in Eq. (4.2), that is the molecular orbitals, is a doable task, with the resulting states being expressed in terms of more complex mathematical expressions compared to the atomic orbitals of hydrogen. Therefore, methods that provide simplified versions of these expressions while maintaining a reasonable degree of physical rigor are highly valuable. In this context, the LCAO, as mentioned earlier, stands out as one of the most useful techniques. It expresses molecular orbitals as finite linear combinations of atomic orbitals for both the L and R nuclei, that is

$$|\psi\rangle = \sum_{i=L,R} \sum_{j=0}^{N} c_{i,j} |\Lambda_{i,j}\rangle, \tag{4.3}$$

where N is a predetermined cutoff on the atomic orbitals to consider, and the set $\{|\Lambda_{L,j}\rangle, |\Lambda_{R,j}\rangle\}$ represents, in this case, the jth excited state of a hydrogen atom localized in the same positions as the L and R nuclei, respectively. Assuming that the set of states $\{|\Lambda_{i,j}\rangle\}$ is known, the goal now is to determine the constants $c_{i,j}$ in the superposition that can accurately reproduce the molecular orbitals. This problem can be formulated in terms of the variational principle [63], which reduces the problem to finding non-trivial solutions of the equation (for more details, please refer to Appendix 9.1)

$$\sum_{\boldsymbol{i}} c_{\boldsymbol{i}} \left(H_{\boldsymbol{m},\boldsymbol{i}} - \langle \hat{H} \rangle S_{\boldsymbol{m},\boldsymbol{i}} \right) = 0, \tag{4.4}$$

where $\boldsymbol{i} \equiv (i, j)$, $H_{\boldsymbol{m},\boldsymbol{i}} \equiv \langle \Lambda_{\boldsymbol{m}} | \hat{H} | \Lambda_{\boldsymbol{i}} \rangle$, $S_{\boldsymbol{m},\boldsymbol{i}} \equiv \langle \Lambda_{\boldsymbol{m}} | \Lambda_{\boldsymbol{i}} \rangle$ and $\langle \hat{H} \rangle$ corresponds to the expected value of the Hamiltonian evaluated under Eq. (4.3). In matrix form, the non-trivial solutions of this set of equations are those satisfying

$$\left| \mathsf{H} - \langle \hat{H} \rangle \mathsf{S} \right| = 0, \tag{4.5}$$

with H and S matrices whose elements are given by $H_{\boldsymbol{m},\boldsymbol{i}}$ and $S_{\boldsymbol{m},\boldsymbol{i}}$.

The beauty of this method lies in its efficiency, specially for simple molecules as is the case of H_2^+. In these cases, it is sufficient to consider only a few atomic orbitals to accurately reproduce the properties of the molecular orbitals to a significant extent. Specifically for H_2^+, we can obtain approximately good descriptions of the two lowest energy molecular orbitals by selecting the atomic orbitals $\{|\Lambda_{i,j}\rangle\} = \{|g_L\rangle, |g_R\rangle\}$, corresponding to the ground state of atomic hydrogen located at the positions of the L and R centers. Therefore, Eq. (4.3) can be written as

$$|\psi\rangle = c_L|g_L\rangle + c_R|g_R\rangle, \tag{4.6}$$

and consequently Eq. (4.5) as

$$\begin{vmatrix} H_{L,L} - \langle\hat{H}\rangle & H_{L,R} - \langle\hat{H}\rangle S_{L,R} \\ H_{L,R} - \langle\hat{H}\rangle S_{L,R} & H_{L,L} - \langle\hat{H}\rangle \end{vmatrix} = 0, \tag{4.7}$$

where we have taken into account that $H_{L,L} = H_{R,R}$, $H_{L,R} = H_{R,L}$, $S_{L,L} = S_{R,R} = 1$ and $S_{L,R} = S_{L,R}$. These elements can be explicitly calculated by using the atomic ground state orbitals. However, for our description, these expressions are not necessary and are given elsewhere [59]. By solving this equation, we find two non-degenerate values of $\langle\hat{H}\rangle$ for which this equation is satisfied. Specifically, we obtain

$$\langle\hat{H}\rangle_{\pm} = \frac{H_{L,L} \pm H_{L,R}}{1 + S_{L,R}} \equiv \mathcal{E}_{\mathrm{b/a}}, \tag{4.8}$$

which introduced in Eq. (4.4) to find the coefficients c_L and c_R, leads to

$$|\psi_{\mathrm{b}}\rangle = \frac{1}{\mathcal{N}_{\mathrm{b}}}\Big[|g_L\rangle + |g_R\rangle\Big], \quad |\psi_{\mathrm{a}}\rangle = \frac{1}{\mathcal{N}_{\mathrm{a}}}\Big[|g_L\rangle - |g_R\rangle\Big], \tag{4.9}$$

respectively for $\mathcal{E}_{\mathrm{b}}$ and $\mathcal{E}_{\mathrm{a}}$.

These states are commonly referred to as bonding ($|\psi_{\mathrm{b}}\rangle$) and antibonding ($|\psi_{\mathrm{a}}\rangle$) states, and they provide a good approximation to the ground and first excited states of the H_2^+ molecule, respectively. Their correspondence with the ground and first excited states, i.e. $\mathcal{E}_{\mathrm{b}} < \mathcal{E}_{\mathrm{a}}$, becomes more evident when explicitly computing the $H_{i,j}$ and $S_{i,j}$ quantities. This is illustrated in Fig. 4.2a, where the two energies are shown as a function of the bond length R. For all values of R, the inequality $\mathcal{E}_{\mathrm{b}} < \mathcal{E}_{\mathrm{a}}$ is satisfied.

One of the main features about Fig. 4.2a is that the bonding orbital provides a stable structure to the molecule around $R \approx 2.5$ a.u., where a minimum in energy is found. This stability is not observed for the antibonding state. A more detailed view of the reason underlying this behavior is provided in panel (b). This plot shows the probability of finding the electron at a given point in space for both the bonding (red curve) and antibonding (blue curve) states. For the bonding states, the electron is delocalized between the two centers of the molecule, resulting in a symmetric charge distribution and, consequently, a net dipole moment equal to zero. This leads to a

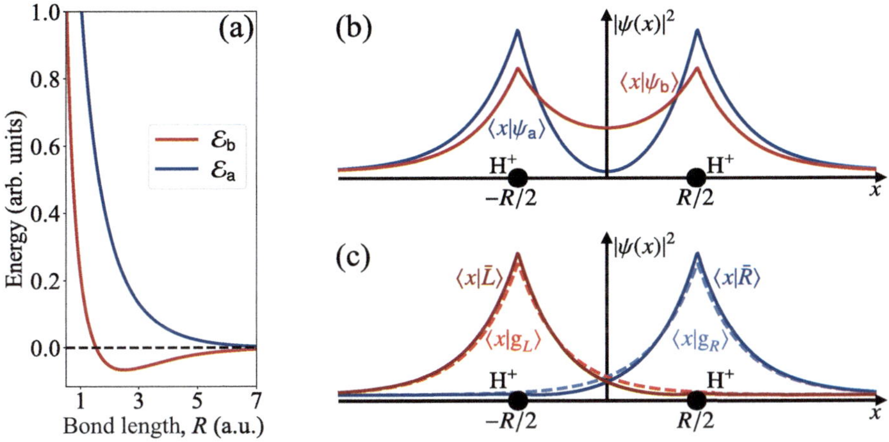

Fig. 4.2 Two lowest energy eigenstates of the H_2^+ molecule according to the LCAO. In **a**, the energy of the ground (bonding, b) and first excited (antibonding, a) states are depicted as a function of the bond length. In **b**, their corresponding wavefunctions are presented alongside a graphical representation of the molecular centers. In **c**, a comparison is made between the localized left states $|\bar{L}\rangle$ and $|\bar{R}\rangle$ and the ground states of the corresponding atomic orbitals $\{|g_L\rangle, |g_R\rangle\}$ located at the respective positions of each molecular center

stable molecule at an equilibrium distance of $R \approx 2.5$ a.u., according to the LCAO. In contrast, this is not the case for antibonding states where the electron is mostly localized either around one center or the other, with almost vanishing probability of being in between.

Furthermore, these states can also be used to construct a set of localized states $\{|\bar{L}\rangle, |\bar{R}\rangle\}$ around the left and right centers, as shown in Fig. 4.2c with the continuous curves. These localized states resemble the atomic orbitals $|g_L\rangle$ and $|g_R\rangle$ presented in panel 4.2c with the dashed curves, although slightly more localized than the former.[3] This set of localized states is defined as

$$|\bar{L}\rangle = \frac{1}{\sqrt{2}}\Big[\,|\psi_b\rangle - |\psi_a\rangle\,\Big], \quad |\bar{R}\rangle = \frac{1}{\sqrt{2}}\Big[\,|\psi_b\rangle + |\psi_a\rangle\,\Big], \tag{4.10}$$

and unlike the set $\{|g_L\rangle, |g_R\rangle\}$, span an orthonormal set that is slightly more localized in the left and right centers compared to the atomic orbitals.

To conclude this section, while the LCAO is effective in modeling the behavior of the ground and first excited states of the H_2^+ molecules, there are certain aspects it does not fully account for. For example, it predicts an equilibrium bond length ($R \approx 2.5$ a.u.) that is slightly greater than the experimental values ($R \approx 2.0$ a.u.). Furthermore, the energies predicted at this equilibrium distance are around 170 kJ for the LCAO, while experimental observations report values of 255 kJ [59]. Despite

[3] In particular, when $R \to \infty$, the set of localized states and the atomic orbitals converge given that $\langle g_L|g_R\rangle \to 0$.

these disagreements, the LCAO method provides a solid foundation for performing calculations in H_2^+ molecules, which can be further refined using more sophisticated methods [59, 64]. Therefore, we employ it in the subsequent discussion.

4.1.2 A Quantum Optical Analysis

The analysis of the HHG process in molecular systems has been extensively studied, from a semiclassical perspective, using various methods. These have mainly involved fully numerical approaches [65–73], in some cases extending beyond the Born-Oppenheimer approximation [74–78], while others considered SFA extensions to the molecular scenario [10, 53–55, 79–82]. These studies have provided valuable insights into the characteristics of the HHG process and its consequences when dealing with molecular systems as the excited medium. They have explored factors such as the dependence of the harmonic generation on the molecular structure, alignment with the applied field's polarization, differences between simple and more complex molecules, and the role of the electron recombination with the same or different centers.

In this section, based on our work in Ref. [48], we take a step further by employing a description using quantized fields, specifically by considering an extension of the SFA to the molecular scenario. Given the similarity of the SFA model used here to that of the atomic description, we omit a semiclassical description of this method in this section and instead refer the reader to Sect. 3.1. Therefore, for the quantum optical analysis, the Schrödinger equation describing the dynamics between a strong-laser field and the H_2^+ molecule within the dipole, Born-Oppenheimer and single-active electron approximations, is given in the length gauge by

$$i\hbar\frac{\partial|\Psi(t)\rangle}{\partial t} = \left[\hat{H}_{\text{mol}} + \mathrm{e}\hat{\mathbf{E}}(t)\cdot\hat{\mathbf{r}} + \hat{H}_{\text{field}}\right]|\Psi(t)\rangle\,, \tag{4.11}$$

where $\hat{H}_{\text{mol}}$ is the molecular Hamiltonian within the Born-Oppenheimer approximation presented in Eq. (4.2), $\hat{\mathbf{E}}(t)$ is the electric field operator within the discrete mode description as provided in Eq. (3.35), and $\hat{H}_{\text{field}}$ represents the free-field Hamiltonian. For this differential equation, we consider the initial condition to be given by

$$|\Psi(t=t_0)\rangle = |\psi_{\text{b}}\rangle \otimes |\alpha_{\mathbf{k}_L}\rangle \bigotimes_{q=2}^{q_c} |0_{\mathbf{k}_q}\rangle, \tag{4.12}$$

that is, we impose the electron to be initially placed in the bonding state and the driving field mode $\mathbf{k}_L$ in a coherent state of large amplitude $\alpha_{\mathbf{k}_L}$. Linear polarization is assumed hereupon. It is worth noting at this point that, in general, the harmonic radiation generated using molecular systems is not linearly polarized. However, for linearly polarized driving fields, the generated ellipticity is very small, which leads to almost linearly polarized harmonics [83, 84].

Introducing here the same unitary transformations as done in the description of HHG in atomic systems presented in Sect. 3.3.2, we arrive at the following differential equation

$$i\hbar\frac{\partial|\Psi'(t)\rangle}{\partial t} = \mathsf{e}\hat{\mathbf{E}}(t)\cdot\hat{\mathbf{r}}(t)\,|\Psi'(t)\rangle\,, \quad \text{with} \quad \hat{\mathbf{r}}(t) = \hat{U}_{\text{sc}}^{\dagger}(t)\hat{\mathbf{r}}\hat{U}_{\text{sc}}(t), \tag{4.13}$$

with $\hat{U}_{\text{sc}}(t) \equiv \hat{\mathcal{T}}\exp[-\int_{t_0}^{t}\mathrm{d}\tau\,\hat{H}_{\text{sc}}(t)]$, where $\hat{H}_{\text{sc}}(t) \equiv \hat{H}_{\text{mol}} + \mathsf{e}\mathbf{E}_{\text{cl}}(t)\cdot\hat{\mathbf{r}}$ in this case. Thus, with this notation, the relation between $|\Psi'(t)\rangle$ and $|\Psi(t)\rangle$ is given as in Eq. (3.86), and the initial condition as

$$|\Psi'(t = t_0)\rangle = |\psi_{\mathsf{b}}\rangle\bigotimes_{q=1}^{q_c}|0_{\mathbf{k}_q}\rangle \equiv |\psi_{\mathsf{b}}\rangle\otimes|\bar{0}\rangle. \tag{4.14}$$

Similarly to what we did for atomic HHG in Sect. 3.3.2, we now introduce the SFA version of the identity. However, we depart from standard SFA assumptions by incorporating not only the ground state contribution to the identity but also that of the first excited state. The rationale behind the inclusion of these two rather than just the ground state lies in the fact that, as mentioned earlier, the HHG spectra in molecular systems showcases interferences due to recombinations happening between different atomic centers. Consequently, as we have shown in Eq. (4.10), for H_2^+ the inclusion of the first two excited states is necessary to span a basis of localized states. Therefore, the approximated version of the identity in this case reads as

$$\hat{\mathbb{1}} \approx |\psi_{\mathsf{b}}\rangle\langle\psi_{\mathsf{b}}| + |\psi_{\mathsf{a}}\rangle\langle\psi_{\mathsf{a}}| + \int \mathrm{d}\mathbf{v}\,|\mathbf{v}\rangle\langle\mathbf{v}|\,, \tag{4.15}$$

where $\{|\mathbf{v}\rangle\}$ is the set of scattering states for the molecular system. Introducing this identity in Eq. (4.13), we arrive to

$$i\hbar\frac{\partial|\Psi'(t)\rangle}{\partial t} = \mathsf{e}\hat{\mathbf{E}}(t)\cdot\hat{\mathbf{r}}(t)\Big[\langle\psi_{\mathsf{b}}|\Psi'(t)\rangle\,|\psi_{\mathsf{b}}\rangle + \langle\psi_{\mathsf{a}}|\Psi'(t)\rangle\,|\psi_{\mathsf{a}}\rangle + \int \mathrm{d}\mathbf{v}\langle\mathbf{v}\rangle\Psi'(t)\,|\mathbf{v}\rangle\Big], \tag{4.16}$$

which we project with respect to the bonding and antibonding states to get the following set of coupled differential equations

$$i\hbar\frac{\partial|\Phi_{\mathsf{b}}(t)\rangle}{\partial t} = \boldsymbol{\mu}_{\mathsf{bb}}(t)\cdot\hat{\mathbf{E}}(t)\,|\Phi_{\mathsf{b}}(t)\rangle + \boldsymbol{\mu}_{\mathsf{ba}}(t)\cdot\hat{\mathbf{E}}(t)\,|\Phi_{\mathsf{a}}(t)\rangle + \int \mathrm{d}\mathbf{v}\boldsymbol{\mu}_{\mathsf{bv}}(t)\cdot\hat{\mathbf{E}}(t)\,|\Phi(\mathbf{v},t)\rangle\,, \tag{4.17}$$

$$i\hbar\frac{\partial|\Phi_{\mathsf{a}}(t)\rangle}{\partial t} = \boldsymbol{\mu}_{\mathsf{ab}}(t)\cdot\hat{\mathbf{E}}(t)\,|\Phi_{\mathsf{b}}(t)\rangle + \boldsymbol{\mu}_{\mathsf{aa}}(t)\cdot\hat{\mathbf{E}}(t)\,|\Phi_{\mathsf{aa}}(t)\rangle + \int \mathrm{d}\mathbf{v}\boldsymbol{\mu}_{\mathsf{av}}(t)\cdot\hat{\mathbf{E}}(t)\,|\Phi(\mathbf{v},t)\rangle\,, \tag{4.18}$$

where we have defined $|\Phi_{\text{i}}(t)\rangle \equiv \langle\psi_{\text{i}}|\Psi'(t)\rangle$, $|\Phi(\mathbf{v},t)\rangle \equiv \langle\mathbf{b}|\Psi'(t)\rangle$ for the quantum optical states. Regarding the matrix elements, we have denoted them as $\boldsymbol{\mu}_{\text{ij}}(t) \equiv \text{e}\langle\psi_{\text{i}}|\hat{\mathbf{r}}(t)|\psi_{\text{j}}\rangle$ and $\boldsymbol{\mu}_{\text{iv}}(t) \equiv \text{e}\langle\psi_{\text{i}}|\hat{\mathbf{r}}(t)|\mathbf{v}\rangle$.

The two equations defined in Eqs. (4.17) and (4.18) are first-order inhomogeneous differential equations with well-defined homogeneous and inhomogeneous parts. As we have seen in previous chapters of this Thesis, a solution to these equations can be written as

$$\begin{aligned}|\Phi_{\text{b}}(t)\rangle = \hat{\mathcal{D}}\big(\chi_{\text{b}}(t,t_0)\big)\,|\Phi_{\text{b}}(t_0)\rangle - \frac{i}{\hbar}\int_{t_0}^{t} dt_1 \hat{\mathcal{D}}\big(\chi_{\text{b}}(t,t_1)\big)\hat{M}_{\text{ba}}(t_1)\,|\Phi_{\text{a}}(t_1)\rangle \\ - \frac{i}{\hbar}\int_{t_0}^{t} dt_1 \int d\mathbf{v}\, \hat{\mathcal{D}}\big(\chi_{\text{b}}(t,t_1)\big)\hat{M}_{\text{b}\mathbf{v}}(t_1)\,|\Phi(\mathbf{v},t_1)\rangle,\end{aligned} \tag{4.19}$$

$$\begin{aligned}|\Phi_{\text{a}}(t)\rangle = \hat{\mathcal{D}}\big(\chi_{\text{a}}(t,t_0)\big)\,|\Phi_{\text{b}}(t_0)\rangle - \frac{i}{\hbar}\int_{t_0}^{t} dt_1 \hat{\mathcal{D}}\big(\chi_{\text{a}}(t,t_1)\big)\hat{M}_{\text{ab}}(t_1)\,|\Phi_{\text{b}}(t_1)\rangle \\ - \frac{i}{\hbar}\int_{t_0}^{t} dt_1 \int d\mathbf{v}\, \hat{\mathcal{D}}\big(\chi_{\text{a}}(t,t_1)\big)\hat{M}_{\text{a}\mathbf{v}}(t_1)\,|\Phi(\mathbf{v},t_1)\rangle,\end{aligned} \tag{4.20}$$

where we defined $\hat{M}_{\text{ij}}(t) \equiv \boldsymbol{\mu}_{\text{ij}}(t)\cdot\hat{\mathbf{E}}(t)$ and $\hat{\mathcal{D}}(\boldsymbol{\chi}_{\text{i}}(t,t_0)) \equiv e^{i\varphi_{\text{i},q}(t)}\prod_{q=1}^{q_c}\hat{D}(\chi_q^{(\text{i})}(t,t_0))$. In this case, the amplitude of the displacements $\chi_q^{(\text{i})}(t,t_0)$ and the phase $\varphi_q^{(\text{i})}(t)$ depend on the specific molecular orbital under consideration. More specifically, they are given by

$$\chi_q^{(\text{i})}(t,t_0) = -\frac{1}{\hbar}\sqrt{\frac{\hbar\omega_{\mathbf{k}_q}}{2\varepsilon_0 V}}\int_{t_0}^{t} d\tau e^{i\omega_{\mathbf{k}_q}\tau}\big(\boldsymbol{\epsilon}_{\mathbf{k}_q}\cdot\boldsymbol{\mu}_{\text{ii}}(\tau)\big), \tag{4.21}$$

$$\varphi_q^{(\text{i})}(t) = \frac{\text{e}^2}{\hbar^2}\frac{\hbar\omega_{\mathbf{k}_q}}{2\varepsilon_0 V}\int_{t_0}^{t} dt_1 \int_{t_0}^{t_1} dt_2 \big(\boldsymbol{\epsilon}_{\mathbf{k}_q}\cdot\boldsymbol{\mu}_{\text{ii}}(t_1)\big)\big(\boldsymbol{\epsilon}_{\mathbf{k}_q}\cdot\boldsymbol{\mu}_{\text{ii}}(t_2)\big)\sin(\omega_{\mathbf{k}_q}(t_1-t_2)). \tag{4.22}$$

From the results obtained for the set of coupled differential equations, we can distinguish three different terms. Specifically, for Eq. (4.19) (Eq. (4.20)) the first term introduces a multimode displacement due to the effective oscillation the electron performs from and to the bonding (antibonding) state. The second term introduces quantum optical fluctuations due to transitions from bonding (antibonding) to antibonding (bonding) states, while the last term introduces quantum optical fluctuations due to the back-action of the ionized photoelectrons.

In the following, similarly to what was done for atomic HHG processes, we invoke the SFA assumptions and consider the contribution of the last term to be much smaller than the other two at all times. This holds true in the regime of weak depletion of the ground state, and when the antibonding state becomes more populated than the continuum ones. We anticipate this approximation to be valid within the parameter regime we are working since, during the HHG dynamics, the electron undergoes transitions where it either recombines with the initial molecular center or with the opposite center, leading to interference effects in the resulting HHG spectra [53–55]. In agreement with Eq. (4.10), a set of localized states approximately describing localization in each center needs contributions from both bonding and antibonding states. For this reason, we keep both bound states in our description. Therefore, in the following, we approximate Eqs. (4.19) and (4.20) as

$$|\Phi_{\mathsf{b}}(t)\rangle \approx \hat{\mathcal{D}}\big(\chi_{\mathsf{b}}(t,t_0)\big)\,|\Phi_{\mathsf{b}}(t_0)\rangle - \frac{i}{\hbar}\int_{t_0}^{t} \mathrm{d}t_1 \hat{\mathcal{D}}\big(\chi_{\mathsf{b}}(t,t_1)\big)\hat{M}_{\mathsf{ba}}(t_1)\,|\Phi_{\mathsf{a}}(t_1)\rangle\,, \tag{4.23}$$

$$|\Phi_{\mathsf{a}}(t)\rangle \approx \hat{\mathcal{D}}\big(\chi_{\mathsf{a}}(t,t_0)\big)\,|\Phi_{\mathsf{a}}(t_0)\rangle - \frac{i}{\hbar}\int_{t_0}^{t} \mathrm{d}t_1 \hat{\mathcal{D}}\big(\chi_{\mathsf{a}}(t,t_1)\big)\hat{M}_{\mathsf{ab}}(t_1)\,|\Phi_{\mathsf{b}}(t_1)\rangle\,. \tag{4.24}$$

As observed from Eqs. (4.23) and (4.24), the solution of each equation depends on the other, as expected from their coupled structure. By combining them, one arrives at the following set of recursive relations

$$\begin{aligned}|\Phi_{\mathsf{b}}(t)\rangle &= \hat{\mathcal{D}}\big(\chi_{\mathsf{b}}(t,t_0)\big)\,|\Phi_{\mathsf{b}}(t_0)\rangle - \frac{i}{\hbar}\int_{t_0}^{t}\mathrm{d}t_1 \hat{\mathcal{D}}\big(\chi_{\mathsf{b}}(t,t_1)\big)\hat{M}_{\mathsf{ba}}(t_1)\hat{\mathcal{D}}\big(\chi_{\mathsf{a}}(t_1,t_0)\big)\,|\Phi_{\mathsf{a}}(t_0)\rangle\\ &\quad + \left(\frac{i}{\hbar}\right)^2\int_{t_0}^{t}\mathrm{d}t_1\int_{t_0}^{t_1}\mathrm{d}t_2 \hat{\mathcal{D}}\big(\chi_{\mathsf{b}}(t,t_1)\big)\hat{M}_{\mathsf{ba}}(t_1)\hat{\mathcal{D}}\big(\chi_{\mathsf{a}}(t_1,t_2)\big)\hat{M}_{\mathsf{ab}}(t_2)\,|\Phi_{\mathsf{b}}(t_2)\rangle\,,\end{aligned} \tag{4.25}$$

$$\begin{aligned}|\Phi_{\mathsf{a}}(t)\rangle &= \hat{\mathcal{D}}\big(\chi_{\mathsf{a}}(t,t_0)\big)\,|\Phi_{\mathsf{a}}(t_0)\rangle - \frac{i}{\hbar}\int_{t_0}^{t}\mathrm{d}t_1 \hat{\mathcal{D}}\big(\chi_{\mathsf{a}}(t,t_1)\big)\hat{M}_{\mathsf{ab}}(t_1)\hat{\mathcal{D}}\big(\chi_{\mathsf{b}}(t_1,t_0)\big)\,|\Phi_{\mathsf{b}}(t_0)\rangle\\ &\quad + \left(\frac{i}{\hbar}\right)^2\int_{t_0}^{t}\mathrm{d}t_1\int_{t_0}^{t_1}\mathrm{d}t_2 \hat{\mathcal{D}}\big(\chi_{\mathsf{a}}(t,t_1)\big)\hat{M}_{\mathsf{ab}}(t_1)\hat{\mathcal{D}}\big(\chi_{\mathsf{b}}(t_1,t_2)\big)\hat{M}_{\mathsf{ba}}(t_2)\,|\Phi_{\mathsf{a}}(t_2)\rangle\,,\end{aligned} \tag{4.26}$$

where each recursive iteration leads to additional transitions between bonding and antibonding states through the operators $\hat{M}_{\mathsf{ab}}(t)$ and its Hermitian conjugate. By further introducing the initial conditions depicted in Eq. (4.14), we arrive at

$$\begin{aligned}|\Phi_{\mathsf{b}}(t)\rangle &= \hat{\mathcal{D}}\big(\chi_{\mathsf{b}}(t,t_0)\big)|\bar{0}\rangle\\ &\quad + \left(\frac{i}{\hbar}\right)^2\int_{t_0}^{t}\mathrm{d}t_1\int_{t_0}^{t_1}\mathrm{d}t_2 \hat{\mathcal{D}}\big(\chi_{\mathsf{b}}(t,t_1)\big)\hat{M}_{\mathsf{ba}}(t_1)\hat{\mathcal{D}}\big(\chi_{\mathsf{a}}(t_1,t_2)\big)\hat{M}_{\mathsf{ab}}(t_2)\,|\Phi_{\mathsf{b}}(t_2)\rangle\,,\end{aligned} \tag{4.27}$$

$$\begin{aligned}|\Phi_{\mathsf{a}}(t)\rangle &= -\frac{i}{\hbar}\int_{t_0}^{t}\mathrm{d}t_1 \hat{\mathcal{D}}\big(\chi_{\mathsf{a}}(t,t_1)\big)\hat{M}_{\mathsf{ab}}(t_1)\hat{\mathcal{D}}\big(\chi_{\mathsf{b}}(t_1,t_0)\big)|\bar{0}\rangle\\ &\quad + \left(\frac{i}{\hbar}\right)^2\int_{t_0}^{t}\mathrm{d}t_1\int_{t_0}^{t_1}\mathrm{d}t_2 \hat{\mathcal{D}}\big(\chi_{\mathsf{a}}(t,t_1)\big)\hat{M}_{\mathsf{ab}}(t_1)\hat{\mathcal{D}}\big(\chi_{\mathsf{b}}(t_1,t_2)\big)\hat{M}_{\mathsf{ba}}(t_2)\,|\Phi_{\mathsf{a}}(t_2)\rangle\,.\end{aligned} \tag{4.28}$$

Let us now proceed with the analysis of the processes described by these equations. In Eq. (4.27), the first term depicts an event in which the electron populates the bonding state exclusively. Due to the HHG dynamics, which are identical to the atomic dynamics discussed in Sect. 3.3.2, each harmonic mode experiences a shift of $\chi_q^{(\mathrm{b})}(t, t_0)$. In contrast, the second term presents a process where at time t_2, the electron transitions from the bonding state to the antibonding state, with the interaction mediated by the operator $\hat{M}_{\mathrm{ab}}(t_2)$. Then, within the time interval $[t_2, t_1]$, where $t_1 \geq t_2$, the quantum optical state experiences a displacement of $\chi_q^{(\mathrm{a})}(t_1, t_2)$ due to the interaction between the electron and the field modes when it resides in the antibonding state. Finally, at time t_1, the electron returns to the bonding state, where it remains until the final time t, yielding an extra displacement $\chi_q^{(\mathrm{b})}(t, t_1)$ on the field modes. In the case of Eq. (4.28), similar dynamics are observed, although with the absence of the first term in Eq. (4.27) due to the imposed initial conditions. Consequently, the only mechanism that allows the electron to be in the antibonding state involves a transition from the bonding state.

In the following, we restrict our equations to first order interaction terms. In other words, we consider scenarios in which the electron undergoes only one transition between the bonding and the antibonding state, or vice versa. This approximation is applicable in situations where $|\boldsymbol{\mu}_{\mathrm{aa}}(t)| > |\boldsymbol{\mu}_{\mathrm{ab}}(t)|$ and $|\boldsymbol{\mu}_{\mathrm{bb}}(t)| > |\boldsymbol{\mu}_{\mathrm{ab}}(t)|$ (it is worth noting that $|\boldsymbol{\mu}_{\mathrm{ba}}(t)| = |\boldsymbol{\mu}_{\mathrm{ab}}(t)|$). In a different phrasing, this approximation is valid when, at all times, the likelihood of transitioning between bonding and antibonding states, or vice versa, is lower than the probability of remaining in the bonding (antibonding) state. To confirm that this condition is satisfied in our analysis, we present in Fig. 4.3 the behavior of $|\boldsymbol{\mu}_{\mathrm{ab}}(t)|$ in comparison to $|\boldsymbol{\mu}_{\mathrm{bb}}(t)|$ (panel (a)) and $|\boldsymbol{\mu}_{\mathrm{aa}}(t)|$ (panel (b)) as a function of time. The figure demonstrates that these conditions hold for all time values. It is important to emphasize that, even though we have employed $R = 2.5$ a.u. in the numerical calculations leading to Fig. 4.3, the same behavior is observed for the set of bond lengths that will be used throughout the remainder of this section.

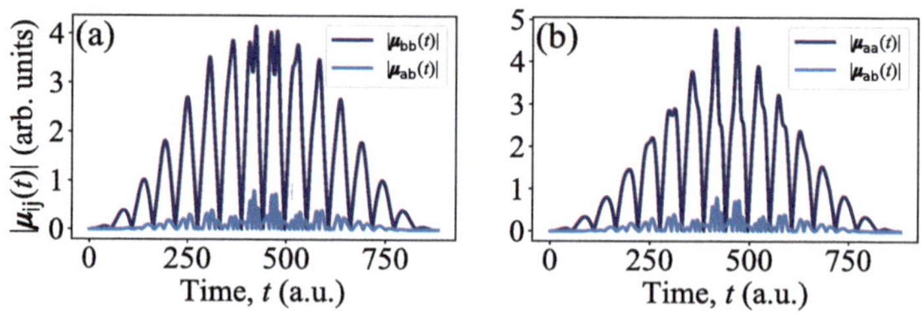

Fig. 4.3 Time-dependent dipole moment matrix elements for different combinations of bonding and antibonding states. In **a**, the absolute value of the matrix element $\mu_{\mathrm{bb}}(t)$ (dark blue curve) is compared to $\mu_{\mathrm{ab}}(t)$ (light blue curve). In **b**, the latter is contrasted with $\mu_{\mathrm{aa}}(t)$. The molecular ion, with a bond length $R = 2.5$ a.u., was excited by a sinusoidal squared laser pulse with a peak intensity of $I = 5 \times 10^{14}$ W/cm^2, central wavelength $\lambda_{\mathbf{k}_L} = 800$ nm, and a duration of $\Delta t \approx 21$ fs (8 optical cycles)

With all of this considered, the final state of the joint electron-light system can be approximated as follows

$$|\Psi'(t)\rangle \approx \frac{1}{\sqrt{\mathcal{N}}}\Big[|\psi_{\mathsf{b}}\rangle \otimes |\Phi_{\mathsf{b}}(t)\rangle + |\psi_{\mathsf{a}}\rangle \otimes |\Phi_{\mathsf{a}}(t)\rangle\Big], \tag{4.29}$$

which can also be expressed in terms of the localized states $\{|\bar{L}\rangle, |\bar{R}\rangle\}$ presented in Eq. (4.10) as

$$\begin{aligned}|\Psi'(t)\rangle &= \frac{1}{\sqrt{2\mathcal{N}}}\Big[|\bar{L}\rangle \otimes \Big(|\Phi_{\mathsf{b}}(t)\rangle - |\Phi_{\mathsf{a}}(t)\rangle\Big) + |\bar{R}\rangle \otimes \Big(|\Phi_{\mathsf{b}}(t)\rangle + |\Phi_{\mathsf{a}}(t)\rangle\Big)\Big]\\ &\equiv \frac{1}{\sqrt{2\mathcal{N}}}\Big[|\bar{L}\rangle \otimes |\Phi_{\bar{L}}(t)\rangle + |\bar{R}\rangle \otimes |\Phi_{\bar{R}}(t)\rangle\Big],\end{aligned} \tag{4.30}$$

with the states $|\Phi_{\mathsf{b}}(t)\rangle$ and $|\Phi_{\mathsf{a}}(t)\rangle$ given by

$$|\Phi_{\mathsf{b}}(t)\rangle \approx \hat{\mathcal{D}}\big(\chi_{\mathsf{b}}(t, t_0)\big)|\bar{0}\rangle, \tag{4.31}$$

$$|\Phi_{\mathsf{a}}(t)\rangle \approx -\frac{i}{\hbar}\int_{t_0}^{t} \mathrm{d}t_1 \hat{\mathcal{D}}\big(\chi_{\mathsf{a}}(t, t_1)\big)\hat{M}_{\mathsf{ab}}(t_1)\hat{\mathcal{D}}\big(\chi_{\mathsf{b}}(t_1, t_0)\big)|\bar{0}\rangle. \tag{4.32}$$

4.1.2.1 Extension to the Many-Molecules Scenario

Up to this point, our analysis has centered on the single-active electron approximation in the context of a single molecular system. However, in practice, we usually encounter multiple molecules independently interacting with the driving electromagnetic field. In this subsection, we expand upon previously discussed results to encompass situations where a total of N_{mol} molecules independently interact with the driving field. In this scenario, the total Hamiltonian, under the Born-Oppenheimer and dipole approximations, can be expressed as follows

$$\hat{H}(t) = \sum_{i=1}^{N_{\text{mol}}}\Big[\hat{H}_{i,\text{mol}} + \mathsf{e}\hat{\mathbf{E}}(t)\cdot\hat{\mathbf{r}}_i\Big] + \hat{H}_{\text{field}}, \tag{4.33}$$

with the subscript i ranging over all the molecules involved in the dynamics. By introducing the same unitary transformations as in the single-molecule analysis, we arrive to the following Schrödinger equation

$$i\hbar\frac{\partial|\bar{\Psi}'(t)\rangle}{\partial t} = \mathsf{e}\sum_{i=1}^{N_{\text{mol}}}\hat{\mathbf{E}}(t)\cdot\hat{\mathbf{r}}_i(t)|\bar{\Psi}'(t)\rangle, \tag{4.34}$$

where $|\bar{\Psi}'(t)\rangle$ represents the joint state of the molecules and the electromagnetic field. In this context, we consider the initial state of the system to be given by

$$|\bar{\Psi}'(t)\rangle = \bigotimes_{i=1}^{N_{\text{mol}}} |\psi_{i,\text{b}}\rangle \otimes |\bar{0}\rangle. \tag{4.35}$$

To solve this equation, the same steps as were taken in the single-molecule scenario can be followed. This involves introducing the (many-molecule) SFA version of the identity operator and projecting the equation with respect to specific states, resulting in a set of coupled differential equations. However, for the sake of clarity, at this stage we intentionally neglect terms accounting for quantum optical fluctuations due to the ionized photoelectrons' back-action. In formal terms, these approximations should be performed at the end of our analysis, as presented in the atomic and single-molecule scenarios. Nevertheless, the results we would obtain by delaying these approximations are fundamentally consistent. Making these approximations at this point proves particularly beneficial to prevent the notation from becoming excessively cumbersome. Therefore, having this in mind and projecting Eq. (4.35) with respect to $|\psi_{\mathbf{m}}\rangle \equiv |\psi_{1,\text{i}}\rangle \otimes |\psi_{2,\text{j}}\rangle \otimes \cdots \otimes |\psi_{N_{\text{mol}},\text{N}}\rangle$, where $\mathbf{m} = \{(1, \text{i}), (2, \text{j}), \cdots, (N_{\text{mol}}, \text{N})\}$ denotes the state of each molecule, we arrive at

$$i\hbar \frac{\partial |\Phi_{\mathbf{m}}(t)\rangle}{\partial t} \approx \text{e} \sum_{\mathbf{n}} \Bigg[\langle \psi_{\mathbf{m}}| \sum_{i=1}^{N_{\text{mol}}} \hat{\mathbf{r}}_i(t) \, |\psi_{\mathbf{n}}\rangle \cdot \hat{\mathbf{E}}(t) \otimes |\Phi_{\mathbf{n}}(t)\rangle \Bigg], \tag{4.36}$$

where the sum over $\mathbf{n}$ encompasses all possible combinations of states in which the molecule can end up. Hereupon, we restrict to scenarios where it is exceedingly unlikely for the molecule to undergo a transition from the bonding to the antibonding state. The primary implication of this restriction is that, after the HHG dynamics, nearly all the molecules ultimately return to the initial state, namely, the bonding state, with the exception of one molecule that effectively performs a transition from the bonding state to the antibonding state. Mathematically, this implies that in the summation over $\mathbf{n}$, we consider only those terms for which $\mathbf{n} = \bar{\text{b}} := \{\mathbf{m} : \text{i} = \text{j} = \cdots = \text{N} = \text{b}\}$ and $\mathbf{n} = \bar{\text{a}}_{\text{k}} := \{\mathbf{m} : \text{i} = \text{j} = \cdots \neq \text{k} \neq \cdots = \text{N} = \text{b} \,\&\, \text{k} = \text{a}\}$, in other words, those cases in which the electrons either return to the bonding state or where at most one of them ends up in the antibonding state. Under this assumption, the system of equations reads

$$i\hbar \frac{\partial |\Phi_{\bar{\text{b}}}(t)\rangle}{\partial t} = N_{\text{mol}} \boldsymbol{\mu}_{\text{bb}}(t) \cdot \hat{\mathbf{E}}(t) \, |\Phi_{\bar{\text{b}}}(t)\rangle + \boldsymbol{\mu}_{\text{ba}}(t) \cdot \hat{\mathbf{E}}(t) \sum_{\text{k}=1}^{N_{\text{mol}}} |\Phi_{\bar{\text{a}}_{\text{k}}}(t)\rangle \,, \tag{4.37}$$

$$i\hbar \frac{\partial |\Phi_{\bar{\text{a}}_{\text{k}}}(t)\rangle}{\partial t} = \Big[\boldsymbol{\mu}_{\text{aa}}(t) + (N_{\text{mol}} - 1) \boldsymbol{\mu}_{\text{bb}}(t) \Big] \cdot \hat{\mathbf{E}}(t) |\Phi_{\bar{\text{a}}_{\text{k}}}(t)\rangle + \boldsymbol{\mu}_{\text{ab}}(t) \cdot \hat{\mathbf{E}}(t) \, |\Phi_{\bar{\text{b}}}(t)\rangle \quad \forall \text{k} \in \{1, 2, \cdots, N_{\text{mol}}\}. \tag{4.38}$$

Although these equations have different prefactors, they exhibit an identical structure to those presented in Eqs. (4.17) and (4.18), albeit without the presence of the continuum terms. Consequently, they can be solved in a similar manner to the single-molecule scenario. By applying the same approximations about neglecting the higher-order terms in the transitions, we obtain

$$|\Phi_{\bar{\mathsf{b}}}(t)\rangle \approx \hat{\mathcal{D}}\big(N_{\text{mol}}\boldsymbol{\chi}_{\mathsf{b}}(t,t_0)\big)|\bar{0}\rangle, \tag{4.39}$$

$$|\Phi_{\bar{\mathsf{a}}_{\mathsf{k}}}(t)\rangle \approx -\frac{i}{\hbar}\int_{t_0}^{t} \mathrm{d}t_1 \hat{\mathcal{D}}\Big(\boldsymbol{\chi}_{\mathsf{a}}(t,t_1) + (N_{\text{mol}}-1)\boldsymbol{\chi}_{\mathsf{b}}(t,t_1)\Big)\hat{M}_{\mathsf{ab}}(t_1) \times \hat{\mathcal{D}}\big(N_{\text{mol}}\boldsymbol{\chi}_{\mathsf{b}}(t_1,t_0)\big)|\bar{0}\rangle, \tag{4.40}$$

which, when dealing with significantly large number of molecules, as commonly encountered in practical scenarios, we can make the approximation $N_{\text{mol}} - 1 \approx N_{\text{mol}}$. Furthermore, considering that the generated displacements in the single-molecule case $\chi_{\mathsf{i}}(t, t_0)$ are typically perturbative, we can make the additional approximation $N_{\text{mol}}\chi_{\mathsf{b}}(t, t_0) - \chi_{\mathsf{a}}(t, t_0) \approx N_{\text{mol}}\chi_{\mathsf{b}}(t, t_0)$. This allows us to rephrase Eq. (4.40) as

$$|\Phi_{\bar{\mathsf{a}}}(t)\rangle \approx -\frac{i}{\hbar}\hat{\mathcal{D}}\big(N_{\text{mol}}\boldsymbol{\chi}_{\mathsf{b}}(t,t_0)\big)\int_{t_0}^{t}\mathrm{d}t_1 e^{\theta_{\mathsf{b}}}\boldsymbol{\mu}_{\mathsf{ab}}(t)\cdot\Big[\hat{\mathbf{E}}(t_1) + \mathbf{E}_{\text{cl}}^{(\mathsf{b})}(t_1)\Big]|\bar{0}\rangle, \tag{4.41}$$

such that the multimode displacement operator can be factored out of the integral, ultimately simplifying the numerical analysis. In this expression, we have defined

$$\theta_{\mathsf{b}} \equiv N_{\text{mol}}^2 \sum_q \frac{1}{2}\Big[\chi_{\mathsf{b}}^{(q)}(t,t_1)\big(\chi_{\mathsf{b}}^{(q)}(t_1,t_0)\big)^* - \big(\chi_{\mathsf{b}}^{(q)}(t,t_1)\big)^*\chi_{\mathsf{b}}^{(q)}(t_1,t_0)\Big], \tag{4.42}$$

$$\mathbf{E}_{\text{cl}}^{(\mathsf{b})}(t) \equiv -iN_{\text{mol}}\sum_q \mathbf{g}(\omega_{\mathbf{k}_q})\Big[\big(\chi_{\mathsf{b}}^{(q)}(t,t_0)\big)^* e^{i\omega_{\mathbf{k}_q}t} - \chi_{\mathsf{b}}^{(q)}(t,t_0)e^{-i\omega_{\mathbf{k}_q}t}\Big]. \tag{4.43}$$

With all this, the joint state of the system reads as follows

$$\begin{aligned} |\Psi'(t)\rangle &= \frac{1}{\sqrt{N}}\Bigg[|\psi_{\bar{\mathsf{b}}}\rangle \otimes |\Phi_{\bar{\mathsf{b}}}(t)\rangle + \sum_{\mathsf{k}=1}^{\mathrm{N_{mol}}}|\psi_{\bar{\mathsf{a}}_{\mathsf{k}}}\rangle \otimes |\Phi_{\bar{\mathsf{a}}_{\mathsf{k}}}\rangle\Bigg] \\ &= \frac{1}{\sqrt{N}}\Big[|\psi_{\bar{\mathsf{b}}}\rangle \otimes |\Phi_{\bar{\mathsf{b}}}(t)\rangle + \sqrt{N_{\text{mol}}}\,|\psi_{\bar{\mathsf{a}}}\rangle \otimes |\Phi_{\bar{\mathsf{a}}}\rangle\Big], \end{aligned} \tag{4.44}$$

where in transitioning from the first to the second equality, we take into account that the quantum optical state $|\Phi_{\bar{\mathsf{a}}_{\mathsf{k}}}\rangle$ maintains the same expression regardless of the molecule whose electron undergoes the transition from the bonding to the anti-bonding state. As a result, we can remove the subscript k. Additionally, we define $|\psi_{\bar{\mathsf{a}}}\rangle = (1/\sqrt{N_{\text{mol}}})\sum_{\mathsf{k}=1}^{N_{\text{mol}}}|\psi_{\bar{\mathsf{a}}_{\mathsf{k}}}\rangle$.

It is important to note that in scenarios where N_{mol} becomes extremely large, the term proportional to $\sqrt{N_{\text{mol}}}$ in Eq. (4.44) can become dominant. Within this regime, the assumption of restricting to first-order terms regarding the bonding-antibonding transitions may not longer hold true. Therefore, in the following we limit our focus to situations where the probability of transitions between bonding and antibonding states in the many-molecule regime, remains lower than those in which all the electrons exclusively occupy the bonding state. These approximations yield results consistent with the expectations derived from semiclassical analyses.

In the following, unless otherwise is stated, the same laser parameters as those outlined in the caption of Fig. 4.3 are employed. However, the bond length of the molecule is used as free parameter.

4.1.2.2 Mean Photon Number of the Different Harmonic Modes

In this subsection, we primarily focus on the analysis of the mean photon number of the different harmonic modes. This analysis provides an efficient way of interpreting the HHG spectra from a quantum optical perspective [85]. Furthermore, it also serve us as a benchmark for evaluating the validity of the approximations we have introduced thus far, ensuring their alignment with the predictions of semiclassical theories.

With this purpose in mind, we commence with the single-molecule regime and make the assumption that we have no knowledge about the state with which the electron has recombined with. In this context, the density matrix describing the quantum optical state at time t reads as follows

$$\hat{\rho}_{\text{field}}(t) = \text{tr}_{\text{elec}}\Big(\big|\Psi'(t)\big\rangle\!\big\langle\Psi'(t)\big| \Big) = \frac{1}{\mathcal{N}}\Big[|\Phi_{\text{b}}(t)\rangle\!\langle\Phi_{\text{b}}(t)| + |\Phi_{\text{a}}(t)\rangle\!\langle\Phi_{\text{a}}(t)| \Big], \tag{4.45}$$

where we have performed a partial trace with respect to the electronic degrees of freedom. As a result, the mean photon number present in the qth harmonic mode can be calculated as

$$\begin{aligned} \langle \hat{a}^{\dagger}_{\mathbf{k}_q} \hat{a}_{\mathbf{k}_q} \rangle &= \frac{1}{\mathcal{N}}\Big[\langle \Phi_{\text{b}}(t)| \, \hat{a}^{\dagger}_{\mathbf{k}_q} \hat{a}_{\mathbf{k}_q} \, |\Phi_{\text{b}}(t)\rangle + \langle \Phi_{\text{a}}(t)| \, \hat{a}^{\dagger}_{\mathbf{k}_q} \hat{a}_{\mathbf{k}_q} \, |\Phi_{\text{a}}(t)\rangle \Big] \\ &\equiv \frac{1}{\mathcal{N}}\Big[\langle \hat{a}^{\dagger}_{\mathbf{k}_q} \hat{a}_{\mathbf{k}_q} \rangle_{\text{b}} + \langle \hat{a}^{\dagger}_{\mathbf{k}_q} \hat{a}_{\mathbf{k}_q} \rangle_{\text{a}} \Big]. \end{aligned} \tag{4.46}$$

The results obtained from the numerical evaluation of Eq. (4.46) are shown in Fig. 4.4, where we considered the interaction of H_2^+ with a laser field linearly polarized along the molecular axis, with peak intensity $I = 5 \times 10^{14}$ W/cm^2, central wavelength $\lambda_{\mathbf{k}_L} = 800$ nm and with a total duration of $\Delta t \approx 21$ fs (8 optical cycles). Specifically, in panels (a) and (b), we separately depict the contributions of $\langle \hat{a}^{\dagger}_{\mathbf{k}_q} \hat{a}_{\mathbf{k}_q} \rangle_{\text{b}}$ (darker blue curve with circular markers) and $\langle \hat{a}^{\dagger}_{\mathbf{k}_q} \hat{a}_{\mathbf{k}_q} \rangle_{\text{a}}$ (lighter blue curve with square markers), for two bond lengths.

For both the bonding and antibonding components, characteristic features of the H_2^+ HHG spectra [8, 52–55] are evident, particularly the presence of several plateau regions. More specifically, three distinguishable plateaus can be roughly identified: one located between the 1st and 13th harmonic mode, another between the 13th and

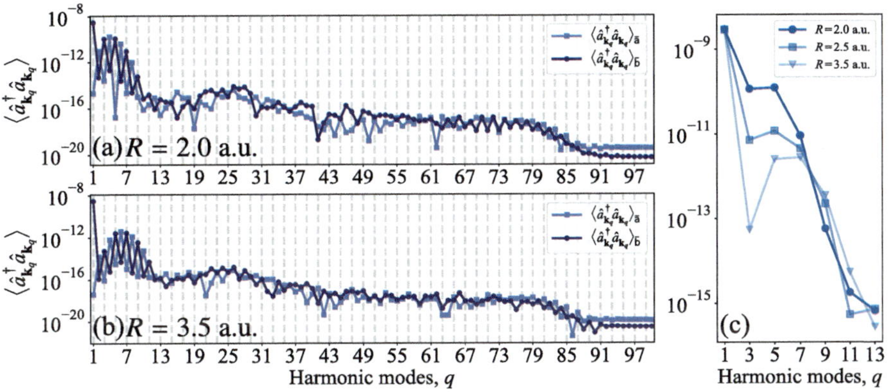

Fig. 4.4 Mean photon number for different harmonic modes under single-molecule dynamics. In panels **a** and **b**, two bond lengths have been considered, specifically $R = 2.0$ a.u. and $R = 3.5$ a.u., respectively. The results exhibit the separate contributions of the bonding state (dark blue curve with circular markers) and antibonding state (light blue curve with square markers). In **c**, the mean photon number for the first seven odd harmonic orders is depicted, with each curve representing a different bond length. These calculations were conducted for an H_2^+ molecule excited with a sinusoidal squared laser pulse of peak intensity $I = 5 \times 10^{14}$ W/cm^2, central wavelength $\lambda_{\mathbf{k}_L} = 800$ nm, and duration $\Delta t \approx 21$ fs (8 optical cycles)

the 41st harmonic mode, and a final one extending until the 80th harmonic mode. The presence of this multiple plateau structure has been attributed to the interference of HHG mechanisms where the electron ends up in the same center where it originally was, or in a different one [53–55]. These interferences have been shown to depend on various factors such as the intensity of the driving field, its orientation with respect to the molecular alignment, and the specific molecule that is considered [8, 51, 52]. In this case, while in the second and third plateau structures the presence of even and odd harmonics cannot be clearly distinguished,[4] in the first plateau structure well-resolved harmonics can be observed. However, the parity of these harmonics depends on the electron's recombination process, as $\langle \hat{a}^\dagger_{\mathbf{k}_q} \hat{a}_{\mathbf{k}_q} \rangle_{\mathrm{b}}$ contributes with odd harmonic orders, while $\langle \hat{a}^\dagger_{\mathbf{k}_q} \hat{a}_{\mathbf{k}_q} \rangle_{\mathrm{a}}$ contributes with even orders. Although odd harmonics are typically expected in parametric HHG processes, the change in parity observed in the second case arises from the electron transitioning from the bonding to the antibonding state, which have opposite parity. As a result, this inversion in the final electronic state is reflected in the mean photon number distribution shown in Fig. 4.4, featuring the presence of even harmonic orders [88, 89].

The bond length also plays a crucial role on the generated harmonic radiation, particularly its intensity. This can be already observed when comparing the difference in mean photon numbers between Fig. 4.4a and b, with the distinction becoming more explicit in panel (c). In this plot, the total mean photon number, as computed

[4] We remind that this is a common feature resulting from the interference of distinct electronic trajectories recombining at different times [86, 87].

in Eq. (4.46), is presented for the first plateau region when considering different values of R. It is observed that, as R increases, the peak of the spectrum for $q > 1$ becomes smaller. This behavior can be better understood by visualizing HHG processes in terms of recombinations with right and left centers. Within this framework, the larger the distance between the two centers, the lower the probability of ionization-recombination mechanisms occurring between them. Consequently, this leads to a reduced efficiency of the HHG process and a smaller mean photon number. However, it is worth noting that this behavior concerning the bond length strongly depends on the molecular alignment with respect to the field's polarization orientation [90].

However, one of the most surprising aspects about Fig. 4.4 is that the relative contribution between $\langle \hat{a}^\dagger_{\mathbf{k}_q} \hat{a}_{\mathbf{k}_q} \rangle_{\mathsf{b}}$ and $\langle \hat{a}^\dagger_{\mathbf{k}_q} \hat{a}_{\mathbf{k}_q} \rangle_{\mathsf{a}}$ show the same order of magnitude. This is because the probability of generating a photon within the single-molecule scenario is almost equal for both the bonding-bonding and bonding-antibonding transition channels. This probability ranges from 10^{-10} until 10^{-16} from the lowest to the highest harmonic orders in panel (a). Nevertheless, when we focus on the electronic population of both energetic states, the probability of finding the electron in the bonding state becomes dominant, as in most cases the electron barely interacts with the field.

If the same analysis is performed in the many-molecule regime, Eq. (4.46) is replaced by

$$\begin{aligned}\langle \hat{a}^\dagger_{\mathbf{k}_q} \hat{a}_{\mathbf{k}_q} \rangle &= \frac{1}{N}\Big[\langle \Phi_{\bar{\mathsf{b}}}(t)| \, \hat{a}^\dagger_{\mathbf{k}_q} \hat{a}_{\mathbf{k}_q} \, |\Phi_{\bar{\mathsf{b}}}(t)\rangle + N_{\text{mol}} \, \langle \Phi_{\bar{\mathsf{a}}}(t)| \, \hat{a}^\dagger_{\mathbf{k}_q} \hat{a}_{\mathbf{k}_q} \, |\Phi_{\bar{\mathsf{a}}}(t)\rangle \Big] \\ &\equiv \frac{1}{N}\Big[\langle \hat{a}^\dagger_{\mathbf{k}_q} \hat{a}_{\mathbf{k}_q} \rangle_{\bar{\mathsf{b}}} + N_{\text{mol}} \langle \hat{a}^\dagger_{\mathbf{k}_q} \hat{a}_{\mathbf{k}_q} \rangle_{\bar{\mathsf{a}}} \Big],\end{aligned} \tag{4.47}$$

whose outcomes from the numerical analysis are presented in Fig. 4.5 when setting $N_{\text{mol}} = 1 \times 10^8$ for the number of molecules[5]. Comparing these results with those of Fig. 4.4, we observe that the contributions of $\langle \hat{a}^\dagger_{\mathbf{k}_q} \hat{a}_{\mathbf{k}_q} \rangle_{\mathsf{b}}$ scale with N^2_{mol}, while the contributions of $\langle \hat{a}^\dagger_{\mathbf{k}_q} \hat{a}_{\mathbf{k}_q} \rangle_{\mathsf{a}}$ scale linearly with N_{mol}. This behavior is consistent with expectations from semiclassical theories [91].

4.1.2.3 Wigner Function for Different Field Modes

In atomic systems, we observed that the non-classical states of light could be obtained through autocorrelation measurements involving the fundamental mode and the generated harmonics. In this context, HHG behaved as a parametric process, where the initial and final state correspond to the ground state of the respective atom. However,

[5] It is important to note that in the numerical analysis, N_{mol} does not correspond exactly to the actual number of molecules, but is a proportionate quantity. This discrepancy arises from the use of arbitrary units in the numerical evaluation of the time-dependent dipole moments $\boldsymbol{\mu}_{\mathsf{ij}}(t)$. As a result, there is a scaling factor between the actual number of molecules and the introduced N_{mol}. However, we estimate this scaling factor to be on the order of $10^{-2} \sim 10^{-3}$.

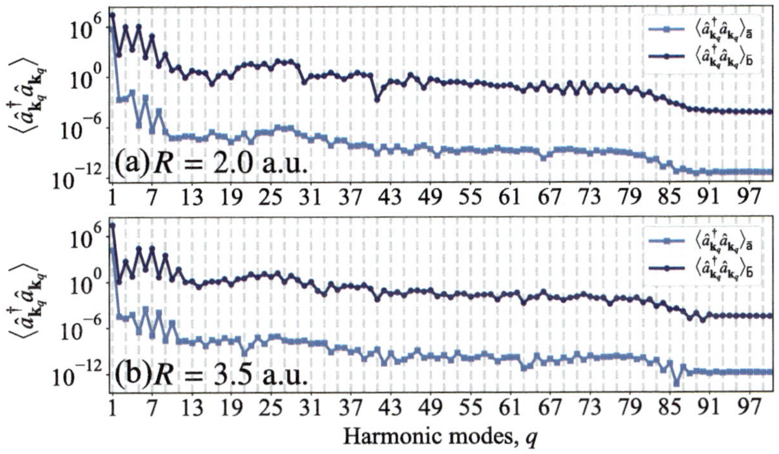

Fig. 4.5 Mean photon number distribution for different harmonic modes when considering $\mathbf{N_{mol}}$ molecules. Both plots maintain a fixed number of molecules, $N_{\text{mol}} = 10^8$. Panels **a** and **b** correspond to bond lengths of $R = 2.0$ a.u. and $R = 3.5$ a.u., respectively. It is observed that recombination events culminating in the bonding state (dark blue curve with circular markers) contribute more significantly to the mean photon number compared to those leading to the antibonding state (light blue curve with squared markers). Specifically, the former scales with N_{mol}^2 while the latter scales with N_{mol}

in the analysis presented here for molecular systems, we observe that the introduction of excited states could potentially lead to non-classical states, as both Eqs. (4.29) and (4.44) are generally entangled states. Nevertheless, the choice of using molecules rather than atoms for an analysis involving the excited states as well is not arbitrary. In the case of molecular systems, indications of the significance of these two lowest energetic states are already observed in semiclassical HHG spectra [8, 51–55], and their importance also depends on the excitation conditions [90]. Thus, it is expected that non-classical features related to the involvement of excited states to be more pronounced in these systems.

In this section, we take the initial steps in characterizing non-classical properties of the generated states. More specifically, we investigate the Wigner function of various harmonic modes under different conditions. This includes situations where we lack knowledge about the final electronic state, as well as scenarios in which the final state of the electron is known. For the first scenario, the quantum optical state is described by Eq. (4.45). Therefore, in accordance with Eq. (2.76), we can express the Wigner function of the qth harmonic mode as

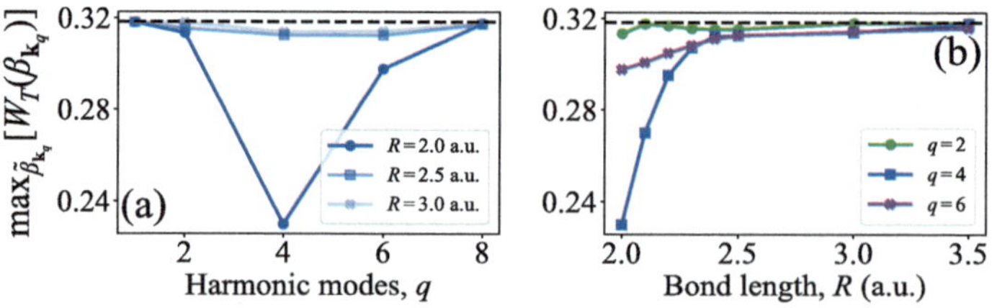

Fig. 4.6 Maximum value of the Wigner function for the final state after the laser-molecule interaction. In **a**, the maximum value is presented as a function of the harmonic mode q for various bond lengths. Panel **b** exhibits this maximum value as a function of bond length for different harmonic orders. The number of molecules considered is $N_{\text{mol}} = 10^9$. Although not visually presented, the resulting Wigner functions exhibit a Gaussian-like behavior

$$
\begin{aligned}
W_T^{(q)}(\beta) &= \frac{2}{\pi}\text{tr}\Big(\hat{\rho}_{\text{field}}(t)\hat{D}_{\mathbf{k}_q}(\beta)\hat{\Pi}\hat{D}_{\mathbf{k}_q}(-\beta)\Big)\\
&= \frac{2}{\pi\mathcal{N}}\Big[\langle\Phi_{\bar{\mathsf{b}}}(t)|\,\hat{D}_{\mathbf{k}_q}(\beta)\hat{\Pi}\hat{D}_{\mathbf{k}_q}(-\beta)\,|\Phi_{\bar{\mathsf{b}}}(t)\rangle\\
&\qquad + N_{\text{mol}}\langle\Phi_{\bar{\mathsf{a}}}(t)|\,\hat{D}_{\mathbf{k}_q}(\beta)\hat{\Pi}\hat{D}_{\mathbf{k}_q}(-\beta)\,|\Phi_{\bar{\mathsf{a}}}(t)\rangle\Big]\\
&\equiv \frac{1}{\mathcal{N}}\Big[W_{\bar{\mathsf{b}}}^{(q)}(\beta) + N_{\text{mol}}\mathcal{N}_{\bar{\mathsf{a}}}W_{\bar{\mathsf{a}}}^{(q)}(\beta)\Big],
\end{aligned}
\tag{4.48}
$$

where in transitioning from the second to the third equality, we explicitly introduce the normalization constant $\mathcal{N}_{\bar{\mathsf{a}}} \equiv \langle\Phi_{\mathsf{i}}(t)|\Phi_{\mathsf{i}}(t)\rangle$, and define $W_{\mathsf{i}}^{(q)}(\beta) \equiv (2/\pi)\langle\Phi_{\mathsf{i}}(t)|\,\hat{D}_{\mathbf{k}_q}(\beta)\hat{\Pi}\hat{D}_{\mathbf{k}_q}(-\beta)\,|\Phi_{\mathsf{i}}(t)\rangle$. As can be observed, Eq. (4.48) consists of two separate contributions: one originating from the bonding component, and the other from the antibonding term. For the former, it follows from its definition in Eq. (4.39), which is a product of coherent states with different amplitudes, that the resulting Wigner distribution for all harmonic modes is a Gaussian with a maximum value π^{-1}. However, due to the influence of the antibonding component, which is generally not a Gaussian state due to the action of the $\hat{M}_{\mathsf{ab}}(t)$ operator, the total Wigner function gets modified.

In situations where the term proportional to $\sqrt{N_{\text{mol}}}$ in Eq. (4.44) does not dominate over the bonding component[6], the Wigner function in Eq. (4.48) still exhibits a Gaussian-like shape, albeit with reduced maximum values. The behavior of the maximum peak is depicted in Fig. 4.6 when considering a driving field analogous to that used in Figs. 4.4 and 4.5. Specifically, in Fig. 4.6a, the maximum value of the obtained Wigner distribution is shown as a function of the harmonic order and for different bond lengths. The dashed black curve represents the upper bound of π^{-1}. Here, the analysis is limited to even harmonic orders since, in accordance with Figs. 4.4 and 4.5, these are the modes for which the antibonding state contributes the most. The more a given harmonic mode contributes to the antibonding component, the lower

[6] As mentioned earlier, this corresponds to cases where the bonding-antibonding transition probability remains smaller than the bonding-bonding ones within the many-molecule scenario.

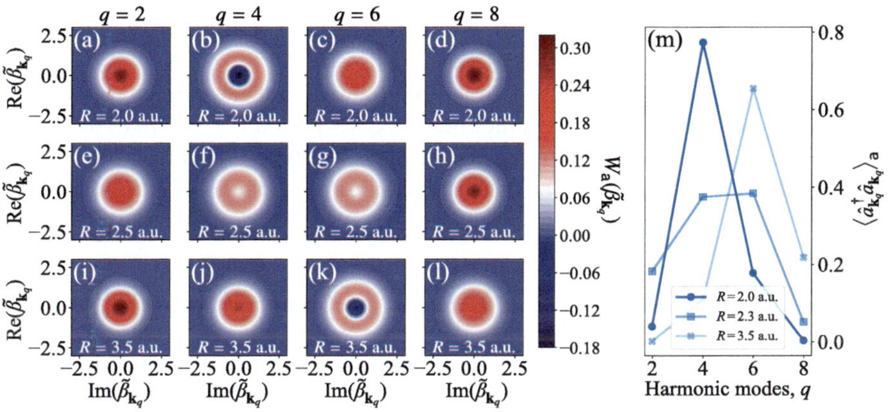

Fig. 4.7 Wigner function of the quantum optical state when conditioning the electron to be found in the antibonding state. In these plots, $\tilde{\beta}_{\mathbf{k}_q} = \beta_{\mathbf{k}_q} - \chi_{\text{a}}^{(q)}(t)$. In panels **a** to **l**, the different rows correspond to a distinct bond length ($R = 2.0$ a.u., 2.5 a.u., and 3.5 a.u. from top to bottom). In contrast, each of the columns depict different harmonic modes (q =2, 4, 6, and 8 from left to right). In panel **m**, the normalized mean photon number for the antibonding component is displayed as a function of the harmonic modes q, with each curve representing different bond lengths R

$\max_{\tilde{\beta}_{\mathbf{k}_q}}[W_T(\tilde{\beta}_{\mathbf{k}_q})]$ becomes. This is evident in the case of $q = 4$ and $R = 2.0$ a.u. in Fig. 4.6a, while for greater bond lengths, the reduction of the Wigner function's maximum value is less pronounced compared to the $R = 2.0$ a.u. case (blue curve with circular markers). In fact, as the bond length increases, the harmonic generation efficiency lowers, which in turn leads to a decrease in the contribution of the antibonding component to the Wigner function. This behavior is presented in panel (b), where we observe that increasing values of R lead to $\max_{\tilde{\beta}_{\mathbf{k}_q}}[W_T(\tilde{\beta}_{\mathbf{k}_q})]$ converging to π^{-1}.

From this analysis, it becomes clearer that the antibonding contribution to Eq. (4.48) exhibits a contrasting behavior compared to the Gaussian-like bonding component. For this reason, we study it separately by projecting Eq. (4.44) onto the antibonding state[7] $|\psi_{\bar{\text{a}}}\rangle$. This can be understood as an alternative operation to the conditioning on HHG operation presented in Sect. 3.3.2.1, as here the electron is used a heralding system. Consequently, the quantum optical state is given by $|\Phi_{\bar{\text{a}}}(t)\rangle$, and its Wigner function by $W_{\text{a}}^{(q)}(\beta)$.

In Fig. 4.7a to l, we display the antibonding component of the Wigner function for different harmonic modes and bond lengths. We restrict our analysis to even harmonic orders since the antibonding contribution is more significant for these modes. In most cases, the resulting Wigner distributions display a Gaussian-like behavior with varying flatness of the central maximum, depending on the harmonic

[7] It is worth noting that, formally, the state with respect to which we are projecting is a dressed version of the antibonding state, as we are working in the interaction picture with respect to the semiclassical Hamiltonian.

mode. However, in specific cases, the resulting function exhibits a distinctive ring-like shape, notably in panels (b), (f), (g) and (k).

When comparing these results with the corresponding mean-photon number distributions shown in Fig. 4.7m, we notice that the cases producing ring-like shaped distributions correspond to situations where the mean-photon number reaches its maximum values. This behavior can be better understood by further expanding Eq. (4.41) as

$$|\Phi_{\bar{a}}(t)\rangle = \hat{\mathcal{D}}\big(N_{\text{mol}}\boldsymbol{\chi}_{\text{b}}(t, t_0)\big)\Bigg[\sum_{q=1}^{q_c} h_1^{(q)}|1_{\mathbf{k}_q}\rangle \bigotimes_{q'\neq q}|0_{\mathbf{k}_q}\rangle + h_2(t)|\bar{0}\rangle\Bigg], \tag{4.49}$$

where we have defined

$$h_1^{(q)}(t) = -\frac{i}{\hbar}\sqrt{N_{\text{mol}}}\int_{t_0}^{t} \mathrm{d}t_1 e^{\bar{\theta}_{\text{b}}(t_1)}\mathbf{g}(\omega_{\mathbf{k}_q})\cdot\boldsymbol{\mu}_{\text{ab}}(t_1)e^{i\omega_{\mathbf{k}_q}t_1}, \tag{4.50}$$

$$h_2(t) = -\frac{i}{\hbar}\sqrt{N_{\text{mol}}}\int_{t_0}^{t} \mathrm{d}t_1 e^{\bar{\theta}_{\text{b}}(t_1)}\mathbf{E}_{\text{cl}}^{(\text{b})}(t)\cdot\boldsymbol{\mu}_{\text{ab}}(t_1). \tag{4.51}$$

From Eq. (4.49), one can observe that the antibonding quantum optical component is given as a superposition of displaced single-photon excitations which are delocalized along the different harmonic modes, and a displaced vacuum state. Consequently, the cases yielding ring-like shaped distribution are associated to scenarios where the displaced single-photon excitations make the most significant contribution to the state. To elaborate further, for a specific harmonic mode, the greater this contribution, and consequently the bigger the corresponding mean photon number, the deeper the central minimum in the Wigner distribution.

This observation offers a plausible explanation for a potential source of noise that contributes to the diminished minima observed in the atomic HHG experimental Wigner functions (see Figs. 3.20 and 3.21). In these experiments, not all the harmonic modes are measured, but some of them are filtered out. Therefore, any remaining correlation with those modes that are not entirely excluded but have been effectively traced out from the measurements could impact the final Wigner distribution, akin to what happens in Fig. 4.7.

Finally, it is worth noting that similar results to those in Fig. 4.6 are obtained if projections with respect to the set of localized states in Eq. (4.10) are considered. Specifically, in these cases, we have the superposition presented in Eq. (4.30), where the bonding term dominates over the antibonding one. Consequently, the leading term in the resulting Wigner functions correspond to a Gaussian distribution with maximum value π^{-1}, which may get reduced due to the influence of the antibonding component.

4.1.2.4 Analysis of the Entanglement Between Light and Matter

As mentioned earlier, the state presented in Eq. (4.44) is generally entangled, as one can find regimes where the overlap $|\langle\Phi_{\bar{b}}(t)|\Phi_{\bar{a}}(t)\rangle \leq 1$, therefore leading to a non-separable superposition. More strikingly, the generated state can be regarded as a hybrid state in two senses. First, it involves entanglement between discrete and continuous variables, the former through the bound states of the molecule and the latter via the quantum optical state. On the other side, it involves entanglement between two naturally distinct systems: matter and light.

These points motivate an analysis of the degree of entanglement between these two subsystems and how this depends on the bond length and the number of independent molecules involved in the HHG process. Since we are dealing with pure states, the entanglement entropy becomes an ideal entanglement measure to probe. Similarly to the ATI analysis performed in Sect. 3.3.1, it is particularly convenient to trace out the field degrees of freedom, as it provides us with an effective qubit system for the electron. More specifically, we get

$$
\begin{aligned}
\hat{\rho}_{\text{elec}}(t) &= \text{tr}_{\text{field}}\big(\left|\Psi'(t)\middle\rangle\!\middle\langle\Psi'(t)\right| \big) \\
&= \frac{1}{\mathcal{N}}\Big[\left|\psi_{\bar{b}}\middle\rangle\!\middle\langle\psi_{\bar{b}}\right| + \sqrt{N_{\text{mol}}}h_2^*(t) \left|\psi_{\bar{b}}\middle\rangle\!\middle\langle\psi_{\bar{a}}\right| \\
&\qquad + N_{\text{mol}}\mathcal{N}_{\bar{b}} \left|\psi_{\bar{a}}\middle\rangle\!\middle\langle\psi_{\bar{a}}\right| + \sqrt{N_{\text{mol}}}h_2(t) \left|\psi_{\bar{a}}\middle\rangle\!\middle\langle\psi_{\bar{b}}\right| \Big].
\end{aligned} \tag{4.52}
$$

The results obtained from the computation of $\mathcal{S}(\hat{\rho}_{\text{elec}}(t))$ are shown in Fig. 4.8. Specifically, in panel (a), the degree of entanglement is shown as a function of the number of molecules for several bond lengths, while in (b), it is shown as a function of the bond length for different values of N_{mol}. As can be observed, the amount of entanglement decreases with the bond length while it increases with the number of molecules. Interestingly, the maximum value of this entanglement measure, i.e. $\mathcal{S}(\hat{\rho}_{\text{elec}}(t)) \approx 0.75$, is obtained for a bond length $R = 2.0$ a.u., corresponding to the experimentally measured equilibrium distance of the molecular ion. Moreover, the obtained trend suggests $\mathcal{S}(\hat{\rho}_{\text{elec}}(t))$ to become even greater for smaller bond lengths. However, one has to bear in mind that, within this regime, the two nuclei will strongly repulse each other and, consequently, nuclear motion needs to be included in the analysis. Finally, it is worth noting that the values of N_{mol} used here match the estimates found in the experiments described in Sect. 3.3.2.1 with atomic gases, suggesting that the calculated effects could be potentially observed experimentally.

4.1.2.5 Analysis of the Entanglement Between the Field Modes

For atomic systems we found that, right after the interaction with the gas jet, the final state of the system was represented as a product of different coherent states. This result was obtained under the standard SFA assumptions, specifically by assuming the ground state depletion to be negligible and by disregarding all other bound

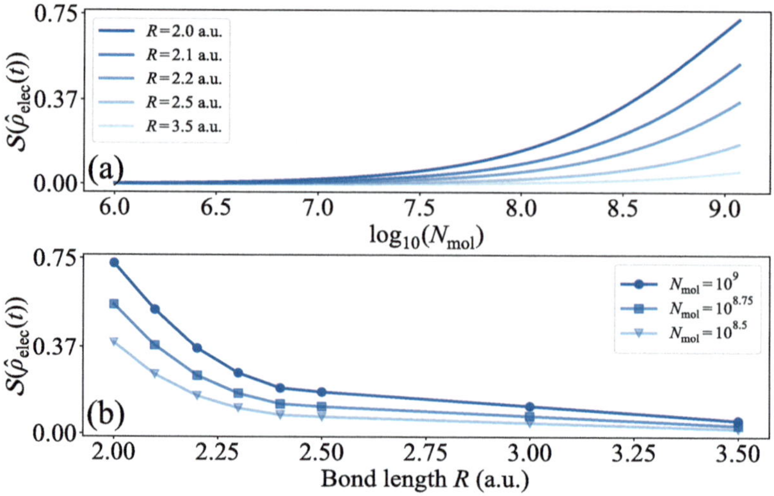

Fig. 4.8 Light-matter entanglement between the electronic degrees of freedom and all the harmonic modes after the laser-molecule interaction. In **a**, the dependence of the entanglement entropy is plotted as a function of the number of molecules. The various curves correspond to different bond lengths R. In **b**, the number of molecules is fixed instead, and the dependence of this entanglement measure is shown as a function of the bond length

states. However, through the conditioning on HHG operation, we were able to theoretically obtain non-classical states of light represented in terms of coherent state superpositions.

For the case of molecules, the situation is slightly different due to the introduction of the first excited state on top of the ground state. Because of this, and when assuming the initial state of the electron to be the bonding state, we encounter two scenarios: when the final state of the electron corresponds to the bonding state it originated from, the final quantum optical state is analogous to that found for atomic interactions; when the final state of the electron corresponds to the antibonding state, we arrive at an entangled state between the field modes where the harmonic excitation is delocalized among the different modes. Thus, we expect to find in this case entanglement features between the field modes even for the case where no conditioning on HHG operation is applied. In this section, we aim to analyze these entanglement features under different scenarios and how they depend on parameters such as the bond length and the number of molecules in the interaction region.

Given that, in general, the number of harmonic modes involved in HHG can easily surpass dozens and even hundreds of orders, the entanglement characterization of the quantum optical state obtained after these processes would certainly benefit from a multipartite entanglement analysis. However, since the study of entanglement for systems involving more than two parties is actually a topic of active research [92, 93],

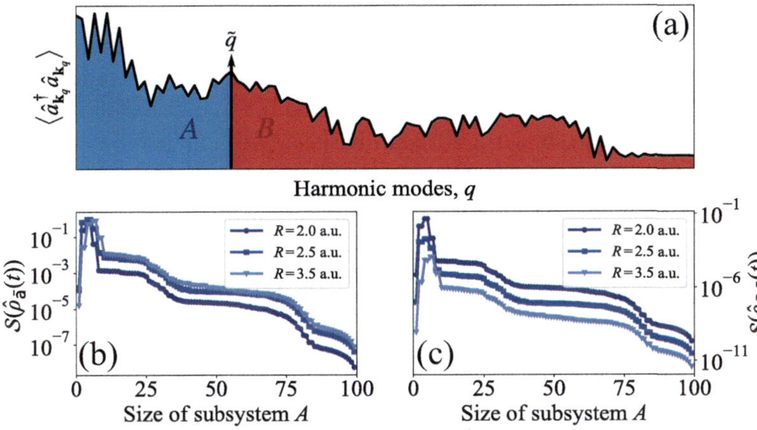

Fig. 4.9 Entanglement between the field modes when the electron is conditioned to be on a specific quantum state after the laser-molecule interaction. In panel **a**, a schematic of the approach is presented where the harmonic modes are divided into two sets, $A := \{q : q \leq \tilde{q}\}$ and $B := \{q : q > \tilde{q}\}$. In the other panels, the entanglement entropy between these two sets is presented as a function of the size of subsystem A. Panel **b** illustrates the entanglement entropy when the electron is conditioned to the antibonding state. Panel **c** showcases this quantity when the electron is conditioned to a localized left or right state for $N_{\text{mol}} = 10^9$. In these two panels, the various curves represent different bond lengths

in our case we restrict to bipartite scenarios. In this case, the entanglement measures presented in Sect. 2.3.4.1 allow us to readily quantify the degree of entanglement of the generated states.

First scenario: conditioning on antibonding states. In this first scenario, we consider cases where the electronic state is conditioned to be found in the state $|\psi_{\bar{\mathrm{a}}}\rangle$, such that the quantum optical state is given by $|\Phi_{\bar{\mathrm{a}}}(t)\rangle$. Thus, in order to perform a bipartite entanglement analysis, we split the frequency modes in two sets, which we denote as A and B. These sets are given by

$$A := \{q : q \leq \tilde{q}\} \quad \text{and} \quad B := \{q : q > \tilde{q}\}, \tag{4.53}$$

and are pictorially represented in Fig. 4.9a, with the black arrow representing the location of the limiting harmonic mode $\tilde{q}$. In terms of these, we can express Eq. (4.49) as

$$\begin{aligned} |\Phi_{\bar{\mathrm{a}}}(t)\rangle = \frac{\hat{\mathcal{D}}\big(\chi_{\mathrm{b}}(t,t_0)\big)}{\mathcal{N}_{\bar{\mathrm{a}}}} \Bigg[\bigg(\sum_{q\in A} h_1^{(q)}(t) |1_{\mathbf{k}_1}\rangle \bigotimes_{q'\neq q} |0_{\mathbf{k}_q}\rangle \bigg) |\bar{0}_B\rangle \\ + |\bar{0}_A\rangle \bigg(\sum_{q\in B} h_1^{(q)}(t) |1_{\mathbf{k}_1}\rangle \bigotimes_{q'\neq q} |0_{\mathbf{k}_q}\rangle \bigg) + h_2(t) |\bar{0}_A\rangle |\bar{0}_B\rangle \Bigg], \end{aligned} \tag{4.54}$$

where, for brevity, we have defined $|\bar{0}_A\rangle \equiv \bigotimes_{q\in A}|0_{\mathbf{k}_q}\rangle$ and $|\bar{0}_B\rangle \equiv \bigotimes_{q\in B}|0_{\mathbf{k}_q}\rangle$. Since this state is pure, the entanglement entropy serves us here as well as an appropriate entanglement measure. However, in order to find computationally inexpensive versions of the Schmidt matrix obtained when tracing out one of the subsystems, it is more convenient to work under a different basis set for both subsystems A and B. More specifically, and utilizing subsystem A as an example, we consider the basis set spanned by the following states

$$\left\{|\tilde{0}_A\rangle \equiv \hat{\mathcal{D}}\big(\chi_{\mathrm{b}}(t,t_0)\big)|\bar{0}_A\rangle,\ |\tilde{1}_A\rangle \equiv \frac{\hat{\mathcal{D}}\big(\chi_{\mathrm{b}}(t,t_0)\big)}{\sqrt{\mathcal{N}_A}}\sum_{q\in A} h_1^{(q)}(t)|1_{\mathbf{k}_q}\rangle \bigotimes_{q'\neq q}|0_{\mathbf{k}_q}\rangle,\ \cdots\right\}, \tag{4.55}$$

where the ellipsis, denotes orthonormal states with respect to the other two, which are already orthonormal ($\mathcal{N}_A$ represents the normalization constant of $|\tilde{1}_A\rangle$). It is worth noting that the first two states in this set correspond, within a displaced frame, to either zero or single-photon excitations on the field modes. Consequently, the additional orthonormal states represented by the ellipsis, obtainable through methods such as the Gram-Schmidt decomposition, may contain more than one excitation in their components. However, in our case, it is sufficient to consider those explicitly presented in Eq. (4.55), as with these, we can rewrite Eq. (4.54) as

$$|\Phi_{\bar{\mathrm{a}}}(t)\rangle = \frac{1}{\sqrt{\mathcal{N}_{\bar{\mathrm{a}}}}}\Big[\sqrt{\mathcal{N}_A}|\tilde{1}_A\rangle|\tilde{0}_B\rangle + \sqrt{\mathcal{N}_B}|\tilde{0}_A\rangle|\tilde{1}_B\rangle + h_2(t)|\tilde{0}_A\rangle|\tilde{0}_B\rangle\Big]. \tag{4.56}$$

After tracing out the degrees of freedom of one of the subsystems, for example, those of subsystem B, we obtain for the reduced density matrix for subsystem A

$$\begin{aligned}\hat{\rho}_{\bar{\mathrm{a}}}^{(A)}(t) &= \mathrm{tr}_B\Big(|\Phi_{\bar{\mathrm{a}}}(t)\rangle\langle\Phi_{\bar{\mathrm{a}}}(t)|\Big)\\ &= \frac{1}{\mathcal{N}_{\bar{\mathrm{a}}}}\Big[\mathcal{N}_A|\tilde{1}_A\rangle\langle\tilde{1}_A| + h_2^*(t)\sqrt{\mathcal{N}_A}|\tilde{1}_A\rangle\langle\tilde{0}_A|\\ &\qquad + h_2(t)\sqrt{\mathcal{N}_A}|\tilde{0}_A\rangle\langle\tilde{1}_A| + \Big(\mathcal{N}_B + |h_2(t)|^2\Big)|\tilde{0}_A\rangle\langle\tilde{0}_A|\Big],\end{aligned} \tag{4.57}$$

which can then be used for computing the entanglement entropy. In Fig. 4.9b, we present the behavior of this quantity as a function of the size of subsystem A, denoted hereafter as $||A|| \equiv \dim(A)$, for different bond lengths. One of the most remarkable aspects of the behavior of this entanglement measure is that it is possible to distinguish the three different plateau regions observed in Figs. 4.4 and 4.5. This is ultimately related to the fact that $S(\hat{\rho}_{\bar{\mathrm{a}}}(t))$ depends on the probability of harmonics generated at specific orders.

In Fig. 4.9b, we also observe the presence of optimal values of $||A||$ for which the degree of entanglement reaches a maximum value. Specifically, for $R = 2.0$, 2.5 and 3.5 a.u., we obtain $\max_{||A||}[S(\hat{\rho}_{\bar{\mathrm{a}}}(t))] \approx 0.69$, 0.99 and 0.78, respectively. Thus, by carefully defining sets A and B, it is possible to generate highly frequency-entangled bipartite states. In this regard, a comparison with Fig. 4.4 reveals that this kind of states can be achieved by defining the set A such that $\tilde{q}$ is a harmonic mode providing the maximum mean-photon number. Beyond this value, and consequently for increasing $||A||$, the entanglement entropy decreases. This reduction occurs because the probability of having displaced Fock states in those modes belonging to subsystem B diminishes when higher harmonic orders are included in its definition.

Notably, we observe that when B is predominantly composed of higher harmonic orders, i.e. for $||A|| \gtrsim 30$, the entanglement increases for larger bond lengths. This emphasizes that, even though recombination events between different centers become less likely as R increases in the laser-matter excitation configuration considered here, they become more efficient in terms of harmonic generation. In other words, as the bond length increases, the probability of generating harmonics through recombination events via the antibonding state becomes more similar between lower harmonic orders (second plateau) and higher harmonic orders (third plateau). This effectively results in increased values in the amount of entanglement.

Second scenario: conditioning on left-like localized and right-like localized states. Alternatively, we can also consider the situation where the final state of the electron is found in the left or right center. In this regard, while for the case of single-molecules the set of localized states is spanned by Eq. (4.10), for the many-molecule scenario this situation becomes more complicated. And this is because we consider a regime where all except one molecule undergo a parametric process. In other words, only one out of these N_{mol} molecules can actually end up in the antibonding state. Consequently, we consider the following many-molecule extension to the localized set of states, which we shall denote as $\{|\bar{L}_{N_{\text{mol}}}\rangle, |\bar{R}_{N_{\text{mol}}}\rangle\}$,

$$|\bar{L}_{N_{\text{mol}}}\rangle \equiv \frac{1}{\sqrt{2}}\Big[|\psi_{\bar{\text{b}}}\rangle - |\psi_{\bar{\text{a}}}\rangle\Big] = \frac{1}{\sqrt{2}}\Bigg[|\psi_{\bar{\text{b}}}\rangle - \frac{1}{\sqrt{N_{\text{mol}}}}\sum_{k=1}^{N_{\text{mol}}}|\psi_{\bar{\text{a}}_k}\rangle\Bigg], \tag{4.58}$$

$$|\bar{R}_{N_{\text{mol}}}\rangle \equiv \frac{1}{\sqrt{2}}\Big[|\psi_{\bar{\text{b}}}\rangle + |\psi_{\bar{\text{a}}}\rangle\Big] = \frac{1}{\sqrt{2}}\Bigg[|\psi_{\bar{\text{b}}}\rangle + \frac{1}{\sqrt{N_{\text{mol}}}}\sum_{k=1}^{N_{\text{mol}}}|\psi_{\bar{\text{a}}_k}\rangle\Bigg]. \tag{4.59}$$

These two states recover the definition of localized states in Eq. (4.10) when $N_{\text{mol}} = 1$, while providing a generalization for $N_{\text{mol}} > 1$. More specifically, they are expressed as a superposition of states where either all electrons are in their respective bonding states, or all except one. In this last case, represented by the $|\psi_{\bar{\text{a}}}\rangle$, the electron excitation is delocalized among all the considered molecules. In Appendix 9.2 we further discuss the properties of this set of localized states, in terms of which Eq. (4.44) reads

$$\begin{aligned}|\Psi'(t)\rangle = \frac{1}{\sqrt{N}}\Big[&|\bar{L}_{N_{\text{mol}}}\rangle \otimes \Big(|\Phi_{\bar{\text{b}}}(t)\rangle - \sqrt{N_{\text{mol}}}\,|\Phi_{\bar{\text{a}}}(t)\rangle\Big)\\ &+ |\bar{R}_{N_{\text{mol}}}\rangle \otimes \Big(|\Phi_{\bar{\text{b}}}(t)\rangle + \sqrt{N_{\text{mol}}}\,|\Phi_{\bar{\text{a}}}(t)\rangle\Big)\Big],\end{aligned} \tag{4.60}$$

and projections of this state with respect to $|\bar{L}_{N_{\text{mol}}}\rangle$ and $|\bar{R}_{N_{\text{mol}}}\rangle$ respectively lead to

$$|\Phi_{\bar{L}/\bar{R}}(t)\rangle = \frac{1}{\sqrt{\mathcal{N}_{\bar{a}}}}\Big[\sqrt{\mathcal{N}_A}|\tilde{1}_A\rangle|\tilde{0}_B\rangle + \sqrt{\mathcal{N}_B}|\tilde{0}_A\rangle|\tilde{1}_B\rangle + \big(\mp 1 + h_2(t)\big)|\tilde{0}_A\rangle|\tilde{0}_B\rangle\Big], \tag{4.61}$$

when expressed in the basis set presented in Eq. (4.55). Thus, the reduced density matrix of the final state can be obtained from Eq. (4.57) upon the substitution $h_2(t) \to \mp 1 + h_2(t)$, respectively for $|\bar{L}_{N_{\text{mol}}}\rangle$ and $|\bar{R}_{N_{\text{mol}}}\rangle$.

In Fig. 4.9c, we present the entanglement entropy for this case, that is $\mathcal{S}(\hat{\rho}_{\bar{L}/\bar{R}}(t))$, as a function of $||A||$ for three different bond lengths. For each of these cases, a similar trend in behavior with respect to $||A||$ is observed, specifically the presence of the three plateau regions, although with overall reduced values for the amount of entanglement. More specifically, the maximum values found for $R = 2.0, 2.5$ and 3.5 a.u. are $\max_{||A||}[\mathcal{S}(\hat{\rho}_{\bar{L}/\bar{R}}(t))] \approx 4.4 \times 10^{-2}, 1.8 \times 10^{-3}$ and 1.11×10^{-4}, respectively, contrasting with the $\sim 10^{-1}$ obtained earlier. Furthermore, the behavior with respect to the bond length also differs from the case of Fig. 4.9b: increasing bond lengths results in a reduced entanglement entropy at all values of $||A||$. This is an expected feature since the importance of $|\Phi_{\bar{a}}(t)\rangle$ in Eq. (4.60) is ultimately determined by the drift in population between the bonding and antibonding states. Thus, the smaller the bond length, the more likely these events become for the laser-matter configuration employed in our analysis. Hence, we observe that the electronic-conditioning operation strongly impacts the amount of entanglement between the field modes in the final state.

Third scenario: absence of knowledge about the electronic state. Finally, we study the entanglement between the field modes in the scenario where we have no knowledge about the final electronic state. Consequently, we trace out the associated degrees of freedom, leading, within the many-molecule regime, to the following density matrix for the quantum optical state

$$\hat{\rho}_{\text{field}}(t) = \frac{1}{\mathcal{N}}\Big[\hat{\rho}_{\bar{b}}(t) + N_{\text{mol}}\mathcal{N}_{\bar{a}}\hat{\rho}_{\bar{a}}(t)\Big], \tag{4.62}$$

which is given as the mixture of the density matrices of two pure states, consequently leading to a mixed state. Thus, and as mentioned in Sect. 2.3.4.1, the entanglement entropy is not a reliable entanglement measure anymore since its validity is constrained to pure states. However, in this case, we can use the logarithmic negativity presented in Eq. (2.129). This measure depends on the negativity, i.e., the sum of all negative values of the partial transpose of $\hat{\rho}_{\text{field}}(t)$ with respect to one of the subsystems.

In this case, we focus on the entanglement between the $\tilde{q}$th harmonic mode and the rest. Thus, we define subsystems $\bar{A}$ and $\bar{B}$ as

$$\bar{A} := \{\tilde{q}\} \quad \text{and} \quad \bar{B} := \{q : \forall q \neq \tilde{q}\}, \tag{4.63}$$

respectively. Then, by performing the partial transpose of Eq. (4.62) with respect to subsystem $\bar{B}$, we arrive at

$$\hat{\rho}_{\text{field}}^{T_{\bar{B}}}(t) = \frac{1}{\mathcal{N}}\Big[\hat{\rho}_{\bar{\text{b}}}(t) + N_{\text{mol}}\mathcal{N}_{\bar{\text{a}}}\hat{\rho}_{\bar{\text{a}}}^{T_{\bar{B}}}(t)\Big], \tag{4.64}$$

where we have taken into account that $\hat{\rho}_{\bar{\text{b}}}(t)$ is a pure separable state, and thus remains unaffected by the partial transpose transformation.

The remaining task in assessing the logarithmic negativity lies in the computation of the negative eigenvalues of Eq. (4.64). Despite the feasibility of exactly computing this entanglement measure for $N_{\text{mol}} \gg 1$, for which we can use the basis sets outlined in Eq. (4.55), challenges arise for relatively small values of N_{mol}, where the overall displacements obtained for both the bonding and antibonding states are notable. Consequently, the formal computation of the negativity necessitates consideration of the full basis which, in the Fock representation, might involve hundreds of states. This imposes substantial numerical demands.

To circumvent these challenges, we propose a numerically manageable lower bound for the logarithmic negativity. In this regard, we take into account that, given three Hermitian matrices A, B and $\mathsf{C} = \mathsf{A} + \mathsf{B}$ with respective eigenvalues $\{\lambda_1^{(\mathsf{A})} > \lambda_2^{(\mathsf{A})} > \cdots > \lambda_n^{(\mathsf{A})}\}$, $\{\lambda_1^{(\mathsf{B})} > \lambda_2^{(\mathsf{B})} > \cdots > \lambda_n^{(\mathsf{B})}\}$ and $\{\lambda_1^{(\mathsf{C})} > \lambda_2^{(\mathsf{C})} > \cdots > \lambda_n^{(\mathsf{C})}\}$, the following relationship holds for their eigenvalues [94, 95]

$$\lambda_{i+j-1}^{(\mathsf{C})} \leq \lambda_i^{(\mathsf{A})} + \lambda_j^{(\mathsf{B})}, \tag{4.65}$$

and, for the potential negative eigenvalues of C, we can then express their absolute value as

$$|\lambda_{i+j-1}^{(\mathsf{C})}| \geq |\lambda_i^{(\mathsf{A})} + \lambda_j^{(\mathsf{B})}|. \tag{4.66}$$

Therefore, if we identify $\mathsf{A} = \hat{\rho}_{\bar{\text{a}}}(t)$, $\mathsf{B} = \hat{\rho}_{\bar{\text{b}}}(t)$ and $\mathsf{C} = \hat{\rho}_{\text{field}}(t)$, we can write for the negativity of C

$$\mathsf{N} = \sum_{\lambda_i^{(\mathsf{C})}<0} |\lambda_i^{(\mathsf{C})}| \geq \sum_{\lambda_i^{(\mathsf{A})}+\lambda_j^{(\mathsf{B})}<0} \left|\min_{i,j}\left(\lambda_i^{(\mathsf{A})} + \lambda_j^{(\mathsf{B})}\right)\right| \geq \left|\min_i \lambda_i^{(\mathsf{A})}\right| \quad \text{with} \quad \min_i \lambda_i^{(\mathsf{A})} < 0, \tag{4.67}$$

where in transitioning to the second inequality, we have taken into account that B does not have, by definition, negative eigenvalues. Consequently, the final term corresponds to the lowest (negative) eigenvalue of A. This derivation then establishes the subsequent lower bound for the logarithmic negativity of our state

$$E_N^{(\bar{q})}\big(\hat{\rho}_{\text{field}}(t)\big) \geq \tilde{E}_N^{(\bar{q})} = \log_2\left(2\left|\min_i \lambda_i^{(\mathsf{A})}\right| + 1\right). \tag{4.68}$$

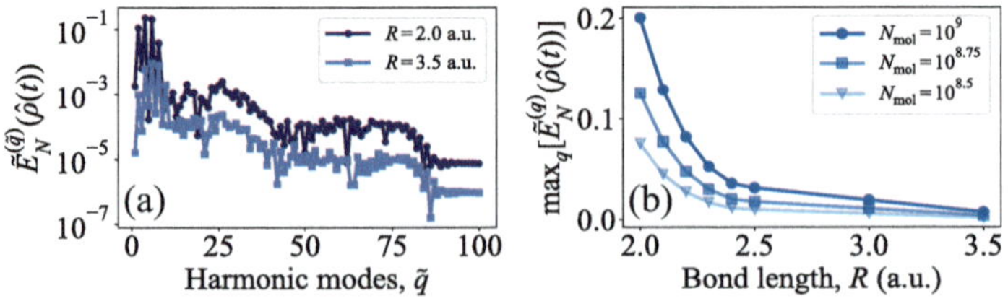

Fig. 4.10 Entanglement between the field modes assuming no knowledge about the final electronic state. In **a**, the degree of entanglement evaluated through the logarithmic negativity between the qth mode and the rest, presented as a function of the mode $\tilde{q}$ for $N_{\text{mol}} = 10^8$. The various curves correspond to different bond lengths. In **b**, the maximum value of this quantity with respect to the harmonic modes, is plotted against the bond length for different numbers of molecules N_{mol}. It is worth noting that, in this figure, the logarithmic negativity is evaluated through a lower bound to the exact quantity (see text for further details)

The results obtained from evaluating the lower bound to Eq. (4.68) are presented in Fig. 4.10 as a function of $\tilde{q}$ for two bond lengths. Similarly to the outcomes of Fig. 4.9, the three plateau regions are distinguishable, albeit with some intermediate maxima and minima. Specifically, in the low harmonic region $\tilde{q} \lesssim 15$, troughs are found for odd harmonic orders, while peaks occur for even values. This behavior arises due to the fact that it is the antibonding component in Eq. (4.62) that provides non-vanishing entanglement features to the state, as explicitly demonstrated in Eq. (4.68). Consequently, since even orders are the states that are most populated for electronic recombinations with the antibonding state, they dominate the entanglement features. For these modes, the logarithmic negativity attains its maximum values. In particular, for $\tilde{q} = 4$ we find $\tilde{E}_N^{(\tilde{q})} \approx 0.2$ for $R = 2.0$ a.u., while for $\tilde{q} = 6$ and $R = 3.5$ a.u. we get $\tilde{E}_N^{(\tilde{q})} \approx 9.29 \times 10^{-3}$. This comparison also reveals that larger bond lengths lead to smaller values of this entanglement measure, as explicitly shown in Fig. 4.10b where $\max_{\tilde{q}}[\tilde{E}_N^{(\tilde{q})}]$ is presented as a function of R for different number of molecules in the interaction region.

4.2 High-Harmonic Generation Processes in Semiconductor Materials

In atomic gas jets, atoms are randomly located in space, separated by large distances and, consequently, they barely interact with each other. In contrast, solid-state systems are characterized by their organized structure, where each atomic nucleus occupies a specific position in space, resulting in an organized, periodic structure commonly referred to as a crystal. These ordered arrangements are possible due to the strong interactions between the atoms in the system, which share their electrons among each other leading to a stable overall structure.

These significant structural differences between atoms and solid-state systems have consequences on the electron dynamics during the HHG process, as well as on the outcomes of the laser-matter interaction. For this reason, in this section we commence with a concise introduction to the fundamentals of electrons interacting with a periodic potential, providing the main ingredients needed to discuss the process of HHG generation in semiconductors from a semiclassical perspective. Through this, we aim to highlight the diverse electron dynamics leading to the generation of high-order harmonics. The effects of these dynamics on the quantum optical state, following the conditioning on HHG operation, are then analyzed. This is achieved by proposing a quantum optical description of the laser-solid interaction built upon the discussed semiclassical description.

4.2.1 An Electron in a Periodic Potential

The interaction of an electron with a periodic potential, as, for instance, that provided by an ideal crystal,[8] can be described by means of the following Hamiltonian

$$\hat{H}_{\text{cr}}(t) = \frac{\hat{\mathbf{p}}^2}{2m_{\text{e}}} + \hat{V}_{\text{cr}}(\hat{\mathbf{r}}), \tag{4.69}$$

where $\hat{V}_{\text{cr}}(\hat{\mathbf{r}})$ is the crystal Hamiltonian satisfying $\hat{V}(\hat{\mathbf{r}}) = \hat{V}(\hat{\mathbf{r}} + \mathbf{R})$. Here, $\mathbf{R}$ represents a vector belonging to a set of vectors spanning what is usually referred to as the Bravais lattice of the crystal, i.e., the minimal set of vectors needed to span the whole crystal structure. This is pictorially represented in Fig. 4.11, where the periodic potential $\hat{V}(\mathbf{r})$ is shown as a function of the electron position, with the circles below representing the atomic nuclei that are separated by the Bravais vector $\mathbf{R}$.

It is worth noting that the Hamiltonian presented in Eq. (4.69) does not account for electron-electron correlations, assuming that these are much smaller than effects of the periodic potential. This assumption is valid for a wide range of materials, where these interactions can be considered negligible. In other particular cases, such as what is usually referred to as strongly correlated materials, electron-electron interactions become extremely influential, leading to exotic phases of matter (see e.g. Ref. [96] and references therein). However, in this section, we focus on the former which further allows us to introduce one of the most crucial concepts in the theory of condensed-matter physics: Bloch's theorem [97–99].

[8] By an ideal crystal, we refer to those systems where no impurities are found, and that break the translation symmetry of the structure.

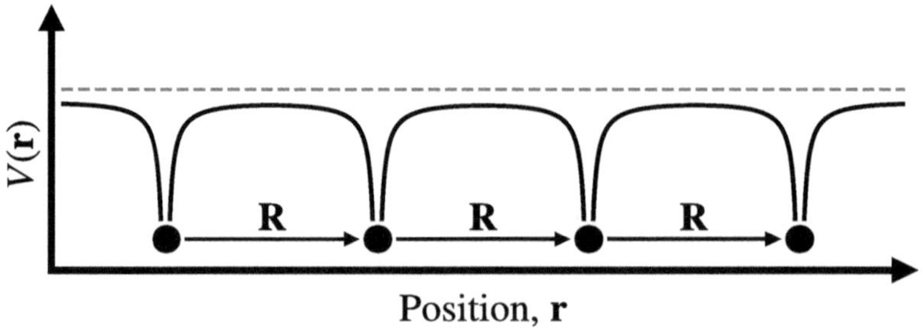

Fig. 4.11 Pictorial representation of a periodic potential. Each circle corresponds to a spatially fixed atom positioned at a constant distance R from the other atoms. The dashed curve represents $V(\mathbf{r}) = 0$

4.2.1.1 Bloch's Theorem

Bloch's theorem states that the eigenstates of Hamiltonians whose form is that of Eq. (4.69), i.e. where the potential to which electrons are subjected satisfies periodicity constraints, can be chosen, in position representation, as

$$\phi_{m,\boldsymbol{k}}(\mathbf{r}) = e^{i\boldsymbol{k}\cdot\mathbf{r}} u_{m,\boldsymbol{k}}(\mathbf{r}), \tag{4.70}$$

where $u_{m,\boldsymbol{k}}(\mathbf{r})$ is a function with the same periodicity as that of the lattice, that is $u_{m,\boldsymbol{k}}(\mathbf{r}) = u_{m,\boldsymbol{k}}(\mathbf{r} + \mathbf{R})$. The states in Eq. (4.70), denoted in Dirac representation as $|\phi_{m,\boldsymbol{k}}\rangle$, are the so-called Bloch states. They are eigenstates of the crystal Hamiltonian with eigenvalues indexed by m, and of the translation operator[9] with eigenvalues $\boldsymbol{k}$, known as crystal momenta. It is essential to note that, despite their similar names, crystal momentum is neither equal nor directly proportional to the (free) electronic momentum. Furthermore, Bloch states are not eigenstates of the momentum operator $\hat{\mathbf{p}}$. Instead, crystal momentum can be considered an extension of the electronic momentum when dealing with periodic potentials.

4.2.1.2 The Tight-Binding Approximation

In general, computing the exact form of the Bloch states in Eq. (4.70), as well as their energy eigenvalues is not a trivial task. However, in cases where the electron-electron correlations are weak, there are some approximations that allow obtaining extremely good estimations for specific solids. An example is the tight-binding approximation [99] which, to some extent, closely resembles the ideas of the LCAO method

[9] The translation operator $\hat{\mathrm{T}}_{\mathbf{R}}$ is defined such that, given a function $f(\mathbf{r})$, it then leads to $\mathrm{T}_{\mathbf{R}} f(\mathbf{r}) = f(\mathbf{r} + \mathbf{R})$.

presented in Sect. 4.1.1 for molecular systems. Thus, our primary objective is to now solve the eigenvalue equation

$$\hat{H}_{\text{cr}}|\phi_{m,\boldsymbol{k}}\rangle = E_m(\boldsymbol{k})|\phi_{m,\boldsymbol{k}}\rangle, \tag{4.71}$$

with $E_m(\boldsymbol{k})$ representing the energy associated to the Bloch state $|\phi_{m,\boldsymbol{k}}\rangle$.

One of the two most crucial points about the tight-binding approximation is to consider the solid system as a set of atoms located at different sites $\mathbf{R}_l = l\mathbf{R}$. Consequently, and as illustrated in Fig. 4.11, the lth atom contributes to the Hamiltonian with the single-atom potential $\hat{V}(\hat{\mathbf{r}} - \mathbf{R}_l)$, such that within this picture we can rewrite Eq. (4.71) as

$$\left[\frac{\hat{\mathbf{p}}^2}{2m_{\text{e}}} + \sum_l \hat{V}(\hat{\mathbf{r}} - \mathbf{R}_l)\right]|\phi_{m,\boldsymbol{k}}\rangle = E_m(\boldsymbol{k})|\phi_{m,\boldsymbol{k}}\rangle. \tag{4.72}$$

The equation above shares similarities with the description of a molecular system under the Born-Oppenheimer approximation, where the behavior of a single electron is influenced by the atomic potentials of all the atoms within the lattice. Therefore, to solve, Eq. (4.72), it is reasonable to consider an LCAO-based ansatz, where the electronic state is expressed as a linear combination of various atomic orbital functions centered around each atomic nucleus. However, it is crucial that this ansatz adheres to the periodicity constrains of the solid-state system. In this context, Wannier states become particularly useful due to their spatial localization [99–102]. This stands in contrast to Bloch states, which are entirely delocalized in space given that they are fully localized in reciprocal space, i.e., the space where the crystal momentum $\boldsymbol{k}$ is defined. The relation between the wavefunctions of these two states is given by

$$w_{m,l}(\mathbf{r}) = \frac{1}{\sqrt{N}}\sum_{\boldsymbol{k}} e^{-i\boldsymbol{k}\cdot\mathbf{R}_m}\phi_{\mathbf{m},\boldsymbol{k}}(\mathbf{r}), \tag{4.73}$$

which, when inverting the transformation, can also be expressed as

$$\phi_{m,\boldsymbol{k}}(\mathbf{r}) = \frac{1}{\sqrt{N}}\sum_l e^{i\boldsymbol{k}\cdot\mathbf{R}_l} w_{m,l}(\mathbf{r} - \mathbf{R}_l). \tag{4.74}$$

As observed in Eq. (4.74), the localization of Wannier functions in space leads to the complete delocalization of Bloch states, which are expressed as a sum of these localized functions spanning the lattice's entire length. This distinction is exemplified in Fig. 4.12, which depicts both functions within a 1D lattice. In panel (b), Wannier states exhibit localization around specific atomic positions, standing in stark contrast with the Bloch states shown in panel (a). Furthermore, values of the crystal momentum satisfying $|\boldsymbol{k}| > 0$, result in an additional modulation onto this real space representation, although not shown in this figure.

Returning to Eq. (4.72), the expansion of Bloch states in terms of Wannier functions offers us an ideal framework for proposing an LCAO-based ansatz for solving

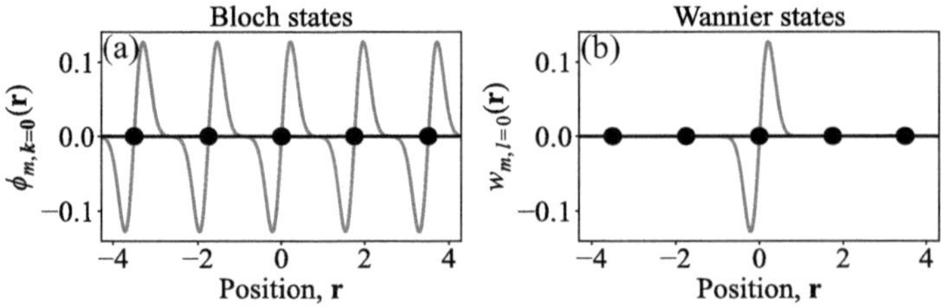

Fig. 4.12 Pictorial representation of Bloch and Wannier states. In **a**, the Bloch wavefunction corresponding to $\boldsymbol{k} = \mathbf{0}$ is depicted as a function of position $\mathbf{r}$. In **b**, the associated Wigner function for $\mathbf{r} = \mathbf{0}$ is presented as a function of the position. The circles represent the atomic locations within the lattice. These figures draw inspiration from Fig. 1 in Ref. [102]

the aforementioned equation. Specifically, denoting the atomic orbitals as $f_m^{(\mathrm{at})}(\mathbf{r})$, we can propose the following ansatz

$$\phi_{m,\boldsymbol{k}}(\mathbf{r}) = \frac{1}{\sqrt{\mathcal{N}}} \sum_{l} e^{i\boldsymbol{k}\cdot\mathbf{R}_l} f_m^{(\mathrm{at})}(\mathbf{r} - \mathbf{R}_l), \tag{4.75}$$

which, by construction, are eigenstates of the crystal Hamiltonian. Thus, the associated energy eigenvalue $E_m(\boldsymbol{k})$ for this ansatz can be computed as

$$E_m(\boldsymbol{k}) = \frac{\langle \phi_{m,\boldsymbol{k}} | \hat{H}_{\mathrm{cr}} | \phi_{m,\boldsymbol{k}} \rangle}{\langle \phi_{m,\boldsymbol{k}} | \phi_{m,\boldsymbol{k}} \rangle}, \tag{4.76}$$

where both numerator and denominator can be expanded as

$$\begin{aligned} \langle \phi_{m,\boldsymbol{k}} | \hat{H}_{\mathrm{cr}} | \phi_{m,\boldsymbol{k}} \rangle &= \frac{1}{\mathcal{N}} \sum_{l,n} e^{i\boldsymbol{k}\cdot(\mathbf{R}_l - \mathbf{R}_n)} \\ &\quad \times \int \mathrm{d}\mathbf{r} \left[f_m^{(\mathrm{at})}(\mathbf{r} - \mathbf{R}_n) \right]^* \langle \mathbf{r} | H_{\mathrm{cr}} | \mathbf{r} \rangle f_m^{(\mathrm{at})}(\mathbf{r} - \mathbf{R}_l), \end{aligned} \tag{4.77}$$

$$\langle \phi_{m,\boldsymbol{k}} | \phi_{m,\boldsymbol{k}} \rangle = \frac{1}{\mathcal{N}} \sum_{l,n} e^{i\boldsymbol{k}\cdot(\mathbf{R}_l - \mathbf{R}_n)} \int \mathrm{d}\mathbf{r} \left[f_m^{(\mathrm{at})}(\mathbf{r} - \mathbf{R}_n) \right]^* f_m^{(\mathrm{at})}(\mathbf{r} - \mathbf{R}_l). \tag{4.78}$$

The second crucial aspect about the tight-binding approximation involves assuming that electrons predominantly localize near the atomic sites, resulting in a minimal overlap between the wavefunctions of adjacent atomic sites l and $l + 1$. Specifically, as the distance between two atomic sites increases, the overlap between their respective wavefunctions diminishes. Consequently, based on this assumption, we approximate Eq. (4.78) as

$$\langle \phi_{m,\boldsymbol{k}} | \phi_{m,\boldsymbol{k}} \rangle \approx \frac{1}{\mathcal{N}} \sum_{n,m} e^{i\boldsymbol{k}\cdot(\mathbf{R}_l - \mathbf{R}_n)} = \frac{N_{\mathrm{at}}}{\mathcal{N}}, \tag{4.79}$$

where we have only kept the leading-order term accounting for the overlap of each atomic wavefunctions located at the different sites with themselves. Here, N_{at} represents the number of atoms in the lattice. If similar approximations are applied to Eq. (4.77), it is possible to rewrite it as [99]

$$\begin{aligned}\langle \phi_{m,\boldsymbol{k}} | \hat{H}_{\text{cr}} | \phi_{m,\boldsymbol{k}} \rangle &= \frac{1}{\mathcal{N}} \sum_{l,n} e^{i\boldsymbol{k}\cdot(\mathbf{R}_l - \mathbf{R}_n)} \big[\delta_{l,n} A_m^{(0)} + \delta_{l,n\pm 1} A_m^{(1)} + \cdots \big] \\ &\approx \frac{1}{\mathcal{N}} \sum_{l,n} e^{i\boldsymbol{k}\cdot(\mathbf{R}_l - \mathbf{R}_n)} \big[\delta_{l,n} A_m^{(0)} + \delta_{l,n\pm 1} A_m^{(1)} \big],\end{aligned} \tag{4.80}$$

where in transitioning from the first to the second equality, we have restricted to nearest neighbor interactions. The coefficients $A_m^{(i)}$ in this expression correspond to the integrals of distinct atomic potentials when evaluated for the atomic orbitals centered around sites l and $l \pm i$. Then, upon explicitly evaluating the sum with respect to the l and n subscripts in this equation and combining it with Eq (4.79) in accordance with Eq. (4.76), we arrive at

$$E_m(\boldsymbol{k}) \approx A_m^{(0)} + \frac{A_m^{(1)}}{\mathcal{N}} \cos(\boldsymbol{k} \cdot \boldsymbol{a}), \tag{4.81}$$

with $\boldsymbol{a}$ the lattice vector, which determines the volume of the corresponding Bravais lattice.

Unlike atomic systems, for which energy levels maintain a constant value, solids exhibit a dispersion relation where different values of $k_i \in [-\pi/a_i, \pi/a_i]$, defining the Brillouin zone, correspond to distinct energy values. Due to their quasi-continuum behavior, rather than referring to these as energy levels, we designate them as *energy bands*, each labeled by the subscript m. In broad terms, there are typically two primary bands distinguished by an energy band-gap E_g at $\boldsymbol{k} = \boldsymbol{0}$. These bands, which in turn may comprise numerous other sub-bands, are the valence band ($m = v$) and conduction band ($m = c$), satisfying $E_v(\boldsymbol{k}) < E_c(\boldsymbol{k})$. Their features are pivotal in determining the material's properties. Generically, we can discern between three different regimes:

- In the case where the energy band-gap E_g is excessively large, leading to the conduction band being initially unpopulated, high-energy photons are required to overcome this substantial gap and populate the conduction band. This condition characterizes insulator materials with significant band-gaps, creating a challenge for electrons to transition to the conduction band.
- As E_g decreases, we enter the realm of semiconductor materials. In this scenario, photons with relatively small energies, typically in the optical domain, can excite an electron from the usually completely filled valence band to the conduction band.
- For very small, or even in some cases vanishing, values of E_g, the conduction band is initially populated without necessarily requiring external energy input. This regime corresponds to metals, where electrons can freely move between valence and conduction bands without the need for external excitations.

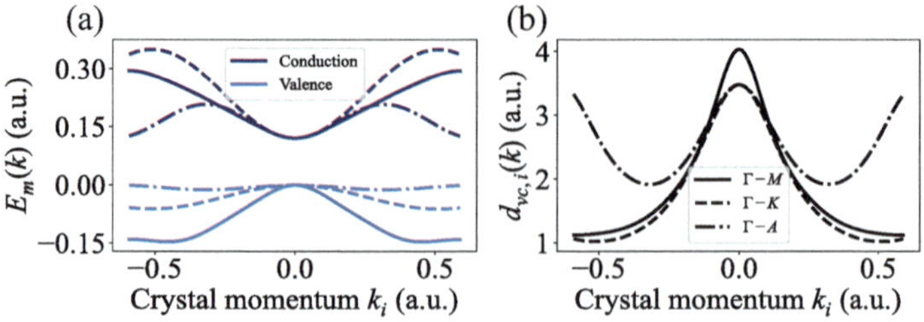

Fig. 4.13 Energy band dispersion relation and transition dipole moments between valence and conduction bands. In **a**, the transition dipole matrix element between valence and conduction bands is plotted for various solid directions. In **b**, the dispersion relation is shown, with each curve representing a distinct solid direction. These plots are conducted for the so-called Γ-point, implying that for each solid direction's function plot, the remaining values of crystal momentum have been set to zero

In the subsequent sections, we focus on semiconductor materials. Specifically, we approximate their band structure in terms of two bands which we refer as the valence and conduction bands, with energy dispersion relations given by

$$E_v(\boldsymbol{k}) = \sum_{i=x,y,z} \sum_{j=0}^{\infty} \alpha^j_{v,i} \cos(jk_i a_i), \tag{4.82}$$

$$E_c(\boldsymbol{k}) = E_g + \sum_{i=x,y,z} \sum_{j=0}^{\infty} \alpha^j_{c,i} \cos(jk_i a_i), \tag{4.83}$$

which can be derived as generalization of the tight-binding approximation, allowing for interactions beyond nearest neighbors. In the case of ZnO, a customary material in the analysis of HHG processes in solids [103], the dispersion relation of these two bands are presented in Fig. 4.13. The coefficients $\{\alpha^j_{m,i}\}$, where i denotes one of the three crystal directions for ZnO (specifically $i = x \equiv \Gamma - M$, $i = y \equiv \Gamma - K$ and $i = z \equiv \Gamma - A$), are detailed in Appendix 9.3, and have been obtained from Refs. [104, 105]. Notably, their computation involves more elaborated numerical methods such as the nonlocal empirical pseudopotential method [106].

4.2.2 *A Bloch-Based Semiclassical Description of High-Harmonic Generation*

In the previous subsection, we briefly covered the fundamental concepts of solid-state physics, specifically delving into the eigenstates of the crystal Hamiltonian and common approximations used when analyzing their band structure. Now, we shift our focus to the scenario where the solid-state system, particularly a semiconductor, interacts with a strong driving field meeting the criteria outlined in the introduction:

- The field intensity remains below the material's damage threshold, typically within the range $I \approx 10^{11} - 10^{12}$ W/cm^2, corresponding to[10] $E_0 \approx 0.1 - 0.6$ V/Å.
- The field's central frequency is significantly lower than typical band-gap energies of semiconductor materials, which correspond to electromagnetic fields within the MIR regime [25, 26].

In this subsection our focus lies on semiclassical descriptions, specifically addressing the light-matter interaction Hamiltonian formulated in the length gauge as

$$\hat{H}(t) = \hat{H}_{\text{cr}} + \mathrm{e}\hat{r}_i E_{\text{cl}}(t), \tag{4.84}$$

where, for the sake of simplicity in our presentation, we proceed by considering a linearly polarized electric field in the direction $\boldsymbol{\epsilon}_i$. Therefore, $\hat{r}_i$ denotes the position operator along the direction defined by $\boldsymbol{\epsilon}_i$, that is, $\hat{r}_i = \boldsymbol{\epsilon}_i \cdot \hat{\mathbf{r}}$. It is important to note that in formulating Eq. (4.84), we have applied the dipole approximation, which we assume to be valid within this context. Specifically, we expect the dipole approximation to hold provided that typical lattice sites are on the order of 1 Å in size, roughly four orders of magnitude smaller than the employed laser field wavelengths.

Compared to individual atoms, electrons in a solid-state system can experience two different kind of transitions. On the one hand, they can be excited to a higher energy band, resulting in an *interband* transition. On the other hand, unlike in atomic systems, electrons can also undergo transitions within the same band, known as *intraband* transitions. These distinct dynamics play a differential role in the HHG process and their outcomes, prompting us to explicitly differentiate between the two. Consequently, we express the position operator as

$$\hat{r}_i = \hat{r}_{i,\text{tra}} + \hat{r}_{i,\text{ter}}, \tag{4.85}$$

where $\hat{r}_{i,\text{tra}}$ accounts for the intraband component of the position operator occurring along the direction specified by $\boldsymbol{\epsilon}_i$, whereas $\hat{r}_{i,\text{ter}}$ pertains the interband one. In the Bloch basis, their matrix elements can be written as

$$\langle \phi_{m,\boldsymbol{k}} | \hat{r}_{i,\text{tra}} | \phi_{l,\boldsymbol{k}} \rangle = i\hbar \delta_{m,l} \frac{\partial}{\partial k_i} \delta(\boldsymbol{k} - \boldsymbol{k}') \tag{4.86}$$

$$\langle \phi_{m,\boldsymbol{k}} | \hat{r}_{i,\text{ter}} | \phi_{l,\boldsymbol{k}} \rangle = \mathrm{e}^{-1} d^{(i)}_{m,l}(\boldsymbol{k}) \delta(\boldsymbol{k} - \boldsymbol{k}'), \tag{4.87}$$

with $d^{(i)}_{m,l}(\boldsymbol{k})$ the dipole matrix element between the valence and conduction bands. This quantity can be computed as

[10] Herein, we employ field strengths to denote the maximum amplitude of the applied field, measured in units of volts per Angstrom (V/Å), following the customary approach within the strong-field solid-state community.

$$\begin{aligned} d_{m,l}^{(i)}(\boldsymbol{k}) = \langle \phi_{m,\boldsymbol{k}} | \hat{r}_i | \phi_{l,\boldsymbol{k}} \rangle &= \frac{1}{E_l(\boldsymbol{k}) - E_m(\boldsymbol{k})} \langle \phi_{m,\boldsymbol{k}} | [\hat{r}_i, \hat{H}_{\text{cr}}] | \phi_{l,\boldsymbol{k}} \rangle \\ &= \frac{i\hbar}{m_{\mathrm{e}}[E_l(\boldsymbol{k}) - E_m(\boldsymbol{k})]} \langle \phi_{m,\boldsymbol{k}} | \hat{p}_i | \phi_{l,\boldsymbol{k}} \rangle, \end{aligned} \tag{4.88}$$

which can be also written as

$$d_{m,l}^{(i)}(\boldsymbol{k}) = \sqrt{\frac{\mathfrak{E}_{p,i}}{\left[2(E_l(\boldsymbol{k}) - E_c(\boldsymbol{k}))^2\right]}}, \tag{4.89}$$

where $\mathfrak{E}_{p,i}$ is the so-called Kane parameter, whose values are tabulated for many materials. Specifically, for ZnO, these can be found in Appendix 9.3, leading to Fig. 4.13b when considering the different ZnO crystal directions.

By introducing this decomposition of the position operator, we can express the Schrödinger equation describing the single-electron dynamics as

$$i\hbar \frac{\partial |\psi(t)\rangle}{\partial t} = \left[\hat{H}_{\text{cr}} + \mathrm{e}\hat{r}_{i,\text{tra}} E_{\text{cl}}(t) + \mathrm{e}\hat{r}_{i,\text{ter}} E_{\text{cl}}(t)\right] |\psi(t)\rangle, \tag{4.90}$$

where the initial state is given by $|\psi(t = t_0)\rangle = |\phi_{v,\boldsymbol{k}_0}\rangle$. In this context, it is essential to note that the many-electron initial state corresponds to a completely filled valence band, with each electron having associated a specific crystal momentum $\boldsymbol{k}$. However, due to the weak electron-electron correlations within the systems under consideration, we treat them as independent entities, and focus on the single-electron dynamics.

Expanding Eq. (4.90) in the Bloch basis introduces a primary challenge, as it leads to a system of coupled differential equations. These equations are interconnected through the crystal momentum $\boldsymbol{k}$ due to the presence of $\hat{r}_{i,\text{tra}}$ which couples different values of $\boldsymbol{k}$ within a band. Therefore, with the aim of simplifying our analysis, it is particularly convenient to operate in the interaction picture with respect to this term, yielding

$$i\hbar \frac{\partial |\bar{\psi}(t)\rangle}{\partial t} = \left[\hat{H}_{\text{cr}}(t) + \mathrm{e}\hat{r}_{i,\text{ter}}(t) E_{\text{cl}}(t)\right] |\bar{\psi}(t)\rangle \quad \text{with} \quad |\psi(t)\rangle = \hat{U}_{\text{tra}}(t) |\bar{\psi}(t)\rangle, \tag{4.91}$$

wherein $\hat{U}_{\text{tra}}(t) \equiv e^{i\mathrm{e}A_{\text{cl}}(t)\hat{r}_{i,\text{tra}}/\hbar}$, such that the time-dependent crystal Hamiltonian and interband component of the position operator are given by

$$\hat{H}_{\text{cr}}(t) \equiv \hat{U}_{\text{tra}}^\dagger(t) \hat{H}_{\text{cr}} \hat{U}_{\text{tra}}(t) \quad \text{and} \quad \hat{r}_{i,\text{ter}}(t) \equiv \hat{U}_{\text{tra}}^\dagger(t) \hat{r}_{i,\text{ter}} \hat{U}_{\text{tra}}(t). \tag{4.92}$$

By transitioning to the interaction picture with respect to the intraband component, we effectively shift our frame to that of the oscillating electron within the field. Thus, rather than working with the crystal momentum $\boldsymbol{k}$, we use the *canonical* crystal momentum $\boldsymbol{K} = \boldsymbol{k} - \mathrm{e}\epsilon_i A_{\text{cl}}(t)$, which spans the basis set $\{|\boldsymbol{K}, m\rangle\}$. Then, we proceed to expand $|\bar{\psi}(t)\rangle$ in this basis set as

$$|\bar{\psi}(t)\rangle = \sum_{m=v,c} \int \mathrm{d}\boldsymbol{K}\, b_m(\boldsymbol{K},t)\,|\boldsymbol{K},m\rangle\,, \tag{4.93}$$

such that, upon its introduction in Eq. (4.90) and subsequent projection of the equation with respect to a specific $|\boldsymbol{K}, m\rangle$, we arrive at

$$\begin{aligned} i\hbar\frac{\partial b_m(\boldsymbol{K},t)}{\partial t} &= E_m\big(\boldsymbol{K} + \mathsf{e}\epsilon_i A_{\text{cl}}(t)\big)b_m(\boldsymbol{K},t) \\ &\quad + \sum_{l=v,c} d^{(i)}_{m,l}\big(\boldsymbol{K} + \mathsf{e}\epsilon_i A_{\text{cl}}(t)\big)b_l(\boldsymbol{K},t) \quad \forall m \in \{v,c\}, \end{aligned} \tag{4.94}$$

which results in a set of coupled differential equations between the two bands. Yet, these equations remain uncoupled with respect to $\boldsymbol{K}$. It is, however, customary to express these equations using a density matrix formalism, describing the evolution of the density matrix population and coherences, which we denote as $n_m(\boldsymbol{K},t) \equiv b^*_m(\boldsymbol{K},t)b_m(\boldsymbol{K},t)$ and $\pi(\boldsymbol{K},t) \equiv b^*_v(\boldsymbol{K},t)b_c(\boldsymbol{K},t)$, respectively. Specifically, upon performing the time derivatives of $n_m(\boldsymbol{K},t)$ and $\pi(\boldsymbol{K},t)$ while considering the relationships derived in Eq. (4.94), we obtain for the evolution of these populations and coherences

$$i\hbar\frac{\partial n_m(\boldsymbol{K},t)}{\partial t} = \theta_m E_{\text{cl}}(t)\Big[d^{(i)}_{c,v}\big(\boldsymbol{K} + \mathsf{e}\epsilon_i A_{\text{cl}}(t)\big)\Big]^* \pi(\boldsymbol{K},t) + \text{c.c.}, \tag{4.95}$$

$$\begin{aligned} i\hbar\frac{\partial \pi(\boldsymbol{K},t)}{\partial t} &= \Big[\mathcal{E}_g\big(\boldsymbol{K} + \mathsf{e}\epsilon_i A_{\text{cl}}(t)\big) + E_{\text{cl}}(t)\xi^{(i)}_g\big(\boldsymbol{K} + \mathsf{e}\epsilon_i A_{\text{cl}}(t)\big)\Big]\pi(\boldsymbol{K},t) \\ &\quad + E_{\text{cl}}(t)d^{(i)}_{c,v}\big(\boldsymbol{K} + \mathsf{e}\epsilon_i A_{\text{cl}}(t)\big)w(\boldsymbol{K},t), \end{aligned} \tag{4.96}$$

where we have defined $\theta_v \equiv 1$ and $\theta_c \equiv -1$, the band-gap energy as $\mathcal{E}_g(\boldsymbol{k}) \equiv E_c(\boldsymbol{k}) - E_v(\boldsymbol{k})$, the difference between the Berry connection $d^{(i)}_{m,m}(\boldsymbol{k})$ of each band as $\xi^{(i)}_g(\boldsymbol{k}) \equiv d^{(i)}_{v,v}(\boldsymbol{k}) - d^{(i)}_{c,c}(\boldsymbol{k})$ and the population difference between bands as $w(\boldsymbol{K},t) \equiv n_v(\boldsymbol{K},t) - n_c(\boldsymbol{K},t)$.

These equations, sometimes referred to as the semiconductor Bloch equations[11] (SBEs) [43], offer us practical advantages over those in Eq. (4.94), as they readily accommodate environmental effects, accounting for additional electron dynamics. In particular, within the realm of HHG processes, it is common to introduce dephasing effects, characterized by the dephasing time T_2, to account for electron-phonon and electron-electron couplings leading to non-radiative electron decay from the conduction to the valence band [104, 105, 108]. These processes explicitly impact the time evolution of the density matrix coherences, resulting in Eq. (4.96) getting modified as

[11] Formally, the semiconductor Bloch equations are derived in the context of electron-hole excitations within a second quantization formalism, as for instance in Ref. [107], and are defined in terms of the crystal momentum rather than the canonical momentum we consider here.

$$i\hbar\frac{\partial \pi(\boldsymbol{K},t)}{\partial t} = \left[\mathcal{E}_g\big(\boldsymbol{K}+\mathsf{e}\epsilon_i A_{\text{cl}}(t)\big) + E_{\text{cl}}(t)\xi_g^{(i)}\big(\boldsymbol{K}+\mathsf{e}\epsilon_i A_{\text{cl}}(t)\big) - \frac{i}{T_2}\right]\pi(\boldsymbol{K},t) + E_{\text{cl}}(t)d_{c,v}^{(i)}\big(\boldsymbol{K}+\mathsf{e}\epsilon_i A_{\text{cl}}(t)\big)w(\boldsymbol{K},t), \tag{4.97}$$

The solution to these differential equations allow us to compute the HHG spectra, in this case expressed in terms of the various currents generated within the solid due to the electron dynamics. The total current encompasses both interband and intraband components, respectively given by

$$j_{i,\text{ter}}(t) = \mathsf{e}\frac{\mathrm{d}}{\mathrm{d}t}\Big[\text{tr}(\hat{r}_{i,\text{ter}}\hat{\rho}(t))\Big], \tag{4.98}$$

$$j_{i,\text{tra}}(t) = -\mathsf{e}\text{tr}\Big(\Big[\hat{r}_{i,\text{tra}}, \hat{H}_{\text{cr}}\Big]\hat{\rho}(t)\Big), \tag{4.99}$$

where $\hat{\rho}(t)$ represents the density matrix of the electron. In terms of these current components, the intensity of the electromagnetic field mode with frequency $\omega_{\mathbf{k}}$ and polarized along the $\boldsymbol{\epsilon}_i$ direction, can be expressed as [108]

$$I_i(\omega_{\mathbf{k}}) \propto \omega_{\mathbf{k}}^2\left|\int_{t_0}^{t} \mathrm{d}\tau e^{i\omega_{\mathbf{k}}\tau}\big[j_{i,\text{tra}}(\tau) + j_{i,\text{ter}}(\tau)\big]\right|^2. \tag{4.100}$$

The currents in Eqs. (4.98) and (4.99) originate from various physical mechanisms, leading to the emission of radiation. Among these, the dominant contributions are mainly attributed to intraband and interband processes [21, 26, 104, 109–111], which are pictorially depicted in Fig. 4.14. The intraband process associates radiation emission with electronic scattering within the non-parabolic energy dispersion profile of a specific band (represented by the dashed green curves in Fig. 4.14), commonly referred to as Bloch oscillations [97]. In contrast, interband processes closely resemble the dynamics described in the three-step model for atoms (see Sect. 3.1.1.2), in which:

1. an electron-hole pair is created through the promotion of an electron from the valence band to the conduction band after the interaction with the driving field (as represented by the dashed red curve in Fig. 4.14);
2. the pair accelerates driven by the field within their respective bands (as indicated by the dashed green curves in Fig. 4.14) and;
3. lastly, a recombination event between the electron and the hole occurs in the valence band, resulting in the emission of a photon whose frequency depends on the energy difference between the bands (illustrated by the dashed purple curves in Fig. 4.14).

In Fig. 4.15, the impact of each current's contribution to the HHG spectrum is illustrated for the case of ZnO when excited by a Gaussian envelope laser pulse. The laser characteristics include a central frequency $\lambda_{\mathbf{k}_L} = 3.25\ \mu\text{m}$, a peak field strength of 0.5 V/Å and a duration of $\Delta t \approx 96$ fs. Specifically, the spectrum is computed for laser fields polarized along the $\Gamma - M$ (panels (a) and (c)) and the $\Gamma - A$ (panels (b)

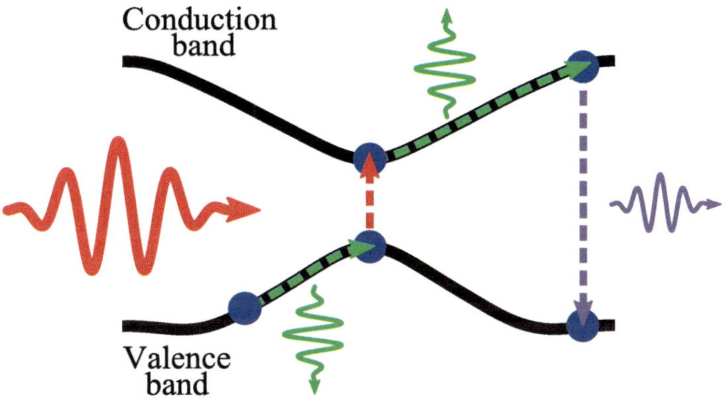

Fig. 4.14 Illustration depicting electron dynamics during the HHG process in solid-state systems. Initially, the excited electron undergoes acceleration within the valence band (lower green dashed arrow), driven by the field, potentially emitting radiation. Due to the interaction with the field, the excited electron can also transition from the valence to the conduction band (vertical red dashed arrow), where it gets accelerated (upper green dashed arrow) due to the driving field's influence, leading to the emission of radiation. Ultimately, the electron can return to the valence band (vertical purple dashed arrow) by emitting a photon with energy equivalent to the energetic difference between the bands

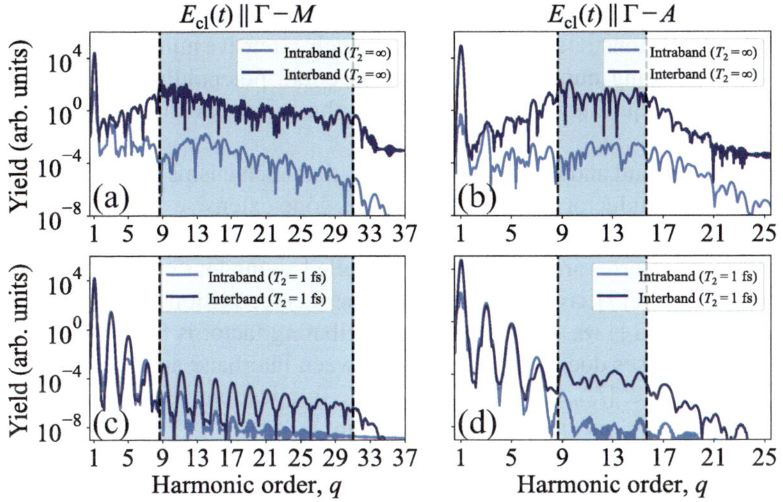

Fig. 4.15 High-harmonic generation spectra of ZnO for laser fields polarized linearly along distinct crystal directions. The laser field features a Gaussian envelope with a central wavelength $\lambda_{\mathbf{k}_L} = 3.25\ \mu\text{m}$, field strength of 0.5 V/Å and a duration of $\Delta t \approx 96$ fs. Specifically, in panels **a** and **c**, we examine the case of the $\Gamma - M$ direction, while for panels **b** and **d**, we consider the case of the $\Gamma - A$ direction. In panels **a** and **b**, infinite dephasing times are assumed, while in panels **c** and **d**, a finite $T_2 = 1$ fs is employed, resulting in well-resolved harmonic peaks

and (d)) crystal directions. In both cases, we observe that the interband contribution (depicted by the dark blue curve) dominates within the non-perturbative regime over the intraband contribution (depicted by the light blue curve). This non-perturbative regime is delineated between the minimum and maximum band-gap energies for each solid direction, illustrated by the blue region enclosed within the black dashed vertical lines in the figure. Conversely, within the perturbative region of the spectra, located to the left of the leftmost vertical line, clear dominance of one contribution over the other is not readily apparent.

While panels (a) and (b) exhibit characteristic features of non-perturbative HHG spectra, such as the plateau region followed by a sharp cutoff, a clear distinction between odd and even harmonics is not evident. In addressing this discrepancy with respect to the well-resolved HHG spectra found experimentally (see for instance Fig. 4.1c), the introduction of dephasing becomes fundamental, as allows recovering the expected clear harmonic spectra. However, achieving these involves using extremely small dephasing times, specifically $T_2 \sim 1$ fs [104, 105], which amount to less than a quarter of an optical cycle. While employing such small dephasing times has been common practice in theoretical approaches, it remains a subject of controversy within the community [20]. In this context, some studies justify their values by suggesting that electrons with high values of crystal momentum could decohere within such time-scales [104]. Others propose additional mechanisms beyond the presented dephasing, such as propagation effects [112]. Recently, M. Kolesik and J. V. Moloney demonstrated in Ref. [113] through extensive numerical computations that issues regarding numerical convergence could potentially become a concern in calculations requiring extremely small dephasing times to show well-resolved harmonic spectra.

Beyond interband and intraband processes, the mechanisms underlying the HHG process in semiconductor materials have undergone extensive scrutiny in the literature [20, 108]. For instance, in the context of non-centrosymmetric materials where Berry connection terms are non-zero, the so-called anomalous current assumes a pivotal role [43, 114–117], contributing to the imperfect recollision mechanism observed in solid-state HHG [118, 119]. Another contributing factor is the so-called mixture current, which arises due to the interplay between interband and intraband position operators [120], i.e. $[\hat{r}_{i,\text{tra}}, \hat{r}_{i,\text{ter}}]$. However, the implications of this mixture current have remain largely unexplored in the literature [108].

4.2.2.1 Beyond Bloch-Based Analyses

While our focus here has been on a Bloch-based analysis of HHG processes, where the quantum state has been expanded in the (accelerated) Bloch basis, alternative approaches based on the use of other basis sets have been also explored in the literature [119]. Given that all the employed basis sets belong to the same Hilbert space, they are necessarily related through unitary transformations. Hence, the use of different bases must not alter physical observables' results. Yet, they could provide alternative, insightful information about the electron dynamics or, in some cases, become helpful for relaxing the requirements of numerical calculations [108, 121]. For instance, in Ref. [122], E. Osika and coworkers employed a mixed representation between Wannier and Bloch states. They expanded the valence band in the real-space localized Wannier basis while retaining the conduction band in the Bloch basis (see Appendix 9.4). This approach allows an atomistic-like interpretation of HHG processes where, unlike atoms, the single electron initiates the dynamics from an initially localized state without necessarily recombining with the same state. This could potentially explain the reduced dependence of the field intensity with the ellipticity of the HHG signal in solids compared to atoms [26].

Recently, in Ref. [123], A. M. Parks and collaborators developed a comprehensive Wannier-based approach, exploring the contributions from spatially short and long trajectories[12] on HHG spectra. Additionally, in Ref. [124], G. G. Brown and colleagues demonstrated that well-resolved harmonic peaks could be observed upon the destructive interference between these spatially long trajectories without requiring extremely short dephasing times. These interferences were attributed to additional intensity-dependent phase factors arising from the laser field's intensity profile.

In another vein, L. Li and coworkers proposed in Ref. [125] a Huygens-Fresnel-based interpretation of the HHG process in solids, deviating from standard particle-based interpretations [119]. Specifically, they showcased that the geometric optics limit aligns with the scenario where the saddle-point approximation can be applied in evaluating the integrals arising from Eq. (4.100). This regime corresponds to the foundation onto which the three-step model is built, and therefore the particle-based interpretation of HHG processes.

4.2.3 A Bloch-Based Quantum Optical Description of High-Har-Monic Generation

While the analysis of HHG processes in solid-state systems from a quantum optical perspective has not been as extensive as in its atomic counterpart, as discussed in Sect. 3.2, recent years have seen notable advancements. In Ref. [126], I. Gonoskov and S. Gräfe developed a novel set of approximations aimed at simplifying the description of light-matter interactions under a quantum optical framework, applicable to arbitrary nuclear potentials. Subsequently, these authors, alongside other collaborators, applied this approach in Ref. [127] to analyze semiconductor materials. Their work revealed that intraband dynamics in solids can induce non-trivial effects such as squeezing and entanglement in the final quantum optical state, attributed to the non-parabolic band structure of solids. However, under typical perturbative HHG conditions where intraband contributions dominate, these effects were found to be negligible.

---

[12] *Spatially* short and long trajectories refer to the distance between the Wannier sites from which the electron ionizes and recombines.

In a parallel yet distinct direction, the groups led by P. Tzallas and M. Lewenstein delved deeper into approaches centered on the conditioning on HHG operation. Notably, in Ref. [128], experimental demonstrations showed that the conditioning on HHG operation (as depicted in Fig. 3.10) can be extended to the interaction of light with semiconductor materials. The theoretical analysis in Ref. [50], employing a Bloch-based framework, showed that these operations could result in the generation of non-classical states of light, akin to those found in atomic systems. However, the properties of the resultant states were dictated by the different electron dynamics within these periodic systems. Moreover, in Ref. [49], employing a Wannier-based approach analogous to that in Ref. [122], it was demonstrated that due to the delocalized nature of HHG processes in solids, electron and light become entangled after electronic recombination in the valence band. Nevertheless, for typical semiconductor materials, this effect was found to be negligible.

In this subsection, our focus centers on the analysis presented in Ref. [50], while those of Ref. [49] are outlined in Appendix 9.4. The approach adopted in the former reference constitutes a direct quantum optical extension of the analysis presented in Sect. 4.2.2. Therefore, assuming identical excitation conditions to its semiclassical counterpart, the light-matter interaction Hamiltonian is expressed as follows

$$\hat{H}(t) = \hat{H}_{\text{cr}} + \mathsf{e}\hat{r}_i\hat{E}(t) + \hat{H}_{\text{field}}, \tag{4.101}$$

where $\hat{H}_{\text{field}}$ denotes the free-field Hamiltonian, while $\hat{E}(t)$ represents the electric field operator given in Eq. (3.35) after performing the scalar product with respect to the polarization mode[13] $\boldsymbol{\epsilon}_i$, i.e., $\hat{E}(t) \equiv \boldsymbol{\epsilon}_i \cdot \hat{\mathbf{E}}(t)$. Similar to Sect. 4.2.2, we further decompose the position operator $\hat{r}_i$ into its intraband and interband components. Consequently, the Schrödinger equation describing the dynamics of the joint state composed by the field modes and the single active electron is expressed as

$$i\hbar\frac{\partial|\Psi(t)\rangle}{\partial t} = \Big[\hat{H}_{\text{cr}} + \mathsf{e}\hat{r}_{i,\text{tra}}\hat{E}(t) + \mathsf{e}\hat{r}_{i,\text{ter}}\hat{E}(t) + \hat{H}_{\text{field}}\Big]|\Psi(t)\rangle, \tag{4.102}$$

with the initial state given by

$$|\Psi(t=t_0)\rangle = |\phi_{v,\boldsymbol{k}_0}\rangle \otimes |\alpha_{\mathbf{k}_L}\rangle \bigotimes_{q=2}^{q_c} |0_{\mathbf{k}_q}\rangle. \tag{4.103}$$

Similar to the analysis conducted in atomic and molecular systems, we implement the series of unitary transformation illustrated in Fig. 3.12 on the electromagnetic field modes. The introduction of these, results in

[13] Formally, the different polarization directions have to be considered at this level, as their population is determined by the quantum optical state. However, for simplicity in the presentation, we restrict our discussion to single polarization modes. Nevertheless, it is important to note that this approach can be extended to encompass all polarization directions.

$$i\hbar\frac{\partial|\tilde{\Psi}(t)\rangle}{\partial t} = \Big[\hat{H}_{\text{cr}} + \mathsf{e}\hat{r}_{i,\text{tra}}\big(E_{\text{cl}}(t)+\hat{E}(t)\big) + \mathsf{e}\hat{r}_{i,\text{ter}}\big(E_{\text{cl}}(t)+\hat{E}(t)\big)\Big]|\tilde{\Psi}(t)\rangle$$
$$\text{with} \quad |\Psi(t)\rangle = \hat{U}_{\text{field}}(t)\hat{D}_{\mathbf{k}_L}(\alpha_{\mathbf{k}_L})|\tilde{\Psi}(t)\rangle, \tag{4.104}$$

for which the initial state satisfies the following condition

$$|\tilde{\Psi}(t_0)\rangle = |\phi_{v,\boldsymbol{k}_0}\rangle \bigotimes_{q=1}^{q_c} |0_{\mathbf{k}_q}\rangle \equiv |\phi_{v,\boldsymbol{k}_0}\rangle \otimes |\bar{0}\rangle. \tag{4.105}$$

However, unlike these other systems, we perform an additional transformation simultaneously affecting the electronic and electromagnetic field modes. This joint transformation is represented by $\hat{\hat{U}}_{\text{tra}}(t) \equiv e^{i\mathsf{e}(A_{\text{cl}}(t)+\hat{A}(t))\hat{r}_{i,\text{tra}}/\hbar}$, and can be understood as the quantum optical counterpart to the transformation defined in Eq. (4.91). Apart from the classical vector potential's action $A_{\text{cl}}(t)$, $\hat{\hat{U}}_{\text{tra}}(t)$ also accounts for the action of its corresponding operator $\hat{A}(t)$. Introducing this unitary transformation into Eq. (4.104) yields

$$i\hbar\frac{\partial|\bar{\Psi}(t)\rangle}{\partial t} = \hat{\hat{U}}^{\dagger}_{\text{tra}}(t)\Big[\hat{H}_{\text{cr}} + \mathsf{e}\hat{r}_{i,\text{ter}}\big(E_{\text{cl}}(t)+\hat{E}(t)\big)\Big]\hat{\hat{U}}_{\text{tra}}(t)|\bar{\Psi}(t)\rangle$$
$$\text{with} \quad |\tilde{\Psi}(t)\rangle = \hat{\hat{U}}_{\text{tra}}(t)|\bar{\Psi}(t)\rangle, \tag{4.106}$$

where the unitary transformation $\hat{\hat{U}}_{\text{tra}}(t)$ simultaneously affects the electronic and quantum optical operators in the equation. Hereafter, we take into account that the coupling factor $\mathrm{g}(\omega_{\mathbf{k}_q})$ in the vector potential's definition is a perturbative quantity. Hence, if we expand the $e^{i\mathsf{e}\hat{A}(t)\hat{r}_{i,\text{tra}}}$ component up to first order in $\mathrm{g}(\omega_{\mathbf{k}_q})$, i.e. $e^{i\mathsf{e}\hat{A}(t)\hat{r}_{i,\text{tra}}} \approx \mathbb{1} + i\mathsf{e}\hbar^{-1}\hat{r}_{i,\text{tra}}\hat{A}(t)$, each term at the right-hand side of Eq. (4.106) takes the form

$$\hat{\hat{U}}^{\dagger}_{\text{tra}}(t)\hat{H}_{\text{cr}}\hat{\hat{U}}_{\text{tra}}(t) \approx \hat{H}_{\text{cr}} - i\mathsf{e}\hbar^{-1}\Big[\hat{r}_{i,\text{tra}}, \hat{H}_{\text{cr}}\Big]\hat{A}(t), \tag{4.107}$$

$$\hat{\hat{U}}^{\dagger}_{\text{tra}}(t)\hat{r}_{i,\text{ter}}\hat{\hat{U}}_{\text{tra}}(t) \approx \hat{r}_{i,\text{ter}} - i\mathsf{e}\hbar^{-1}\Big[\hat{r}_{i,\text{tra}}, \hat{r}_{i,\text{ter}}\Big]\hat{A}(t), \tag{4.108}$$

$$\hat{\hat{U}}^{\dagger}_{\text{tra}}(t)\hat{E}(t)\hat{\hat{U}}_{\text{tra}}(t) \approx \hat{E}(t) - i\mathsf{e}\hbar^{-1}\hat{r}_{i,\text{tra}}\Big[\hat{A}(t), \hat{E}(t)\Big] \approx \hat{E}(t), \tag{4.109}$$

where in Eq. (4.109) we have neglected second order terms in $\mathrm{g}(\omega_{\mathbf{k}_q})$. Regarding the other two equations, Eq. (4.107) introduces the impact of the intraband current on the quantum optical fluctuations, while Eq. (4.108) captures the effect of the mixture current. In the subsequent analysis, for a direct comparison with the semiclassical approach in Sect. 4.2.2, we specifically focus on the influence of interband and intraband currents on the quantum optical state. Therefore, we exclude the influence of the mixture current, which, it is worth stressing, remains uncertain within the scope of semiclassical theories.

Under these conditions, we can approximate Eq. (4.106) as

$$i\hbar\frac{\partial|\bar{\Psi}(t)\rangle}{\partial t} = \Big[\hat{H}_{\text{cr}}(t) + \mathsf{e}\hat{r}_{i,\text{ter}}(t)\big(E_{\text{cl}}(t) + \hat{E}(t)\big) + i\mathsf{e}\hbar^{-1}\hat{v}_{i,\text{tra}}(t)\hat{A}(t)\Big]|\bar{\Psi}(t)\rangle, \tag{4.110}$$

where we have defined $\hat{v}_{i,\text{tra}} \equiv [\hat{r}_{i,\text{tra}}, \hat{H}_{\text{tra}}]$. It is important to note that $\hat{H}_{\text{cr}}(t)$ and $\hat{r}_{i,\text{ter}}$ are defined as in Eq. (4.92), with $\hat{v}_{i,\text{tra}}(t)$ obtained similarly by doing the replacement $\hat{r}_{i,\text{ter}} \rightarrow \hat{v}_{i,\text{tra}}$ in the aforementioned equation. As a final step before explicitly solving the resulting differential equation, we transition to the interaction picture with respect to $\hat{H}_{\text{sc}}(t) \equiv \hat{H}_{\text{cr}}(t) + \mathsf{e}\hat{r}_{i,\text{ter}}(t)E_{\text{cl}}(t)$, yielding

$$i\hbar\frac{\partial|\bar{\Psi}'(t)\rangle}{\partial t} = \Big[\mathsf{e}\hat{\bar{r}}_{i,\text{ter}}(t)\hat{E}(t) - i\mathsf{e}\hbar^{-1}\hat{\bar{v}}_{i,\text{tra}}(t)\hat{A}(t)\Big]|\bar{\Psi}'(t)\rangle \\ \text{with} \quad |\bar{\Psi}(t)\rangle = \hat{U}_{\text{sc}}(t, t_0)|\bar{\Psi}'(t)\rangle, \tag{4.111}$$

where $\hat{\bar{r}}_{i,\text{ter}}(t) \equiv \hat{U}^{\dagger}_{\text{sc}}(t, t_0)\hat{r}_{i,\text{ter}}(t)\hat{U}_{\text{sc}}(t, t_0)$ and $\hat{\bar{v}}_{i,\text{tra}}(t) \equiv \hat{U}^{\dagger}_{\text{sc}}(t, t_0)\hat{v}_{i,\text{tra}}(t)\hat{U}_{\text{sc}}(t, t_0)$.

To solve this equation, and in virtue of the semiclassical analysis performed earlier, we expand the state $|\bar{\Psi}'(t)\rangle$ into the crystal momentum eigenbasis, that is, we write

$$|\bar{\Psi}'(t)\rangle = \sum_{l=c,v}\int \mathrm{d}\boldsymbol{K}\, |\boldsymbol{K}, l\rangle \otimes |\Phi_l(\boldsymbol{K}, t)\rangle\,, \tag{4.112}$$

with $|\Phi_l(\boldsymbol{K}, t)\rangle$ denoting the quantum optical state conditioned on finding the electron in the state $|\boldsymbol{K}, l\rangle$. Introducing this expansion inside Eq. (4.111), and after projecting with respect to $|\boldsymbol{K}, m\rangle$, we arrive at

$$i\hbar\frac{\partial|\Phi_m(\boldsymbol{K}, t)\rangle}{\partial t} = \sum_{l=v,c}\int \mathrm{d}\boldsymbol{K}'\Big[\mathsf{e}\langle\boldsymbol{K}, m|\hat{\bar{r}}_{i,\text{ter}}(t)|\boldsymbol{K}', l\rangle\hat{E}(t)\,|\Phi_l(\boldsymbol{K}', t)\rangle \\ - i\mathsf{e}\hbar^{-1}\langle\boldsymbol{K}, m|\hat{\bar{v}}_{i,\text{tra}}(t)|\boldsymbol{K}', l\rangle\hat{A}(t)\,|\Phi_l(\boldsymbol{K}', t)\rangle\Big], \tag{4.113}$$

which is valid for both the valence and the conduction bands.

Before proceeding further, it is first convenient to elaborate on the matrix elements involving the operators $\hat{\bar{r}}_{i,\text{ter}}$ and $\hat{\bar{v}}_{i,\text{tra}}$, whose form will determine the strategy followed in order to solve the differential equation above. To facilitate this analysis, we take into account that, in agreement with Eq. (4.94), we can express

$$\hat{U}_{\text{sc}}(t)\,|\boldsymbol{K}, m\rangle = b_v(\boldsymbol{K}, t|m)\,|\boldsymbol{K}, v\rangle + b_c(\boldsymbol{K}, t|m)\,|\boldsymbol{K}, c\rangle\,, \tag{4.114}$$

given that this unitary operator does not couple the different values of the crystal momentum. In this expression, $b_l(\boldsymbol{K}, t|m)$ represents the probability amplitude of being in the energy band l provided that the electron initially was in band m. Consequently, for the intraband contributions appearing in Eq. (4.113), we can write

$$
\begin{aligned}
\int \mathrm{d}\boldsymbol{K}' \langle \boldsymbol{K}, m|\hat{\hat{r}}_{i,\text{ter}}(t)|\boldsymbol{K}', l\rangle
&= \int \mathrm{d}\boldsymbol{K}'\big[b_v^*(\boldsymbol{K}, t|m)\,\langle \boldsymbol{K}, v| + b_c^*(\boldsymbol{K}, t|m)\,\langle \boldsymbol{K}, c|\,\big]\hat{r}_{i,\text{ter}}(t)\\
&\qquad \times \big[b_v(\boldsymbol{K}, t|l)\,|\boldsymbol{K}', v\rangle + b_c(\boldsymbol{K}, t|l)\,|\boldsymbol{K}', c\rangle\,,\big]\,|\Phi_l(\boldsymbol{K}', t)\rangle\\
&= \sum_{\mathsf{m}=v,c}\sum_{\mathsf{l}=v,c} b^*_{\mathsf{m}}(\boldsymbol{K}, t|m) b_{\mathsf{l}}(\boldsymbol{K}, t|l) d_{\mathsf{m},\mathsf{l}}\big(\boldsymbol{K} + \mathsf{e}\epsilon_i A_{\text{cl}}(t)\big)\,|\Phi_{\mathsf{l}}(t)\rangle\\
&\equiv \mathsf{e}^{-1} M^{(\text{ter})}_{m,l}(\boldsymbol{K}, t)\,|\Phi_l(\boldsymbol{K}, t)\rangle\,,
\end{aligned}
\tag{4.115}
$$

where in going from the first to the second equality, we have used Eq. (4.87) alongside the fact that $|\boldsymbol{K}, m\rangle = e^{-i\mathsf{e}A_{\text{cl}}(t)\hat{r}_{i,\text{tra}}}|\phi_{m,\boldsymbol{k}}\rangle$, with $\boldsymbol{k} = \boldsymbol{K} + \mathsf{e}\epsilon_i A_{\text{cl}}(t)$. Thus, we observe that these matrix elements are diagonal in the crystal momentum basis.

Regarding the matrix element involving the $\hat{\hat{v}}_{i,\text{tra}}(t)$ operator, we break it down into two parts. More specifically, given that $\hat{v}_{i,\text{tra}}(t)$ is defined in terms of the commutator between $\hat{r}_{i,\text{tra}}$ and $\hat{H}_{\text{cr}}$, we examine each element within the commutator individually. For one of these, we find

$$
\begin{aligned}
\int \mathrm{d}\boldsymbol{K}' \langle \boldsymbol{K}, m|\hat{\hat{r}}_{i,\text{tra}}(t)\hat{\hat{H}}_{\text{cr}}(t)|\boldsymbol{K}', l\rangle
&= \int \mathrm{d}\boldsymbol{K}'\big[b_v^*(\boldsymbol{K}, t|m)\,\langle \boldsymbol{K}, v| + b_c^*(\boldsymbol{K}, t|m)\,\langle \boldsymbol{K}, c|\,\big]\hat{r}_{i,\text{tra}}(t)\hat{H}_{\text{cr}}\\
&\qquad \times \big[b_v(\boldsymbol{K}, t|l)\,|\boldsymbol{K}', v\rangle + b_c(\boldsymbol{K}, t|l)\,|\boldsymbol{K}', c\rangle\,,\big]\,|\Phi_l(\boldsymbol{K}', t)\rangle\\
&= \int \mathrm{d}\boldsymbol{K}'\big[b_v^*(\boldsymbol{K}, t|m)\,\langle \boldsymbol{K}, v| + b_c^*(\boldsymbol{K}, t|m)\,\langle \boldsymbol{K}, c|\,\big]\hat{r}_{i,\text{tra}}(t)\\
&\qquad \times \big[E_v\big(\boldsymbol{K}' + \mathsf{e}\epsilon_i A_{\text{cl}}(t)\big) b_v(\boldsymbol{K}, t|l)\,|\boldsymbol{K}', v\rangle\\
&\qquad\quad + E_c\big(\boldsymbol{K}' + \mathsf{e}\epsilon_i A_{\text{cl}}(t)\big) b_c(\boldsymbol{K}, t|l)\,|\boldsymbol{K}', c\rangle\,\big]\,|\Phi_l(\boldsymbol{K}', t)\rangle\\
&= \sum_{\mathsf{m}=c,v}\sum_{\mathsf{l}=c,v}\int \mathrm{d}\boldsymbol{K}' b^*_{\mathsf{m}}(\boldsymbol{K}, t|m)\langle \boldsymbol{K}, \mathsf{m}|\hat{r}_{i,\text{tra}}|\boldsymbol{K}', \mathsf{l}\rangle\\
&\qquad \times b_{\mathsf{m}}(\boldsymbol{K}', t|l) E_{\mathsf{m}}\big(\boldsymbol{K}' + \mathsf{e}\epsilon_i A_{\text{cl}}(t)\big)\,|\Phi_l(\boldsymbol{K}', t)\rangle\\
&= i\hbar \sum_{\mathsf{m}=c,v} b^*_{\mathsf{m}}(\boldsymbol{K}, t|m)\frac{\partial}{\partial k_i}\Big[b_{\mathsf{m}}(\boldsymbol{K}, t|l) E_{\mathsf{m}}\big(\boldsymbol{K} + \mathsf{e}\epsilon_i A_{\text{cl}}(t)\big)\,|\Phi_l(\boldsymbol{K}, t)\rangle\Big],
\end{aligned}
\tag{4.116}
$$

where in transitioning from the third to the fourth equality we used Eq. (4.86).

For the other term of the commutator, i.e. $\hat{\hat{H}}_{\text{cr}}(t)\hat{\hat{r}}_{i,\text{tra}}(t)$, a similar analysis can be followed, yielding

$$
\begin{aligned}
\int \mathrm{d}\boldsymbol{K}' \langle \boldsymbol{K}, m|\hat{\hat{H}}_{\text{cr}}(t)\hat{\hat{r}}_{i,\text{tra}}|\boldsymbol{K}', l\rangle\,|\Phi_l(\boldsymbol{K}', t)\rangle
&= i\hbar \sum_{\mathsf{m}=c,v} b^*_{\mathsf{m}}(\boldsymbol{K}, t|m) E_{\mathsf{m}}\big(\boldsymbol{K} + \mathsf{e}\epsilon_i A_{\text{cl}}(t)\big)\frac{\partial}{\partial k_i}\Big[b_{\mathsf{m}}(\boldsymbol{K}, t|l)\,|\Phi_l(\boldsymbol{K}, t)\rangle\Big],
\end{aligned}
\tag{4.117}
$$

such that when combining Eqs. (4.116) and (4.117), we arrive at

$$
\begin{aligned}
\int \mathrm{d}\boldsymbol{K}'\langle \boldsymbol{K}, m|\hat{v}_{i,\mathrm{tra}}(t)|\boldsymbol{K}', l\rangle\, |\Phi_l(\boldsymbol{K}', t)\rangle
&= i\hbar \sum_{\mathrm{m}=c,v} b_{\mathrm{i}}^*(\boldsymbol{K}, t|m) b_{\mathrm{i}}(\boldsymbol{K}, t|l) \frac{\partial}{\partial k_i}\Big[E\mathrm{m}\big(\boldsymbol{K} + \mathrm{e}\epsilon_i A_{\mathrm{cl}}(t)\big)\Big]\, |\Phi_l(\boldsymbol{K}, t)\rangle \\
&\equiv i\hbar \mathrm{e}^{-1} M_{m,l}^{(\mathrm{tra})}(\boldsymbol{K}, t)\, |\Phi_l(\boldsymbol{K}, t)\rangle\,, \qquad (4.118)
\end{aligned}
$$

which, as Eq. (4.115), is diagonal in the crystal momentum basis.

Introducing Eqs. (4.115) and (4.118) into Eq. (4.113), and explicitly distinguishing between projections onto the valence and conduction bands, we obtain the following set of coupled differential equations

$$
\begin{aligned}
i\hbar \frac{\partial}{\partial t}|\Phi_v(\boldsymbol{K}, t)\rangle &= \Big[M_{v,v}^{(\mathrm{ter})}(\boldsymbol{K}, t)\hat{E}(t) + M_{v,v}^{(\mathrm{tra})}(\boldsymbol{K}, t)\hat{A}(t)\Big]\, |\Phi_v(\boldsymbol{K}, t)\rangle \\
&\quad + \Big[M_{v,c}^{(\mathrm{ter})}(\boldsymbol{K}, t)\hat{E}(t) + M_{v,c}^{(\mathrm{tra})}(\boldsymbol{K}, t)\hat{A}(t)\Big]\, |\Phi_c(\boldsymbol{K}, t)\rangle\,, \qquad (4.119) \\
i\hbar \frac{\partial}{\partial t}|\Phi_c(\boldsymbol{K}, t)\rangle &= \Big[M_{c,v}^{(\mathrm{ter})}(\boldsymbol{K}, t)\hat{E}(t) + M_{c,v}^{(\mathrm{tra})}(\boldsymbol{K}, t)\hat{A}(t)\Big]\, |\Phi_v(\boldsymbol{K}, t)\rangle \\
&\quad + \Big[M_{c,c}^{(\mathrm{ter})}(\boldsymbol{K}, t)\hat{E}(t) + M_{c,c}^{(\mathrm{tra})}(\boldsymbol{K}, t)\hat{A}(t)\Big]\, |\Phi_c(\boldsymbol{K}, t)\rangle\,, \qquad (4.120)
\end{aligned}
$$

which are, however, decoupled in $\boldsymbol{K}$. In these equations, the elements $M_{m,l}^{(\mathrm{ter})}(\boldsymbol{K}, t)$ and $M_{m,l}^{(\mathrm{tra})}(\boldsymbol{K}, t)$ account for the influence of the interband and intraband currents, respectively, on the quantum optical state. Specifically, the diagonal elements $M_{m,m}^{(\mathrm{ter})}(\boldsymbol{K}, t)$ and $M_{m,m}^{(\mathrm{tra})}(\boldsymbol{K}, t)$ exactly resemble the time-dependent interband polarization and the intraband current, specifically in situations where the electron begins and ends up the dynamics in the band m. In contrast, the off-diagonal terms represent those contributions in scenarios where the electron ends up in a different state from where it was initially.

In this context, a main distinction between $M_{m,m}^{(\mathrm{ter})}(\boldsymbol{K}, t)$ and $M_{m,m}^{(\mathrm{tra})}(\boldsymbol{K}, t)$ terms and the currents defined in Eqs. (4.99) and (4.98) is that the former have been computed in a noiseless scenario, while the latter can potentially incorporate dephasing effects. However, we introduce these effects phenomenologically when evaluating the matrix elements $M_{v,v}^{(\mathrm{ter})}(\boldsymbol{K}, t)$ and $M_{v,v}^{(\mathrm{tra})}(\boldsymbol{K}, t)$ elements, which play a fundamental influence in the equations below, as

$$
M_{v,v}^{(\mathrm{ter})}(\boldsymbol{K}, t) = \pi(\boldsymbol{K}, t) d_{cv}\big(\boldsymbol{K} + \mathrm{e}\epsilon_i A_{\mathrm{cl}(t)}(t)\big) + \text{c.c.}, \qquad (4.121)
$$

$$
M_{v,v}^{(\mathrm{tra})}(\boldsymbol{K}, t) = \sum_{m=c,v} n_m(\boldsymbol{K}, t) \frac{\partial}{\partial K_i}\big[E_{\mathrm{i}}\big(\boldsymbol{K} + \mathrm{e}\epsilon_i A_{\mathrm{cl}}(t)\big)\big], \qquad (4.122)
$$

where the $n_m(\boldsymbol{K}, t)$ and $\pi(\boldsymbol{K}, t)$ are obtained by numerically solving Eqs. (4.95) and (4.97), with the dephasing time T_2 used as a parameter. Specifically, we expect that in the weak-coupling limit—indeed the case here, given $\text{g}(\omega_{\mathbf{k}_q}) \ll 1$—between the quantum optical degrees of freedom and the time-dependent interband and intraband operators depicting the electron dynamics, the results derived through this phenomenological approach to converge with those obtained by solving the corresponding Liouville equation. The rationale behind this statement lies in the eventual decoupling of the final joint state of the electron and the electromagnetic field in both approaches. Furthermore, it is noteworthy that the method employed here resembles the use of non-Hermitian Hamiltonians, a strategy already applied in semiclassical analyses [129], yielding similar outcomes to those achieved when solving Eqs. (4.95) and (4.97).

Focusing our attention into the system of coupled differential equations presented in Eqs. (4.119) and (4.120), it is worth noting that they mirror those obtained in the molecular analysis in Sect. 4.1.2, specifically Eqs. (4.17) and (4.18). Consequently, they yield analogous solutions. However, for clarity in notation, we begin by solving the homogeneous part of these equations. Specifically, for the energy band m, we have

$$i\hbar\frac{\partial}{\partial t}|\Phi_m^{(\text{hom})}(\boldsymbol{K}, t)\rangle = \Big[M_{m,m}^{(\text{ter})}(\boldsymbol{K}, t)\hat{E}(t) + M_{m,m}^{(\text{tra})}(\boldsymbol{K}, t)\hat{A}(t)\Big]|\Phi_m^{(\text{hom})}(\boldsymbol{K}, t)\rangle, \tag{4.123}$$

where it is particularly convenient to expand the electric field and the vector potential operators in terms of their creation and annihilation operators

$$\hat{E}(t) = -i\sum_{q=1}^{q_c}\text{g}(\omega_{\mathbf{k}_q})\Big[f_q(t)\hat{a}^\dagger_{\mathbf{k}_q} - f_q^*(t)\hat{a}_{\mathbf{k}_q}\Big], \tag{4.124}$$

$$\hat{A}(t) = -\int \text{d}t\,\hat{E}(t) = i\sum_{q=1}^{q_c}\text{g}(\omega_{\mathbf{k}_q})\Big[F_q(t)\hat{a}^\dagger_{\mathbf{k}_q} - F_q^*(t)\hat{a}_{\mathbf{k}_q}\Big], \tag{4.125}$$

where in these expressions we have defined $f_q(t) \equiv f(t)e^{i\omega_{\mathbf{k}_q}t}$ and $F_q(t) \equiv \int \text{d}t f_q(t)$. These functions are shown in Fig. 4.16 and, as observed, both tend to zero as $t \to \pm\infty$. Introducing them in Eq. (4.124) leads to

$$\begin{aligned} i\hbar\frac{\partial}{\partial t}|\Phi_m^{(\text{hom})}(\boldsymbol{K}, t)\rangle = i\sum_{q=1}^{q_c}\text{g}(\omega_{\mathbf{k}_q})\Big[&\Big(-f_q(t)M_{m,m}^{(\text{ter})}(\boldsymbol{K}, t) + F_q(t)M_{m,m}^{(\text{tra})}(\boldsymbol{K}, t)\Big)\hat{a}^\dagger_{\mathbf{k}_q}\\ &-\Big(-f_q^*(t)M_{m,m}^{(\text{ter})}(\boldsymbol{K}, t) + F_q^*(t)M_{m,m}^{(\text{tra})}(\boldsymbol{K}, t)\Big)\hat{a}_{\mathbf{k}_q}\Big]|\Phi_m^{(\text{hom})}(\boldsymbol{K}, t)\rangle. \end{aligned} \tag{4.126}$$

In this case the solution to the homogeneous equation is given by

$$|\Phi_m^{(\text{hom})}(\boldsymbol{K}, t)\rangle = \hat{\mathcal{D}}\big(\boldsymbol{\chi}_m(\boldsymbol{K}, t, t_0)\big)|\Phi_m^{(\text{hom})}(\boldsymbol{K}, t_0)\rangle, \tag{4.127}$$

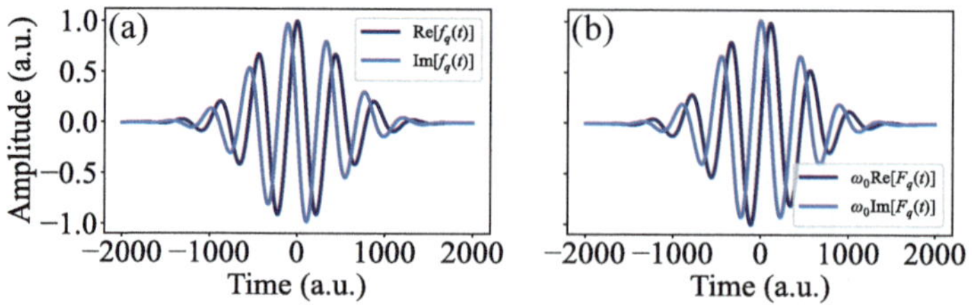

Fig. 4.16 Field envelopes for the electric end vector potential operators. The real (dark blue curve) and imaginary (light blue curve) components of the envelope functions corresponding to the **a** electric field operator and **b** vector potential operator are shown when $\omega_{\mathbf{k}_q} = \omega_{\mathbf{k}_L}$ (similar trends are observed for higher frequencies). Both functions tend towards zero as $t \to \pm\infty$

where $\hat{\mathcal{D}}(\boldsymbol{\chi}_m(\boldsymbol{K}, t, t_0))$ is a multimode displacement operator identical to the one obtained for atomic and molecular systems, with the displacements $\chi_q^{(m)}(\boldsymbol{K}, t, t_0)$ determined in this case by

$$\chi_q^{(m)}(\boldsymbol{K}, t, t_0) = \frac{1}{\hbar}\sqrt{\frac{\hbar\omega_{\mathbf{k}_q}}{2\varepsilon_0 V}} \int_{t_0}^{t} \mathrm{d}\tau \Big[- f_q(t) M_{m,m}^{(\mathrm{ter})}(\boldsymbol{K}, t) + F_q(t) M_{m,m}^{(\mathrm{tra})}(\boldsymbol{K}, t)\Big], \tag{4.128}$$

that is, the displacement found on the field modes explicitly depends on the contributions arising from the interband and intraband currents.

Analogous to the solutions of Eqs. (4.17) and (4.18) given in Eqs. (4.25) and (4.26), employing the result above enables us to express a solution to Eqs. (4.119) and (4.120) through the following set of recursive relations

$$\begin{aligned} |\Phi_v(\boldsymbol{K}, t)\rangle &= \hat{\mathcal{D}}\big(\boldsymbol{\chi}_v(\boldsymbol{K}, t, t_0)\big) |\Phi_v(\boldsymbol{K}, t_0)\rangle \\ &\quad - \frac{i}{\hbar}\int_{t_0}^{t} \mathrm{d}t_1 \hat{\mathcal{D}}\big(\boldsymbol{\chi}_v(\boldsymbol{K}, t, t_1)\big) \hat{\mathcal{M}}_{v,c}(\boldsymbol{K}, t_1) |\Phi_c(\boldsymbol{K}, t_1)\rangle, \end{aligned} \tag{4.129}$$

$$\begin{aligned} |\Phi_c(\boldsymbol{K}, t)\rangle &= \hat{\mathcal{D}}\big(\boldsymbol{\chi}_c(\boldsymbol{K}, t, t_0)\big) |\Phi_c(\boldsymbol{K}, t_0)\rangle \\ &\quad - \frac{i}{\hbar}\int_{t_0}^{t} \mathrm{d}t_1 \hat{\mathcal{D}}\big(\boldsymbol{\chi}_c(\boldsymbol{K}, t, t_1)\big) \hat{\mathcal{M}}_{c,v}(\boldsymbol{K}, t_1) |\Phi_v(\boldsymbol{K}, t_1)\rangle, \end{aligned} \tag{4.130}$$

which combined with the initial expansion for the joint state of the system in Eq. (4.112), while undoing the transformations presented in Eqs. (4.106) and (4.111), yields

$$|\tilde{\Psi}(t)\rangle = \hat{\hat{U}}_{\mathrm{tra}}(t)\hat{U}_{\mathrm{sc}}(t, t_0) \int \mathrm{d}\boldsymbol{K}\Big[|\boldsymbol{K}, v\rangle \otimes |\Phi_v(\boldsymbol{K}, t)\rangle + |\boldsymbol{K}, c\rangle \otimes |\Phi_c(\boldsymbol{K}, t)\rangle\Big], \tag{4.131}$$

that is, a generally entangled state between the field and electronic degrees of freedom. Before proceeding further, we impose the initial conditions corresponding to finding the electron at time t_0 in the valence band with crystal momentum $\boldsymbol{K}$.

Given that in our case we assume that $t_0 \to -\infty$, then it follows that $A_{\text{cl}}(t_0) = 0$ and $\hat{A}(t_0) = \hat{0}$ (see Fig. 4.16b). Consequently, in this asymptotic regime, the crystal momentum and its canonical counterpart coincide. Thus, provided that $\hat{\hat{U}}_{\text{tra}}(t_0) = \mathbb{1}$ and $\hat{U}_{\text{sc}}(t_0, t_0) = \mathbb{1}$, we conclude that

$$|\Phi_v(\boldsymbol{k}, t_0)\rangle = \delta(\boldsymbol{k} - \boldsymbol{k}_0)|\bar{0}\rangle \quad \text{and} \quad |\Phi_c(\boldsymbol{k}, t_0)\rangle = 0. \tag{4.132}$$

This implies that, at the final time $t \to \infty$ where the crystal momentum and the canonical crystal momentum coincide as well, given that $\lim_{t\to\infty} A_{\text{cl}}(t) = 0$, we can express Eqs. (4.129) and (4.130), up fo first order terms in $g(\omega_{\mathbf{k}_q})$ as

$$|\Phi_v(\boldsymbol{k}, t)\rangle \approx \delta(\boldsymbol{k} - \boldsymbol{k}_0)\hat{\mathcal{D}}\big(\boldsymbol{\chi}_v(\boldsymbol{k}, t, t_0)\big)|\bar{0}\rangle, \tag{4.133}$$

$$|\Phi_c(\boldsymbol{k}, t)\rangle \approx -\delta(\boldsymbol{k} - \boldsymbol{k}_0)\frac{i}{\hbar}\int_{t_0}^{t} \mathrm{d}t_1 \hat{\mathcal{D}}\big(\boldsymbol{\chi}_c(\boldsymbol{k}, t, t_1)\big)\hat{\mathcal{M}}_{c,v}(\boldsymbol{k}, t_1)\hat{\mathcal{D}}\big(\boldsymbol{\chi}_v(\boldsymbol{k}, t_1, t_0)\big)|\bar{0}\rangle, \tag{4.134}$$

such that the joint state reads as

$$\begin{aligned}|\tilde{\Psi}(t)\rangle = \hat{\hat{U}}_{\text{tra}}(t)\hat{U}_{\text{sc}}(t, t_0)\Big[&\hat{\mathcal{D}}\big(\boldsymbol{\chi}_v(\boldsymbol{k}_0, t, t_0)\big)\,|\boldsymbol{k}_0, v\rangle \otimes |\bar{0}\rangle \\ &- \frac{i}{\hbar}\int_{t_0}^{t} \mathrm{d}t_1 \hat{\mathcal{D}}\big(\boldsymbol{\chi}_c(\boldsymbol{k}_0, t, t_1)\big)\hat{\mathcal{M}}_{c,v}(\boldsymbol{k}_0, t_1)\hat{\mathcal{D}}\big(\boldsymbol{\chi}_v(\boldsymbol{k}_0, t_1, t_0)\big)\,|\boldsymbol{k}_0, c\rangle \otimes |\bar{0}\rangle\Big],\end{aligned} \tag{4.135}$$

where we have taken into account that in the asymptotic limit $\boldsymbol{K} = \boldsymbol{k}_0$.

It is essential to highlight that the unitary transformation $\hat{\hat{U}}_{\text{tra}}(t)$ entangles both the electronic and quantum optical degrees of freedom. This entanglement arises due to the non-diagonal nature of both $\hat{A}(t)$ and $\hat{r}_{i,\text{tra}}$ in the coherent state and the Bloch basis, respectively. However, within our discrete field mode description, where the field's envelope is accounted for using the weight function $f(t)$ in the definition of the electric field operator, leads to $\hat{\hat{U}}_{\text{tra}}(t) \to \mathbb{1}$ as $t \to \infty$. Consequently, at the final time t, we can express

$$\begin{aligned}|\tilde{\Psi}(t)\rangle = \hat{U}_{\text{sc}}(t, t_0)\Big[&\hat{\mathcal{D}}\big(\boldsymbol{\chi}_v(\boldsymbol{k}_0, t, t_0)\big)\,|\boldsymbol{k}_0, v\rangle \otimes |\bar{0}\rangle \\ &- \frac{i}{\hbar}\int_{t_0}^{t} \mathrm{d}t_1 \hat{\mathcal{D}}\big(\boldsymbol{\chi}_c(\boldsymbol{k}_0, t, t_1)\big)\hat{\mathcal{M}}_{c,v}(\boldsymbol{k}_0, t_1)\hat{\mathcal{D}}\big(\boldsymbol{\chi}_v(\boldsymbol{k}_0, t_1, t_0)\big)\,|\boldsymbol{k}_0, c\rangle \otimes |\bar{0}\rangle\Big],\end{aligned} \tag{4.136}$$

which is still an entangled state between light and matter.

Finally, we point our focus onto processes where the electron is ultimately found in the valence band. This involves the application of the projector $\hat{P}_v = \int \mathrm{d}\boldsymbol{K}\, |\boldsymbol{K}, v\rangle\langle\boldsymbol{K}, v|$, which, along with $\hat{P}_c = \int \mathrm{d}\boldsymbol{K}\, |\boldsymbol{K}, c\rangle\langle\boldsymbol{K}, c|$, spans a POVM set. Thus, after tracing out the electronic degrees of freedom while working in the regime of weak depletion of the valence band (equivalently the conduction band), we can approximate the final quantum optical state as

$$\begin{aligned}\hat{\rho}_{\text{field}}^{(v)} &= \text{tr}_{\text{elec}}\Big(\hat{P}_v|\tilde{\Psi}(t)\rangle\langle\tilde{\Psi}(t)|\Big) \approx \hat{\mathcal{D}}\big(\boldsymbol{\chi}_v(\boldsymbol{k}_0, t, t_0)\big)|\bar{0}\rangle\langle\bar{0}|\hat{\mathcal{D}}^\dagger\big(\boldsymbol{\chi}_v(\boldsymbol{k}_0, t, t_0)\big)\\ &\equiv |\Phi_v(t)\rangle\langle\Phi_v(t)|\,,\end{aligned} \tag{4.137}$$

which indeed results in a product state, similar to those observed in atomic and molecular systems. However, it is worth noting that retaining terms of the form $\langle \boldsymbol{K}, m|\,\hat{U}_{\text{sc}}(t, t_0)\,|\boldsymbol{K}', c\rangle$, which become negligible in the regime of weak depletion, would introduce light-matter entanglement. This aspect is further explored in Appendix 9.4 through a more convenient Wannier-based perspective that allows for a discrete treatment of electronic states (see Fig. 4.12). In the specific excitation conditions of ZnO leading to Fig. 4.15, the degree of entanglement is almost negligible, suggesting that in those instances Eq. (4.137) holds true.

Before delving into the analysis of the various quantum optical observables and quantum information measures, it is worth highlighting that the state in Eq. (4.137) has been derived under a single-active electron model. However, as mentioned earlier, in reality the ground state of a semiconductor material consists of a fully occupied valence band where each electron possesses a well-defined crystal momentum. While electron-electron correlations in semiconductor materials are weak, their impact can be approximated to some extent through effective couplings, such as dephasing. Within this framework, each electron can be considered independently, each having a distinct initial crystal momentum $\boldsymbol{k}$ spanning the entire Brillouin zone. Consequently, it is possible to extend Eq. (4.137) to the many-electron regime through

$$|\Phi_v(t)\rangle \approx \hat{\mathcal{D}}\bigg(N_z \int \mathrm{d}\boldsymbol{K}\,\boldsymbol{\chi}_v(\boldsymbol{K}, t, t_0)\bigg)|\bar{0}\rangle \equiv \hat{\mathcal{D}}\big(\bar{\boldsymbol{\chi}}_v(\boldsymbol{K}, t, t_0)\big)|\bar{0}\rangle, \tag{4.138}$$

where N_z denotes the number of Brillouin zones excited by the laser field. This quantity fundamentally relies on the experimental conditions at which the excitation takes place: the laser beam width, the crystal alignment concerning the field's polarization and, in case HHG takes place during transmission, the sample's width. Unless otherwise specified, in subsequent sections, the selected N_z values are within the range of 10^6 to 10^7. For these values, we observe the emergence of non-classical states upon the conditioning on HHG operation.

4.2.3.1 Mean Photon Number of the Different Harmonic Modes

Similar to the analysis conducted on molecular systems in Sect. 4.1.2, the computation of the mean-photon number of the harmonic modes serves as a qualitative validation method for the proposed quantum optical theoretical approach. The results when computing this quantity should, to some extent [85], resemble the semiclassical spectra presented in Fig. 4.15. Considering the state in Eq. (4.138), the mean-photon number of the qth harmonic mode is given by

$$\langle \hat{a}^{\dagger}_{\mathbf{k}_q} \hat{a}_{\mathbf{k}_q} \rangle = \langle \Phi_v(t) | \hat{a}^{\dagger}_{\mathbf{k}_q} \hat{a}_{\mathbf{k}_q} | \Phi_v(t) \rangle = N_z^2 \left| \int \mathrm{d}\boldsymbol{K} \, \chi_q^{(v)}(\boldsymbol{K}, t, t_0), \right|^2 \tag{4.139}$$

which comprises both interband and intraband contributions, as apparent from the definition of $\chi_q^{(v)}(\boldsymbol{K}, t, t_0)$ given in Eq. (4.128). However, for a more precise comparison with Fig. 4.15, we selectively deactivate one of the contributions while retaining the other.

The results from the computation of Eq. (4.139) are presented in Fig. 4.17 for a ZnO crystal under the excitation of a linearly polarized laser field aligned along different crystal orientations with $\lambda_{\mathbf{k}_L} = 3.25$ µm, peak field strength of 0.5 V/Å and a duration of $\Delta t \approx 96$ fs. Specifically, panels (a) and (c) correspond to the scenario where the field is linearly polarized along the $\Gamma - M$ direction, while panels (b) and (d) depict the case where it is polarized parallel to the $\Gamma - A$ direction. As can be observed, akin to Fig. 4.15, the mean photon number replicates the key aspects of the HHG spectra: it exhibits a plateau region spanning from the lowest to the maximum energy band-gap (shaded blue region), followed by a sharp cutoff. In all instances,

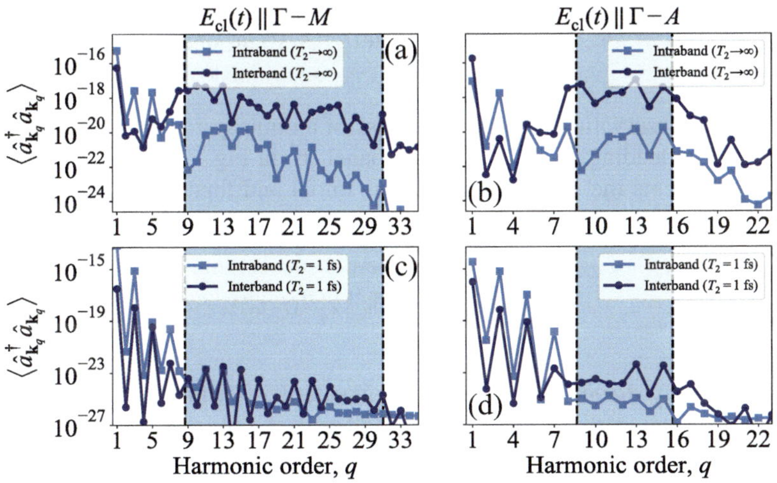

Fig. 4.17 Mean photon numbes calculated for the harmonic modes after the laser-solid interaction. For this figure, we consider a ZnO material under the influence of a linearly polarized laser pulse directed along the $\Gamma - M$ direction in panels **a** and **c**, and the $\Gamma - A$ direction in panels **b** and **d**. Laser settings comprise a peak field strength of 0.5 V/Å, a central wavelength of $\lambda_{\mathbf{k}_L} = 3.25$ µm, and a duration of $\Delta t \approx 96$ fs. In panels **a** and **b**, the illustration assumes an ideal solid scenario without dephasing effects ($T_2 \to \infty$). Conversely, panels **c** and **d** incorporate scenarios with dephasing effects ($T_2 = 1$ fs). The blurred blue region signifies the minimum and maximum energy bandgaps in the specified solid direction

the interband component (dark blue curve) prevails over the intraband component (light blue curve) within the non-pertubative part of the spectrum. In the perturbative one, this is not the generally the case. Furthermore, diminished dephasing times lead to decreased overall harmonic intensities (observed in panels (c) and (d)) and enhanced resolution of harmonic peaks. Consequently, we conclude that, from a qualitative point of view, the presented quantum optical approach effectively captures the primary characteristics of the harmonic spectra.

4.2.3.2 Wigner Function of the Fundamental Mode and Its Fidelity

In the regime of weak depletion of the valence band, we found that the quantum optical state, conditioned on the presence of electrons in this band, can be written as in Eq. (4.138). While this state remains classical, it serves as a foundation for potentially distilling a non-classical state through the conditioning on HHG operation, as detailed in Sect. 3.3.2.1. It is worth highlighting that, when considering semiconductor materials as the driven system, the implementation of this operation is indeed feasible, as demonstrated by N. Tsatrafyllis and colleagues in Ref. [128].

Hence, conditioning Eq. (4.138) on the generation of harmonic radiation results, up to a normalization factor, in the emergence of the following massively frequency-entangled state

$$|\bar{\Phi}_v(t)\rangle = \hat{D}_{\mathbf{k}_L}(\alpha_{\mathbf{k}_L})\left[\bigotimes_{q=1}^{q_c}|\bar{\chi}_q^{(v)}(t,t_0)\rangle - \xi_v(t,t_0)\bigotimes_{q=1}^{q_c}|0_{\mathbf{k}_q}\rangle\right], \tag{4.140}$$

similar to the observations done for the case of atomic systems, upon reversing the transformation leading to panel (c) from panel (b) in Fig. 3.12. In this expression, $\xi_v(t,t_0)$ represents the overlap between the initial and final optical states after the solid-matter interaction, that is

$$\xi_v(t,t_0) = \prod_{q=1}^{q_c}\langle 0_{\mathbf{k}_q}|\chi_q^{(v)}(t,t_0)\rangle, \tag{4.141}$$

which ultimately dictates the degree of entanglement in the resulting state: when $\xi_v(t,t_0) \to 0$, the state becomes separable due to the absence of the second term in the superposition. Otherwise, whenever $0 < \xi_v(t,t_0) < 1$, the state is generally entangled.

In this and subsequent subsections, we analyze various quantum optical properties and quantum information measures of the state described in Eq. (4.140) to asses its non-classical behavior. Due to the presence of additional electron dynamics in solids, absent in atomic systems, we anticipate their pivotal role in shaping the non-classical properties of the obtained state. To commence, we investigate the Wigner function

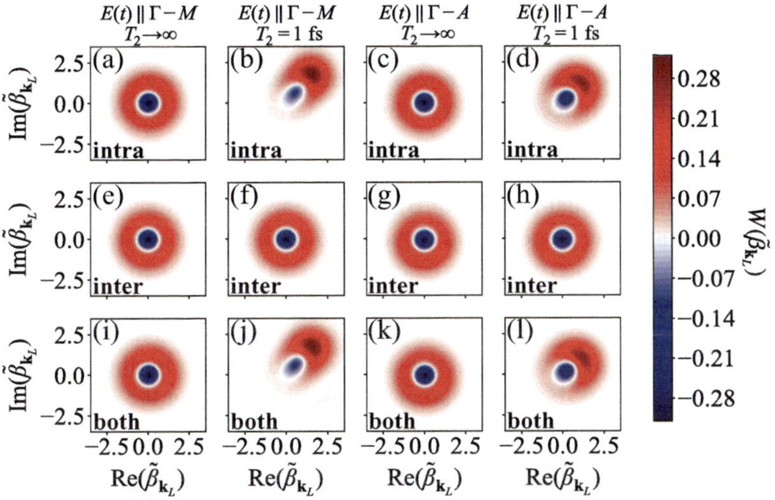

Fig. 4.18 Wigner function representations of the IR component after the laser solid interaction. Each column represents distinct values of the dephasing time, T_2, alongside different orientations of the applied linearly polarized laser field, specifically aligned with the $\Gamma - M$ and $\Gamma - A$ directions. The first two rows focus individually on the intraband and interband contributions, while the last row displays their combined effect. The laser parameters used in this case correspond to $\lambda_{\mathbf{k}_L} = 3.25$ µm, field strength of 0.5 V/Å and a duration of $\Delta t \approx 96$ fs

of the IR mode obtained after measuring the harmonic radiation. In this case, the resulting state is expressed as follows

$$|\bar{\Phi}_{v,\text{IR}}(t)\rangle = \hat{D}_{\mathbf{k}_L}(\alpha_{\mathbf{k}_L})\Big[|\bar{\chi}^{(v)}_{\mathbf{k}_L}(t,t_0)\rangle - \tilde{\xi}_v|0_{\mathbf{k}_L}\rangle\Big], \tag{4.142}$$

where $\tilde{\xi}_v \equiv \xi_{v,\text{IR}}|\xi_{\text{HH}}|^2$ with $\xi_{v,\text{IR}} \equiv \langle 0_{\mathbf{k}_L}|\bar{\chi}^{(v)}_{\mathbf{k}_L}(t,t_0)\rangle$ and $\xi_{v,\text{HH}} \equiv \prod_{q=2}^{q_c}\langle 0_{\mathbf{k}_L}|\bar{\chi}^{(v)}(t,t_0)\rangle$.

The Wigner function of the state given in Eq. (4.142), representing an unbalanced coherent state superposition, can be calculated using Eq. (2.76). The results of this evaluation are shown in Fig. 4.18 under varying conditions. Each row, from left to right, explores scenarios where the laser field is polarized along different crystal directions, considering two distinct dephasing times: $T_2 \to \infty$ and $T_2 = 1$ fs. Additionally, for different crystal directions, we employed varying numbers of Brillouin zones N_z. Specifically, for the $\Gamma - M$ cases (first two columns starting from the left hand side), we used $N_z = 6.6 \times 10^6$, while for the $\Gamma - A$ scenario, we utilized $N_z = 1.6 \times 10^7$. It is important to note that larger values of these parameters result in more dispersed and unbalanced coherent state superpositions, leading to $\tilde{\xi}_v \to 0$ and, consequently, Gaussian-like Wigner functions. Conversely, smaller values of N_z produce distributions resembling those of displaced single-photon excitations, as explicitly demonstrated in Eq. (3.104).

Upon examining the various columns in Fig. 4.18, two primary observations emerge. Firstly, as T_2 diminishes, we witness more unbalanced coherent state superpositions, leading to reduced Wigner negativities. This trend arises due to the increasing magnitude of $|\xi^{(v)}_{\mathbf{k}_L}|$. Secondly, excitation scenarios where the laser field is polarized along the $\Gamma - M$ direction yield more extensive displacements, even when employing a smaller value of N_z compared to the $\Gamma - A$ scenario. Specifically, when considering an equal number of Brillouin zones, excitations along the $\Gamma - M$ direction generate more disperse and unbalanced coherent state superpositions compared to the $\Gamma - A$ counterpart. This discrepancy can be attributed to the band structures illustrated in Fig. 4.13a. As observed there, the electron mobility along the valence band of the $\Gamma - A$ direction involves smaller energy exchanges than what is observed for the $\Gamma - M$ direction, as the valence band is flatter for the former. Consequently, fewer electrons, hence a smaller N_z, oscillating along the $\Gamma - M$ direction are required to achieve similar displacements as those observed along the $\Gamma - A$ direction.

The first two rows of Fig. 4.18 individually examine intraband and interband contributions, respectively, by intentionally deactivating the other. The third row illustrates the combined effect, obtained as the sum of interband and intraband influences. A comparison between the rows reveals that, across all scenarios, the primary contribution predominantly originates from intraband terms, which further increases as T_2 decreases. This predominance and its relation to dephasing times can be elucidated as follows. Intraband contributions primarily arise from Bloch oscillations, involving electron's acceleration within a specific band under the field's influence. Diminishing dephasing times implies that electrons spend less time in the conduction band after excitation due to electron-phonon or electron-electron interactions. Consequently, the electron transitions out of the conduction band more rapidly after excitation, through a non-radiative decay process. This shift in behavior accentuates the significance of intraband phenomena as T_2 diminishes compared to interband effects, whose influence becomes smaller. This can be further observed in panels (g) and (h) where the Wigner function becomes slightly more homogeneous as T_2 decreases.

The findings depicted in Fig. 4.18 underscore one of the advantages of solid-state systems towards the potential generation of non-classical states: the ability to tune the generated coherent state superposition through different solid orientations. However, this adjustability can also be achieved through alternative means, such as modifying the electric field strength applied to excite the solid. In Fig. 4.19, we explore this tunability by examining the fidelity[14] of Eq. (4.142) with respect to (1) a coherent state of the form $\hat{D}_{\mathbf{k}_L}(\alpha_{\mathbf{k}_L})|\chi^{(q)}_{\mathbf{k}_L}(t, t_0)\rangle$, and (2) a displaced Fock state $\hat{D}_{\mathbf{k}_L}(\alpha_{\mathbf{k}_L})|1_{\mathbf{k}_L}\rangle$. These cases represent the two limiting cases found in this conditioning approach, as demonstrated in Eqs. (3.103) and (3.104).

In Fig. 4.19, we present the fidelity when computed for ZnO excited along the $\Gamma - M$ direction. In panel (a), we illustrate its behavior as a function of the dephasing

[14] Given a state $|\psi\rangle$, its fidelity with respect to a state $|\varphi\rangle$ is defined as $\mathcal{F}(|\varphi\rangle, |\psi\rangle) \equiv |\langle\phi|\psi\rangle|^2$, and can be interpreted as a measure of the distance between the two states [130].

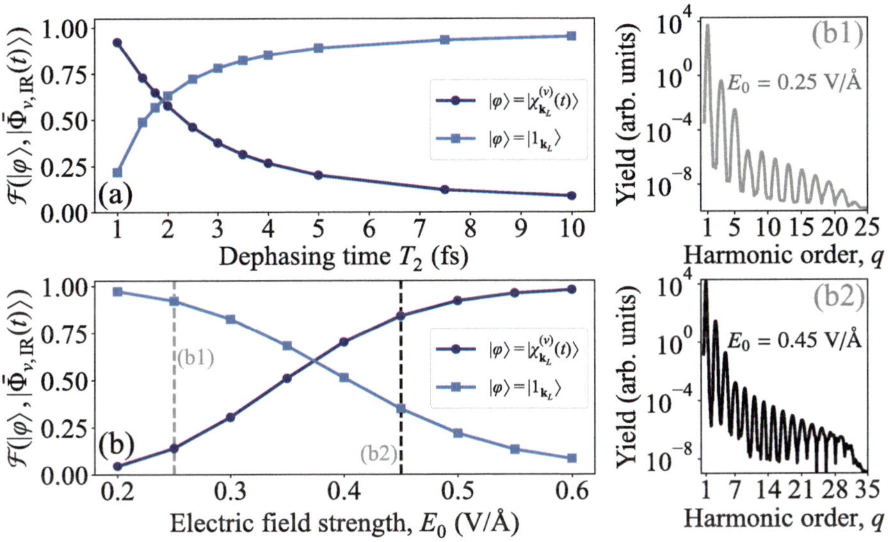

Fig. 4.19 Fidelity of the IR state after the laser-solid interaction with respect to different states. The light blue curve with squared markers indicates the fidelity concerning $\hat{D}_{\mathbf{k}_L}(\alpha_{\mathbf{k}_L})|1_{\mathbf{k}_L}\rangle$, while the dark blue curve with circular markers represents $\hat{D}_{\mathbf{k}_L}(\alpha_{\mathbf{k}_L})|\chi^{(q)}_{\mathbf{k}_L}(t, t_0)\rangle$. For simplicity, they are referenced as $|1_{\mathbf{k}_L}\rangle$ and $|\chi^{(q)}_{\mathbf{k}_L}(t, t_0)\rangle$ in the legend. Panel **a** shows the fidelity against the dephasing time T_2, while **b** illustrates its dependence with the electric field strength E_0. The semiclassical spectra in panels **b1** and **b2** correspond to $E_0 = 0.25$ V/Å and $E_0 = 0.45$ V/Å, respectively, indicated by the vertical dashed lines in panel (**b**). Both panels **a** and **b** assume a dephasing time $T_2 = 1$ fs, with the linearly polarized laser field aligned along the $\Gamma - M$ direction

time for a constant $E_0 = 0.5$ V/Å, while in panel (b) we display its variation with respect to the electric field strength for a fixed $T_2 = 1$ fs. In the latter case, we limit the field strengths within the range of 0.2 to 0.6 V/Å since higher values typically exceed the damage threshold of many materials [20, 23, 24]. The results reveal that as T_2 increases or E_0 decreases, the fidelity with respect to the displaced Fock state increases (light blue curve), while for the coherent state it decreases (dark blue curve). Although dephasing times are typically beyond experimental control as they are determined by the material properties, adjusting the electric field strength provides a means to influence the fidelity of the generated states.

Furthermore, within the considered ranges—which are consistent with experimental implementations—the fidelity of the generated state is highly tunable. For instance, the fidelity of the Fock (coherent) state changes from approximately $\mathcal{F} \approx 0.98$ (≈ 0.05) at $E_0 = 0.2$ V/Å, to around $\mathcal{F} \approx 0.09$ (≈ 0.98) at $E_0 = 0.6$ V/Å. However, it is worth remarking that this analysis has been based on $N_z = 6.6 \times 10^6$, indicating that factors such as the beam width in the interaction regions and the sample's width (in the case of HHG in transmission setups), also play an instrumental role in controlling the fidelity of the generated states.

4.2.3.3 Entanglement Between the Field Modes

As mentioned earlier, the state described in Eq. (4.140) generally represents a massively entangled state where all the frequency modes excited during the HHG process exhibit correlations with each other. In this subsection, our focus lies in assessing the degree of entanglement within this state between the qth harmonic mode and all the other modes ($\forall q' \neq q$). Since we are dealing with states involving continuous variables, the linear entropy, as defined in Eq. (2.127), serves as a suitable measure of entanglement due to its easier computation when compared to the entanglement entropy (2.123). Specifically, it involves the computation of the purity of the reduced state $\hat{\rho}_q$, the latter given by

$$\hat{\rho}_q(t) = \frac{\text{tr}_{q' \neq q}\big(|\bar{\Phi}_v(t)\rangle\langle\bar{\Phi}_v(t)|\big)}{\text{tr}\big(|\bar{\Phi}_v(t)\rangle\langle\bar{\Phi}_v(t)|\big)}. \tag{4.143}$$

The outcomes for this entanglement measure are depicted in Fig. 4.20 as a function of the harmonic mode q. Notably, distinct features akin to those observed in typical harmonic spectra are found. Specifically, there is a plateau in frequency extending

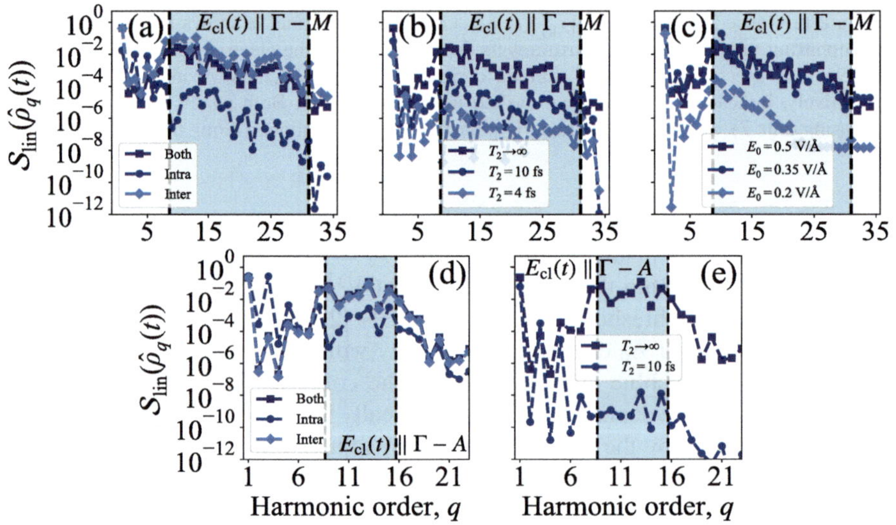

Fig. 4.20 Linear entropy as a function of the harmonic modes after the laser-solid interaction. The upper row considers the scenario where the field is polarized along the $\Gamma - M$ crystal direction, while the lower row shows the case where the laser is aligned with the $\Gamma - A$ direction. Panels **a** and **d** explore the separate contributions of interband dynamics (lightest blue curve with diamond markers) and intraband dynamics (darker blue curve with circular markers), as well as their combination (darkest blue curve with squared markers). These cases maintain $E_0 = 0.5$ V/Å and assume $T_2 \to \infty$. Panels **b** and **e** consider various dephasing times while keeping $E_0 = 0.5$ V/Å. Lastly, in panel **c**, different field strength values are considered while setting $T_2 \to \infty$

across the non-perturbative region between the minimum and maximum energy band-gaps (highlighted with the blue shaded region). This plateau is followed by a sharp cutoff, located at the frequency corresponding to the maximum energy difference between the bands. The first row of this figure illustrates laser-solid interactions under various conditions, driven by a linearly polarized laser field parallel to the $\Gamma - M$ direction. The second row showcases the scenario where the field is aligned with the $\Gamma - A$ direction.

In panels (a) and (d), the individual contributions of interband (lightest blue curve with diamond markers), intraband (darker blue curve with circular markers), and their combined effect (darkest blue curve with square markers) on the linear entropy are presented for the case $T_2 \to \infty$. Interestingly, unlike the observations made for the Wigner function, the interband contribution here emerges as the dominant one. Specifically, for $q = 1$, the entanglement reaches a peak value of $\hat{S}_{\text{lin}}(\hat{\rho}_q) \approx 0.44$ in panel (a), while $\hat{S}_{\text{lin}}(\hat{\rho}_q) \approx 0.23$ in panel (d). This dominance aligns with the understanding that, within the perturbative range of the spectrum, interband effects play a more critical role than intraband contributions in shaping the harmonic plateau, particularly in terms of their intensity. Therefore, considering that the degree of entanglement heavily depends on the population of these harmonic modes, the interband contribution significantly influences this quantum information measure.

As the dephasing time increases, the harmonic yield diminishes for constant field strength (in this case $E_0 = 0.5$ V/Å), resulting in a reduction in the degree of entanglement, as illustrated in panels (b) and (e). Conversely, when considering a fixed dephasing value, such as $T_2 \to \infty$, panel (c) shows the degree of entanglement for three distinct field strength values. As E_0 decreases, the cutoff frequency at which the linear entropy abruptly decreases, also diminishes. This behavior mirrors the pattern observed in the harmonic spectrum, similar to the trends seen in panels (b1) and (b2) of Fig. 4.19, and further discussed in Ref. [26].

4.3 Discussion and Outlook

In this chapter, we have conducted a quantum optical analysis of high-harmonic generation in both molecular and solid-state systems. The additional dynamics experienced by the electron in these media when excited by a strong driving field, in contrast to the atomic case, leave their imprint on the resultant light states after the interaction. More broadly, we observed that under certain scenarios, these additional electron dynamics are sufficient to generate states exhibiting non-classical features.

For the case of molecules, our focus was on the H_2^+ molecule, representing the simplest system in this context. Unlike atomic systems, here an electron can ionize from one molecular center and recombine with the other, a process inducing interference effects in the resultant harmonic spectra [8, 51–55]. While our quantum optical model captures these features, it also unveils the presence of light-matter entanglement due to these mechanisms. Furthermore, the inversion of symmetry the electron potentially undergoes during recombination can lead to the generation of frequency-entangled states. These persist regardless of conditioning on HHG operations, although employing heralding measurements on the electronic state can enhance entanglement and other non-classical features on the field modes.

In this regard, one could contemplate the following possibilities to further expand our work:

- **Introducing the conditioning on HHG operations.** The analysis conducted in this chapter concerning molecular ions did not involve the introduction of the conditioning on HHG operations. While not essential to observe non-classical features in the state after the laser-molecule interaction, employing such operations could potentially strengthen these features. Therefore, evaluating the extent to which these operations benefit or affect the assessed non-classical measures is crucial from both fundamental and practical standpoints, particularly towards potential experimental implementations.
- **Development of entanglement witnesses.** Delving into experimental implementations, the entanglement measures employed in this chapter might not be optimal in the context of strong-field physics. This stems from their dependence on tomographic reconstructions which, in the context of ATI analyses, have shown strong susceptibility to experimental noise and additional correlations resulting from the laser-matter interaction [131]. Consequently, the development of entanglement witnesses derived from experimentally available measures, and that avoid tomography of the generated states, becomes pivotal in this context [132, 133].
- **Enhanced optical configurations.** While our analysis focused on H_2^+ under a linearly polarized field aligned with the molecular axis, employing more intricate configurations could amplify the studied non-classical features. Notably, in Ref. [90], it was demonstrated that exciting H_2^+ with lasers polarized along different directions, enhances the exchange of electronic population between the ground and first excited states. Consequently, we anticipate that employing these diverse configurations will accentuate some of the evaluated entanglement measures.
- **Asymmetric molecules and fields.** Building upon the prior point, studies have indicated that in the case of asymmetric molecules, where the composing centers within the molecule belong to different atomic species, the population of excited and ground states can become comparable [134]. As a result, the inclusion of excited states, on top of the ground states, becomes pivotal in the context of HHG [88]. Moreover, this necessity can be also emphasized by introducing perturbative ultraviolet fields in conjunction with the strong IR field [89]. Therefore, exploring how these asymmetries, either in the molecular system or in the driving field, influence the non-classical properties of the state after the interaction, is of considerable interest given the role excited states exert on them.

For the case of solid-state systems, our focus laid on the analysis of semiconductor materials, characterized by a band-gap that separates valence and conduction bands, the latter typically unpopulated initially. In these systems, several electron dynamics shape the HHG spectrum, with significant relevance attributed to transitions between bands and electronic acceleration within a single band. In this context, our analysis underscored the pivotal influence of these distinct electron dynamics on the potential non-classical properties of the state following the conditioning on HHG operation. On one hand, the analysis of the Wigner function of the final IR field unveiled the predominant role of intraband dynamics in determining their overall shape and the presence of negative regions. On the other hand, concerning the entanglement among field modes, interband dynamics exerted a more pronounced impact. Furthermore, our study highlighted that material properties such as inherent dephasing times or the strength of the applied radiation, significantly influence these non-classical properties.

In this context, potential follow-ups to our work could include the following:

- **Experimental implementations.** Solid-state systems could offer distinct advantages over atomic gas jets when performing conditioning on HHG operations due to their inherent scales. Experimental setups commonly utilize samples with widths of 500 μm to observe HHG outcomes [21, 26]. Confirming these theoretically predicted results in experimental setups could unveil the potential for constructing non-classical light sources on smaller scales.
- **Increased influence of the conduction band states.** Our analysis relied on the assumption of weak depletion of the valence band states. This allowed us to derive a quantum optical state after the interaction akin to that obtained in atomic systems, with the conditioning on HHG operations yielding similar outcomes. While this assumption generally holds for semiconductor materials like the ZnO employed here, recent demonstrations using sophisticated pulse envelopes have shown substantial depletions within the valence band [135]. In atomic systems, deviations from weak depletion revealed squeezing and entanglement features without the need for conditioning on HHG operations [136] (refer to Fig. 3.33). Thus, understanding the extension of these findings to solids becomes intriguing given their natural scales highlighted in the previous point.
- **Applications in materials characterization.** We observed that intrinsic material properties, particularly dephasing processes, significantly impact the observed non-classical characteristics. Specifically, they lead to broader coherent state superpositions after conditioning on HHG operations and diminish entanglement between field modes. Therefore, a comparison between these distinct non-classical measures, alongside the alterations induced by varying field strengths, might serve as a means to evaluate some of the material properties. In this direction, more extensive discussions on the influence of other electron dynamics such as the anomalous current induced due to non-vanishing Berry connections [43, 114–117], also becomes crucial.
- **Extension to strongly-correlated materials.** Our analysis thus far has focused on semiconductor materials exhibiting weak electron-electron correlations, primarily assessed through finite dephasing times. However, in the case of strongly correlated materials, these correlations become of paramount importance as they underpin the unique properties exhibited by these materials [1]. Within this context, HHG has proven to offer fundamental insights into the emergence of these properties [32, 46]. Therefore, it is crucial to explore how these emergent properties in the system

influence the quantum optical states, and whether their generation could be useful for further characterization of these materials. In this context, C. S. Lange and colleagues recently conducted analytical studies focusing on HHG processes within the Mott insulating phase of the Fermi-Hubbard model. Their work demonstrates the generation of non-classical light [137].

References

1. Zong A, Nebgen BR, Lin S-C, Spies JA, Zuerch M (2023) Nat Rev Mater 8:224. https://doi.org/10.1038/s41578-022-00530-0
2. Cireasa R, Boguslavskiy AE, Pons B, Wong MCH, Descamps D, Petit S, Ruf H, Thiré N, Ferré A, Suarez J, Higuet J, Schmidt BE, Alharbi AF, Légaré F, Blanchet V, Fabre B, Patchkovskii S, Smirnova O, Mairesse Y, Bhardwaj VR (2015) Nat Phys 11:654. https://doi.org/10.1038/nphys3369
3. Ayuso D, Ordonez AF, Smirnova O (2022) Phys Chem Chem Phys 24:26962. https://doi.org/10.1039/D2CP01009G
4. Ayuso D, Neufeld O, Ordonez AF, Decleva P, Lerner G, Cohen O, Ivanov M, Smirnova O (2019) Nat Photonics 13:866. https://doi.org/10.1038/s41566-019-0531-2
5. Ayuso D, Ordonez AF, Decleva P, Ivanov M, Smirnova O (2021) Nat Commun 12:3951. https://doi.org/10.1038/s41467-021-24118-4
6. Ayuso D, Ordonez AF, Ivanov M, Smirnova O (2021) Optica 8:1243. https://doi.org/10.1364/OPTICA.423618
7. Mayer N, Ayuso D, Decleva P, Khokhlova M, Pisanty E, Ivanov M, Smirnova O (2024) Nat Photon 18:1155-1160. https://doi.org/10.1038/s41566-024-01499-8
8. Lein M, Hay N, Velotta R, Marangos JP, Knight PL (2002) Phys Rev A 66:023805. https://doi.org/10.1103/PhysRevA.66.023805
9. Kanai T, Minemoto S, Sakai H (2005) Nature 435:470. https://doi.org/10.1038/nature03577
10. Lein M (2007) J Phys B: Atlc. Mol Opt Phys 40:R135. https://doi.org/10.1088/0953-4075/40/16/R01
11. Torres R, Kajumba N, Underwood JG, Robinson JS, Baker S, Tisch JWG, de Nalda R, Bryan WA, Velotta R, Altucci C, Turcu ICE, Marangos JP (2007) Phys Rev Lett 98:203007. https://doi.org/10.1103/PhysRevLett.98.203007
12. Morishita T, Le A-T, Chen Z, Lin CD (2008) Phys Rev Lett 100:013903. https://doi.org/10.1103/PhysRevLett.100.013903
13. Itatani J, Levesque J, Zeidler D, Niikura H, Pépin H, Kieffer JC, Corkum PB, Villeneuve DM (2004) Nature 432:867. https://doi.org/10.1038/nature03183
14. Smirnova O, Mairesse Y, Patchkovskii S, Dudovich N, Villeneuve D, Corkum P, Ivanov MY (2009) Nature 460:972. https://doi.org/10.1038/nature08253
15. Haessler S, Caillat J, Boutu W, Giovanetti-Teixeira C, Ruchon T, Auguste T, Diveki Z, Breger P, Maquet A, Carré B, Taïeb R, Saliéres P (2010) Nat Phys 6:200. https://doi.org/10.1038/nphys1511
16. Kraus PM, Mignolet B, Baykusheva D, Rupenyan A, Horný L, Penka EF, Grassi G, Tolstikhin OI, Schneider J, Jensen F, Madsen LB, Bandrauk AD, Remacle F, Wörner HJ (2015) Science 350:790. https://doi.org/10.1126/science.aab2160
17. Ciappina MF, Becker A, Jaroń-Becker A (2007) Phys Rev A 76:063406. https://doi.org/10.1103/PhysRevA.76.063406
18. Laarmann T, Shchatsinin I, Stalmashonak A, Boyle M, Zhavoronkov N, Handt J, Schmidt R, Schulz CP, Hertel IV (2007) Phys Rev Lett 98:058302. https://doi.org/10.1103/PhysRevLett.98.058302

19. Ciappina MF, Becker A, Jaroń-Becker A (2008) Phys Rev A 78:063405. https://doi.org/10.1103/PhysRevA.78.063405
20. Goulielmakis E, Brabec T (2022) Nat Photonics 16:411. https://doi.org/10.1038/s41566-022-00988-y
21. Schubert O, Hohenleutner M, Langer F, Urbanek B, Lange C, Huttner U, Golde D, Meier T, Kira M, Koch SW, Huber R (2014) Nat Photonics 8:119. https://doi.org/10.1038/nphoton.2013.349
22. You YS, Reis DA, Ghimire S (2017) Nat Phys 13:345. https://doi.org/10.1038/nphys3955
23. Kim H, Han S, Kim YW, Kim S, Kim S-W (2017) ACS Photonics 4:1627. https://doi.org/10.1021/acsphotonics.7b00350
24. Garratt D, Misiekis L, Wood D, Larsen EW, Matthews M, Alexander O, Ye P, Jarosch S, Ferchaud C, Strüber C, Johnson AS, Bakulin AA, Penfold TJ, Marangos JP (2022) Nat Commun 13:1. https://doi.org/10.1038/s41467-022-31008-w
25. Chin AH, Calderón OG, Kono J (2001) Phys Rev Lett 86:3292. https://doi.org/10.1103/PhysRevLett.86.3292
26. Ghimire S, DiChiara AD, Sistrunk E, Agostini P, DiMauro LF, Reis DA (2011) Nat Phys 7:138. https://doi.org/10.1038/nphys1847
27. Vampa G, Hammond T, Thiré N, Schmidt B, Légaré F, McDonald C, Brabec T, Klug D, Corkum P (2015) Phys Rev Lett 115:193603. https://doi.org/10.1103/PhysRevLett.115.193603
28. Luu TT, Garg M, Kruchinin SY, Moulet A, Hassan MT, Goulielmakis E (2015) Nature 521:498. https://doi.org/10.1038/nature14456
29. Lanin AA, Stepanov EA, Fedotov AB, Zheltikov AM (2017) Optica 4:516. https://doi.org/10.1364/OPTICA.4.000516
30. You YS, Cunningham E, Reis DA, Ghimire S (2018) J Phys B: Atlc. Mol Opt Phys 51:114002. https://doi.org/10.1088/1361-6455/aac11d
31. Lakhotia H, Kim HY, Zhan M, Hu S, Meng S, Goulielmakis E (2020) Nature 583:55. https://doi.org/10.1038/s41586-020-2429-z. Note number: 7814
32. Silva REF, Blinov IV, Rubtsov AN, Smirnova O, Ivanov M (2018) Nat Photonics 12:266. https://doi.org/10.1038/s41566-018-0129-0
33. Roy A, Bera S, Saha K (2020) Phys Rev Res 2:043133. https://doi.org/10.1103/PhysRevResearch.2.043133
34. Orthodoxou C, Zaïr A, Booth GH (2021) NPJ Quantum Mater 6:1. https://doi.org/10.1038/s41535-021-00377-8
35. Kanega M, Ikeda TN, Sato M (2021) Phys Rev Res 3:L032024. https://doi.org/10.1103/PhysRevResearch.3.L032024
36. Murakami Y, Takayoshi S, Koga A, Werner P (2021) Phys Rev B 103:035110. https://doi.org/10.1103/PhysRevB.103.035110
37. Murakami Y, Uchida K, Koga A, Tanaka K, Werner P (2022) Phys Rev Lett 129:157401. https://doi.org/10.1103/PhysRevLett.129.157401
38. Uchida K, Mattoni G, Yonezawa S, Nakamura F, Maeno Y, Tanaka K (2022) Phys Rev Lett 128:127401. https://doi.org/10.1103/PhysRevLett.128.127401
39. Bauer D, Hansen KK (2018) Phys Rev Lett 120:177401. https://doi.org/10.1103/PhysRevLett.120.177401
40. Jürß C, Bauer D (2019) Phys Rev B 99:195428. https://doi.org/10.1103/PhysRevB.99.195428
41. Silva REF, Jiménez-Galán A, Amorim B, Smirnova O, Ivanov M (2019) Nat Photonics 13:849. https://doi.org/10.1038/s41566-019-0516-1
42. Jiménez-Galán A, Silva REF, Smirnova O, Ivanov M (2020) Nat Photonics 14:728. https://doi.org/10.1038/s41566-020-00717-3
43. Chacón A, Kim D, Zhu W, Kelly SP, Dauphin A, Pisanty E, Maxwell AS, Picón A, Ciappina MF, Kim DE, Ticknor C, Saxena A, Lewenstein M (2020) Phys Rev B 102:134115. https://doi.org/10.1103/PhysRevB.102.134115
44. Baldelli N, Bhattacharya U, González-Cuadra D, Lewenstein M, Graß T (2022) ACS Omega 7:47424. https://doi.org/10.1021/acsomega.2c07169

45. Pattanayak A, Pujari S, Dixit G (2022) Sci Rep 12:6722. https://doi.org/10.1038/s41598-022-10465-9
46. Alcalá J, Bhattacharya U, Biegert J, Ciappina M, Elu U, Graß T, Grochowski PT, Lewenstein M, Palau A, Sidiropoulos TPH, Steinle T, Tyulnev I (2022) Proc Natl Acad Sci 119:e2207766119. https://doi.org/10.1073/pnas.2207766119
47. Ripoll JJG (2002) Quantum information and quantum optics with superconducting circuits. Cambridge University Press, Cambridge. https://doi.org/10.1017/9781316779460
48. Rivera-Dean J, Stammer P, Maxwell AS, Lamprou T, Pisanty E, Tzallas P, Lewenstein M, Ciappina MF (2024) Phys Rev A 109:033706. https://doi.org/10.1103/PhysRevA.109.033706
49. Rivera-Dean J, Stammer P, Maxwell AS, Lamprou T, Ordóñez AF, Pisanty E, Tzallas P, Lewenstein M, Ciappina MF (2023) Entanglement and non-classical states of light in a strong-laser driven solid-state system. https://doi.org/10.48550/arXiv:2211.00033, arXiv:2211.00033 [physics, physics:quant-ph]
50. Rivera-Dean J, Stammer P, Maxwell AS, Lamprou T, Ordóñez AF, Pisanty E, Tzallas P, Lewenstein M, Ciappina MF (2024) Nonclassical states of light after high-harmonic generation in semiconductors: a Bloch-based perspective. . https://doi.org/10.1103/PhysRevB.109.035203
51. Lein M, Hay N, Velotta R, Marangos JP, Knight PL (2002) Phys Rev Lett 88:183903. https://doi.org/10.1103/PhysRevLett.88.183903
52. Lein M, Corso PP, Marangos JP, Knight PL (2003) Phys Rev A 67:023819. https://doi.org/10.1103/PhysRevA.67.023819
53. de Morisson Faria CF (2007) Phys Rev A 76:043407. https://doi.org/10.1103/PhysRevA.76.043407
54. Suárez N, Chacón A, Pérez-Hernández JA, Biegert J, Lewenstein M, Ciappina MF (2017) Phys Rev A 95:033415. https://doi.org/10.1103/PhysRevA.95.033415
55. Suárez Rojas N (2018) Strong-field processes in atoms and polyatomic molecules, type Doctoral thesis, school Universitat Politécnica de Catalunya. https://upcommons.upc.edu/handle/2117/114001
56. Ito K, Hall RI, Ukai M (1996) J Chem Phys 104:8449. https://doi.org/10.1063/1.471594
57. Sansone G, Kelkensberg F, Pérez-Torres JF, Morales F, Kling MF, Siu W, Ghafur O, Johnsson P, Swoboda M, Benedetti E, Ferrari F, Lépine F, Sanz-Vicario JL, Zherebtsov S, Znakovskaya I, L'Huillier A, Ivanov MY, Nisoli M, Martín F, Vrakking MJJ (2010) Nature 465:763. https://doi.org/10.1038/nature09084
58. Finkelstein BN, Horowitz GE (1928) Z Phys 48:118. https://doi.org/10.1007/BF01351582
59. Atkins P, Friedman R (2015) An introduction to molecular structure. In: Molecular quantum mechanics. Oxford University Press, Oxford, chapter 8, pp 249–286
60. Poll JD, Karl G (1966) Can J Phys 44:1467. https://doi.org/10.1139/p66-122
61. Silvera IF (1980) Rev Mod Phys 52:393. https://doi.org/10.1103/RevModPhys.52.393
62. Born M, Oppenheimer R (1927) Ann Phys 389:457. https://doi.org/10.1002/andp.19273892002
63. Atkins P, Friedman R (2015) Techniques of approximation. In: Molecular quantum mechanics. Oxford University Press, Oxford, chapter 6, pp 168–205
64. Atkins P, Friedman R (2015) The calculation of electronic structure. In: Molecular quantum mechanics. Oxford University Press, Oxford, chapter 6, pp 287–341
65. Ivanov MY, Corkum PB (1993) Phys Rev A 48:580. https://doi.org/10.1103/PhysRevA.48.580
66. Zuo T, Chelkowski S, Bandrauk AD (1993) Phys Rev A 48:3837. https://doi.org/10.1103/PhysRevA.48.3837
67. Yu H, Bandrauk AD (1995) J Chem Phys 102:1257. https://doi.org/10.1063/1.468913
68. Moreno P, Plaja L, Roso L (1997) Phys Rev A 55:R1593. https://doi.org/10.1103/PhysRevA.55.R1593
69. Kopold R, Becker W, Kleber M (1998) Phys Rev A 58:4022. https://doi.org/10.1103/PhysRevA.58.4022

70. Alon OE, Averbukh V, Moiseyev N (1998) Phys Rev Lett 80:3743. https://doi.org/10.1103/PhysRevLett.80.3743
71. Bandrauk AD, Yu H (1999) Phys Rev A 59:539. https://doi.org/10.1103/PhysRevA.59.539
72. Lappas DG, Marangos JP (2000) J Phys B: AtIc, Mol Opt Phys 33:4679. https://doi.org/10.1088/0953-4075/33/21/313
73. Averbukh V, Alon OE, Moiseyev N (2001) Phys Rev A 64:033411. https://doi.org/10.1103/PhysRevA.64.033411
74. Kreibich T, Lein M, Engel V, Gross EKU (2001) Phys Rev Lett 87:103901. https://doi.org/10.1103/PhysRevLett.87.103901
75. Morales F, Riviére P, Richter M, Gubaydullin A, Ivanov M, Smirnova O, Martín F (2014) J Phys B: AtIc, Mol Opt Phys 47:204015. https://doi.org/10.1088/0953-4075/47/20/204015
76. Ahmadi H, Vafaee M, Maghari A (2016) J Phys B: AtIc, Mol Opt Phys 49:035602. https://doi.org/10.1088/0953-4075/49/3/035602
77. Vafaee M, Ahmadi H, Maghari A (2016) J Phys B: AtIc, Mol Opt Phys 50:025601. IOP Publishing. https://doi.org/10.1088/1361-6455/50/2/025601
78. Ahmadi H, Vafaee M, Maghari A (2016) Phys Rev A 94:033415. https://doi.org/10.1103/PhysRevA.94.033415
79. Kjeldsen TK, Madsen LB (2004) J Phys B: AtIc, Mol Opt Phys 37:2033. https://doi.org/10.1088/0953-4075/37/10/003
80. Milošević DB (2006) Phys Rev A 74:063404. https://doi.org/10.1103/PhysRevA.74.063404
81. Chirilć CC, Lein M (2006) Phys Rev A 73:023410. https://doi.org/10.1103/PhysRevA.73.023410
82. Suárez N, Chacón A, Ciappina MF, Wolter B, Biegert J, Lewenstein M (2016) Phys Rev A 94:043423. https://doi.org/10.1103/PhysRevA.94.043423
83. Yuan K-J, Lefebvre C, Chelkowski S, Lu H, Bandrauk AD (2020) Phys Rev A 101:023411. https://doi.org/10.1103/PhysRevA.101.023411
84. Habibović D, Becker W, Milošević DB (2021) Eur Phys J D 75:122. https://doi.org/10.1140/epjd/s10053-021-00133-4
85. Sundaram B, Milonni PW (1990) Phys Rev A 41:6571. https://doi.org/10.1103/PhysRevA.41.6571
86. Schafer KJ, Kulander KC (1997) Phys Rev Lett 78:638. https://doi.org/10.1103/PhysRevLett.78.638
87. Saliéres P, Antoine P, de Bohan A, Lewenstein M (1998) Phys Rev Lett 81:5544. https://doi.org/10.1103/PhysRevLett.81.5544
88. Bian X-B, Bandrauk AD (2010) Phys Rev Lett 105:093903. https://doi.org/10.1103/PhysRevLett.105.093903
89. Bian X-B, Bandrauk AD (2011) Phys Rev A 83:023414. https://doi.org/10.1103/PhysRevA.83.023414
90. Han Y-C, Madsen LB (2013) Phys Rev A 87:043404. https://doi.org/10.1103/PhysRevA.87.043404
91. Eberly JH, Fedorov MV (1992) Phys Rev A 45:4706. https://doi.org/10.1103/PhysRevA.45.4706
92. Walter M, Gross D, Eisert J (2016) Multipartite entanglement, in quantum information. Wiley, chapter 14, pp 293–330. https://doi.org/10.1002/9783527805785.ch14
93. Bengtsson I, Zyczkowski K (2016) A brief introduction to multipartite entanglement. arXiv:1612.07747 [quant-ph]
94. Weyl H (1912) Math Ann 71:441. https://doi.org/10.1007/BF01456804
95. Fulton W (2000) Eigenvalues, invariant factors, highest weights, and Schubert calculus. https://doi.org/10.48550/arXiv:math/9908012. arXiv:math/9908012
96. Dagotto E (2005) Science 309:257. https://doi.org/10.1126/science.1107559
97. Bloch F (1929) Z Phys 52:555. https://doi.org/10.1007/BF01339455
98. Ashcroft NW, Mermin ND (1976) Electron levels in a periodic potential: general properties. In: Solid state physics. Cengage Learning, USA, chapter 8, pp 131–150

99. Haug H, Koch SW (2004) Periodic lattice of atoms. In: Quantum theory of the optical and electronic properties of semiconductors. World Scientific Publishing Co. Pte. Ltd., chapter 3, pp 41–44
100. Wannier GH (1937) Phys Rev 52:191. https://doi.org/10.1103/PhysRev.52.191
101. Kohn W (1959) Phys Rev 115:809. https://doi.org/10.1103/PhysRev.115.809
102. Marzari N, Mostofi AA, Yates JR, Souza I, Vanderbilt D (2012) Rev Mod Phys 84:1419. https://doi.org/10.1103/RevModPhys.84.1419
103. Park J, Subramani A, Kim S, Ciappina MF (2022) Adv Phys: X 7:2003244. https://doi.org/10.1080/23746149.2021.2003244
104. Vampa G, McDonald C, Orlando G, Klug D, Corkum P, Brabec T (2014) Phys Rev Lett 113:073901. https://doi.org/10.1103/PhysRevLett.113.073901
105. Vampa G, McDonald CR, Orlando G, Corkum PB, Brabec T (2015) Phys Rev B 91:064302. https://doi.org/10.1103/PhysRevB.91.064302
106. Goano M, Bertazzi F, Penna M, Bellotti E (2007) J Appl Phys 102:083709. https://doi.org/10.1063/1.2794380
107. Haug H, Koch SW (2004) Semiconductor bloch equations. In: Quantum theory of the optical and electronic properties of semiconductors. World Scientific Publishing Co. Pte. Ltd., chapter 12, pp 211–234
108. Yue L, Gaarde MB (2022) JOSA B 39:535. https://doi.org/10.1364/JOSAB.448602
109. Golde D, Meier T, Koch SW (2008) Phys Rev B 77:075330. https://doi.org/10.1103/PhysRevB.77.075330
110. Golde D, Meier T, Koch SW (2009) Phys Status Solidi C 6:420. https://doi.org/10.1002/pssc.200880309
111. Ghimire S, DiChiara AD, Sistrunk E, Ndabashimiye G, Szafruga UB, Mohammad A, Agostini P, DiMauro LF, Reis DA (2012) Phys Rev A 85:043836. https://doi.org/10.1103/PhysRevA.85.043836
112. Kilen I, Kolesik M, Hader J, Moloney JV, Huttner U, Hagen MK, Koch SW (2020) Phys Rev Lett 125:083901. https://doi.org/10.1103/PhysRevLett.125.083901
113. Kolesik M, Moloney JV (2023) Phys Rev B 108:115433. https://doi.org/10.1103/PhysRevB.108.115433
114. Liu H, Li Y, You YS, Ghimire S, Heinz TF, Reis DA (2017) Nat Phys 13:262. https://doi.org/10.1038/nphys3946
115. Banks HB, Wu Q, Valovcin DC, Mack S, Gossard AC, Pfeiffer L, Liu R-B, Sherwin MS (2017) Phys Rev X 7:041042. https://doi.org/10.1103/PhysRevX.7.041042
116. Luu TT, Wörner HJ (2018) Nat Commun 9:916. https://doi.org/10.1038/s41467-018-03397-4
117. Avetissian HK, Mkrtchian GF (2020) Phys Rev B 102:245422. https://doi.org/10.1103/PhysRevB.102.245422
118. Yue L, Gaarde MB (2020) Phys Rev Lett 124:153204. https://doi.org/10.1103/PhysRevLett.124.153204
119. Li L, Lan P, Zhu X, Lu P (2023). Rep Prog Phys. https://doi.org/10.1088/1361-6633/acf144
120. Wilhelm J, Grössing P, Seith A, Crewse J, Nitsch M, Weigl L, Schmid C, Evers F (2021) Phys Rev B 103:125419. https://doi.org/10.1103/PhysRevB.103.125419
121. Silva REF, Martín F, Ivanov M (2019) Phys Rev B 100:195201. https://doi.org/10.1103/PhysRevB.100.195201
122. Osika EN, Chacón A, Ortmann L, Suárez N, Pérez-Hernández JA, Szafran B, Ciappina MF, Sols F, Landsman AS, Lewenstein M (2017) Phys Rev X 7:021017. https://doi.org/10.1103/PhysRevX.7.021017
123. Parks AM, Ernotte G, Thorpe A, McDonald CR, Corkum PB, Taucer M, Brabec T (2020) Optica 7:1764. https://doi.org/10.1364/OPTICA.402393
124. Brown GG, Jiménez-Galán A, Silva REF, Ivanov M (2024) Phys Rev Res 6:043005. https://doi.org/10.1103/PhysRevResearch.6.043005
125. Li L, Lan P, Zhu X, Lu P (2021) Phys Rev Lett 127:223201. https://doi.org/10.1103/PhysRevLett.127.223201

126. Gonoskov I, Gräfe S (2021) J Chem Phys 154:234106. https://doi.org/10.1063/5.0048930
127. Gonoskov I, Sondenheimer R, Hünecke C, Kartashov D, Peschel U, Gräfe S (2024) Phys Rev B 109:125110. https://doi.org/10.1103/PhysRevB.109.125110
128. Tsatrafyllis N, Kühn S, Dumergue M, Foldi P, Kahaly S, Cormier E, Gonoskov I, Kiss B, Varju K, Varro S, Tzallas P (2019) Phys Rev Lett 122:193602. https://doi.org/10.1103/PhysRevLett.122.193602
129. Wang G, Du T-Y (2021) Phys Rev A 103:063109. https://doi.org/10.1103/PhysRevA.103.063109
130. Nielsen MA, Chuang IL (2010) Distance measures for quantum information. In: Quantum computation and quantum information: 10th anniversary edition. Cambridge University Press, Cambridge, UK, chapter 9, pp 399–424
131. Bourassin-Bouchet C, Barreau L, Gruson V, Hergott J-F, Quéré F, Saliéres P, Ruchon T (2020) Phys Rev X 10:031048. https://doi.org/10.1103/PhysRevX.10.031048
132. Gühne O, Tóth G (2009) Phys Rep 474:1. https://doi.org/10.1016/j.physrep.2009.02.004
133. Caspar P, Oudot E, Sekatski P, Maring N, Martin A, Sangouard N, Zbinden H, Thew R (2022) Quantum 6:671. https://doi.org/10.22331/q-2022-03-22-671
134. Ben-Itzhak I, Gertner I, Heber O, Rosner B (1993) Phys Rev Lett 71:1347. https://doi.org/10.1103/PhysRevLett.71.1347
135. Holkundkar AR, Bandyopadhyay JN (2023) Phys Rev A 108:063503. https://doi.org/10.1103/PhysRevA.108.063503
136. Stammer P, Rivera-Dean J, Maxwell AS, Lamprou T, Argüello-Luengo J, Tzallas P, Ciappina MF, Lewenstein M (2024) Entanglement and squeezing of the optical field modes in high harmonic generation. https://doi.org/10.1103/PhysRevLett.132.143603
137. Lange CS, Hansen T, Madsen LB (2024) Phys Rev A 109:033110. https://doi.org/10.1103/PhysRevA.109.033110

Chapter 5
Non-Classical States of Light for Device-Independent Quantum Key Distribution

> *"Quantum mechanics... delivers much, but does not really bring us any closer to the secret of the Old One. I, at any rate, am convinced that He does not play dice."*
>
> – Albert Einstein, in a letter to Max Born, December 1926.

Communication stands as one of the most precious facets of human interaction, forming the bedrock upon which relationships, knowledge exchange, and societal progress are established. However, effective communication necessitates a shared language among the involved parties. Without this shared understanding, communication becomes ineffective, leading to a loss of the intended message. A prime example illustrating this challenge is found in Egyptian hieroglyphs, whose decryption remained within the realm of speculation until the discovery of the Rosetta Stone in 1799 [1].

While the primary purpose of Egyptian hieroglyphs was not necessarily secrecy, certain communication contexts demand confidentiality. In scenarios involving two subjects, typically referenced as Alice and Bob, communicating privately, cryptography assumes a pivotal role. By means of cryptographic protocols, Alice and Bob encrypt, transmit, and decrypt their messages using specific algorithms to ensure the information remains confidential from an eavesdropper, usually denoted as Eve. For the encryption, the use of a *key* becomes crucial, that is, a secret method used by Alice and Bob to encode and decode their message. Thus, the use of provably secure encryption mechanisms is ideally paramount in order to guarantee the message's secrecy during transmission from Alice to Bob.

J. Rivera-Dean, *Non-classical States of Light*, Springer Theses,
https://doi.org/10.1007/978-3-031-73769-5_5

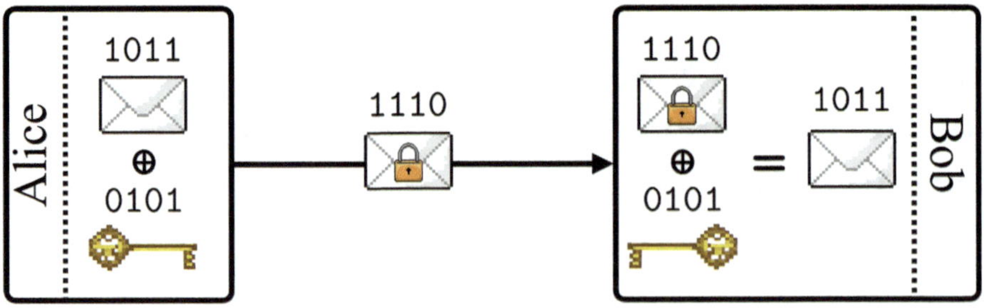

Fig. 5.1 Illustration depicting the one-time pad encryption technique. In this encryption process, Alice converts the message into binary format and combines it through binary addition with a secret key shared with Bob. The resulting encrypted message is transmitted from Alice to Bob through an untrusted channel, safeguarding it from eavesdroppers who lack the key. Upon reception at Bob's end, Bob performs binary addition between the encrypted message and the shared secret key to unveil the original message

The first provably secure encryption method was the *one-time pad*, originally attributed to Frank Miller in 1882 [2], although it is also commonly referred to as the Vernam cipher due to Gilbert Vernam, who patented it in 1917 [3]. According to this method [4], pictorially represented in Fig. 5.1, Alice and Bob posses a secret random key K, exclusively known to them and encoded in binary format. To ensure privacy during communication when sending the message M to Bob, encoded in binary format as well, Alice performs binary addition on M and K, sending the result $M \oplus K$ to Bob. Upon receiving the encoded message, Bob performs the same operation with the same key, i.e. $M \oplus K \oplus K$, to retrieve the secret message. As demonstrated by Claude E. Shannon in 1949 [4, 5], this methodology is provably secure provided that: (1) the key is genuinely random, (2) it matches the length of the original message, (3) it remains confidential, and (4) it is not reused for other message exchanges.

Beyond the one-time pad algorithm, one of the most used encryption methods is the so-called Rivest-Shamir-Adleman (RSA) cryptosystem [6], which generates keys that are computationally difficult to guess. By *computationally difficult*, it implies that obtaining the secret key requires solving problems for which no known efficient algorithm exists. In the case of RSA, this involves the factorization of two large prime numbers. However, the computational difficulty does not inherently guarantee provable security: if a potential hacker develops an algorithm capable of efficiently solving these problems, such as Shor's algorithm for prime number factorization [7, 8], the security of the employed protocol becomes compromised. It is important to note that the potential hacker would also require the means to implement the algorithm, which, in the case of Shor's algorithm, necessitates a fault-tolerant quantum computer [9]. Thus, the advent of quantum technology compromises the security of existing key generation protocols.

One could consider two approaches to address this security vulnerability: one focuses on developing new encryption methods capable of resisting attacks from quantum computers, known as *post-quantum cryptography*. The other approach involves developing physically secure methods for key distribution. In this second context, the field of quantum key distribution (QKD) takes utmost importance. QKD aims to generate and distribute provably secure cryptographic keys analogously to the one-time pad encryption scheme, utilizing the principles of quantum mechanics, wherein Alice and Bob leverage quantum states and measurements to generate the secret key. Several reasons underscore the rationale behind QKD's potential usefulness in this realm, under the assumption of the correctness of quantum mechanics [4, 9]:

- the inherent randomness associated to quantum states and the measurements performed on them;
- the no-cloning theorem [10, 11], which establishes the impossibility of cloning a quantum state, a stark contrast to classical bits' behavior;
- the impossibility of Eve obtaining information without disturbing the state used by Alice and Bob [10];
- the capability of QKD to generate new keys for each message exchange, eliminating vulnerabilities associated with key reuse, although its efficiency heavily relies on the system utilized for this purpose.

In this chapter, our focus revolves around a specific variant of QKD known as Device-Independent Quantum Key Distribution (DIQKD). In DIQKD, the security analysis of the protocol employed by Alice and Bob operates under the premise that they cannot trust the operations conducted by their measurement devices. Our exploration, based on Refs. [12, 13], is centered on a photonic-based DIQKD protocol reliant on the utilization of (non-classical) quantum dot single-photon sources. We delve into the prerequisites these sources must meet to ensure the establishment of a secret key between Alice and Bob. Subsequently, we investigate whether employing more generic sources, specifically not solely dependent on qubit systems, might aid in alleviating the stringent requirements of DIQKD.

5.1 A Brief Introduction to Quantum Key Distribution

As mentioned earlier, QKD aims to generate a secure key between Alice and Bob by leveraging the principles of quantum mechanics. However, contingent upon the correctness of quantum mechanics, the security of QKD typically relies on certain assumptions regarding the cryptographic protocol's functioning. This section introduces an example of a QKD protocol, which we recur throughout the chapter. This example serves to familiarize readers with the core concepts within QKD and elucidates the common assumptions made during the analysis of its security.

5.1.1 An Example of a Quantum Key Distribution Protocol: the Ekert 91 Protocol

In general terms, typical quantum key distribution protocols rely on the generation of quantum states, onto which measurements are performed to obtain some outcomes. In its most simplest form, two parties, Alice and Bob, are involved in the key generation and, depending on the specific process followed by them, we can distinguish between two types of protocols [14]. On one hand, there are the *prepare-and-measure* protocols, where one party generates quantum states within their system, transmits them through a channel, and the other party subsequently measures them. The first example in this regard is the commonly known as BB84 protocol [15], named after its founders C. H. Benett and G. Brassard, which has undergone various adaptations and extensions [16–18]. On the other hand, *entanglement-based* protocols involve a station, usually positioned between Alice and Bob, sharing an entangled state with them, upon which they perform measurements (see Fig. 5.2). An example is the E91 protocol [19], named after its founder A. Ekert.

In practical implementations, these two protocols demand significantly different resources. However, from a theoretical perspective, security proofs for the entanglement-based protocol are also valid for the prepare-and-measure one. This assertion is based on the premise that transitioning from an entanglement-based protocol to the prepare-and-measure approach is feasible by considering that the source of entangled states resides within Alice's station, for instance. Therefore, assuming the secure nature of Alice's station, where no information is leaked without their

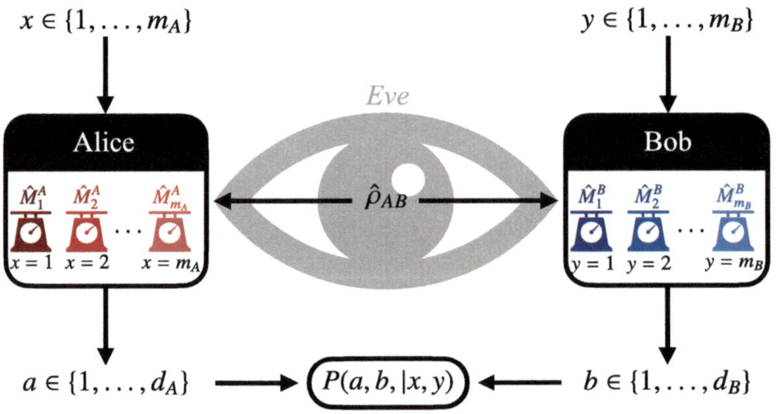

Fig. 5.2 Standard setup of quantum key distribution protocols. Upon receiving their corresponding portion of the quantum state $\hat{\rho}_{AB}$, Alice and Bob randomly select an input x and y out of m_A and m_B choices. These settings correspond to the application of measurement operators $\hat{M}_x^A$ and $\hat{M}_y^B$ on their respective share of the quantum state. As a result, they observe outcomes a and b, respectively. The probabilities of obtaining specific outcomes, given particular measurements, are described by the set of correlations $\{P(a, b|x, y)\}$ between their measurement choices and observed results

permission, prevents Eve from acting on the source, resulting in fewer vulnerabilities to exploit. Consequently, in subsequent discussions, our focus is limited on the entanglement-based approach, for which the E91 protocol serves as cornerstone.

In accordance with the E91 protocol [4, 9, 19, 20], Alice and Bob are located in two spatially separated stations, with a source positioned between them capable of generating maximally entangled states. For the purpose of our discussion, we confine ourselves to qubits and represent the state produced by the source as

$$|\psi\rangle = \frac{1}{\sqrt{2}}\Big[|0_A\rangle \otimes |0_B\rangle + |1_A\rangle \otimes |1_B\rangle\Big], \tag{5.1}$$

where qubit $|i_A\rangle$ ($i \in [0, 1]$) is transmitted to Alice, and qubit $|j_B\rangle$ ($j \in [0, 1]$) is sent to Bob through a designated quantum channel. Within their respective laboratories, Alice and Bob possess devices capable of performing measurements, $m_A \in \mathbb{W}$ for Alice and $m_B \in \mathbb{W}$ for Bob, on their respective shares of the state. We denote the operators representing these measurements as $\hat{M}_x^A$ for Alice and $\hat{M}_y^B$ for Bob, where $x \in \{1, \ldots, m_A\}$ and $y \in \{1, \ldots, m_B\}$. It is assumed that both parties have access to machines capable of generating genuinely random numbers, determining the values of x and y used by Alice and Bob for measuring their qubits.

As a result of these measurements, Alice and Bob obtain outcomes a and b, respectively, where $a \in \{1, \ldots, o_A\}$ and $b \in \{1, \ldots, o_B\}$, with o_A and o_B the total number of outcomes obtained by Alice and Bob. In the following, we assume that $o_A = d_A$ and $o_B = d_B$, with d_A and d_B corresponding to the dimensions of their respective Hilbert spaces, denoted as $\mathcal{H}_A$ and $\mathcal{H}_B$. In this specific case involving qubits, we have $d_A = d_B = 2$. Furthermore, we set $m_A = 2$ and $m_B = 3$ [20], with the measurement operators given by

$$\hat{M}_1^A = \hat{\sigma}_z, \quad \hat{M}_2^A = \hat{\sigma}_x, \tag{5.2}$$

$$\hat{M}_1^B = \frac{1}{\sqrt{2}}\big(\hat{\sigma}_z - \hat{\sigma}_x\big), \quad \hat{M}_2^B = \frac{1}{\sqrt{2}}\big(\hat{\sigma}_z + \hat{\sigma}_x\big), \quad \hat{M}_3^B = \hat{\sigma}_z, \tag{5.3}$$

where $\{\hat{\sigma}_x, \hat{\sigma}_y, \hat{\sigma}_z\}$ is the set of Pauli operators.

It is noteworthy that, in principle, Alice and Bob could have the capability to perform other measurements beyond those in Eqs. (5.2) and (5.3), generically represented as a superposition between different Pauli operators. However, when considering the measurements specified in Eqs. (5.2) and (5.3), in conjunction with the quantum state they are sharing (see Eq. (5.1)), they can establish nonlocal correlations (refer to Sect. 2.3.4.2) which serve as a fundamental cornerstone of QKD [21]. These correlations manifest in the conditional probability distributions $P(a, b|x, y)$, depicting the likelihood of Alice and Bob obtaining outcomes a and b, respectively, given that measurements x and y are applied. As presented in Eq. (2.135), these probability distributions are given by the Born rule [22] which, in this particular case, can be written as

$$P(a, b|x, y) = \langle\psi| \left(\hat{M}_{a|x} \otimes M_{b|y}\right) |\psi\rangle, \tag{5.4}$$

where $|\psi\rangle$ is given in Eq. (5.1), and where $\{\hat{M}_{a|x}\}$ ($\{\hat{M}_{b|y}\}$) represents a POVM decomposition of the measurement operator $\hat{M}_x^A$ ($\hat{M}_y^B$), where each $\hat{M}_{a|x}$ ($\hat{M}_{b|y}$) denotes the operator corresponding to Alice (Bob) performing measurement x (y) and obtaining outcome a (b). As a consequence, if these probability distributions exhibit nonlocal behavior, they cannot be described by a local-hidden variable model as formulated in Eq. (2.132). The primary implication is that Eve would be unable to access these potential hidden variables, which, if accessible, would determine the outcomes of their measurements in conjunction with the measurement inputs considered by Alice and Bob.

Furthermore, a primary advantage of nonlocality is the ability for Alice and Bob to quantitatively assess the degree of nonlocality in their experiments using the Bell inequalities outlined in Eq. (2.136). In the context of the considered protocol, Alice and Bob can utilize the outputs of the measurements characterized by the inputs $x \in \{1, 2\}$ and $y \in \{1, 2\}$ to evaluate the following expression

$$\langle\hat{\mathcal{B}}\rangle = |\langle\hat{M}_1^A \hat{M}_2^B\rangle + \langle\hat{M}_2^A \hat{M}_1^B\rangle + \langle\hat{M}_2^A \hat{M}_2^B\rangle - \langle\hat{M}_1^A \hat{M}_1^B\rangle|, \tag{5.5}$$

for which behaviors $\boldsymbol{P} = \{P(a, b, |x, y)\}$ belonging to the local set $\mathcal{L}$ (see Fig. 2.14), and therefore admitting a local-hidden variable model, satisfy

$$S \equiv \left|\langle\hat{\mathcal{B}}\rangle\right| \leq 2, \tag{5.6}$$

leading to the commonly known as Clauser-Horne-Shimony-Holt (CHSH) inequality [23]. Conversely, nonlocal behaviors, i.e. those that are in $\mathcal{Q}$ but not in $\mathcal{L}$,[1] violate this inequality, thus yielding

$$2 < S \leq 2\sqrt{2}, \tag{5.7}$$

where the upper bound, generally known as the Tsirelson bound [24], is obtained for Eq. (5.1) when applying measurements $x \in \{1, 2\}$ and $y \in \{1, 2\}$ presented in Eqs. (5.2) and (5.3).

Therefore, through the examination of Eq. (5.5), Alice and Bob can verify the presence of nonlocal correlations in their shared data. Specifically, if their evaluation yields the $2\sqrt{2}$ bound, they can confidently assert that Eve lacks knowledge of the generated key, which they can compute from the outcomes acquired when jointly measuring $x = 1$ and $y = 3$, as they yield same outcomes for both parties.

[1] In the following, we refer to the set $\mathcal{Q} \cap \mathcal{L}$ as the *nonlocal* set $\mathcal{NL}$.

Thus, by publicly communicating the inputs chosen in each round of the protocol—defined by Alice and Bob receiving their share of the quantum state and performing measurements on it—they can construct a secret *raw* key.

5.1.2 *Classical Postprocessing Employed on the* Raw *Key*

In the preceding subsection, we presented a scenario wherein Alice and Bob received a maximally entangled state, subsequently generating a key through local operations and classical communication. However, practical implementations encounter numerous sources of noise that can introduce errors into the generated key. For instance, the central station responsible for generating the states shared by both parties might fail to produce maximally entangled states. Additionally, the generated quantum state could experience decoherence as they traverse the quantum channel transmitting them to Alice and Bob's stations. Notably, these imperfections could also stem from a potential interference by Eve.

After completing n protocol rounds, resulting in a raw key of length $l_{\text{raw}} < n$, Alice and Bob need to evaluate its security. This assessment provides an error estimation within the key: surpassing a predefined threshold prompts protocol termination; otherwise, Alice and Bob proceed to what is known as *classical postprocessing*. This postprocessing aims to distill a cryptographic key smaller than the raw key obtained in the protocol, i.e. $l_{\text{key}} < l_{\text{raw}}$, but of enhanced security. Thus, the entire process involves three primary steps: parameter estimation, error correction and privacy amplification [9, 14].

- **Parameter estimation.** In this phase, Alice and Bob focus on estimating the error rate, given that potential sources of noise may have caused discrepancies in their obtained outputs. To achieve this, they select a given number l_{PE} of elements from the raw key, and publicly announce them. By comparing these elements, they derive an estimate of the error rate, essential for limiting the information potentially acquired by Eve. Depending on this estimation, Alice and Bob determine whether to continue or abort the protocol.
- **Error correction.** Using the shorter key of length $l_{\text{raw}} - l_{\text{PE}}$ obtained from the preceding step, Alice and Bob focus on error elimination. Various methods for achieving this exist (refer to Ref. [14] and references therein), leading to a reduced-size key $l_{\text{EC}} < l_{\text{raw}}$, where the errors have been effectively removed.
- **Privacy amplification.** To further secure the generated key, Alice and Bob execute a privacy amplification process [14] aimed at minimizing Eve's potential knowledge of the secret key. The outcome is a shorter key of length $l_{\text{key}} < l_{\text{EC}}$ but of enhanced security. This refined key serves as the cryptographic foundation for encrypting their messages.

5.1.3 Device Independent Quantum Key Distribution

In the description of the E91 protocol provided in Sect. 5.1.1, certain underlying assumptions have been made about its operation. These assumptions are [25]:

- Alice and Bob are placed in distant regions in completely shielded stations, from which no information can be leaked without their permission;
- they both possess authentic random number generators that determine the measurement inputs x and y;
- they can communicate through an authenticated classical channel, that is, ensuring sender identity verification and message integrity during transmission;
- quantum mechanics is correct;
- the inner workings of the measurement are inaccessible to Eve, and therefore Alice and Bob completely trust them.

Given this last assumption, these protocols often fall under the category of *device-dependent*. Yet, it opens an opportunity for Eve to exploit and acquire information about the key without being detected [26]. This vulnerability has led to various attacks, which in photonic implementations have taken advantage of exploiting sub-optimal efficiencies of the employed photodetectors [27–29], or by inducing the desired outcome in them through the use of relatively intense fields manipulated by Eve [30–32].

In this context, *device-independent* quantum key distribution takes a further step, aiming to study the security of a protocol while assuming no knowledge about the inner workings of the devices employed by Alice and Bob [33]. Essentially, in DIQKD, the measurement apparatuses used by Alice and Bob are treated as black boxes, as illustrated in Fig. 5.3, where they introduce measurement inputs x and y, and respectively obtain outcomes a and b. Consequently, the security of DIQKD protocols fundamentally relies on the set of correlations $\{P(a, b|x, y)\}$ established by Alice and Bob without making any assumption about their origin.

The paramount quantity used to assess the security of a DIQKD protocols is the *key rate*, denoted hereupon as $r = l_{\text{key}}/n$, defining the number of secret bits Alice and Bob are able to generate per protocol round. As a first step to ensure the protocol's security, this quantity is evaluated in the asymptotic regime $n \to \infty$, where it is assumed that Alice and Bob have access to an infinitely large number of rounds. However, even in this limit, the estimation of key rates remains challenging, and what is typically studied are either lower or upper bounds to this quantity. In this chapter, we are interested in analyzing the minimal requirements a given physical device has to satisfy in order to ensure security of DIQKD protocols. Thus, our interest focus towards the seek of lower bounds to the key rate.

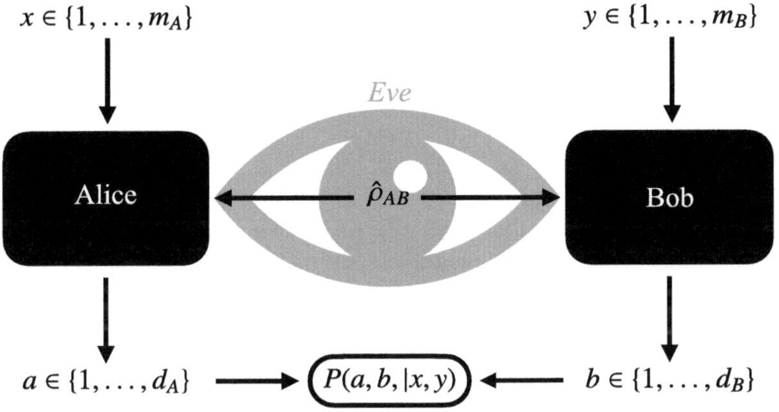

Fig. 5.3 Standard setup of device-independent quantum key distribution protocols. In contrast to Fig. 5.2, device-independent quantum key distribution protocols are established on the premise of distrust towards the physical devices engaged in state preparation and measurements. This skepticism is symbolically depicted by representing Alice's and Bob's stations as black boxes, concealing the inner workings of the devices. Within this framework, they introduce randomly chosen inputs x and y, and receive corresponding outcomes a and b, lacking any insight into the device mechanisms. Consequently, the security of these protocols relies on the set of correlations $\{P(a, b|x, y)\}$ derived from multiple iterations of the experiment

5.1.3.1 Lower Bounding the Key Rate

In scenarios where public communication occurs in only one direction, for instance from Alice to Bob, the key rate is typically lower bounded by the Devetak-Winter bound [34], given by

$$r_{\text{DW}} = H(A|x = m, E) - H(A|B, x = m, y = m + 1), \tag{5.8}$$

where $H(\cdot|\cdot)$ represents the conditional entropy. This bound showcases the difference between $H(A|x = m, E)$, representing Eve's lack of knowledge about Alice outcomes for the key generation input $x = m$, and $H(A|B, x = m, y = m + 1)$, indicating Bob's lack of knowledge about Alice's outcomes when they introduce inputs $x = m$ and $y = m + 1$ in their respective black boxes. Consequently, a DIQKD protocol is said to be secure whenever $r > 0$, implying that Eve's ignorance about Alice's outcomes is greater than that of Bob.

Given the set of correlations $\{P(a, b|x = m, y = m + 1)\}$, the conditional entropy between Alice and Bob $H(A|B, x = m, y = m + 1)$ can be straightforwardly computed as follows [35]

$$
\begin{aligned}
H(A|B, x = m, y = m + 1)& \\
= -\sum_{a,b} P(a, b|x = m, y = m + 1) \log_d \big(P(a, b|x = m, y = m + 1)\big)& \\
+ \sum_{a,b} P(b|y = m + 1) \log_d \big(P(b|y = m + 1)\big),& \qquad (5.9)
\end{aligned}
$$

where the base of the logarithm is relative to Alice's, or equivalently Bob's when $d_A = d_B$, number of outcomes. Nevertheless, estimating $H(A|x = m, E)$, which represents the von Neumann entropy conditioned on Eve's side information, remains a challenging task. Consequently, it is a common practice to explore alternative approaches aiming to establish manageable lower bounds on this quantity. Some of these approaches utilize analytical formulas, while others require the performance of numerical optimization, potentially with Semidefinite Programming (SDP) techniques. In the following, we present the ones used in forthcoming sections.

Analytical bound based on the CHSH inequality violation. As mentioned in Sect. 5.1.1, one of the crucial steps undertaken by Alice and Bob in typical QKD protocols, involves utilizing the generated statistics from a subset of the implemented measurements to compute the score of a Bell inequality. For the case $d_A = d_B = 2$ and $m = 2$, we can use the CHSH inequality. In this context, the score S attained by Alice and Bob, can be used to obtain a lower bound on $H(A|x = m, E)$, as shown by A. Acín and collaborators in Ref. [33]. More specifically, this lower bound is expressed as follows

$$
H(A|x = m, E) \geq 1 - h\left(\frac{1 + \sqrt{(S/2)^2 - 1}}{2}\right), \qquad (5.10)
$$

where $h(\cdot)$ denotes the binary entropy. As it can be seen from this expression, when the CHSH inequality is maximally violated such that $S = 2\sqrt{2}$, the term at the right hand side of the inequality becomes 1, signifying that Eve possesses no information regarding Alice's output. Conversely, for $S = 2$, the resulting lower bound of 0 indicates the opposite scenario.

Analytical bounds adding noisy preprocessing techniques. In certain scenarios, the bound introduced in Eq. (5.10) can undergo slight enhancement by incorporating noisy preprocessing of the data before its conversion to a viable key. Specifically, Alice has the capability to execute a bit-flip operation on the acquired outcomes with a probability q. Therefore, in this context Eq. (5.10) transforms into [36]

$$
H(A|x = m, E) \geq 1 - h\left(\frac{1 + \sqrt{(S/2)^2 - 1}}{2}\right) + h\left(\frac{1 + \sqrt{1 - q(q-1)(8 - S^2)}}{2}\right), \qquad (5.11)
$$

which, contingent on the specific value of q, this process can result in an increasing of Eve's lack of knowledge about Alice's outputs compared to that of Bob.

The min-entropy. The two methods presented thus far can be analytically assessed based on the set of correlations $\{P(a, b|x, y)\}$ established between Alice and Bob. In contrast, evaluating the min-entropy typically necessitates addressing an optimization problem. The min-entropy is formally defined as [37]

$$H_{\min}(A|x = m, E) = -\log_d \left[G(A|x = m, E)\right], \tag{5.12}$$

where $G(A|x = m, E)$ denotes the guessing probability [38]. This probability represents Eve's likelihood of correctly guessing Alice's outcome when Alice employs input $x = m$. In our context, the guessing probability can be lower bounded through the following optimization problem

$$\begin{aligned} G(A|x = x^*, E) = \ \text{maximize} \quad & \sum_{a=1}^{d} \langle \hat{M}_{a|m} \hat{Z}_a \rangle \\ \text{subject to} \quad & \langle \hat{M}_{a|x} \hat{M}_{b|y} \rangle = P(a, b|x, y) \quad \forall a, b, x, y; \\ & \langle \hat{M}_{a|x} \rangle = P(a|x),\ \langle \hat{M}_{b|y} \rangle = P(b|y) \quad \forall a, b, x, y; \\ & \sum_{a=1}^{d} \hat{M}_{a|x} = \sum_{b=1}^{d} \hat{M}_{b|y} = \mathbb{1} \quad \forall x, y; \\ & \sum_{a} \hat{Z}_a = \mathbb{1}, \quad \hat{Z}_a \geq 0 \ \ \forall a; \\ & [\hat{M}_{a|x}, \hat{M}_{b|y}] = [\hat{M}_{a|x}, \hat{Z}_c] = [\hat{Z}_c, \hat{M}_{b|y}] = 0 \\ & \qquad \forall a, b, c, x, y; \end{aligned} \tag{5.13}$$

where it is worth noting that the minimization problem can be derived by negating the objective function. Additionally, in the formulation above, the sets $\{\hat{M}_{a|x}\}$ and $\{\hat{M}_{b|y}\}$ correspond to the POVM sets associated with the measurements $\hat{M}_x^A$ and $\hat{M}_y^B$ performed by Alice and Bob, respectively.

A convergent hierarchy for the von Neumann entropy. In more recent studies [39, 40], P. Brown and colleagues introduced a convergent hierarchy for the conditional von Neumann entropy. In its most relaxed form, this lower bound can be expressed as follows (for an in-depth derivation, we refer the interested reader to Ref. [40])

$$H(A|x = m, E) \geq c_M + \sum_{i=1}^{M-1} \frac{w_i}{t_i \ln d} \sum_{a=1}^{d} f(t_i, \hat{M}_{a|m}), \tag{5.14}$$

where w_i and t_i correspond to the ith Gauss-Radau quadratures containing a total of M nodes, while $f(t_i, \hat{M}_{a|x})$ is defined in terms of the following optimization problem

$$f(t_i, \hat{M}_{a|m}) = \ \text{minimize} \quad \left\langle \hat{M}_{a|m} (\hat{Z}_{a,i} + \hat{Z}_{a,i}^{\dagger} + (1 - t_i) \hat{Z}_{a,i}^{\dagger} \hat{Z}_{a,i}) + t_i \hat{Z}_{a,i} \hat{Z}_{a,i}^{\dagger} \right\rangle \tag{5.15}$$

$$\begin{aligned} \text{subject to } & \langle \hat{M}_{a|x}\hat{M}_{b|y}\rangle = P(a,b|x,y) \ \ \forall a,b,x,y; \\ & \langle \hat{M}_{a|x}\rangle = P(a|x),\ \langle \hat{M}_{b|y}\rangle = P(b|y) \ \ \forall a,b,x,y; \\ & \hat{M}_{a|x} > 0,\ \hat{M}_{a|y} > 0 \ \ \forall a,b,x,y; \\ & [\hat{M}_{a|x},\hat{M}_{b|y}] = [\hat{M}_{a|x},\hat{Z}_{a,i}] = [\hat{Z}_{a,i},\hat{M}_{b|y}] = 0, \\ & [\hat{M}_{a|x},\hat{M}_{b|y}] = [\hat{M}_{a|x},\hat{Z}^{\dagger}_{a,i}] = [\hat{Z}^{\dagger}_{a,i},\hat{M}_{b|y}] = 0 \\ & \quad \forall a,b,c,x,y, \end{aligned} \tag{5.16}$$

with the lower bound in Eq. (5.14) referred to as a *convergent hierarchy* given that increased values of M result in tighter bounds on the von Neumann entropy.

5.1.3.2 SDP Optimization and the Navascués-Pironio-Acín hierarchy

Broadly,[2] SDP optimization entails the minimization of an objective function formulated in relation to a positive semidefinite matrix Γ, subjected to a defined set of linear constraints. Mathematically, an SDP can be formally expressed as [41]

$$\begin{aligned} & \text{minimize } \operatorname{tr}(\mathsf{C}\Gamma) \\ & \text{subject to } \operatorname{tr}(\mathsf{A}_i\Gamma) = b_i,\ i = 1,\dots,n \\ & \qquad \Gamma \succeq 0, \end{aligned} \tag{5.17}$$

where C, Γ and $\{\mathsf{A}_i\}$ are Hermitian matrices of equal size, with n denoting the number of linear constraints. This form of the SDP is commonly referred to as the primal problem. Corresponding to this, we can formulate its dual problem as

$$\begin{aligned} & \text{maximize } \sum_{i=1}^{n} b_i y_i \\ & \text{subject to } \sum_{i=1}^{n} y_i \mathsf{A}_i \preceq \mathsf{C}, \end{aligned} \tag{5.18}$$

which is defined on the constraints of the primal problem. Feasible points of the primal and the dual, respectively, provide upper and lower bounds to the solution of the optimization problem which, up to some mild conditions, can be thought as sharing the same value. This particular characteristic is in fact harnessed by numerous standard numerical solvers used for SDP optimization problems such as `Mosek` [42, 43], the primary solver utilized throughout this chapter. Specifically, the

---

[2] For those readers less familiar with the topic, we suggest reviewing Section 2.3.4.2 before delving into this section.

algorithm implemented by this solver is the interior point method [44] which, in broad terms, involves a sequence of iterative steps between Eqs. (5.17) and (5.18) until achieving convergence.

The problems presented in Eqs. (5.12) and (5.14) entail an optimization over the set of bounded operators $\mathrm{B}(H)$. However, they can be relaxed to an SDP optimization problem by identifying the variables of the function to optimize with the elements of a positive semidefinite matrix Γ as

$$\Gamma_{i,j} = \mathrm{tr}(\hat{\rho}\hat{O}_i\hat{O}_j), \tag{5.19}$$

where $\hat{O}_i$ and $\hat{O}_j$ represent linear combinations of various products of elements from the POVM sets $\{\hat{M}_{a|x}\}_a$ and $\{\hat{M}_{b|y}\}_b$. By definition, these variables naturally involve the correlations established between Alice and Bob and, depending on the specific problem, with Eve as well. Hence, to properly assess the security of DIQKD protocols, it is important for these correlations to remain *physical*, which in our case implies belonging to the quantum set $\mathcal{Q}$. Thus, a reasonable question to pose at this point is *to what extent does a generic finite-size matrix with elements given by Eq.* (5.19) *depict a behavior within the quantum set?* The Navascués-Pironio-Acín (NPA) hierarchy [45, 46] offers a pathway to address this inquiry based on the fact that, if a behavior $\boldsymbol{P}$ belongs to the quantum set, then a positive semidefinite matrix Γ can be identified, fulfilling

$$\sum_{i,j} c_{i,j}\Gamma_{i,j} = 0 \quad \text{if} \quad \sum_{i,j} c_{i,j}\hat{O}_i\hat{O}_j = 0, \tag{5.20}$$

$$\sum_{i,j} c_{i,j}\Gamma_{i,j} = \sum_{a,b} d_{a,b}P(a,b|x,y) \quad \text{if} \quad \sum_{i,j} c_{i,j}\hat{O}_i\hat{O}_j = \sum_{a,b} d_{a,b}\hat{M}_{a|x}\hat{M}_{b|y}. \tag{5.21}$$

The conditions in the form of Eqs. (5.20) and (5.21) are the ones explicitly appearing in both Eqs. (5.13) and (5.14). However, in those scenarios, the problem is formulated differently as the objective is to find a matrix Γ that satisfies these constraints while minimizing a specified objective function. In this context, depending on the definition of the $\Gamma_{i,j}$ elements in Eq. (5.19), different levels of approximation—referred to as NPA levels—to the quantum set $\mathcal{Q}$ are obtained, denoted in the following as $\mathcal{Q}^{(k)}$. This is pictorially represented in Fig. 5.4.

For instance, the first level of approximation $\mathcal{Q}^{(1)}$ is obtained by considering in Eq. (5.19) only those operators $\hat{O}_i$ and $\hat{O}_j$ belonging to the set $\mathcal{O}^{(1)} := \{\hat{M}_{a|x}\} \cup \{\hat{M}_{b|y}\}$ composed of the local POVM elements acting locally on Alice and Bob's stations. As higher-order correlations, such as those of the form $\hat{M}_{a|x}\hat{M}_{b|y}$, are not explicitly included in our moment matrix, they remain unfixed in the SDP optimization. This leads to correlations that do not necessarily belong to the quantum set $\mathcal{Q}$, but to a broader set $\mathcal{Q}^{(1)}$, consequently satisfying $\mathcal{Q}^{(1)} \supseteq \mathcal{Q}$.

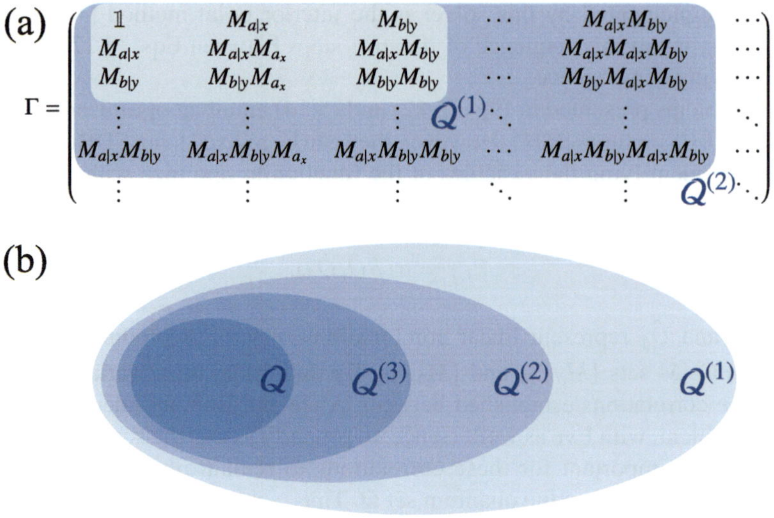

Fig. 5.4 Pictorial representation of the different NPA levels. Each NPA level, indexed by k, encompasses variables (displayed in the first row or column in panel (a)) involving up to kth order monomials. These levels represent successive approximations towards the set of quantum correlations, illustrated in panel (b) as $Q^{(k)}$ for the kth NPA level. As the NPA level increases, denoting a higher k, the resulting set $Q^{(k)}$ becomes closer to the set of quantum correlations Q

Continuing with this reasoning, if we consider the set of operators $O^{(2)} := \{\hat{M}_{a|x}\} \cup \{\hat{M}_{b|y}\} \cup \{\hat{M}_{a|x}\hat{M}_{b|y}\}$, we obtain a more constrained $Q^{(2)}$ set than $Q^{(1)}$, satisfying $Q^{(1)} \supseteq Q^{(2)} \supseteq Q$. This is pictorially shown in Fig. 5.4 (a), where the different approximations to the quantum set shown in (b) involve the inclusion of higher-order moments in the Γ matrix.

As a result, the NPA hierarchy delineates a sequence of convergent sets. The higher the level, which fixes the size of the moment matrix Γ utilized in the SDP optimization, the closer the optimization results align with behaviors belonging within the quantum set Q. However, practical limitations in numerical methods prevent the use of arbitrarily large matrices due to memory constraints. Therefore, it becomes imperative to confine the analysis to specific NPA levels. The choice of the NPA level depends on the specific problem under consideration, in our case Eqs. (5.12) and (5.14), providing tighter lower bounds to $H(A|x = m, E)$. In certain scenarios, hybrid levels can be defined, where all elements of the $O^{(k)}$ set are considered along with few elements from $O^{(k+1)}$.

5.1.3.3 Beyond Lower Bounds: Upper Bounding the Key Rate

While lower bounds to the key rate analyze the minimal requirements necessary to assert the security of DIQKD protocols, upper bounds present an optimistic scenario by providing the best-case estimation for the extractable key rate from a protocol [47–50]. Since the von Neumann conditional entropy $H(A|x = m, E)$ is typically challenging to evaluate directly, obtaining tight upper and lower bounds offers an alternative and relatively comprehensive characterization of DIQKD protocols.

Throughout this chapter, our primary focus is on analyzing lower bounds using the techniques presented in Sect. 5.1.3.1. However, for the sake of assessing the computed bounds' accuracy, we occasionally compare our results with certain upper bound methods from the literature. Of notable interest is the methodology proposed by M. Farkas and colleagues in Ref. [50]. As numerically studied in Ref. [51] by K. Lukanowski and collaborators, the upper bounds derived through this approach align closely with the lower bounds computed using Eq. (5.14).

More specifically, Ref. [50] proposes an upper bound to the key rate using the so-called *Convex-Combination* (CC) attack (see Fig. 5.5). This attack involves Eve sending a set of local correlations ($\boldsymbol{P}^{\mathcal{L}}_{AB}$) to Alice and Bob with a probability denoted as $q^{\mathcal{L}}$, while with probability $q^{\mathcal{NL}}$, Eve sends a set of nonlocal correlations ($\boldsymbol{P}^{\mathcal{NL}}_{AB}$) to them. Thus, the aim of this attack is to optimize the probability of sending $\boldsymbol{P}^{\mathcal{L}}_{AB}$, enhancing Eve's ability to accurately predict Alice's outcomes. However, the combination of these two set of correlations must replicate the statistics observed by Alice and Bob, represented in Fig. 5.5 as $\boldsymbol{P}^{\text{obs}}_{AB}$. Therefore, the CC attack can be formulated as the following linear optimization problem [51]

$$\begin{aligned} \mathbf{q}_{CC}[\boldsymbol{P}^{\mathcal{NL}}_{AB}, \boldsymbol{P}^{\text{obs}}_{AB}] = \underset{\mathbf{q}}{\text{argmax}} \quad & \sum_i q_i^{\mathcal{L}} \\ \text{subject to} \quad & \mathbf{q}^{\mathcal{L}} \cdot \boldsymbol{P}^{\mathcal{L}}_{AB} + \mathbf{q}^{\mathcal{NL}} \cdot \boldsymbol{P}^{\mathcal{NL}}_{AB} = \boldsymbol{P}^{\text{obs}}_{AB}, \\ & \sum_i q_i^{\mathcal{L}} + \sum_i q_i^{\mathcal{NL}} = 1, \\ & \forall_{i,j} : 0 \leq q_i^{\mathcal{L}}, q_i^{\mathcal{NL}} \leq 1, \end{aligned} \tag{5.22}$$

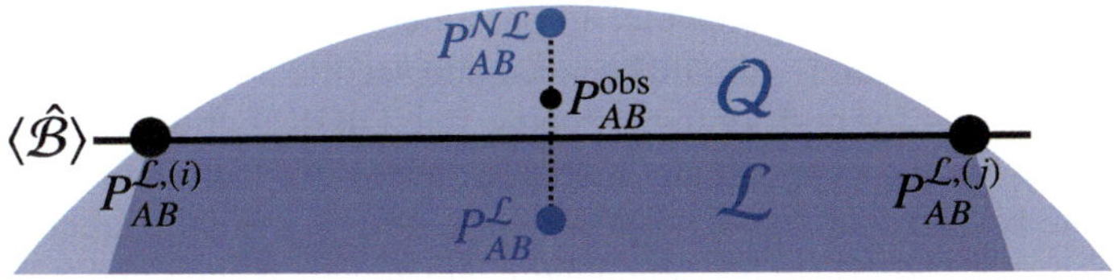

Fig. 5.5 Pictorial representation of the Convex-Combination attack. In the CC attack, Eve randomly transmits either a behavior $\boldsymbol{P}^{\mathcal{L}}_{AB}$ from the set of local correlations $\mathcal{L}$ or $\boldsymbol{P}^{\mathcal{NL}}_{AB}$ from the set of nonlocal correlations $\mathcal{NL}$ to Alice and Bob. This results in a convex combination involving local and non-local correlations, tailored to match the observed statistics by Alice and Bob, denoted here as $\boldsymbol{P}^{\text{obs}}_{AB}$

such that when $\sum_i q_i^{\mathcal{L}} = 1$ ($\sum_i q_i^{\mathcal{NL}} = 0$), Eve sends local correlations and therefore has complete knowledge about the key, while in the opposite scenario Eve is completely uncertain about it.

5.2 DIQKD Using Quantum-Dot Based Single-Photon Sources

In the preceding subsection, we have seen that the successful performance of a DIQKD protocol necessitates the shared correlations between Alice and Bob to fall within the nonlocal set, that is, to violate a Bell inequality. However, in order to guarantee in practice that these correlations cannot be reproduced by a hidden-variable description, the devices used by both parties must meet a series of requirements which, in case they do not, introduce what are usually referred to as *loopholes* [52, 53].

Two primary loopholes are typically identified in the literature within the realm of DIQKD. The first, known as the *locality loophole*, involves a breach on locality regarding the measurements performed by Alice and Bob. Specifically, this implies the possibility of information about Alice's input reaching Bob's station before the performance of measurement y. In QKD contexts, where Alice's and Bob's stations are often located far apart, the use of photonic-based implementations becomes advantageous in addressing this loophole, owing to the speed of light's upper limit on information transmission. The milestone study in Ref. [54] by A. Aspect and collaborators, effectively closed this loophole by physically separating Alice and Bob by a time interval t, equivalent to the time taken by light to travel from the central station to their respective stations. By adjusting their measurement settings within a time frame $t' < t$—while the state sent by the central stations was in transit to both Alice and Bob—they ensured the closure of the locality loophole.

The second loophole is the commonly known as *detection loophole*, stemming from the natural imperfection of physical devices, which in some cases might fail at detection when measurements are performed. Thus, it is possible to think of a local-hidden variable model determining whether an incoming signal is detected or not [55]. Consequently, in order to claim security of the specific protocol, and therefore close the detection loophole, all possible outcomes—including those that lead to a failure in detection—should be taken into account in the security proof [25].

The requirements for performing loophole-free DIQKD protocols impose stringent constraints on the experimental resources necessary for establishing a secure cryptographic key. Recently, two proposals have emerged in this direction utilizing heralding schemes [56, 57], where entanglement between Alice's and Bob's stations is achieved through heralding measurements [58, 59] (see Sect. 2.3.3.4). In these experimental proposals, light was employed for transmission, allowing for countering the effect of transmission losses through the use of heralding measurements, while material systems were utilized in the local stations to enhance detection efficiency.

However, despite these advantages, the use of material measurement devices typically demands advanced setups and suffers from low repetition rates, leading to keys of reduced length due to the time required to run a protocol round. For instance, in Ref. [56], D. P. Nadlinger and collaborators achieved a positive key rate for distances between Alice and Bob of 2 m. In contrast, W. Zhang and coworkers showed in Ref. [57] the presence of Bell inequality violations between parties at 400 m, that were theoretically large enough to establish a secure key. Nonetheless, the number of runs in their experiment was insufficient to achieve this goal. In this context, employing photonic implementations utilizing single-photon sources may potentially alleviate the challenges of long runtimes due to their high-repetition rates [60–62]. Furthermore, integrating these with heralding-based schemes could effectively address as well the requirements for long-distance transmission [63, 64].

In this section based on our work in Ref. [12], we explore a purely photonic scheme utilizing single-photon sources, expanding upon the initial proposal by J. Kołodyński and colleagues in Ref. [65]. This idea was further developed by E. M. González-Ruíz and coworkers in Ref. [66]. In this subsequent work, the authors introduced a detailed model of quantum-dot single-photon sources and examined the scheme's resilience to various imperfections in both the sources and measurement devices concerning the violation of Bell inequalities. Consequently, here we extend this proposal to analyze its applicability to DIQKD protocols.

5.2.1 The Central Heralding Station Scheme

According to the proposal of J. Kołodyński and coworkers in Ref. [65], pictorially presented in Fig. 5.6, Alice and Bob generate two single-photons each, with both photons having orthogonal polarizations, by means of a single-photon source (SPS) and an optical delay line [66]. One of these photons is ideally kept by each of the parties, while the other is sent to a central heralding station (CHS) that performs an heralding measurement. As a result of this measurement, two outcomes can be obtained: a positive one indicating that the photons kept by Alice and Bob are entangled, and a negative one indicating the opposite. As we demonstrate by the end of this subsection, the positive outcomes from the CHS herald the generation of a tunable pure entangled state, which does not depend on the transmission noise between the CHS and the measurement stations.

Here, we offer a practically oriented description of the scheme, concentrating solely on the final states acquired after positive heralding. For an exhaustive analysis encompassing all potential outcomes derived from the CHS, readers are directed to E. M. González-Ruíz's Master thesis in Ref. [67]. Hence, we commence by examining the initial joint state of Alice and Bob, which, under ideal conditions, is given by

$$|\psi\rangle = \hat{a}^{\dagger}_{s,H}\hat{a}^{\dagger}_{s,V}\hat{b}^{\dagger}_{s,H}\hat{b}^{\dagger}_{s,V}|\bar{0}\rangle, \tag{5.23}$$

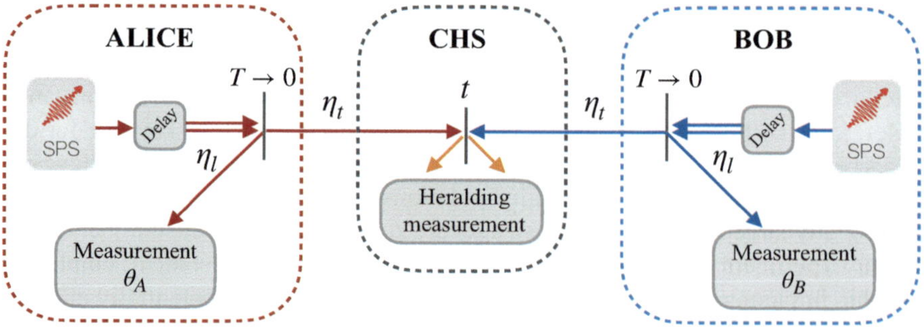

Fig. 5.6 Schematic representation of the central heralding station photonic setup. Each of Alice and Bob's setups includes a single-photon source (SPS) that, via an optical delay circuit, generates two simultaneous photons with orthogonal polarizations. One photon from each lab is directed through a low-transmittance beam splitter ($T \to 0$) to a CHS, where a Bell state measurement is performed. In between, the photons travel through a medium of transmission efficiency η_t. The entanglement of the remaining photons' polarization states between Alice and Bob, post-CHS measurement, is adjustable based on the beam splitter's transmittance t at the CHS. The overall local efficiency at Alice and Bob's stations is represented by η_l. Upon receiving positive feedback from the CHS, Alice and Bob implement some measurement on their respective share of the state

where $\hat{a}^\dagger_{s,H}$ ($\hat{a}^\dagger_{s,V}$) denotes the creation operator that characterizes the single-photon source, exciting a field mode linearly polarized along the H (V) direction on Alice's side. Correspondingly, $\hat{b}^\dagger_{s,H}$ ($\hat{b}^\dagger_{s,V}$) represents the analogous operation on Bob's side. Therefore, $|\bar{0}\rangle$ denotes the vacuum state across all these modes.

For these setups, it is particularly convenient to work in the Heisenberg picture and examine the evolution of the creation operators across the different optical elements. As depicted in Fig. 5.6, both Alice's and Bob's photons initially encounter a beam splitter of transmission efficiency T. If we denote the creation operators representing the reflected modes as $\hat{a}^\dagger_{H/V}$, while those characterizing the transmitted modes as $\hat{a}^\dagger_{t,H/V}$, the relationship between these and $\hat{a}^\dagger_{s,H/V}$ can be expressed as

$$\hat{a}^\dagger_{s,H/V} = i\sqrt{1-T}\hat{a}^\dagger_{H/V} + \sqrt{T}\hat{a}_{t,H/V}, \tag{5.24}$$

where an analogous expression is found for Bob's modes. Subsequent to this beam splitter, the reflected modes are directed to Alice's and Bob's measurement devices, while the transmitted modes are directed towards the CHS. It is worth noting that, with the aim of having just two photons, each coming from different stations, directed to the CHS, it is required for the aforementioned beam splitter to have a low transmittance.

Before being transmitted through a given channel to the CHS, Alice's and Bob's photons first traverse a half-wave plate (not shown in Fig. 5.6), which affects the transmission of the different polarization modes. Specifically, denoting the transmitted modes after the half-wave plate as $\hat{\tilde{a}}^\dagger_{t,H/V}$, their relationship with $\hat{a}_{t,H/V}$ reads as follows

$$\hat{a}^{\dagger}_{t,H} = \frac{1}{\sqrt{2}}\left(\hat{\tilde{a}}^{\dagger}_{t,H} + \hat{\tilde{a}}^{\dagger}_{t,V}\right), \tag{5.25}$$

$$\hat{a}^{\dagger}_{t,V} = \frac{1}{\sqrt{2}}\left(\hat{\tilde{a}}^{\dagger}_{t,H} - \hat{\tilde{a}}^{\dagger}_{t,V}\right). \tag{5.26}$$

After passing through a medium with transmission efficiency η_t, where a photon is absorbed by the medium with probability $(1 - \eta_t)$, both the transmitted photons from Alice and Bob encounter a beam splitter of transmittance t which ideally renders them indistinguishable. Denoting the modes after this beam splitter as $\hat{d}_{H/V}$ and $\hat{\tilde{d}}_{H/V}$, their relationship with the input modes $\hat{\tilde{a}}^{\dagger}_{t,H/V}$ and $\hat{\tilde{b}}^{\dagger}_{t,H/V}$ can be expressed as follows

$$\hat{\tilde{a}}^{\dagger}_{t,H/V} = i\sqrt{1-t}\,\hat{d}^{\dagger}_{H/V} + \sqrt{t}\,\hat{\tilde{d}}^{\dagger}_{H/V}, \tag{5.27}$$

$$\hat{\tilde{b}}^{\dagger}_{t,H/V} = \sqrt{t}\,\hat{d}^{\dagger}_{H/V} + i\sqrt{1-t}\,\hat{\tilde{d}}^{\dagger}_{H/V}, \tag{5.28}$$

such that, by combining all these transformations, we can rewrite the initial $\hat{a}^{\dagger}_{s,H/V}$ and $\hat{b}^{\dagger}_{s,H/V}$ operators as

$$\hat{a}^{\dagger}_{s,H} = i\sqrt{1-T}\hat{a}^{\dagger}_{H} + \sqrt{\frac{\eta_t T}{2}}\left[i\sqrt{1-t}\,\hat{d}^{\dagger}_{H} + \sqrt{t}\,\hat{\tilde{d}}^{\dagger}_{H} + i\sqrt{1-t}\,\hat{d}^{\dagger}_{V} + \sqrt{t}\,\hat{\tilde{d}}^{\dagger}_{V}\right], \tag{5.29}$$
$$+ \sqrt{1-\eta_t}\hat{a}^{\dagger}_{\text{env},H}$$

$$\hat{a}^{\dagger}_{s,V} = i\sqrt{1-T}\hat{a}^{\dagger}_{V} + \sqrt{\frac{\eta_t T}{2}}\left[i\sqrt{1-t}\,\hat{d}^{\dagger}_{H} + \sqrt{t}\,\hat{\tilde{d}}^{\dagger}_{H} - i\sqrt{1-t}\,\hat{d}^{\dagger}_{V} - \sqrt{t}\,\hat{\tilde{d}}^{\dagger}_{V}\right] \tag{5.30}$$
$$+ \sqrt{1-\eta_t}\hat{a}^{\dagger}_{\text{env},V},$$

$$\hat{b}^{\dagger}_{s,H} = i\sqrt{1-T}\hat{b}^{\dagger}_{H} + \sqrt{\frac{\eta_t T}{2}}\left[\sqrt{t}\,\hat{d}^{\dagger}_{H} + i\sqrt{1-t}\,\hat{\tilde{d}}^{\dagger}_{H} + \sqrt{t}\,\hat{d}^{\dagger}_{V} + i\sqrt{1-t}\,\hat{\tilde{d}}^{\dagger}_{V}\right], \tag{5.31}$$
$$+ \sqrt{1-\eta_t}\hat{b}^{\dagger}_{\text{env},H}$$

$$\hat{b}^{\dagger}_{s,V} = i\sqrt{1-T}\hat{b}^{\dagger}_{V} + \sqrt{\frac{\eta_t T}{2}}\left[\sqrt{t}\,\hat{d}^{\dagger}_{H} + i\sqrt{1-t}\,\hat{\tilde{d}}^{\dagger}_{H} - \sqrt{t}\,\hat{d}^{\dagger}_{V} - i\sqrt{1-t}\,\hat{\tilde{d}}^{\dagger}_{V}\right] \tag{5.32}$$
$$+ \sqrt{1-\eta_t}\hat{b}^{\dagger}_{\text{env},V},$$

where we have included the probability amplitude of the corresponding photon getting absorbed by the media through the $\hat{a}^{\dagger}_{\text{env},H/V}$ and $\hat{b}^{\dagger}_{\text{env},H/V}$ operators.

Thus, introducing these terms into the initial state presented in Eq. (5.23), we describe all possible paths that the two photons generated via the SPSs and the optical delay lines can take. However, we are interested in cases where only one of the photons generated by Alice and Bob respectively arrives at the central beam splitter with transmittance t. After passing through this optical element, which ideally renders them indistinguishable, they encounter two polarized beam splitters (PBS) as depicted in Fig. 5.7. These PBSs direct the photons to one of four possible detectors based on their polarization. Consequently, the unnormalized state representing this scenario is given by

$$\begin{aligned}|\psi\rangle = i\sqrt{\frac{\eta_t(1-T)T}{2}}&\Big[\hat{a}_H^\dagger\Big(i\sqrt{1-t}\,(\hat{d}_H^\dagger - \hat{d}_V^\dagger) + \sqrt{t}\,(\hat{\tilde{d}}_H^\dagger - \hat{\tilde{d}}_V^\dagger)\Big)\\ &+ \hat{a}_V^\dagger\Big(i\sqrt{1-t}\,(\hat{d}_H^\dagger + \hat{d}_V^\dagger) + \sqrt{t}\,(\hat{\tilde{d}}_H^\dagger + \hat{\tilde{d}}_V^\dagger)\Big)\Big]\\ &\times\Big[\hat{b}_H^\dagger\Big(\sqrt{t}\,(\hat{d}_H^\dagger - \hat{d}_V^\dagger) + i\sqrt{1-t}\,(\hat{\tilde{d}}_H^\dagger - \hat{\tilde{d}}_V^\dagger)\Big)\\ &+ \hat{a}_V^\dagger\Big(\sqrt{t}\,(\hat{d}_H^\dagger + \hat{d}_V^\dagger) + i\sqrt{1-t}\,(\hat{\tilde{d}}_H^\dagger + \hat{\tilde{d}}_V^\dagger)\Big)\Big],\end{aligned} \tag{5.33}$$

denoting all possible combinations of detections in the D_1, D_2, D_3 and D_4 detectors outlined in Fig. 5.7a. Among these combinations, we are particularly interested in those resulting in simultaneous clicks either on detectors D_1 and D_4, or on D_2 and D_3. These events correspond respectively to keeping those terms proportional to $\hat{d}_H^\dagger\hat{\tilde{d}}_V^\dagger$, or to $\hat{d}_V^\dagger\hat{\tilde{d}}_H^\dagger$ in the state above. These scenarios, referred to as *positive heralding* from the CHS station, lead to the following state for the photons kept by Alice and Bob

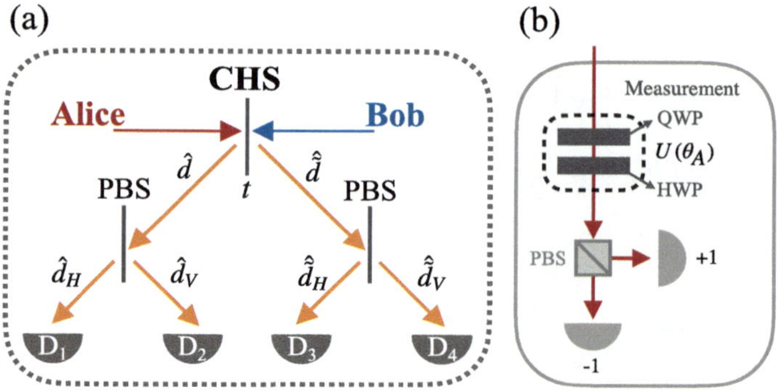

Fig. 5.7 Schematic representation of the central heralding station and the measurement devices. In **a** we illustrate the optical components for the heralding measurement within the CHS. The two PBSs and the four detectors, labeled as D_i where $i = 1, 2, 3, 4$, enable the performance of a Bell state measurement in the polarization basis using incoming photons from Alice's and Bob's stations after passing through the beam splitter of transmittance t. In **b**, the measurement setup at Alice's station is depicted. The HWP and QWP determine the measurement basis by properly setting the relative angle θ_A. Photons with orthogonal polarizations are directed by a PBS to two detectors, registering outcomes as ± 1, respectively

$$|\psi_{1,4}\rangle \propto \Big[(1-2t)\big(\hat{a}_H^\dagger \hat{b}_H^\dagger - \hat{a}_V^\dagger \hat{b}_V^\dagger\big) + \big(\hat{a}_V^\dagger \hat{b}_H^\dagger - \hat{a}_H^\dagger \hat{b}_V^\dagger\big)\Big]|\bar{0}\rangle, \tag{5.34}$$

$$|\psi_{2,3}\rangle \propto \Big[(1-2t)\big(\hat{a}_H^\dagger \hat{b}_H^\dagger - \hat{a}_V^\dagger \hat{b}_V^\dagger\big) + \big(\hat{a}_H^\dagger \hat{b}_V^\dagger - \hat{a}_V^\dagger \hat{b}_H^\dagger\big)\Big]|\bar{0}\rangle, \tag{5.35}$$

which for $0 < t < 1$, these are entangled states whose entanglement features can be adjusted by controlling the transmittance t of the beam splitter located at the CHS. Besides this property, it is worth remarking that the amount of entanglement in the generated state remains independent of the transmission efficiency η_t, which solely determines the success probability of a heralding event. In practical terms, the amount of CHSH violation obtained through this scheme does not depend on the distance between Alice and Bob, although this determines the probability with which they observe a positive heralding event.

5.2.2 *Imperfections on the Scheme When Using Quantum Dots as Single-Photon Sources*

In the preceding subsection, we detailed the central heralding scheme outlined in Ref. [65] under ideal conditions. However, in reality, various sources of noise can influence the correlations established by Alice and Bob. For instance, irrespective of the specific sources utilized for generating their photons, the measurements conducted on their respective share of the quantum state within their stations inevitably suffer from losses. In the following, we denote these as *local losses* characterized by the efficiency η_l.

The presence of limited efficiency, leading to $\eta_l < 1$, may be attributed to various factors, including imperfections in the employed detector. In our scenario, we consider the measurement system illustrated in Fig. 5.7b, where the received photons undergo unitary transformations based on a tunable parameter θ, and are detected using non-photon number resolving detectors. These detectors solely register the arrival of photons without distinguishing the specific photon count. Consequently, their action can be described using the following POVM set

$$\{\hat{M}_0 = |0\rangle\langle 0|\,,\, \hat{M}_1 = \mathbb{1} - |0\rangle\langle 0|\}, \tag{5.36}$$

which either remain inactive or register the presence of radiation, respectively. Thus, in conjunction with the quarter-wave plate (QWP), the half-wave plate (HWP), and the PBS preceding the photodetectors in Fig. 5.7 (b), the measurements conducted by Alice and Bob given inputs x and y, can be respectively described by the POVM sets

$$\Big\{\hat{M}_{(i,j)|x} \equiv \hat{U}^\dagger(\theta_x)\hat{M}_i^{(H)} \otimes \hat{M}_j^{(V)}\hat{U}(\theta_x)\Big\}_{i,j=0,1} \quad \text{with } x \in \{1,2\}, \tag{5.37}$$

$$\Big\{\hat{M}_{(m,n)|y} \equiv \hat{U}^\dagger(\theta_y)\hat{M}_m^{(H)} \otimes \hat{M}_n^{(V)}\hat{U}(\theta_y)\Big\}_{m,n=0,1} \quad \text{with } y \in \{1,2,3\}, \tag{5.38}$$

where $\hat{U}(\theta_x)$ ($\hat{U}(\theta_y)$) denotes the unitary operation resulting from the combined action of the QWP, the HWP and the PBS. The superscripts in $\hat{M}_a^{(i)}$ represent each of the detectors within a station, in this case Alice's one (see Fig. 5.7b). Depending on which detector registers a click, Alice and Bob obtain outcomes $a, b = \pm 1$. However, in cases where no detector registers a click, they obtain a third outcome represented as $a, b = \emptyset$. Furthermore, there might be instances where both detectors provide a simultaneous click.

These measurements, combined with the quantum state $\hat{\rho}$ shared by Alice and Bob conditioned to positive heralding from the CHS, establish the set of correlations $\{P(a, b|x, y)\}$. However, in practical scenarios, the states $\hat{\rho}$ differ from the ideal descriptions detailed in the preceding subsection, that is, Eqs. (5.34) and (5.35). For instance, we can encounter scenarios where both of Alice's photons are reflected at the $T \to 0$ beam splitter, while both of Bob's photons are transmitted, resulting in positive heralding. Consequently, to surpass the idealized scenario and circumvent the detection loophole, we consider all possible statistics $P_{k,l,m,n}$ as depicted in Table 5.1. Here, the indices represent the number of photons detected at the CHS (k), at Alice's and Bob's stations (l and m respectively), and that remain undetected (n). Moreover, to simplify the analysis to two outcomes, we assume that non-detection events for Alice or Bob are assigned a +1 outcome, following the approach in Refs. [25, 66]. It is worth noting that these non-detection events also depend on the probability with

Table 5.1 Probability contributions to the total probability of detection $P(a, b|x, y)$. In this table, the outcome $a, b = \emptyset$ indicates the absence of detected photons, while $a = \pm 1$ ($b = \pm 1$) denotes a classical binary output. The four sub-indices represent the number of photons detected at the CHS, Alice's and Bob's stations, and those that remained undetected, respectively. The probabilities are categorized between those involving four photons (above the middle line) and five photons (below the middle line), where the fifth photon arises due to imperfect purity of the single-photon source ($g^{(2)} \neq 0$). When more than one photon is detected at the local stations, we specify whether they were detected at the same (SD) or different (DD) detectors. As we do not consider number-resolving detectors, in the former case, Alice and Bob identify these events as correct. However, if photons click in different detectors, they recognize that an error occurred and hence assign the outcome as Ø

$P(a = \pm 1, b = \pm 1\|x, y)$	$P(a = \pm 1, b = \emptyset\|x, y)$	$P(a = \emptyset, b = \pm 1\|x, y)$	$P(a = \emptyset, b = \emptyset\|x, y)$
P_{2110}	$P_{2200,SD}$	$P_{2020,SD}$	P_{2002}
	P_{2101}	P_{2011}	$P_{2200,DD}$
			$P_{2020,DD}$
P_{2111}	P_{2102}	P_{2012}	P_{2003}
$P_{2210,SD}$	$P_{2201,SD}$	$P_{2021,SD}$	$P_{2201,DD}$
$P_{2120,SD}$	$P_{2120,DD}$	$P_{2210,DD}$	$P_{2021,DD}$
	$P_{2300,SD}$	$P_{2030,SD}$	$P_{2300,DD}$
			$P_{2030,DD}$

which Alice's and Bob's detectors fail at measuring radiation, which is contingent upon the local efficiency η_l. This probability is thus represented by $(1 - \eta_l)$.

Besides the local efficiency, the imperfections of the specific single-photon source under consideration also pose an important source of noise. Two critical factors come into play here: the indistinguishability between the photons generated by the source and the purity of the source itself. Concerning the former, this can be quantified using the Hong-Ou-Mandel visibility V_{HOM}, where a higher visibility indicates a greater degree of indistinguishability between the two photons. The precise nature of the visibility is contingent upon the decoherence mechanisms affecting the employed source. In the case of our quantum dot single-photon sources, our focus centers on dephasing effects caused by photon interaction with the quantum dot, representing the primary source of noise for the relevant timescales [68]. Consequently, if we denote the dephasing rate as γ_d, according to Ref. [66] the visibility can be expressed as

$$V_{\text{HOM}} = \frac{\gamma}{\gamma + 2\gamma_d}, \tag{5.39}$$

where γ represents the spontaneous decay rate of the quantum dot.

Additionally, achieving indistinguishability between the photons generated by both Alice and Bob stands as another crucial requirement in the proposed method. While slow drifts might cause their respective emitters to intermittently fall in and out of resonance over time, we assume that all photons are equally indistinguishable. We further consider that the limited visibility solely arises from a fast dephasing rate γ_d, an assumption requiring two remote quantum dots to have a similar visibility as that of a single quantum dot. This has been indeed recently demonstrated in Ref. [69], albeit even if this were not the case, in Ref. [66] the protocol's performance demonstrates reduced dependence on the cross visibility between Alice's and Bob's sources compared to their intrinsic visibility.[3]

Regarding the second limiting factor, which involves the purity of the utilized source, the state produced by the quantum dot single-photon source is generally described by the following density matrix [70].

$$\hat{\rho}_s = \sum_{k=0} P_k \, |k\rangle\langle k| \,, \tag{5.40}$$

where P_k represents the probability of generating k photons. In this context, the purity of the source can be characterized by means of the time-zero second-order autocorrelation function, which can be expressed as

$$g^{(2)} \equiv g^2(0) \approx \frac{2P_2}{(P_1 + 2P_2)^2}, \tag{5.41}$$

[3] This is pictorially shown in Fig. 4b of Ref. [66].

where we have assumed the probability of generating more than two photons to be negligible. Furthermore, all the analysis presented in the following subsections has been done by restricting to first order terms with respect to P_2, that is $P_2^2 \approx 0$. This assumption is based on the consideration that the probability of the two sources simultaneously producing two photons is small. Hence, in the most extreme situation, the maximum number of photons we may encounter in the setup under consideration is five. These probabilities are reflected in the distributions presented in the lower section of Table 5.1.

5.2.3 *DIQKD in the Asymptotic Limit*

The imperfections outlined in the previous subsection significantly impact the generation of a secure key in a DIQKD scenario. In this subsection, we examine how these influence the lower bounds on the key rate by employing the various methods described in Sect. 5.1.3.1. Specifically, we compare the analytical lower bounds provided by Eqs. (5.10) and (5.11) with the numerical bound derived in Eq. (5.14).

5.2.3.1 A Numerical Implementation for Optimizing the Convergent Hierarchy for the Von Neumann Entropy

Following the operations conducted by the CHS and the local measurements performed at Alice's and Bob's station, as depicted in Fig. 5.7, the heralding scheme illustrated in Fig. 5.6 relies on the parameters $\boldsymbol{\theta} := (t, \theta_{x=1}, \theta_{x=2}, \theta_{y=1}, \theta_{y=2}, \theta_{y=3})$. Given some fixed imperfections of the employed sources and of the measurement devices, these parameters can be optimized with aims of maximizing the achievable key rate. Thus, the optimization problem addressed here can be generally formulated as

$$r_{\text{lb}} = \max_{\boldsymbol{\theta}} \left[H(A|x = m, E) - H(A|B, x = m, y = m + 1) \right], \tag{5.42}$$

where, here, $H(A|x = m, E)$ is lower bounded using either the analytical expressions in Eqs. (5.10) and (5.11), or the numerical one in Eq. (5.14).

On one hand, when using Eqs. (5.10) and (5.11), the optimization can be straightforwardly conducted as these bounds depend on the CHSH score $S(\boldsymbol{\theta})$ which directly depends on the optimization variables. Employing *local optimization* methods[4], in our case the Nelder-Mead algorithm [71], allows for the direct maximization of Eq. (5.42). It is worth noting that when utilizing Eq. (5.11), an extra parameter

---

[4] Local optimization methods begin with a local guess, which is presumed to be a solution in proximity to the optimal solution. The algorithm then iterates across the search space, making incremental changes to the local guess to enhance the objective function. This iterative process continues until the algorithm converges to a local optimum, which signifies a solution superior to any of its neighboring solutions.

q must be introduced to account for the probability of Alice performing a bitflip operation on the obtained outcomes.

On the other hand, the computational process for Eq. (5.42) deviates from the straightforward optimization when Eq. (5.14) is employed to lower bound the key rate. In this scenario, the dependence of $H(A|x = m, E)$ on the set of parameters $\boldsymbol{\theta}$ arises through the constraints of the SDP primal problem in Eq. (5.14). Consequently, this case involves two distinct optimization processes:

- an SDP optimization, which in this case was implemented in `Python` by means of the `ncpol2sdpa` package [72] and optimized using the `Mosek` solver [43];
- a local optimization on the $\boldsymbol{\theta}$ variables, which we perform using the Nelder-Mead algorithm [71] implemented through the `minimize` function available in the `Scipy` package [73].

One of the main drawbacks of using local optimization methods for optimizing with respect to $\boldsymbol{\theta}$ is its high dependence on the initial guess, often leading to convergence on local minima. Thus, generating a diverse set of initial guesses pertinent to the problem becomes essential in order to minimize the impact of local minima. In our case, we refine our search problem by considering the physical constraints governing our parameters. Firstly, as t represents a transmittance parameter, it adheres to $t \in [0, 1]$. Secondly, as θ_x (θ_y) denotes a rotation angle of the employed linear optical elements in the measurement system, it satisfies $\theta_x \in [0, 2\pi]$ ($\theta_y \in [0, 2\pi]$). With these constraints in place, the subsequent step involves identifying relevant initial guesses that yield nonlocal probability distributions. This procedure, illustrated in Fig. 5.8, entails exploring the viability of the following linear problem using a specific behavior $\boldsymbol{P}$ derived from a random selection of $\boldsymbol{\theta}$

$$\begin{aligned} &\text{minimize} \quad 1 \\ &\text{subject to} \sum_i q_i P_i^{\mathcal{L}} = \boldsymbol{P}, \quad \sum_i q_i = 1, \quad q_i > 0 \ \forall i, \end{aligned} \tag{5.43}$$

where the set of probabilities $\{P_i^{\mathcal{L}}\}$ represents the vertices of the local polytope $\mathcal{L}$ in Fig. 5.8. Consequently, the constraint in this problem mandates that the behavior $\boldsymbol{P}$ can be expressed as a convex combination of the vertices outlining the local polytope, that is, demand $\boldsymbol{P}$ to be within $\mathcal{L}$. Failure to meet this constraint renders the minimization infeasible, indicating that $\boldsymbol{P}$ exhibits nonlocal behavior.

By iterating this method a given number N of times, we can generate, for predetermined values of η_l, V_{HOM} and $g^{(2)}$, a set of $\tilde{N}$ (with $\tilde{N} < N$) nonlocal initial guesses for the local optimization. Hence, the total optimization process can be viewed as a local optimization, employing the initial guesses $\boldsymbol{\theta}$ that render Eq. (5.43) infeasible, where Eq. (5.14) is used as the objective function. Further details regarding this methodology are elaborated in Appendix 10.1.

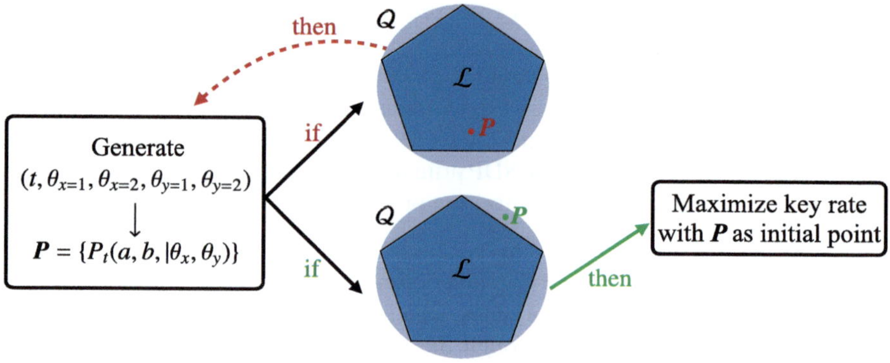

Fig. 5.8 Scheme depicting the numerical optimization approach used for the quantum dot-based central heralding station setup. Initially, a point is randomly generated, adhering to a set of conditions dictated by the physical implementation, and its associated set of probability distributions $\boldsymbol{P}$ is computed. Subsequently, we verify if the generated point falls within the local set $\mathcal{L}$. If it does, a new point is generated. However, if it lies outside the local set, we employ it to execute the SDP and maximize the key rate

5.2.3.2 Comparison Between Numerical Methods and Analytical Formulas

The numerical method outlined in Eq. (5.14) can be understood, to some extent, as a broader extension of the analytical bounds to the key rate in Eqs. (5.10) and (5.11). While the latter two methods utilized the CHSH score S to estimate Eve's knowledge about Alice's outcomes, which in certain scenarios might not be the optimal Bell inequality to use [74, 75], the numerical approach in Eq. (5.14) effectively optimizes the protocol's security with respect to all possible Bell inequalities. As a result, it is expected the convergent hierarchy to the von Neumann entropy in Eq. (5.14) to outperform the considered analytical bounds.

In Fig. 5.9, we showcase key rates derived from the aforementioned formulas, obtained under the ideal scenario of $V_{\text{HOM}} = 1.0$ and $g^{(2)} = 0$. This reflects a state generated by the quantum dot single-photon source that is pure and features indistinguishable photons. Throughout these calculations, we have set the transmittance of the beam splitters located at Alice's and Bob's stations to $T = 10^{-3}$, a fixed value unless specified otherwise. This choice makes the probability these beam splitters transmitting simultaneously two photons from each station negligible, enabling an exploration of the minimal experimental requisites for obtaining a secret key in the asymptotic regime.

In the aforementioned plot, the light blue curves correspond to the bounds obtained for Eqs. (5.10) (dashed curve) and (5.11) (solid curve), respectively. Conversely, the darker curves correspond to those obtained through Eq. (5.14) with (solid curve) and without (dashed curve) noisy preprocessing. It is worth noting that the latter can be obtained by substituting Alice's POVM set from $\{\hat{M}_{0|m}, \hat{M}_{1|m}\}$ to $\{(1-q)\hat{M}_{0|m} +$

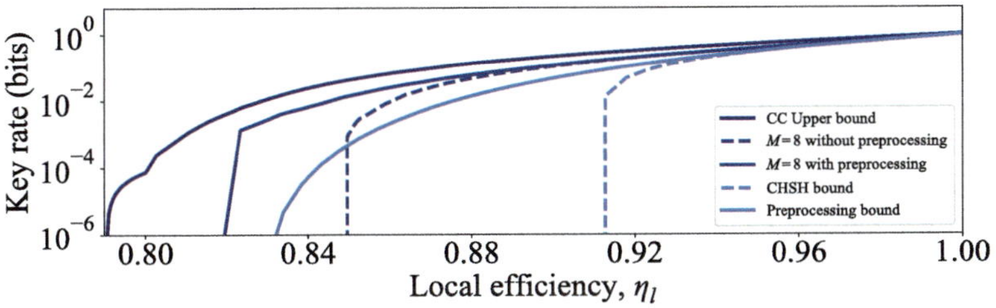

Fig. 5.9 Dependence of the key rate on the detection efficiency for the ideal scenario. In this scenario, parameters are set as follows: $T = 10^{-3}$, $\eta_l = 1.0$, $V_{\text{HOM}} = 1.0$, and $g^{(2)} = 0$. The medium-blue curves represent the key rate obtained by optimizing Eq. (5.14) with $M = 8$, being M the number of Gauss-Radau quadrature nodes. The solid (dashed) curve illustrates the rate obtained by implementing (not implementing) preprocessing strategies. With the lightest blue color, the results obtained using analytical formulas from Eqs. (5.10) (dashed curve) and (5.11) (solid curve) are displayed. Generally, employing optimization techniques yields better bounds on the key rate compared to the corresponding analytic bounds. These findings align with those reported in Ref. [40]. Finally, the darkest solid curve depicts an upper bound on the key rate obtained by employing the so-called CC attack, extracted from Ref. [65]

$q\hat{M}_{1|m}, q\hat{M}_{0|m} + (1-q)\hat{M}_{1|m}\}$ to account for the noisy preprocessing operation. Additionally, the darkest curve represents the upper bound obtained through the CC attack outlined in Sect. 5.1.3.3, sourced from the data[5] presented in Ref. [51].

The figure highlights how the improved bound for the relative entropy between Alice and Eve leads to an enhanced key rate value and lowers the threshold where this quantity turns positive, that is, the critical η_l. Specifically, without noisy preprocessing operations, we note that the key rate starts being positive at $\eta_l > 0.909$ for the analytical bound in Eq. (5.10), while for Eq. (5.14), it begins at $\eta_l > 0.848$. However, as η_l increases, the discrepancy between these results diminishes, converging to $r \approx 1$ bit at $\eta_l = 1$. The use of noisy preprocessing strategies reduces this critical value for η_l even further, resulting in positive key rates at $\eta_l > 0.832$ with Eq. (5.11), while for $\eta_l > 0.818$ with Eq. (5.14) when combined with noisy preprocessing. Notably, this value is quite close to those obtained via the CC attack, which leads to a critical local efficiency of $\eta_l > 0.791$.

5.2.3.3 Effects of the Single-Photon Source Imperfections on the Key Rate

While the bounds in the previous section were determined under an ideal scenario, imperfections in the employed sources inherently compromise the security of the established key. This stems from the fact that Alice and Bob cannot fully trust the devices they use, making it impossible to discern whether these imperfections originate from the source or are introduced by Eve to gain knowledge about the established

[5] Our gratitude to K. Łukanowski and J. Kołodyński for providing this data.

key. Notably, in Ref. [66] it was demonstrated that both $V_{\text{HOM}} < 1$ and $g^{(2)} > 0$ affect the quality of the generated state and consequently impact the attained CHSH score. Thus, it is then expected these imperfections to critically affect the key rate. In this direction, state-of-the-art values for these parameters in quantum dot single-photon sources are approximately $V_{\text{HOM}} \approx 0.96$ and $g^{(2)} \leq 0.05$ [61, 62]. Hence, these are the reference values we utilize in the subsequent analysis.

In Fig. 5.10, we present the key rate values computed using Eq. (5.14), both with (solid curves) and without (dashed curves) noisy preprocessing. Panel (a) illustrates the key rate trends with respect to η_l when setting $g^{(2)} = 0.00$, while varying the Hong-Ou-Mandel visibility from $V_{\text{HOM}} = 0.95$ (lightest curves) to $V_{\text{HOM}} = 1.00$ (darkest curves). Conversely, in panel (b), we fix $V_{\text{HOM}} = 1.00$ and alter the time zero second-order autocorrelation function from $g^{(2)} = 0.05$ (lightest curve) to $g^{(2)} = 0.00$ (darkest curve).

In both scenarios, incorporating noisy preprocessing improves the asymptotic key rate values. To elaborate, in Fig. 5.10a, considering both cases with and without noisy preprocessing, we respectively achieve $\eta_l > 0.918$ and $\eta_l > 0.938$ for $V_{\text{HOM}} = 0.975$, and $\eta_l > 0.949$ and $\eta_l > 0.958$ for $V_{\text{HOM}} = 0.95$. When examining panel (b), for the same order of conditions, we obtain $\eta_l > 0.872$ and $\eta_l > 0.894$ for $g^{(2)} = 0.01$, whereas $\eta_l > 0.935$ and $\eta_l > 0.955$ for $g^{(2)} = 0.05$. Hence, these observations indicate that within an asymptotic regime where Alice and Bob have access to an unlimited number of protocol rounds, state-of-the-art values indeed pertain the attainment of positive key rates. However the primary limiting factor is the local efficiency within Alice's and Bob's stations.

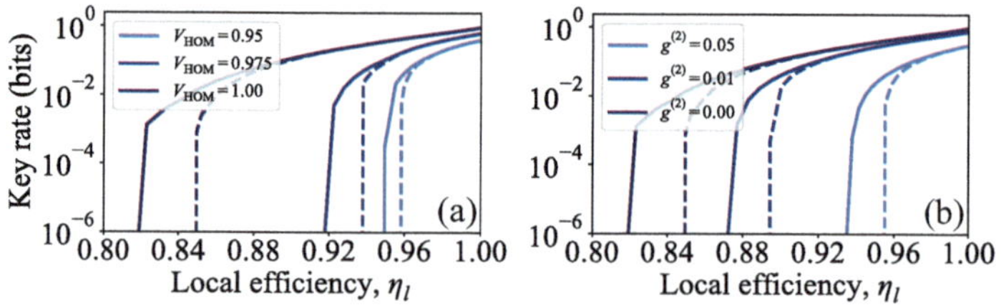

Fig. 5.10 Dependence of the asymptotic key rate on the detection efficiency for imperfect scenarios. Panel **a** illustrates the key rate's dependency on the visibility V_{HOM}, with $g^{(2)}$ set to 0. In panel **b**, the key rate's dependence on the time-zero second-order autocorrelation function $g^{(2)}$ is displayed, with V_{HOM} set to 1. In both cases, $T = 10^{-3}$. The solid curves represent the achieved key rate when utilizing a noisy preprocessing strategy for key generation, while the dashed curves depict scenarios without the application of preprocessing. All curves are generated using Eq. (5.14) with $M = 8$

5.2.4 *DIQKD in the Finite Statistics Regime*

In practical scenarios, Alice and Bob are constrained by a finite number of protocol rounds n, as they aim to derive a secret key within a predefined time frame. This limitation naturally restricts both the attainable values of n and the accessible statistics, consequently imposing limitations on the length l of the secret key they can derive after error correction and privacy amplification. This limitation is illustrated in Fig. 5.11, where the key rate $r = l/n$ is depicted as a function of the number of rounds. Here, we observe that if the number of protocol rounds is insufficiently large, Alice and Bob are not able to distill a secret key. The number of protocol rounds accessible within a time frame is intricately linked to the specifics of the physical implementation, such as the experiment's repetition rate or the spatial separation between Alice and Bob, which significantly influences the successful heralding events at the CHS. Thus, conducting a finite statistics analysis is crucial to comprehensively assess the potential superiority of this photonic-based platform over existing ones.

5.2.4.1 The Entropy Accumulation Theorem

In the analysis of finite statistics effects, the Entropy Accumulation Theorem (EAT) [76–78] has a paramount role. This theorem provides a security proof of the protocol against the so-called *coherent* attacks, in which we allow Eve to simultaneously interact with all the states used by Alice and Bob in the protocol rounds [79].

In brief [53, 76], the EAT considers the scenario depicted in Fig. 5.12. Here, each protocol round i, is modeled in terms of a quantum channel $\mathcal{M}_i$ which receives a *memory* system $\hat{R}_{i-1}$ generated in previous protocol rounds. This memory system is entangled with that of Eve, and produces some classical data: the outputs Alice and Bob obtain from their measurements, denoted as O_i; and the inputs they have

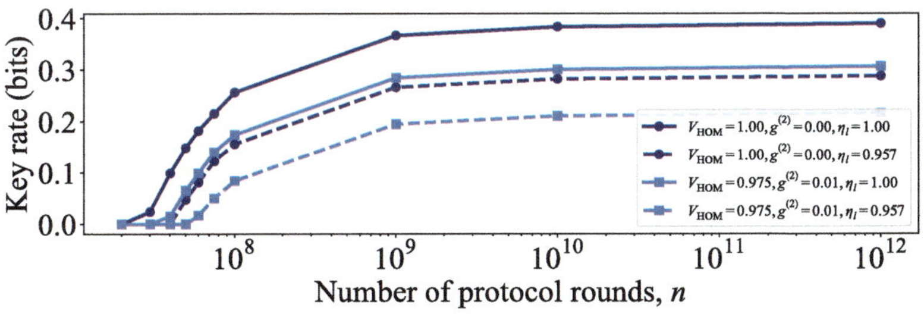

Fig. 5.11 Dependence of the key rate on the number of rounds. When considering finite statistics, Alice and Bob are limited to a finite number of rounds, denoted as n. As n increases sufficiently, the achieved rates converge toward the asymptotic case. In this plot, contrary to prior graphs, the dark curves correspond to $T = 0.0622$, while the lighter ones represent $T = 0.0106$

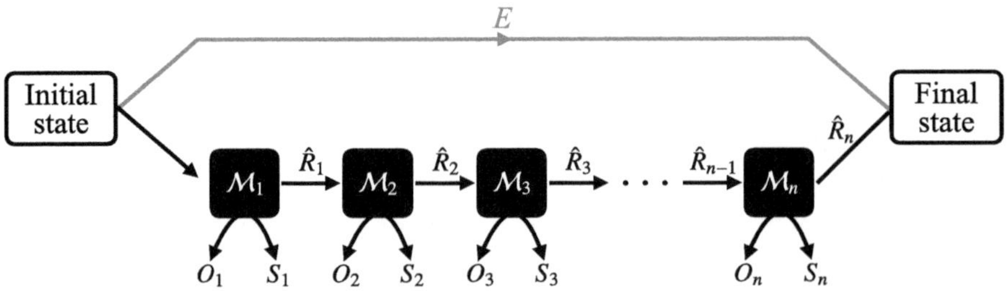

Fig. 5.12 Pictorial representation of the Entropy Accumulation Theorem setup. Each protocol round is represented by a quantum channel $\mathcal{M}_i$. This channel takes in a memory system $\hat{R}_{i-1}$ and generates another memory system $\hat{R}_i$. Additionally, it produces classical data, specifically the outputs O_i obtained by Alice and Bob in the corresponding round of the protocol, along with supplementary side information S_i

introduced into their devices along with potential supplementary side information, denoted as S_i. Additionally, another memory system $\hat{R}_i$ is generated and passed to subsequent protocol rounds. From the sets $\{O_i\}$ and $\{S_i\}$, Alice and Bob can determine the extent to which they share nonlocal correlations, and specifically from $\{O_i\}$ they can distill a secret key.

In this context, the EAT proposes a lower bound on the leaked information of O_i by considering the worst-case scenario in each protocol round. For instance, at the ith protocol round, the state $\hat{\sigma}_i$ inputted to the quantum channel $\mathcal{M}_i$, involves Eve's state and the memory state $\hat{R}_{i-1}$ obtained in the previous round, leading to outputs O_i and S_i. Thus, if we abbreviate the action of the channel $(\mathcal{M}_i \otimes \mathbb{1})(\hat{\sigma}_i)$ as simply $\mathcal{M}_i(\hat{\sigma})$, the quantity of interest in this protocol round is given by

$$\min_{\hat{\sigma}_i} \left[H(O_i|S_i, E)_{\mathcal{M}_i(\hat{\sigma}_i)} \right], \tag{5.44}$$

such that, roughly speaking (we refer the reader to Refs. [76–78] for more technical details) the conditional entropy for the entire protocol can be expressed as

$$H(O^n|S^n, E) \geq \sum_{i=1}^{n} \min_{\hat{\sigma}_i} \left[H(O_i|S_i, E)_{\mathcal{M}_i(\hat{\sigma}_i)} \right] - O(\sqrt{n}), \tag{5.45}$$

where O^n and S^n, respectively, represent all the outcomes and side information obtained after the n rounds of the protocol. Here, a minimization is performed such that we keep the worst case entropy for each protocol round.

It is worth noting that the first term in Eq. (5.45) scales with the number of protocol rounds n while the second, $O(\sqrt{n})$, representing correction factors from finite size analysis, scales with $\sqrt{n}$. Thus, this expression denotes the length of the secure key achievable by Alice and Bob after n protocol rounds. Consequently, in the asymptotic limit $n \to \infty$, the second term scales as $(\sqrt{n}/n) \to 0$, leading the finite size key to converge to the asymptotic one analyzed in Sect. 5.2.3. This convergence is observed in Fig. 5.11 where, for $n > 10^{10}$ the key rate stabilizes to a constant value. However, it is important to note that, in this figure, the values of the transmission efficiency T have been adjusted compared to those in Fig. 5.10, where $T = 10^{-3}$, for reasons explained in the following subsection.

5.2.4.2 Performance of the CHS Photonic-Based Implementation

To assess the performance of the photonic-based implementation illustrated in Fig. 5.6 within a finite size regime, we employ a single-round von Neumann entropy bound similar to the approach in Refs. [56, 80]. This enables the use of Theorem 1 in Ref. [81] to derive an upper bound on the length of the secure key achievable using this protocol. For the sake of completeness, this theorem is presented in Appendix 10.2 along with some details regarding our numerical implementation.

In photonic-based experimental implementations, the number of rounds n typically hinges on the rate at which the single photon-source generates photons, and the duration Alice and Bob allocate for the protocol. For standard quantum-dot-based sources, this rate typically falls around $\nu = 75$ MHz [60–62]. However, not every generated photon contributes to successful protocol rounds; only those resulting in positive heralding from the CHS are utilized for the key generation. This probability, denoted as the probability of heralding P_h, is contingent upon the transmission efficiency between the CHS and Alice's and Bob's station, as well as on the events depicted in Table 5.1. This efficiency, in turn, depends on the distance between the stations and can generally be expressed as

$$\eta_t = \eta_{\text{conv}} \exp[-\frac{L}{2L_0}], \tag{5.46}$$

where L represents the distance between Alice and Bob, assumed to be equidistant from the CHS, L_0 is the attenuation length of the media through which light propagates, and η_{conv} accounts for the efficiency conversion to telecommunication wavelengths. This conversion is indeed necessary because InGaAs quantum dots typically emit at 947 nm, while telecommunication wavelengths operate at either 1300 nm or 1550 nm. Consequently, we set $\eta_{\text{conv}} = 0.5$, in agreement with Refs. [82, 83]. Therefore, for optical fibers functioning at telecommunication wavelengths, a standard value for L_0 is 22 km.

Furthermore, the probability of heralding not only relies on η_t; it is also contingent upon the transmission efficiency T of the initial beam splitter at Alice's and Bob's stations. Low values of T, in the order of 10^{-3}, increase the probability of single photon propagation from each of the parties' stations to the CHS. However, this renders $P_h \propto 10^{-5}$, inducing highly isolated heralding events that trigger robust CHSH violations. Conversely, higher T values amplify the probability of successful heralding but compromise the security of the resulting key. Therefore, in our finite size analysis, we aim to optimize the T parameter to strike a balance between security and transmission distance.

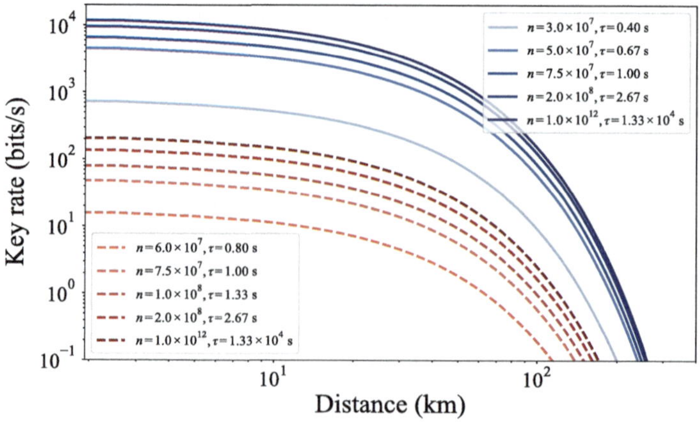

Fig. 5.13 Finite-size key rate in bits per second as a function of the distance between Alice's and Bob's stations. The illustration considers a single-photon source emitting photons at a rate of $\nu = 75$ MHz. The blue solid curves depict the ideal scenario, assuming $\eta_l = 1.0$, $V_{\mathrm{HOM}} = 1.0$, and $g^{(2)} = 0.00$. In contrast, the red dashed curves portray a more realistic scenario with $\eta_l = 0.957$, $V_{\mathrm{HOM}} = 0.975$, and $g^{(2)} = 0.01$. Each curve corresponds to a distinct number of rounds n (as indicated in the legend), where the time required for these rounds is calculated as $\tau = n/\nu$. Optimal transmissivity values for the beam splitter T were determined for each set of experimental parameters, yielding $T = 0.0622$ for the blue solid curves and $T = 0.0106$ for the red dashed curves

In Fig. 5.13 we present the outcomes of our analysis. Here, the key is presented in bits per second, that is as $r = \nu l$, with $\nu = 75$ MHz, as a function of the distance between Alice's and Bob's stations, considering single-photon sources with varying characteristics. The blue solid curves portray the ideal scenario with $\eta_l = 1.00$, $V_{\mathrm{HOM}} = 1.00$ and $g^{(2)} = 0.00$. Conversely, the red dashed curves depict a more realistic scenario where $V_{\mathrm{HOM}} = 0.975$ and $g^{(2)} = 0.01$, attainable with current technologies [61, 62]. In this last setup, the local efficiency is set to $\eta_l = 0.957$ to yield positive key rates.

Each curve in Fig. 5.13 represents a distinct number of protocol rounds used by Alice and Bob to generate the key, as specified in the legends. The legend also includes the time $\tau = n/\nu$, indicating the duration for a single-photon source emitting photons at a rate $\nu = 75$ MHz to execute n protocol rounds. For both types of curves—solid and dashed—we observe that below $n \leq 10^7$ (lightest curves) extracting a positive key considering the protocol to be 10^{-6}-sound and 10^{-3}-complete (refer to Appendix 10.2 for definitions) is not feasible. However, reaching the asymptotic regime already occurs for $n = 10^{12}$ (darkest curves). Although conducting this amount of protocol rounds requires Alice and Bob approximately 10^4 seconds, it is noteworthy that for $\tau \approx 1$ second, they can procure secure keys of around 10^2 bits in length, even with suboptimal sources and detectors, for relatively short distances between them.

As a final point, it is worth noting that achieving the results in Fig. 5.13 requires a transmissivity of the initial beam splitters on the order of 10^{-2}. Specifically, for the blue solid curves, the optimal transmissivity was $T = 0.0622$, whereas for the dashed red curves, it was $T = 0.0105$. This enables obtaining heralding probabilities $P_h \approx 10^{-2}$, facilitating the generation of secure keys generated over longer distances between both parties. Particularly, in the scenario where the devices used by Alice and Bob are imperfect (as represented by the red dashed curves), the key rate reduces to 0.1 bits/s over a range of distances $L \in [113, 170]$ km (limits obtained for $n = 6 \times 10^7$ and $n = 10^{12}$ rounds, respectively). In the ideal case, a similar key rate is observed for $L \in [200, 261]$ km. Therefore, as highlighted earlier, an increase in the heralding probability enables the establishment of secure rates at longer distances, albeit at the expense of reducing the length of the key per second. Alternatively, another approach could be to increase the repetition rate of the single-photon source, although this might require more effort from an experimental standpoint.

5.3 DIQKD Beyond Qubits

The results from the preceding section have revealed that, while DIQKD protocols are feasible, they strain our current technological capabilities. To enable device-independent communication, we must overcome these limitations or explore alternative scenarios to relax the potential requirements. With this aim, in this section based in our work in Ref. [13], we examine whether utilizing higher-dimensional systems beyond qubits, particularly focusing on qutrit systems shared between Alice and Bob, could offer advantages. Specifically, we explore upper and lower bounds on the key rate and assess how these bounds are affected by an increased number of measurement inputs and outputs, as well as variations in the dimensions of the shared quantum system.

In this context, other quantum communication scenarios have demonstrated potential advantages in utilizing higher-dimensional systems by Alice and Bob. For instance, in the realm of device-dependent quantum key distribution, the use of qutrits in the E91 protocol exhibited enhanced noise robustness compared to their qubit counterparts [84, 85]. Additionally, for measurement-device-independent QKD, qutrits displayed higher secret key rates under similar error rates compared to qubits, albeit with lower resilience to transmission noises [86]. Finally, in device-independent randomness generation, the utilization of qutrits enabled the extraction of a greater number of random bits, i.e. showing a reduced correlation with Eve, surpassing the capabilities of qubits [87].

5.3.1 Optimizing the Lower Bounds to the Key Rate

In contrast to Sect. 5.2, here Alice and Bob are allowed to perform m and $m + 1$ measurements, respectively, with $m \geq 2$. This setup allows each of them to obtain one of $d \geq 2$ outputs per measurement. To facilitate this, we consider Alice and Bob to share a $(d \times d)$-dimensional entangled state represented as

$$\hat{\rho} = V\,|\psi\rangle\langle\psi| + \frac{1-V}{d}\hat{\mathbb{1}}, \tag{5.47}$$

where $0 \leq V \leq 1$ denotes the visibility. This parameter quantifies the level of noise present in the state, with $V = 1$ indicating a pure state while $V = 0$ yielding a maximally mixed state.

Consequently, by considering parameterized POVM sets for Alice's and Bob's measurements, which we denote as $M_x^A(\boldsymbol{\theta}) := \{\hat{\Pi}(\boldsymbol{\theta})_{a|x}\}_a$ and $M_y^B(\boldsymbol{\theta}) := \{\hat{\Pi}(\boldsymbol{\theta})_{b|y}\}_b$ respectively, such that $\hat{\Pi}_{a|x}\hat{\Pi}_{a'|x} = \delta_{a,a'}$ (same for Bob's operators), our aim is to optimize the parameters $\boldsymbol{\theta}$ and the state $\hat{\rho}$ for a fixed visibility, to maximize the key rate. However, in the case of $d > 2$, where no analytical bound to $H(A|x = m, E)$ is known, a numerical analysis using either Eqs. (5.12) or (5.14) becomes essential. Nevertheless, the utilization of an optimization method akin to the one detailed in Sect. 5.2.3.1 becomes computationally challenging, particularly for $d = 3$. This challenge arises due to two primary reasons: firstly, the numerical assessment of Eq. (5.14) incurs substantial computational cost owing to the large Γ matrices required for the SDP; secondly, employing highly general POVM sets for the measurements implemented by Alice and Bob (refer to Appendix 10.3 for additional details) significantly increase the number of parameters in the optimization process, rendering the method outlined in Sect. 5.2.3.1 inefficient and computationally time-consuming.

These two points highlighted in the previous paragraph can be mitigated by adopting the alternative approach illustrated in Fig. 5.14a, which utilizes Eq. (5.12) in place of Eq. (5.14). While the min-entropy provides less precise bounds for $H(A|x = m, E)$ compared to the latter, it results in significantly smaller SDP matrices. Thus, for a fixed NPA level, the min-entropy becomes more efficient than the convergent hierarchy to the von Neumann entropy.

The optimization method depicted in Fig. 5.14a comprises two distinct steps: an initial SDP optimization followed by a subsequent local optimization concerning the parameters $\boldsymbol{\theta}$. To illustrate the method's functionality, let us consider the ith step of this optimization procedure. At this stage, the value of η remains fixed, and an initial state $\hat{\rho}_i$ is chosen alongside the parameter $\boldsymbol{\theta}_i$, defining the POVM sets $\{M_x^A(\boldsymbol{\theta}_i)\}_x$ and $\{M_y^B(\boldsymbol{\theta}_i)\}_y$. These, combined with the Born rule in Eq. (2.135), define a behavior $\boldsymbol{P}_{AB}(\boldsymbol{\theta}_i)$. This behavior, can then be incorporated into the constraints of Eq. (5.12) to obtain a lower bound on $H(A|x = m, E)$, which we shall denote $H_{\min}^{(i)}(A|x = m, E)$. In practice, the SDP problem in Eq. (5.12) was formulated using `Mathematica` due to its efficiency in handling the SDP constraints. Subsequently, the optimization was performed in `Matlab` using `Yalmip` [88] for defining the problem and `Mosek` as the solver [42].

One of the fundamental advantages of utilizing Eq. (5.12) for the SDP optimization lies in its clarity when addressing the associated dual problem. As specified in Eq. (5.18), the dual problem is formulated as a maximization over the equality constraints, which, in our specific scenario, can be expressed as follows

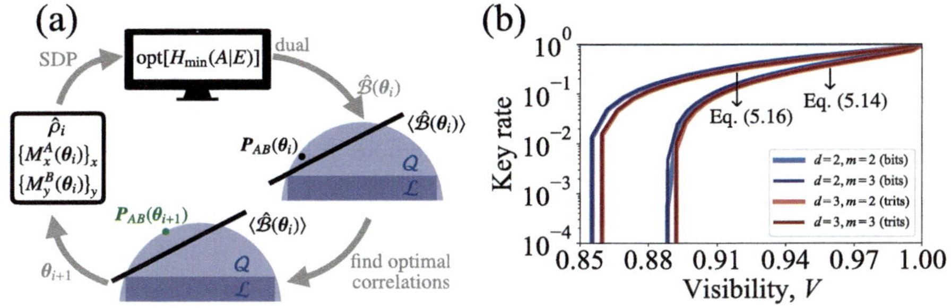

Fig. 5.14 Scheme of the numerical optimization for lower bounds on the key rate and the obtained results. In **a**, a visual representation of the methodology used to optimize Eq. (5.12) concerning the parameters $\boldsymbol{\theta}$ is depicted. Given a state and a sequence of measurements conducted by Alice and Bob, these establish a set of constraints enabling the optimization of $H_{\min}(A|E)$ via Semidefinite Programming. The dual of this SDP provides a Bell inequality, subsequently optimized using local optimization methods with respect to $\boldsymbol{\theta}$, deriving a new state and a set of measurements applicable in subsequent steps. In **b**, lower bounds on the key rate are presented when utilizing Eqs. (5.12) and (5.14); the latter is applied with $M = 16$. Four distinct scenarios are considered based on the values of d ($d = 2$ in blue and $d = 3$ in red) and m (light colors for $m = 2$ and dark colors for $m = 3$)

$$\langle\hat{\mathcal{B}}(\boldsymbol{\theta}_i)\rangle = \sum_{x,y,a,b} c_{x,y}^{a,b}\langle\hat{\Pi}_{a|x}(\boldsymbol{\theta}_i)\hat{\Pi}_{b|y}(\boldsymbol{\theta}_i)\rangle, \tag{5.48}$$

with the set of constants $\{c_{x,y,}^{a,b}\}$ representing the optimized variables of the dual problem. A direct comparison with Eq. (2.136) leads us to conclude that this equation effectively formulates a Bell expression, enabling us to construct the subsequent Bell operator

$$\hat{\mathcal{B}}(\boldsymbol{\theta}_i) = \sum_{x,y,a,b} c_{x,y}^{a,b}\hat{\Pi}_{a|x}(\boldsymbol{\theta}_i) \otimes \hat{\Pi}_{b|y}(\boldsymbol{\theta}_i), \tag{5.49}$$

where it is worth noting that this identification does not necessarily hold for Eq. (5.14), as it is defined as a sum of M SDP problems, each with different associated dual problems.

The maximum eigenvalue, in absolute value, of the operator in Eq. (5.49) corresponds to the maximal Bell inequality violation obtained by Eq. (5.48). Therefore, denoting the eigenvalues and eigenvectors of this Bell operator as $\{\lambda(\boldsymbol{\theta})\}$ and $\{|\varphi(\boldsymbol{\theta})\rangle\}$, we can then perform a search over the parameter $\boldsymbol{\theta}$ to determine an optimal Bell operator. This optimal operator achieves the maximum Bell inequality violation consistent with the $\{c_{x,y}^{a,b}\}$ coefficients. Consequently, the optimization problem we attempt to solve can be expressed as

$$\lambda(\boldsymbol{\theta}_{i+1}) = \underset{\theta}{\text{optimize}}\left[\lambda(\boldsymbol{\theta})\right], \tag{5.50}$$

which has associated the eigenstate $|\varphi(\boldsymbol{\theta}_{i+1})\rangle$ that, combined with Eq. (5.47), leads to the updated $\hat{\rho}_{i+1}$. The optimization in Eq. (5.50) corresponds to the local optimization phase of the method. In contrast to Sect. 5.2.3.1, where the objective function depended on an SDP problem, here it does not, enabling an efficient evaluation. In practice, we utilized built-in methods from the `Matlab Global Optimization` package [89]. Specifically, we employed the `Multistart` function, which proficiently launches optimization processes from multiple initial points, ensuring the retention of the optimal solution.

The updated parameters $\boldsymbol{\theta}_{i+1}$, alongside the updated $\hat{\rho}_{i+1}$ and the parametrization considered for the POVM sets, define a new behavior $\boldsymbol{P}_{AB}(\boldsymbol{\theta}_{i+1})$. This updated behavior can be re-entered into Eq. (5.12), resulting in $H_{\text{min}}^{(i+1)}(A|x = m, E)$. Consequently, these two iterative steps—the SDP and the local optimization—can be evaluated repeatedly until the condition $|H_{\text{min}}^{(i)}(A|x = m, E) - H_{\text{min}}^{(i+1)}(A|x = m, E)| < \epsilon$ is met, indicating convergence of the method. We set the value of ϵ to 10^{-4} in our case, approximately two orders of magnitude above the precision of the `Mosek` solver. Upon achieving convergence, an additional optimization is conducted, aiming to minimize $H(A|B, x = m, y = m + 1)$. Since analytical expressions exist for this quantity, it can be optimized using a local optimization technique analogous to those utilized in the evaluation of Eq. (5.50). Furthermore, the optimal values of $\boldsymbol{\theta}_i$ and $\hat{\rho}_i$ can be utilized to evaluate Eq. (5.14), which indeed provides improved bounds on the key rate compared to the min-entropy. This distinction is illustrated in Fig. 5.14b, where the leftmost curves employ Eq. (5.14) for $M = 16$, while the rightmost ones utilize Eq. (5.12). Each color in this figure represents a distinct combination of d and m. The lighter curves correspond to $m = 2$, with blue representing $d = 2$ (in bits) and red indicating $d = 3$ (in trits). Conversely, the darker curves portray the case for $m = 3$.

In addition to comparing Eqs. (5.12) and (5.14), the aforementioned figure illustrates that cases with $d = 3$ exhibit higher critical visibilities—where the key rate reaches zero—compared to those with $d = 2$. This suggests that, in the realm of DIQKD, an increase in the dimension d of the states used by Alice and Bob in the protocol compromises the noise robustness of the key rate. Specifically, the critical visibilities obtained are approximately $V_{\text{crit}}^{(d=2)} \approx 0.888$ and $V_{\text{crit}}^{(d=3)} \approx 0.892$ when employing Eq. (5.12), whereas $V_{\text{crit}}^{(d=2)} \approx 0.855$ and $V_{\text{crit}}^{(d=3)} \approx 0.860$ when using Eq. (5.14). Moreover, although an increase in the number of measurement settings appears to enhance the obtained bounds, the improvement achieved is almost negligible.

5.3.2 *Upper Bounds on the Key Rate Based on the Convex-Com-Bination Attack*

Similar to Sect. 5.2.3.2, here we explore the CC attack to establish upper bounds on the key rate. As depicted in Eq. (5.22), this attack involves linear optimization

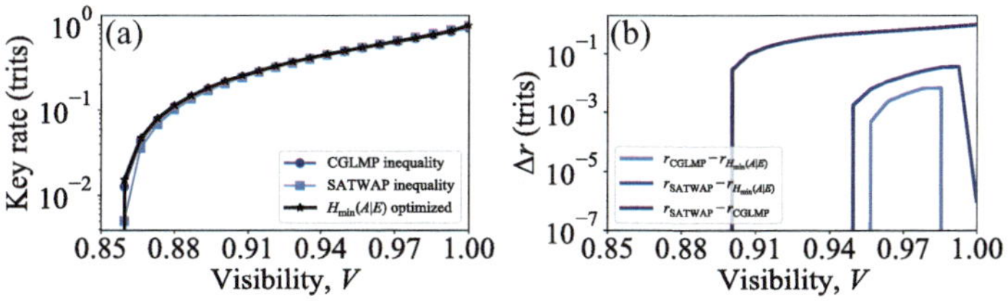

Fig. 5.15 Comparison between the optimization method and qutrit Bell inequalities. In **a**, the key rate for the $d = 3, m = 2$ scenario is depicted as a function of visibility across different methodologies. Specifically, it is computed using the CGLMP inequality (blue curve with circular markers), the maximal violation concerning Salavrakos' inequality (orange curve with squared markers), and the method outlined here based on the optimization shown in Fig. 5.14 (green curve with star markers). In **b**, the disparity among these key rates is illustrated as a function of visibility

of the probability of Eve sharing local or nonlocal correlations, wherein the selection of nonlocal correlations significantly influences Eve's ability to predict Alice's outcome. In qubit systems, it was demonstrated in Ref. [51] that using maximally entangled states for nonlocal correlations—those maximizing the violation of the CHSH inequality—provides optimal upper bounds to the key rate. However, for qutrits this optimality is not assured, necessitating the search of optimal inequalities.

In this context, Fig. 5.15a shows the lower bounds on the key rate calculated using the method described in the preceding subsection. It compares the numerical bounds to Eq. (5.12) obtained in Fig. 5.14b, with those found using the states and measurements that maximally violate two distinct Bell inequalities in the qutrit scenario: the CGLMP inequality formulated by D. Collins and colleagues in Ref. [90] (depicted by the blue curve) maximally violated by the so-called CGLMP state; and the SATWAP inequality (shown with lighter blue curve), established by A. Salavrakos and coworkers in Ref. [91] and optimally violated by maximally entangled states. Panel (b) displays the disparity between these rates. Notably, our optimization method provides improvement over these two inequalities for $V < 0.95$. However, for higher values, it tends to become suboptimal, yet yielding differences on the order of 10^{-2} concerning both inequalities. Consequently, these marginal disparities suggest that these inequalities might present an optimal choice for performing the CC attack.

However, the choice between the two presented inequalities remains unclear. This uncertainty is underscored in Fig. 5.15b, indicating that the SATWAP inequality provides superior lower bounds on the key rate for $V > 0.90$, while the trend reverses for $V \leq 0.90$. In this direction, one significant advantage about the SATWAP inequality is that it allows derivation of analytical bounds that extend to arbitrary dimensions d. This was demonstrated by N. Parker Sánchez in Ref. [92], where, utilizing maximally entangled states and the noise model in Eq. (5.47), the following upper bound on the key rate was obtained

$$r_{\mathrm{ub}} = \frac{1+(d-1)V}{d}\log_d\left(1+(d-1)V\right) + \frac{(d-1)(1-V)}{d}\log_d(1-V) - \frac{1-V}{1-2/I_d^{\max}}, \quad (5.51)$$

where $I_d^{\max}$ represents the violation of the CGLMP inequality for a given dimension d achieved by maximally entangled states of the form $(1/\sqrt{d})\sum_{i=1}^{d}|i\rangle_A \otimes |i\rangle_B$.

Alternatively, utilizing the state that maximally violates the CGLMP inequality, which we denote the CGLMP state, does not permit for analytical expressions and therefore requires numerical evaluation of Eq. (5.22). For the specific optimization details, we refer readers to N. Parker Sánchez's comprehensive bachelor thesis [92], whereas here we focus on discussing the obtained results, summarized in Fig. 5.16. Panel (a) displays both $H(A|x=m,E)$ (darkest blue curve) and $H(A|B,x=m,y=m+1)$ (lightest blue curve) for the maximally entangled (solid curve) and the CGLMP (dashed curve). It is observed that while the latter provides better bounds for the relative entropy between Alice and Eve, the former leads to reduced values for the relative entropy between Alice and Bob. The combination of these two, leads to the key rate depicted in panel (b) where, for all visibilities resulting in a positive key rate, the maximally entangled state (darkest blue curve) outperforms the CGLMP state. Moreover, as Eq. (5.51) extends to arbitrary dimensions, panel (c) illustrates the behavior of upper bounds concerning the dimension d. Contrary to the lower bounds presented in Fig. 5.14b, these upper bounds display lower critical visibilities with increasing d. Specifically, in the limit $d \to \infty$, from Eq. (5.51) it is possible to obtain for the critical visibility $V_{\mathrm{crit}}^{(d\to\infty)} \to 0.7539$ [92].

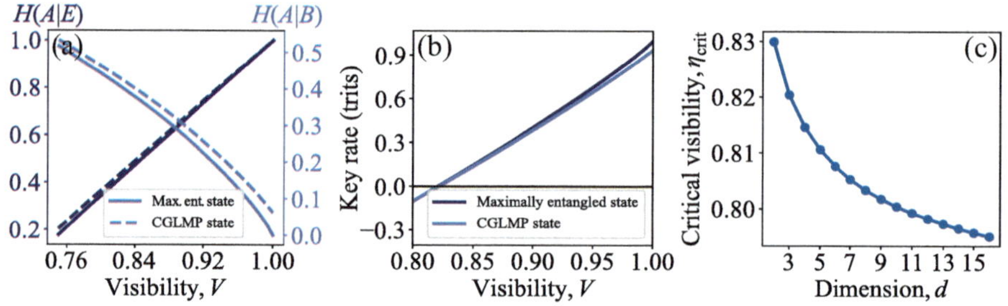

Fig. 5.16 Analytical bounds on the key rate derived from the CC attack for arbitrary dimensions. Panel **a** displays bounds on $H(A|x=m,E)$ (darker curve) alongside the corresponding values of $H(A|B,x=m,y=m+1)$ (lighter curve). This comparison is presented for scenarios where Eve's nonlocal state in the CC attack corresponds to either the maximally entangled state (solid curve) or the CGLMP state (dashed curve). In panel **b**, these two quantities from the previous plot are combined for the evaluation of the key rate. Panel **c** presents the critical visibilities obtained through Eq. (5.51) as a function of the dimension d

5.3.3 Comparison Between the Bounds

Combining the lower and upper bounds computed in Sects. 5.3.1 and 5.3.2, we obtain Fig. 5.17. Here, the number of measurements has been fixed to $m = 2$, while exploring dimensions of $d = 2$ (blue curves) and $d = 3$ (red curves). However, numerical limitations prevent us from evaluating dimensions beyond $d = 3$ as these computations exceed our current numerical capabilities. Different line styles represent the particular method employed for computing the bound.

In panel (a), a direct comparison between the key rates for upper and lower bounds is depicted. As mentioned earlier, notably, unlike the lower bounds, the critical visibilities derived from the upper bounds decrease as d increases. Specifically, the upper bounds yield $V_{\text{crit,ub}}^{(d=3)} \approx 0.820$ and $V_{\text{crit,ub}}^{(d=2)} = 0.830$, while the best-scenario for lower bounds presents $V_{\text{crit,lb}}^{(d=3)} \approx 0.860$ and $V_{\text{crit,lb}}^{(d=2)} = 0.855$ using Eq. (5.14). However, this contrasting trend between lower and upper bounds is not apparent when exclusively examining the bounds to the relative entropy between Alice and Eve, shown in panel (b).

In this case, the upper bounds for $d = 3$ are evaluated using the CGLMP state since, as depicted in Fig. 5.16b, it provides better bounds on $H(A|x = m, E)$ than maximally entangled states. Specifically, for $d = 3$, the critical visibilities for the upper and lower bounds are $V_{\text{crit,ub}}^{(d=3)} \approx 0.687$ and $V_{\text{crit,lb}}^{(d=3)} \approx 0.691$, respectively. For $d = 2$, we obtain $V_{\text{crit,ub}}^{(d=2)} \approx 0.712$ and $V_{\text{crit,lb}}^{(d=2)} \approx 0.713$, respectively.

Moreover, it is observed that the condition $H_{\text{ub}}^{(d=3)}(A|x = x^*, E) \geq H_{\text{ub}}^{(d=2)}(A|x = x^*, E)$ holds true across all visibilities for upper bounds but not for lower bounds. Particularly for the latter, $H_{\text{lb}}^{(d=3)}(A|x = x^*, E)$ and $H_{\text{lb}}^{(d=2)}(A|x = x^*, E)$ become equal at around $V \approx 0.795$, the former assessed in terms of trits and the latter in bits. However, at this point, $H(A|B, x = x^*, y = y^*)$ already surpasses $H(A|x = m, E)$,

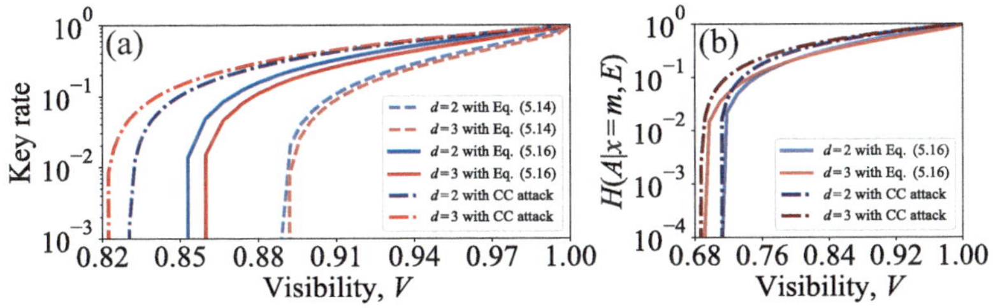

Fig. 5.17 Comparison between lower and upper bounds. In panel **a**, a comparison is made between the lower and upper bounds of the key rates for $(d = 3, m = 2)$ depicted by the red curves and $(d = 2, m = 2)$ illustrated by the blue curves. The dash-dotted curves represent the upper bounds of the key rate. The lower bounds are displayed using solid and dashed curves. Specifically, the lower bounds obtained through Eq. (5.12) are represented by the dashed curves, while the lower bounds derived from Eq. (5.14) are displayed with the solid curves. In panel **b**, a comparison is made between the lower bounds (solid curves) and upper bounds (dash-dotted curves) for $H(A|x = x^*, E)$ concerning $d = 3$ (red curves) and $d = 2$ (blue curves) as a function of visibility. The upper bounds are determined using the CGLMP state for the CC attack

and therefore, this enhancement does not reflect on the key rate. In summary, while an increase in d does not necessarily guarantee improved key rate resilience concerning visibility, it does hold promise for enhanced randomness generation.

5.4 Discussion and Outlook

In this chapter, we have conducted a comprehensive device-independent security analysis of quantum key distribution protocols, examining their resilience against various sources of noise. We focused on realistic photonic-based implementations of DIQKD, leveraging quantum dot single-photon sources. We studied the protocol's security using parameters commonly employed in their characterization. Our findings indicate that achievable parameter values, within the scope of near-future experimental implementations, allow for the extraction of key rates in asymptotic regimes. Additionally, employing a finite statistics analysis, we delved into the temporal and spatial requirements necessary for establishing a secure communication rate between Alice and Bob, which surpass recent DIQKD experimental implementations based on hybrid light-matter approaches [56, 57].

Furthermore, we investigated whether employing larger dimensional systems shared between Alice and Bob offer improvements in the noise resilience of DIQKD protocols. This involved a study of both upper and lower bounds on the key rate in the asymptotic regime when using qubits and qutrits, resulting in varying outcomes. Lower bounds revealed that increasing dimensions did not lead to improved noise resilience for the key rate, while upper bounds demonstrated the opposite trend. However, both lower and upper bounds showcased an advantage in increasing dimensionality concerning visibility constraints for device-independent randomness generation.

Moreover, unlike the numerically evaluated lower bounds, the methods used for computing the upper bounds could be extended to arbitrary dimensions. In this scenario, the enhancement achieved for the critical visibilities when analyzing the key rate, reached a plateau at approximately 75%, a limited improvement compared to the 82% attained for qubits. Consequently, these findings suggest that expanding experimental setups to higher dimensions beyond qubits may not necessarily justify the associated increase in experimental complexity.

Regarding the analysis provided in this chapter, one could contemplate the following possibilities to further expand our work:

- **Relaxation of the device-independent requirements.** As observed along this chapter, the device-independent prerequisites in the studied QKD protocols impose stringent constraints on the required experimental parameters for secret key extraction. Relaxing these prerequisites leads to more feasible experimental implementations. This relaxation may involve exploring options such as one-sided DIQKD [93], where one party distrusts their device while the other does not; or

measurement-device-independent QKD [94], which aims to mitigate risks associated with potentially compromised measurement devices used by Alice and Bob. An extension of our work would involve investigating how these alternative protocols alleviate the noise requirements considered for the quantum dot single-photon sources analyzed in this chapter.
- **Noisy preprocessing and finite statistics when using larger dimensions.** In this chapter, it was observed that increasing the dimensionality of states employed by Alice and Bob does not significantly enhance, if at all, resilience against visibility losses in the asymptotic regime without additional preprocessing operations performed by Alice [36]. Exploring noisy preprocessing strategies for higher dimensions could be insightful, considering that the expanded number of outcomes in the qutrit regime allows for more varied bitflip strategies than in the qubit case. Moreover, it would be of interest to analyze whether potential improvements due to the use of higher dimensional systems can be found when considering a finite statistics analysis. However, this exploration would necessitate examining plausible experimental setups that permit the utilization of qutrits [95, 96] with sufficient tunability to engineer the optimal states derived from the optimization in Fig. 5.14a.

References

1. Singh S (2000) The Code Book: The Science of Secrecy from Ancient Egypt to Quantum Cryptography. Anchor Books, New York, US
2. Bellovin SM (2011) Cryptologia 35:203
3. Vernam GS (1919) Secret Signaling System, uS Patent No. 1310719A
4. Wolf R (2021) In: Quantum key distribution: an introduction with exercises, Lecture notes in physics. Springer International Publishing, Cham, pp 1–12
5. Shannon CE (1949) The Bell Syst Tech J 28:656
6. Rivest RL, Shamir A, Adleman L (1978) Commun ACM 21:120
7. Shor P (1994) In: Proceedings 35th annual symposium on foundations of computer science, pp 124–134
8. Shor PW (1999) SIAM Rev 41:303
9. Renner R, Wolf R (2023) AIAA J 61:1895
10. Nielsen MA, Chuang IL (2010) Quantum information theory. In: Quantum computation and quantum information: 10th anniversary edition. Cambridge University Press, Cambridge, UK, Chap. 12, pp 528–607
11. Wootters WK, Zurek WH (1982) Nature 299:802
12. González-Ruiz EM, Rivera-Dean J, Cenni MFB, Sørensen AS, Acín A, Oudot E (2024) Opt Express 32:13181
13. Rivera-Dean J, Steffinlongo A, Parker-Sánchez N, Acín A, Oudot E (2024) Device-independent quantum key distribution beyond qubits. arXiv:2402.00161 [quant-ph]
14. Wolf R (2021) In: Quantum key distribution: an introduction with exercises, Lecture notes in physics. Springer International Publishing, Cham, pp 91–116
15. Bennett CH, Brassard G (2014) Theoretical computer science theoretical aspects of quantum cryptography – celebrating 30 years of BB84, vol 560, 7
16. Bruß D (1998) Phys Rev Lett 81:3018
17. Bechmann-Pasquinucci H, Gisin N (1999) Phys Rev A 59:4238
18. Scarani V, Acín A, Ribordy G, Gisin N (2004) Phys Rev Lett 92:057901

19. Ekert AK (1991) Phys Rev Lett 67:661
20. Acín A, Massar S, Pironio S (2006) New J Phys 8:126
21. Brunner N, Cavalcanti D, Pironio S, Scarani V, Wehner S (2014) Rev Mod Phys 86:419
22. Born M (1926) Z Phys 37:863
23. Clauser JF, Horne MA, Shimony A, Holt RA (1969) Phys Rev Lett 23:880
24. Cirel'son BS (1980) Lett Math Phys 4:93
25. Pironio S, Acín A, Brunner N, Gisin N, Massar S, Scarani V (2009) New J Phys 11:045021
26. Scarani V, Kurtsiefer C (2014) Theoretical computer science theoretical aspects of quantum cryptography – celebrating 30 years of BB84, vol 560, 27
27. Makarov V, Anisimov A, Skaar J (2006) Phys Rev A 74:022313
28. Qi B, Fung C-HF, Lo H-K, Ma X (2007) Quantum Inf Comput 7:73
29. Zhao Y, Fung C-HF, Qi B, Chen C, Lo H-K (2008) Phys Rev A 78:042333
30. Makarov V (2009) New J Phys 11:065003
31. Lydersen L, Wiechers C, Wittmann C, Elser D, Skaar J, Makarov V (2010) Nat Photonics 4:686
32. Gerhardt I, Liu Q, Lamas-Linares A, Skaar J, Kurtsiefer C, Makarov V (2011) Nat Commun 2:349
33. Acín A, Brunner N, Gisin N, Massar S, Pironio S, Scarani V (2007) Phys Rev Lett 98:230501
34. Devetak I, Winter A (2005) Proceedings of the Royal Society A: Mathematical. Phys Eng Sci 461:207
35. Nielsen MA, Chuang IN (2010) Entropy and information. In: Quantum computation and quantum information: 10th anniversary edition. Cambridge University Press, Cambridge, UK, Chap 11, pp 500–527
36. Ho M, Sekatski P, Tan E-Z, Renner R, Bancal J-D, Sangouard N (2020) Phys Rev Lett 124:230502
37. Konig R, Renner R, Schaffner C (2009) IEEE Trans Inf Theory 55:4337
38. Acín A, Massar S, Pironio S (2012) Phys Rev Lett 108:100402
39. Brown P, Fawzi H, Fawzi O (2021) Nat Commun 12:575
40. Brown P, Fawzi H, Fawzi O (2024) Quantum 8:1445. https://doi.org/10.22331/q-2024-08-27-1445
41. Boyd S, Lieven V (2004) Convex optimization problems. In: Convex optimization. Cambridge University Press, Cambridge, UK, Chap 4, pp 127–214
42. MOSEK ApS (2019) The MOSEK optimization toolbox for MATLAB manual. Version 9.0
43. MOSEK ApS (2022) MOSEK optimizer API for python 9.3.20
44. Boyd S, Lieven V (2004) Interior-point methods. In: Convex optimization. Cambridge University Press, Cambridge, UK, Chap 11, pp 561–630
45. Navascués M, Pironio S, Acín A (2007) Phys Rev Lett 98:010401
46. Navascués M, Pironio S, Acín A (2008) New J Phys 10:073013
47. Kaur E, Wilde MM, Winter A (2020) New J Phys 22:023039
48. Christandl M, Ferrara R, Horodecki K (2021) Phys Rev Lett 126:160501
49. Arnon-Friedman R, Leditzky F (2021) IEEE Trans Inf Theory 67:6606
50. Farkas M, Balanzó-Juandó M, Łukanowski K, Kołodyński J, Acín A (2021) Phys Rev Lett 127:050503
51. Łukanowski K, Balanzó-Juandó M, Farkas M, Acín A, Kołodyński J (2023) Quantum 7:1199
52. Larsson J-A (2014) J Phys A: Math Theor 47:424003
53. Wolf R. (2021) In: Quantum key distribution: an introduction with exercises, Lecture notes in physics. Springer International Publishing, Cham, pp 159–182
54. Aspect A, Dalibard J, Roger G (1982) Phys Rev Lett 49:1804
55. Pearle PM (1970) Phys Rev D 2:1418
56. Nadlinger DP, Drmota P, Nichol BC, Araneda G, Main D, Srinivas R, Lucas DM, Ballance CJ, Ivanov K, Tan EY-Z, Sekatski P, Urbanke RL, Renner R, Sangouard N, Bancal J-D (2022) Nature 607:682
57. Zhang W, van Leent T, Redeker K, Garthoff R, Schwonnek R, Fertig F, Eppelt S, Rosenfeld W, Scarani V, Lim CC-W, Weinfurter H (2022) Nature 607:687
58. Brunner N, Young AB, Hu C, Rarity JG (2013) New J Phys 15:105006

59. Máttar A, Brask JB, Acín A (2013) Phys Rev A 88:062319
60. Wang H, He Y-M, Chung T-H, Hu H, Yu Y, Chen S, Ding X, Chen M-C, Qin J, Yang X, Liu R-Z, Duan Z-C, Li J-P, Gerhardt S, Winkler K, Jurkat J, Wang L-J, Gregersen N, Huo Y-H, Dai Q, Yu S, Höfling S, Lu C-Y, Pan J-W (2019) Nat Photon 13:770
61. Uppu R, Pedersen FT, Wang Y, Olesen CT, Papon C, Zhou X, Midolo L, Scholz S, Wieck AD, Ludwig A, Lodahl P (2020) Sci Adv 6:eabc8268
62. Tomm N, Javadi A, Antoniadis NO, Najer D, Löbl MC, Korsch AR, Schott R, Valentin SR, Wieck AD, Ludwig A, Warburton RJ (2021) Nat Nanotechnol 16:399
63. Gisin N, Pironio S, Sangouard N (2010) Phys Rev Lett 105:070501
64. Máttar A, Acín A (2016) Phys Scr 91:043003
65. Kołodyński J, Máttar A, Skrzypczyk P, Woodhead E, Cavalcanti D, Banaszek K, Acín A (2020) Quantum 4:260
66. González-Ruiz EM, Das SK, Lodahl P, Sørensen AS (2022) Phys Rev A 106:012222
67. González-Ruíz EM (2019) Study of InGaAs quantum dots as single-photon sources for device-independent quantum key distribution. Master thesis, University of Copenhagen
68. Muljarov EA, Zimmermann R (2004) Phys Rev Lett 93:237401
69. Zhai L, Nguyen GN, Spinnler C, Ritzmann J, Löbl MC, Wieck AD, Ludwig A, Javadi A, Warburton RJ (2022) Nat Nanotechnol 17:829
70. Bozzio M, Vyvlecka M, Cosacchi M, Nawrath C, Seidelmann T, Loredo JC, Portalupi SL, Axt VM, Michler P, Walther P (2022) npj Quantum Inf 8:1
71. Nelder JA, Mead R (1965) Comput J 7:308
72. Wittek P (2015) ACM Trans Math Soft 41:21:1
73. Virtanen P, Gommers R, Oliphant TE, Haberland M, Reddy T, Cournapeau D, Burovski E, Peterson P, Weckesser W, Bright J, Van Der Walt SJ. SciPy 1.0 Contributors (2020) Nat Meth 17:261
74. Sekatski P, Bancal J-D, Valcarce X, Tan EY-Z, Renner R, Sangouard N (2021) Quantum 5:444
75. Woodhead E, Acín A, Pironio S (2021) Quantum 5:443
76. Arnon-Friedman R, Dupuis F, Fawzi O, Renner R, Vidick T (2018) Nat Commun 9:459
77. Dupuis F, Fawzi O (2019) IEEE Trans Inf Theory 65:7596
78. Dupuis F, Fawzi O, Renner R (2020) Commun Math Phys 379:867
79. Wolf R (2021) In: Quantum key distribution: an introduction with exercises, Lecture notes in physics. Springer International Publishing, Cham, pp 117–157
80. Liu W-Z, Li M-H, Ragy S, Zhao S-R, Bai B, Liu Y, Brown PJ, Zhang J, Colbeck R, Fan J, Zhang Q, Pan J-W (2021) Nat Phys 17:448
81. Tan EY-Z, Sekatski P, Bancal J-D, Schwonnek R, Renner R, Sangouard N, Lim CC-W (2022) Quantum 6:880
82. Da Lio B, Faurby C, Zhou X, Chan ML, Uppu R, Thyrrestrup H, Scholz S, Wieck AD, Ludwig A, Lodahl P, Midolo L (2022) Adv Quantum Technol 5:2200006
83. Zahidy M, Mikkelsen MT, Müller R, Da Lio B, Krehbiel M, Wang Y, Bart N, Wieck AD, Ludwig A, Galili M, Forchhammer S, Lodahl P, Oxenløwe LK, Bacco D, Midolo L (2024) npj Quantum Inf 10:2. https://doi.org/10.1038/s41534-023-00800-x
84. Fuchs CA, Gisin N, Griffiths RB, Niu C-S, Peres A (1997) Phys Rev A 56:1163
85. Durt T, Cerf NJ, Gisin N, Żukowski M (2003) Phys Rev A 67:012311
86. Jo Y, Son W (2016) Phys Rev A 94:052316
87. Borkała JJ, Jebarathinam C, Sarkar S, Augusiak R (2022) Entropy 24:350
88. Lofberg J (2004) In: 2004 IEEE international conference on robotics and automation (IEEE Cat. No.04CH37508), pp 284–289
89. The MathWorks Inc. (2022) Global optimization toolbox (R2022a)
90. Collins D, Gisin N, Linden N, Massar S, Popescu S (2002) Phys Rev Lett 88:040404
91. Salavrakos A, Augusiak R, Tura J, Wittek P, Acín A, Pironio S (2017) Phys Rev Lett 119:040402
92. Parker i Sánchez N (2024) Upper bounds on key rates in device-independent quantum key distribution with qudits based on convex-combination attacks, Bachelor thesis, Universitat de Barcelona

93. Branciard C, Cavalcanti EG, Walborn SP, Scarani V, Wiseman HM (2012) Phys Rev A 85:010301
94. Lo H-K, Curty M, Qi B (2012) Phys Rev Lett 108:130503
95. Thew RT, Acín A, Zbinden H, Gisin N (2004) Phys Rev Lett 93:010503
96. Malik M, Erhard M, Huber M, Krenn M, Fickler R, Zeilinger A (2016) Nat Photon 10:248

Chapter 6
Conclusions

Nunca es tarde si la dicha es buena.

– Refranero español

Non-classical states form the cornerstone of quantum technology, delineating the potential advantages that quantum mechanics could deliver over conventional technology in our daily lives. However, elucidating practical implementations for these technologies necessitates addressing two fundamental queries: "*how?*" and "*why?*". In this Thesis, we have indirectly tackled these questions within the context of Photonics by (1) investigating how strong-field physics processes can potentially lead to non-classical states of light, and (2) delineating the advantages of non-classical states light states in the context of key distribution.

Yet, it is crucial to remark that this Thesis has not centered on the application of strong-field-based non-classical light states for specific quantum information science applications. Rather, it stands as an initial step, laying the groundwork: we have explored how strong-field physics might give rise to states of interest for quantum technology applications. The next stride is, thus, to combine these two domains: merging the ultrafast and energetic facets of attosecond science with results from quantum information science to propel both fields into unprecedented energy, time and sensitivity scales.

In this final chapter, we briefly recapitulate the two parts constituting this Thesis. Differing from the *Discussion and outlook* sections presented at the conclusion of the main chapters composing this Thesis, our aim here is to motivate potential synergies between them. For that purpose, we opt for an inverted order in their presentation, compared to the sequence followed in previous chapters.

J. Rivera-Dean, *Non-classical States of Light*, Springer Theses,
https://doi.org/10.1007/978-3-031-73769-5_6

Non-classical States of Light for Quantum Key Distribution

In this part of the Thesis, covered in Chap. 5, we delved into the feasibility of device-independent quantum key distribution protocols using a photonic-based setup. Our analysis highlighted the constraints posed by certain parameters defining the source of non-classical states, specifically quantum-dot-based single-photon sources, when attempting to establish a secure key between Alice and Bob. In a first step, our investigation was conducted under the assumption that Alice and Bob execute an infinite number of protocol rounds. Within this framework, we observed that the primary practical limitation revolved around the local efficiency within Alice's and Bob's stations. Notably, for current state-of-the-art quantum dot single-photon sources, the required detection efficiency hovered around 90%, significantly challenging the capabilities of existing detector systems, although potentially achievable for near-term implementations.

As a subsequent step, we explored a more pragmatic scenario where Alice and Bob execute a finite number of rounds, often constrained by the allocated time for generating a secret key. This finite statistics analysis also establishes bounds on the maximum achievable distances between Alice and Bob for generating secure keys. Within this context, the principal limiting factor emerged as the transmission efficiency of the utilized channel, governing the probability with which Alice and Bob successfully share an entangled state. A lower probability of successful transmission reduces the likelihood of a protocol round succeeding. For the examined photonic setup, we observed the potential to establish secure keys at distances spanning hundreds of kilometers.

Finally, we examined the impact of increased dimensions within the non-classical states shared by Alice and Bob on noise constraints concerning key rate acquisition, devoid of any specific physical implementation. Our observations revealed discordance between lower and upper bounds regarding the potential for improvements. The primary conclusion drawn was that despite potential advantages, the experimental challenges involved in obtaining the necessary states might outweigh the prospective enhancements. Nevertheless, our analysis agreed on one aspect: increased dimensionality presents an advantage in the context of device-independent randomness generation.

Non-classical States of Light After Strong Laser-Matter Interactions

In the past few years, there has been a growing interest in quantum optical representations of strong-field phenomena. In this part of the Thesis, we delved into this area. In Chap. 3, attention was on the quantum optical analysis of strong-fields with atomic media, while in Chap. 4 with more intricate systems such as diatomic molecules and semiconductor materials. The primary outcome of these analyses highlighted that strong-field processes within these systems, could be used to generate non-classical light states across a broad spectrum of frequencies, characterized by high photon numbers. Additionally, under specific conditions, the generation of hybrid entangled states involving both light and matter was demonstrated to be feasible.

Specifically, in Chap. 3, the high-photon number nature was leveraged to drive nonlinear optical processes in crystals using non-classical states in the form of coherent state superpositions, marking an unprecedented achievement in optics. While this signifies a notable application in the realm of optics, there exists potential for exploiting this property in quantum technological environments. For instance, recent research in Ref. [1] demonstrated that these states exhibited enhanced metrological capabilities compared to conventional even and odd cat states, showcasing better resilience against transmission noise for quantum metrology applications. Hence, the field of quantum metrology emerges as a promising arena for the utilization of these states.

Moreover, their high-photon count coupled with their extensive frequency entanglement position them as potential candidates for quantum communication protocols. Specifically, their high-photon nature might mitigate, to some extent, the impact of transmission losses—a critical factor, as previously discussed, that significantly influences the feasibility of quantum communication protocols. Additionally, their massively frequency-entangled nature might be suitable for quantum network protocols, enabling communication among multiple parties rather than just two. Furthermore, these states could be employed, together with more general measurements in Alice's and Bob's stations, to enhance the local noise resilience of quantum communication protocols.

Conclusively, in the second part of Chap. 4, we delved into the circumstances conducive to generating these states using semiconductor materials. These systems constitute a cornerstone in today's technology, pervading various aspects of our daily lives, from smartphones to specialized medical equipment. Thus, their versatility and capacity to scale technology down to handheld sizes make them invaluable. Presently, achieving practicable technology necessitates scalability, and based on our theoretical analysis, this appears to pave the way for implementing strong-field-based non-classical light sources on a tabletop scale, akin to the quantum-dot single-photon sources discussed in Chap. 5. Furthermore, recent developments in analog-based quantum simulation [2] may allow us to better understand the mechanisms underlying high-harmonic generation in semiconductors, as well as in quantum materials. This could contribute to the design and search for materials with optimal characteristics for the fabrication of such non-classical state sources. Additionally, it could pave the way for the first proof-of-principle experiments in the direction of Ref. [3], where non-classical light is used to drive strong-field processes.

In summary, this Thesis can be understood as a contribution to the fields of strong-field physics and quantum information science, as well as to their emerging but quickly evolving intersection. We look forward to seeing how this promising field evolves in the future.

References

1. Stammer P, Martos TF, Lewenstein M, Rajchel-Mieldzioć (2024) https://doi.org/10.1088/2058-9565/ad7881
2. Argüello-Luengo J, Rivera-Dean J, Stammer P, Maxwell AS, Weld DM, Ciappina MF, Lewenstein M (2024) PRX Quantum 5:010328. https://doi.org/10.1103/PRXQuantum.5.010328
3. Even Tzur M, Birk M, Gorlach A, Krüger M, Kaminer I, Cohen O (2023) Nature Photonics 17:501. https://doi.org/10.1038/s41566-023-01209-w

Chapter 7
Additional Material of Chapter 2

7.1 Transformation of the Momentum Operator with Respect to $\hat{U}_{\text{VG}}(t)$

According to the Baker-Hausdorff lemma (see e.g. Ref. [1] and references therein), we can write

$$\hat{O} = e^{i\beta\hat{Z}}\hat{X}e^{-i\beta\hat{Z}} = \hat{X} + i\beta[\hat{Z},\hat{X}] + \frac{(i\beta)^2}{2!}[\hat{Z},[\hat{Z},\hat{X}]] + \frac{(i\beta)^3}{3!}[\hat{Z},[\hat{Z},[\hat{Z},\hat{X}]]] + \cdots, \tag{7.1}$$

and in particular for the case of $\hat{U}_{\text{VG}}(t) \equiv e^{-iQ\mathbf{A}(\mathbf{r},t)\cdot\mathbf{r}/\hbar}$ acting on the momentum operator $\hat{\mathbf{p}}$, we get

$$\begin{aligned}\hat{U}^{\dagger}_{\text{VG}}(t)\hat{\mathbf{p}}\hat{U}_{\text{VG}}(t) = \hat{\mathbf{p}} &+ \left(\frac{-iQ}{\hbar}\right)[\mathbf{A}(\hat{\mathbf{r}},t)\cdot\hat{\mathbf{r}},\hat{\mathbf{p}}] \\ &+ \frac{1}{2!}\left(\frac{-iQ}{\hbar}\right)^2[\mathbf{A}(\hat{\mathbf{r}},t)\cdot\hat{\mathbf{r}},[\mathbf{A}(\hat{\mathbf{r}},t)\cdot\hat{\mathbf{r}},\hat{\mathbf{p}}]] + \cdots.\end{aligned} \tag{7.2}$$

Further expanding the first-order commutator, we get

$$\begin{aligned}[\mathbf{A}(\hat{\mathbf{r}},t)\cdot\hat{\mathbf{r}},\hat{\mathbf{p}}] &= \mathbf{A}(\hat{\mathbf{r}},t)[\hat{\mathbf{r}},\hat{\mathbf{p}}] + [\mathbf{A}(\hat{\mathbf{r}},t),\hat{\mathbf{p}}]\hat{\mathbf{r}} \\ &= i\hbar\mathbf{A}(\hat{\mathbf{r}},t) + [\mathbf{A}(\hat{\mathbf{r}},t),\hat{\mathbf{p}}]\hat{\mathbf{r}},\end{aligned} \tag{7.3}$$

which introduced in Eq. (7.2) yields

$$\begin{aligned}\hat{U}^{\dagger}_{\text{VG}}(t)\hat{\mathbf{p}}\hat{U}_{\text{VG}}(t) = \hat{\mathbf{p}} + Q\mathbf{A}(\hat{\mathbf{r}},t) &- \frac{iQ}{\hbar}[\mathbf{A}(\hat{\mathbf{r}},t),\hat{\mathbf{p}}]\hat{\mathbf{r}} \\ &+ \frac{1}{2!}\left(\frac{-iQ}{\hbar}\right)^2[\mathbf{A}(\hat{\mathbf{r}},t)\cdot\hat{\mathbf{r}},[\mathbf{A}(\hat{\mathbf{r}},t),\hat{\mathbf{p}}]\hat{\mathbf{r}}] + \cdots\end{aligned} \tag{7.4}$$

J. Rivera-Dean, *Non-classical States of Light*, Springer Theses,
https://doi.org/10.1007/978-3-031-73769-5_7

and that we represented in the main text as

$$\hat{U}_{\mathrm{VG}}^{\dagger}(t)\hat{\mathbf{p}}\hat{U}_{\mathrm{VG}}(t) = \hat{\mathbf{p}} + Q\mathbf{A}(\hat{\mathbf{r}}, t) - \frac{iQ}{\hbar}[\mathbf{A}(\hat{\mathbf{r}}, t), \hat{\mathbf{p}}]\hat{\mathbf{r}} + \mathcal{O}\Big([\mathbf{A}(\hat{\mathbf{r}}, t), \hat{\mathbf{p}}]\hat{\mathbf{r}}\Big), \quad (7.5)$$

where we defined

$$\mathcal{O}\Big([\mathbf{A}(\hat{\mathbf{r}}, t), \hat{\mathbf{p}}]\hat{\mathbf{r}}\Big) \equiv -\frac{iQ}{\hbar}[\mathbf{A}(\hat{\mathbf{r}}, t), \hat{\mathbf{p}}]\hat{\mathbf{r}} + \frac{1}{2!}\left(\frac{-iQ}{\hbar}\right)^2 [\mathbf{A}(\hat{\mathbf{r}}, t) \cdot \hat{\mathbf{r}}, [\mathbf{A}(\hat{\mathbf{r}}, t), \hat{\mathbf{p}}]\hat{\mathbf{r}}] + \cdots, \quad (7.6)$$

that is, the term involving first and higher-order commutators with respect to $[\mathbf{A}(\hat{\mathbf{r}}, t), \hat{\mathbf{p}}]\hat{\mathbf{r}}$. Note that, under the dipole approximation where $\mathbf{A}(\hat{\mathbf{r}}, t) \approx \mathbf{A}(t)$ such that we discard the spatial dependence of the vector potential, then $[\mathbf{A}(t), \hat{\mathbf{p}}]\hat{\mathbf{r}} = 0$ which consequently leads to $\mathcal{O}([\mathbf{A}(t), \hat{\mathbf{p}}]\hat{\mathbf{r}}) = 0$.

7.2 Poynting's Theorem and the Energy Carried by the Electromagnetic Field

With the aim of formally deriving the energy carried out by the electromagnetic field, we consider a charge Q subjected to an electromagnetic field characterized by the electric field $\mathbf{E}(\mathbf{r}, t)$ and the magnetic field $\mathbf{B}(\mathbf{r}, t)$. In this context, the infinitesimal amount work $\mathrm{d}W$ performed by the electromagnetic field on the aforementioned charge during an infinitesimal time interval $\mathrm{d}t$, can be formulated as

$$\mathrm{d}W = \mathbf{F}(\mathbf{r}, t) \cdot \mathbf{v}\, \mathrm{d}t = Q\big(\mathbf{E}(\mathbf{r}, t) + \mathbf{v} \times \mathbf{B}(\mathbf{r}, t)\big) \cdot \mathbf{v}\, \mathrm{d}t = Q\mathbf{E}(\mathbf{r}, t) \cdot \mathbf{v}\, \mathrm{d}t, \quad (7.7)$$

where in transitioning to the second equality, we introduced the Lorentz force. In the case where a single particle with charge Q is replaced by a charge density $\hat{\rho}(\mathbf{r})$, the aforementioned expression can be redefined as

$$\mathrm{d}W = \left(\int_V \mathrm{d}\mathbf{r}\, \mathbf{E}(\mathbf{r}, t) \cdot \rho(\mathbf{r})\mathbf{v}\right)\mathrm{d}t = \left(\int_V \mathrm{d}\mathbf{r}\, \mathbf{E}(\mathbf{r}, t) \cdot \mathbf{J}(\mathbf{r}, t)\right)\mathrm{d}t, \quad (7.8)$$

with $\mathbf{J}(\mathbf{r}, t) \equiv \rho(\mathbf{r})\mathbf{v}$ denoting the current density.

From Eq. (7.8), it follows that the work conducted by the electromagnetic field per unit time and volume is expressed as $\mathbf{E}(\mathbf{r}, t) \cdot \mathbf{J}(\mathbf{r}, t)$. This can be further rewritten in terms of the electric and magnetic fields, considering that the presence of a charge current modifies Eq. (2.4) to

$$\nabla \times \mathbf{B}(\mathbf{r}, t) = \mu_0\left(\mathbf{J}(\mathbf{r}, t) + \varepsilon_0 \frac{\partial \mathbf{E}(\mathbf{r}, t)}{\partial t}\right), \quad (7.9)$$

where we have taken into account that $c = 1/\sqrt{\mu_0 \varepsilon_0}$. From this expression, we can rewrite

$$\begin{aligned}\mathbf{E}(\mathbf{r},t)\cdot\mathbf{J}(\mathbf{r},t) &= \frac{1}{\mu_0}\mathbf{E}(\mathbf{r},t)\cdot\big(\nabla\times\mathbf{B}(\mathbf{r},t)\big) - \epsilon_0\mathbf{E}(\mathbf{r},t)\cdot\frac{\partial\mathbf{E}(\mathbf{r},t)}{\partial t}\\ &= \frac{1}{\mu_0}\mathbf{B}(\mathbf{r},t)\cdot\big(\nabla\times\mathbf{E}(\mathbf{r},t)\big) - \frac{1}{\mu_0}\nabla\cdot\big(\mathbf{E}(\mathbf{r},t)\times\mathbf{B}(\mathbf{r},t)\big)\\ &\quad - \epsilon_0\mathbf{E}(\mathbf{r},t)\cdot\frac{\partial\mathbf{E}(\mathbf{r},t)}{\partial t},\end{aligned} \tag{7.10}$$

which using Eq. (2.3) can be further expressed as follows

$$\begin{aligned}\mathbf{E}(\mathbf{r},t)\cdot\mathbf{J}(\mathbf{r},t) = &-\frac{1}{2\mu_0}\frac{\partial}{\partial t}|\mathbf{B}(\mathbf{r},t)|^2 - \frac{\epsilon_0}{2}\frac{\partial}{\partial t}|\mathbf{E}(\mathbf{r},t)|^2\\ &-\frac{1}{\mu_0}\nabla\cdot\big(\mathbf{E}(\mathbf{r},t)\times\mathbf{B}(\mathbf{r},t)\big).\end{aligned} \tag{7.11}$$

As a result, upon introducing the expression above into Eq. (7.8), we derive the rate of work conducted by the electromagnetic field per unit time on the specified charge density

$$\begin{aligned}\frac{\mathrm{d}W}{\mathrm{d}t} = &-\frac{\partial}{\partial t}\int_V \mathrm{d}\mathbf{r}\,\frac{1}{2}\left(\frac{1}{\mu_0}|\mathbf{B}(\mathbf{r},t)|^2 + \epsilon_0|\mathbf{E}(\mathbf{r},t)|^2\right)\\ &-\frac{1}{\mu_0}\int_V \mathrm{d}\mathbf{r}\,\nabla\cdot\big(\mathbf{E}(\mathbf{r},t)\times\mathbf{B}(\mathbf{r},t)\big),\end{aligned} \tag{7.12}$$

with this expression constituting what is commonly known as Poynting's theorem. Here, the first term on the left-hand side signifies the total energy stored within the electromagnetic field, while the second term denotes the rate at which energy exits the volume V.

Reference

1. Gerry C, Knight P (2005) Field quantization. In: Introductory quantum optics. Cambridge University Press, Cambridge, UK, pp 10–42. Chap. 1

Chapter 8
Additional Material of Chapter 3

8.1 Numerical Computation of the ATI Wigner Functions

8.1.1 The Saddle-Point Approximation

The Wigner functions presented in Eq. (3.76) exhibit two different kind of contributions. On one hand, there are semiclassical-based terms provided by the $M(p, t_1)$ probability amplitudes. On the other hand, there are quantum-optical-based terms arising from all the other terms within the expression. Both types of contributions introduce corresponding phase factors to the integrals: the semiclassical terms contribute with the semiclassical action $S(p, t, t_1)$ from Eq. (3.15), which scales with $\sqrt{U_p}$, while the quantum optical terms provide a phase dependent on the displacement, scaling with $|\delta_{\mathbf{k}_L}(p, t, t_1)|^2$. For the laser parameters used in Fig. 3.15, we have $\sqrt{U_p} \approx 10$ while $|\delta_{\mathbf{k}_L}(p, t, t_1)| \approx 10^{-1}$. Therefore, we expect the semiclassical phase will play a dominant role in the phase of the integrand. To simplify, we rewrite our integral as follows

$$W(\tilde{\beta}_{\mathbf{k}_L}) = \int_{t_0}^{t} \mathrm{d}t_1 \int_{t_0}^{t} \mathrm{d}t_2 \tilde{M}(p, t_1, t_2, \tilde{\beta}_{\mathbf{k}_L}) e^{\frac{i}{\hbar} S(p,t,t_1)} e^{-\frac{i}{\hbar} S(p,t,t_2)}, \tag{8.1}$$

where we have explicitly separated the dominant phase terms from the rest, condensing the remaining components into the complex function $\tilde{M}(p, t_1, t_2, \tilde{\beta})$. As this function changes relatively slowly compared to the highly oscillatory term, we utilize the saddle-point approximation to compute the integrals. Thus, we approximate Eq. (8.1) as follows

J. Rivera-Dean, *Non-classical States of Light*, Springer Theses,
https://doi.org/10.1007/978-3-031-73769-5_8

$$W(\tilde{\beta}_{\mathbf{k}_L}) \simeq \sum_{t_1 \in \{t_{\text{ion}}\}} \sum_{t_2 \in \{t_{\text{ion}}\}} \sqrt{\frac{2\pi}{i|\partial^2 S(\mathbf{p}, t, t_1)/\partial t_1^2\big|_{t_1=t_{\text{ion}}}|}} \sqrt{\frac{2\pi i}{|\partial^2 S(\mathbf{p}, t, t_1)/\partial t_2^2\big|_{t_2=t_{\text{ion}}}|}} \times \tilde{M}(p, t_{1,\text{ion}}, t_{2,\text{ion}}, \tilde{\beta}) e^{\frac{i}{\hbar} S(p,t,t_{1,\text{ion}})} e^{-\frac{i}{\hbar} S(p,t,t_{2,\text{ion}})}. \tag{8.2}$$

In this last expression, t_{ion} represents the ionization times calculated from the saddle-point equation in Eq. (3.24). As the two phases are identical, the ionization times t_{ion} are the same for t_1 and t_2, although all possible combinations must be considered in the summation.

8.1.2 Using a Numerical Integrator

Unlike the Wigner function plots presented in Fig. 3.15, the plots in Fig. 3.16 correspond to a different set of laser parameters, and notably, they take into account the phenomenological contribution of N_{at} atoms participating in the process. Consequently, the conditions for applying the saddle-point approximation, as done in the previous section, are no longer met. Thus, to generate these plots, we adopted a fully numerical approach, where the integration is carried out using numerical methods. Specifically, we utilized the `nquad` integration routine provided by `SciPy` [1] to perform the double integration described in Eq. (3.76). To enhance the computational efficiency of the code, we incorporated the `numba` package [2], which accelerates the evaluation of the various functions required for computing the integral.

In general, the functions that appear within the integral are complex. In principle, the `nquad` function does not allow the evaluation of complex integrands, meaning that the real and imaginary parts of the integrand must be evaluated separately. However, since the Wigner function is real, as it describes a quasiprobability distribution, we initially verified within a grid that this property was satisfied as a sanity check. Nonetheless, for generating the plots, we avoided the integration over the imaginary part. With this in mind, the process of generating the plots consists of two parts

1. First, we generate a 2D grid of points $\{x_0, \ldots, x_m\}$ and $\{y_0, \ldots, y_n\}$, with $n \neq m$ in general, defining the real and imaginary parts of β, i.e. $\beta \equiv x + iy$. For each of these points, we numerically calculate the integral presented in Eq. (3.76). The evaluation was carried out in a single 2 GHz-CPU core, and the same process was executed in parallel for various values of the canonical momentum across the remaining cores. With an initial grid of 10×40 points, evaluating the Wigner function takes approximately a day (for $p = 0$ a.u. and $N_{\text{at}} = 10^4$). It is important to note that the size of the grid and its limits must be adjusted accordingly depending on the values of the canonical momentum p and the number of atoms N_{at} under consideration. Larger values of these two parameters necessitate bigger matrices with expanded limits, and thus increased computational resources. For the given grid size and the specified parameters ($p = 0$, $N_{\text{at}} = 10^4$ and a 5-cycle linearly polarized pulse with $E_0 = 0.053$ a.u., $\omega_{\mathbf{k}_L} = 0.057$ a.u. and a sinusoidal

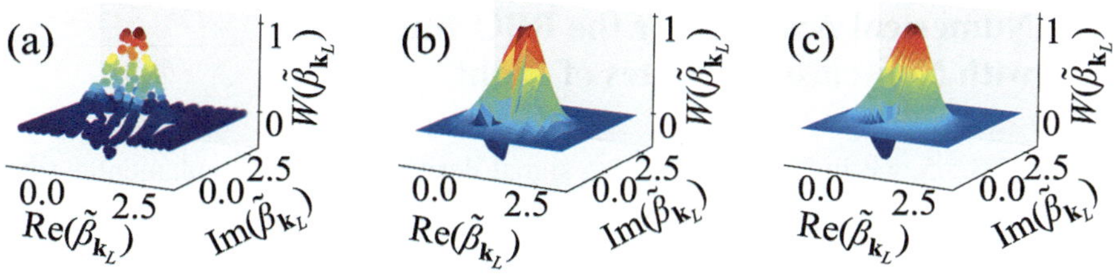

Fig. 8.1 Evaluation and smoothening of the numerically calculated Wigner functions. Evolution of the Wigner function (for $p = 0$ a.u. and $N_{\text{at}} \approx 10^4$) throughout different processing steps. Initially, we considered a grid of size 20×40 for evaluating the Wigner function as per Eq. (3.76), resulting in panels (**a**) and (**b**). In (**a**), we display a scatter plot while in (**b**) we present a surface plot of the obtained data. To further enhance the smoothness of these functions, we applied an interpolation scheme over a grid of size 500×500 within the range defined by the initial grid. The corresponding output is shown in (**c**). It is important to note that the Wigner function has been normalized to its maximum value in these plots

squared envelope), we obtain Fig. 8.1a which illustrates the values of the Wigner function at each evaluated point. The corresponding surface plot is displayed in panel (b). It is worth noting that, in this figure, the Wigner function has been normalized to its maximum value within the studied region.

2. The second part involves the smoothening of the Wigner function plot. This can be accomplished through either the *hard* but exact method or the *easy* but less accurate method. The first approach entails expanding the integration grid, introducing more points in the evaluation of the Wigner function. As a result, this yields a more exact smoothened plot due to the increased number of data points. However, the primary drawback here is that a larger number of points requires greater computational resources. The alternative, which is less exact but more flexible, is to employ an interpolation method on the points already calculated for the initial grid. In this manner, we can artificially augment the number of evaluation points (for example, transitioning from a grid of 10×40 to one of size 500×500) within the same limits without the need to recompute the numerical integration detailed in Eq. (3.76). To implement this interpolation, we utilized the `interpolate.griddata` function provided by the `SciPy` package. This function allows interpolation of two-dimensional data using various methods. It is important to note that the accuracy of the interpolation method is contingent on the fineness of the initial grid. As such, different values of momentum p and the number of atoms N_{at} necessitate varying numbers of initial evaluation points for this approach to be valid. Following the smoothening process, we obtain Fig. 8.1c.

8.2 Numerical Analysis for the BBO Interaction with Non-classical States of Light

In Sect. 3.3.2.5, we investigated the $S_{2\omega}$ signal derived from the fundamental component. Here, we conduct an explicit evaluation of the interaction between the fundamental mode and the non-linear crystal, analyzing the resulting second harmonic mode. Consequently, the Schrödinger equation describing these dynamics is given, in the interaction picture with respect to the free-field Hamiltonian, by

$$i\hbar\frac{\partial|\Phi(t)\rangle}{\partial t} = \Big[\kappa\Big(\hat{a}^2_{\mathbf{k}_L}\hat{a}^\dagger_{2\mathbf{k}_L} + (\hat{a}^\dagger_{\mathbf{k}_L})^2\hat{a}_{2\mathbf{k}_L}\Big)\Big]|\Phi(t)\rangle\,, \tag{8.3}$$

where $\hat{a}^\dagger_{\mathbf{k}_L}$ and $\hat{a}^\dagger_{2\mathbf{k}_L}$ denote the creation operators acting on the field modes of frequency $\omega_{\mathbf{k}_L}$ and $2\omega_{\mathbf{k}_L}$, respectively. The parameter κ represents the coupling constant of the interaction, which is proportionally related to the second-order susceptibility of the crystal under consideration.

We solve this equation under two different scenarios, depending on the initial state. On one hand, we consider the case where no conditioning on HHG operation is applied to the fundamental mode, such that the initial state is

$$|\Phi(t=t_0)\rangle = \Big|\frac{1}{2}(\alpha_{\mathbf{k}_L} + \chi_{\mathbf{k}_L})\big(f_+e^{i\omega t_0} + f_-e^{i\omega t_0}\big)\Big\rangle \otimes |0_{2\mathbf{k}_L}\rangle\,, \tag{8.4}$$

while, on the other hand, we consider the case where conditioning is indeed applied, such that the initial state reads

$$|\Phi(t=t_0)\rangle = |\mathrm{GCAT}(t_0,\tau)\rangle \otimes |0_{2\mathbf{k}_L}\rangle\,. \tag{8.5}$$

For these two initial states, the Schrödinger equation described in Eq. (8.3) was numerically solved using built-in functions from the `QuTiP` package [3, 4] available in Python. Subsequently, the obtained results were utilized to compute the $S_{2\omega}$ function, representing the second harmonic mode, expressed as

$$S_{2\omega} \propto \int \mathrm{d}\tau\big\langle\hat{E}^2_{2\mathbf{k}_L}(t)\big\rangle, \tag{8.6}$$

where $\hat{E}_{2\mathbf{k}_L}(t)$ denotes the electric field operator component acting on the second harmonic mode. For the sake of simplicity, in this case we focus on monochromatic fields by setting $f_+ = f_- = 1$, which further simplifies the numerical analysis.

In Fig. 8.2, the outcomes of this computation are presented, where we fixed $|\alpha_{\mathbf{k}_L}| = 1.0$. To ensure significantly relevant intensities in the second harmonic mode, we specifically chose $\kappa = 10$. From a technical standpoint, this choice was made to expedite the computational time of the numerical simulation. Utilizing the `qutip` package involves expressing both states and operators in the Fock basis,

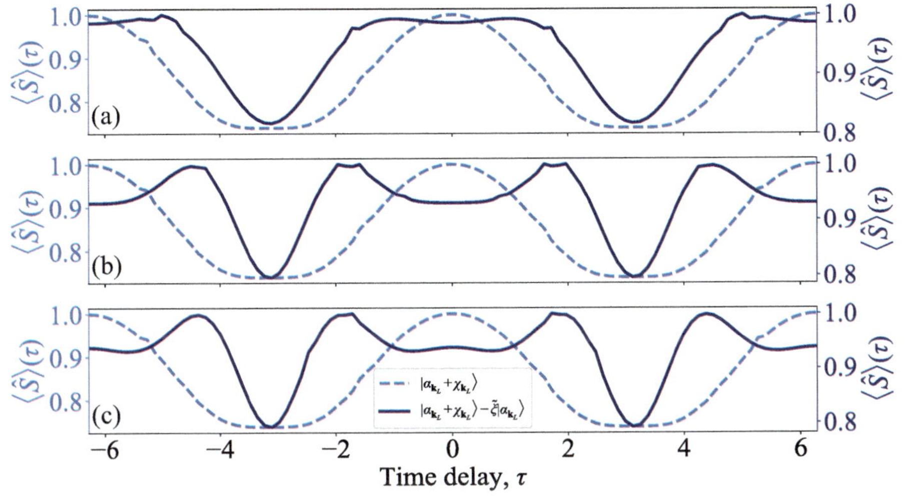

Fig. 8.2 Second harmonic generation signal obtained from numerical analysis of nonlinear optical interactions. These results were generated with $|\alpha_{\mathbf{k}_L}| = 1.0$ and $\kappa = 10$. Each panel represents a different value of $|\chi_{\mathbf{k}_L}|$. Specifically, panels (**a**) to (**c**) respectively depict $|\chi_{\mathbf{k}_L}| = 0.5$, $|\chi_{\mathbf{k}_L}| = 0.1$, and $|\chi_{\mathbf{k}_L}| = 0.01$. The light blue dashed curves illustrate the scenario where the initial state of the fundamental mode is given by Eq. (8.4), while the dark blue solid curves correspond to cases where it follows Eq. (8.5)

which encompasses an infinite number of elements. Therefore, truncation to a specific value, denoted as $n_{\max}$, becomes necessary. Setting $n_{\max} = 60$ resulted in satisfactory convergence for the obtained results.

In panels (a) to (c), we consider decreasing values of $|\chi_{\mathbf{k}_L}|$. Specifically, in (a), $|\chi_{\mathbf{k}_L}| = 0.5$, in (b), $|\chi_{\mathbf{k}_L}| = 0.1$, and in (c), $|\chi_{\mathbf{k}_L}| = 0.01$. Although these plots depict $(2\omega)\tau \in [-2\pi, 2\pi]$, the integration limits in Eq. (8.6) were set to $\pm 10\pi/(2\omega)$. The light blue dashed curves represent the scenario where the initial state is defined by Eq. (8.4), while the dark solid curves correspond to the initial state specified in Eq. (8.5). It is observed that while the initial state in Eq. (8.4) exhibits a sinusoidal-like behavior, the initial state in Eq. (8.5) demonstrates a modulation of this behavior, that becomes more pronounced as $|\chi_{\mathbf{k}_L}|$ decreases. Notably, these observed modulations align with the intra-cycle ones depicted in Fig. 3.29.

Furthermore, the non-classical nature of the probe field induces similar properties in the 2ω field, as illustrated in Fig. 8.3 showcasing the Wigner function of the second harmonic mode under various conditions. In this case, since the Wigner function analysis was done through built-in functions of the `qutip` package, the numerical analysis becomes more efficient, facilitating closer emulation of the experimental scenario. For this figure, we set the mean photon number of the fundamental mode to be in the order of 144 photons. This implies that $n_{\max} \gg 100$, which in our case was indeed set to $n_{\max} = 500$, although convergence tests were done with $n_{\max} = 600$.

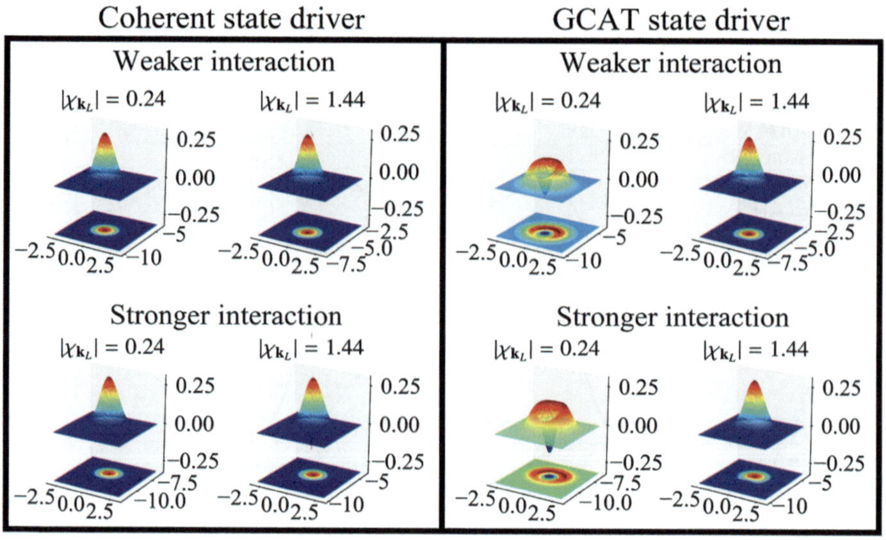

Fig. 8.3 Wigner function of the second harmonic mode obtained from numerical analysis of nonlinear optical interactions. The left box illustrates the scenario using Eq. (8.4) as the initial state, while the right box corresponds to the use of Eq. (8.5). In both rows, the total interaction time with the crystal is $\Delta t = \pi$, with κ values set to 0.010 for the upper row and 0.015 for the lower row. For all cases, the driving field's photon count is set to approximately 144, and the delay in the MZI is $\tau = 0$

Values beyond these led to matrices that were sufficiently big to exceed the memory capabilities of the employed hardware (a MacBook Air M1 2020 with 8 GPU cores).

In Fig. 8.3, the value of κ along with the interaction time were calibrated to yield mean photon numbers within the range of a few to several photons for the resulting second harmonic mode. While excitations with a coherent state fail to induce non-classical features in the second harmonic state (left box), this differs for excitations with the coherent state superposition (right box). Specifically, we observe that stronger interactions–achieved via higher κ and longer interaction times–result in more pronounced non-classical features. Conversely, these features become less pronounced as $|\chi_{\mathbf{k}_L}|$ increases. It is essential to note that these calculations were conducted under the scenario of $\tau = 0$. Deviations from this temporal configuration lead to suboptimal interference at the output of the Mach-Zehnder interferometer, consequently resulting in less pronounced non-classical features compared to those observed in Fig. 8.3.

References

1. Virtanen P, Gommers R, Oliphant TE, Haberland M, Reddy T, Cournapeau D, Burovski E, Peterson P, Weckesser W, Bright J, van der Walt SJ, Brett M, Wilson J, Millman KJ, Mayorov N, Nelson ARJ, Jones E, Kern R, Larson E, Carey CJ, Polat İ, Feng F, Moore EW, VanderPlas J, Laxalde D, Perktold J, Cimrman R, Henriksen I, Quintero EA, Harris CR, Archibald AM, Ribeiro AH, Pedregosa F, van Mulbregt P (2020) SciPy 1.0 Contributors. Nat Methods 17:261. https://doi.org/10.1038/s41592-019-0686-2
2. Lam SK, Pitrou A, Seibert S (2015) Proceedings of the second workshop on the LLVM compiler infrastructure in HPC, pp 1–6
3. Johansson JR, Nation PD, Nori F (2012) Comput Phys Commun 183:1760. https://doi.org/10.1016/j.cpc.2012.02.021
4. Johansson JR, Nation PD, Nori F (2013) Comput Phys Commun 184:1234. https://doi.org/10.1016/j.cpc.2012.11.019

Chapter 9
Additional Material of Chapter 4

9.1 The Variational Principle

The variational principle presented here is commonly known as the Rayleigh-Ritz method, adapted from the exposition provided in Ref. [1]. According to this method, we expand our ansatz $|\psi\rangle$ into the elements (potentially not all) of a given basis set $|\phi_i\rangle$

$$|\psi\rangle = \sum_i c_i |\phi_i\rangle, \tag{9.1}$$

where the coefficients c_i are treated as free parameters which, for the sake of simplicity, we assume to be real. Therefore, the primary objective of the method is to optimize this set of parameters to minimize the energy's mean value. To elaborate further, if we denote the Hamiltonian operator of the system as $\hat{H}$, its mean value can be expressed as

$$\langle \hat{H} \rangle = \frac{\langle \psi | \hat{H} | \psi \rangle}{\langle \psi | \psi \rangle} = \frac{\sum_{i,j} c_i c_j H_{i,j}}{\sum_{i,j} c_i c_j S_{i,j}}, \tag{9.2}$$

where we have denoted $H_{i,j} \equiv \langle \phi_i | \hat{H} | \phi_j \rangle$ and $S_{i,j} \equiv \langle \phi_i | \phi_j \rangle$.

Since our objective is to minimize this quantity concerning the coefficients c_k, we aim to equate the derivative of $\langle \hat{H} \rangle$ with respect to these coefficients to zero, expressed as

J. Rivera-Dean, *Non-classical States of Light*, Springer Theses,
https://doi.org/10.1007/978-3-031-73769-5_9

$$
\begin{aligned}
\frac{\partial\langle\hat{H}\rangle}{\partial c_k} &= \frac{\sum_i c_i[H_{i,k}+H_{k,i}]}{\sum_{i,j} c_i c_j S_{i,j}} - \frac{\left(\sum_i c_i[S_{i,k}+S_{k,i}]\right)\sum_{i,j} c_i c_j H_{i,j}}{\left(\sum_{i,j} c_i c_j S_{i,j}\right)^2} \\
&= \frac{\sum_i c_i[H_{i,k}+H_{k,i}]}{\sum_{i,j} c_i c_j S_{i,j}} - \frac{\left(\sum_i c_i[S_{i,k}+S_{k,i}]\right)\langle\hat{H}\rangle}{\sum_{i,j} c_i c_j S_{i,j}} \\
&= \frac{\sum_i c_i[H_{i,k}-\langle\hat{H}\rangle S_{i,k}]}{\sum_{i,j} c_i c_j S_{i,j}} + \frac{\sum_i c_i[H_{k,i}-\langle\hat{H}\rangle S_{k,i}]}{\sum_{i,j} c_i c_j S_{i,j}} = 0,
\end{aligned}
\tag{9.3}
$$

an equation that is satisfied whenever the numerators of this expression vanish, that is

$$
\sum_i c_i[H_{i,k} - \langle\hat{H}\rangle S_{i,k}] = 0, \tag{9.4}
$$

leading to a system of equations whose condition for the existence of a solution is determined by

$$
\left|\mathsf{H} - \langle\hat{H}\rangle\mathsf{S}\right| = 0, \tag{9.5}
$$

where H and S are the matrices whose elements are respectively given by $H_{i,j}$ and $S_{i,j}$. In this expression, $|\cdot|$ denotes the determinant.

9.2 Properties of the Many-Body Left and Right-Like Localized States

In Sect. 4.1.2.5, we explored an extension to the many-molecule regime involving a set of localized left and right states. Specifically, for the case of left-localized states, we particularly examined

$$
|\bar{L}_{N_{\mathrm{mol}}}\rangle \equiv \frac{1}{\sqrt{2}}\Big[|\psi_{\bar{\mathrm{b}}}\rangle - |\psi_{\bar{\mathrm{a}}}\rangle\Big] = \frac{1}{\sqrt{2}}\Big[|\psi_{\bar{\mathrm{b}}}\rangle - \frac{1}{\sqrt{N_{\mathrm{mol}}}}\sum_{k=1}^{N_{\mathrm{mol}}}|\psi_{\bar{\mathrm{a}}_k}\rangle\Big], \tag{9.6}
$$

where all molecules except one are in a bonding state, an excitation that is delocalized along them. In this section, we explore specific properties of this state, focusing on the probability of locating one of these electrons in a left-localized state.

For that purpose, we rewrite Eq. (9.6) as follows

$$
|\bar{L}_{N_{\mathrm{mol}}}\rangle = \frac{1}{\sqrt{2}}\sum_{i=1}^{N_{\mathrm{mol}}}\left(\frac{1}{N_{\mathrm{mol}}}|\psi_{\mathrm{b}_i}\rangle - \frac{1}{\sqrt{N_{\mathrm{mol}}}}|\psi_{\mathrm{a}_i}\rangle\right)\bigotimes_{j\neq i}^{N_{\mathrm{mol}}}|\psi_{\mathrm{b}_j}\rangle, \tag{9.7}
$$

thus, if we project this expression with respect to $|\bar{L}_i\rangle$, we impose the ith electron to be found in a left-localized state. This leads to the following state

$$\langle \bar{L}_i|\bar{L}_{N_{\text{mol}}}\rangle = \frac{1}{2}\bigg[\frac{1}{N_{\text{mol}}} + \frac{1}{\sqrt{N_{\text{mol}}}}\bigg]\bigotimes_{j\neq i}^{N_{\text{mol}}}|\psi_{\mathsf{b}_j}\rangle \tag{9.8}$$

$$+ \frac{1}{2}\sum_{j\neq i}^{N_{\text{mol}}}\bigg[\frac{1}{N_{\text{mol}}}|\psi_{\mathsf{b}_j}\rangle - \frac{1}{\sqrt{N_{\text{mol}}}}|\psi_{\mathsf{a}_j}\rangle\bigg]\bigotimes_{k\neq i,j}^{N_{\text{mol}}}|\psi_{\mathsf{b}_k}\rangle \tag{9.9}$$

$$= \frac{1}{2N_{\text{mol}}}\Big[N_{\text{mol}} + \sqrt{N_{\text{mol}}}\Big]\bigotimes_{j\neq i}^{N_{\text{mol}}}|\psi_{\mathsf{b}_j}\rangle \tag{9.10}$$

$$- \frac{1}{2}\frac{\sqrt{N_{\text{mol}}-1}}{\sqrt{N_{\text{mol}}}}\bigg[\frac{1}{\sqrt{N_{\text{mol}}-1}}\sum_{j\neq i}^{N_{\text{mol}}}|\psi_{\mathsf{a}_j}\rangle\bigotimes_{k\neq i,j}^{N_{\text{mol}}}|\psi_{\mathsf{b}_k}\rangle\bigg], \tag{9.11}$$

from which we observe that the probability of finding the ith electron in a left localized state without caring the state onto which the other molecules are found is given by

$$P_{\bar{L}_i} = \bigg(\frac{1}{2N_{\text{mol}}}\Big[N_{\text{mol}} + \sqrt{N_{\text{mol}}}\Big]\bigg)^2 + \frac{1}{4}\frac{N_{\text{mol}}-1}{N_{\text{mol}}}. \tag{9.12}$$

In Fig. 9.1, the dark blue curve illustrates the behavior of the probability described by Eq. (9.12) concerning N_{mol}. Conversely, the lighter blue curve represents the probability of encountering the ith molecule in a right-localized state $|\bar{R}_i\rangle$. As anticipated, when $N_{\text{mol}} = 1$, the latter probability reduces to zero while the former reaches 1. With an increase in the number of molecules, the probability of discovering the ith molecule in a left-localized state diminishes, owing to the excitation toward an anti-bonding state becoming delocalized across the N_{mol} molecules involved in the HHG processes. Nonetheless, we observe that, for all values of N_{mol}, the dark blue curve consistently surpasses the lighter one. This signifies that the probability of finding the

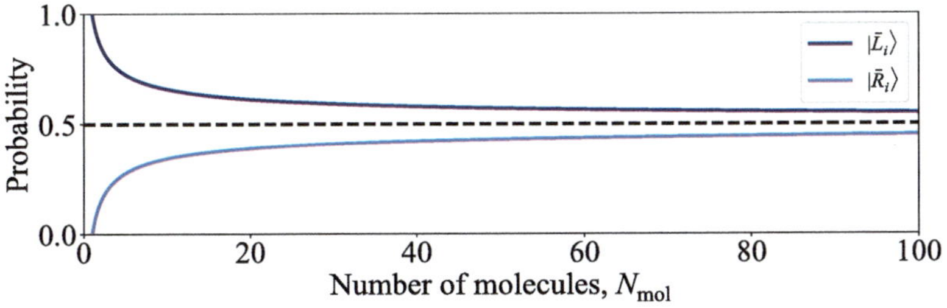

Fig. 9.1 Probability of finding the ith molecule in a left or right localized state for the $|\bar{L}_{\text{N}_{\text{mol}}}\rangle$ state. The dark blue curve represents the probability of locating the ith molecule in a left-localized state, while the light blue curve denotes the probability of finding it in a right-localized state

ith molecule in a left-localized state consistently exceeds the probability of finding it in a right-localized state. Hence, this justifies the notation employed in referencing the states outlined in Eq. (9.7).

9.3 About the ZnO Band Structure Parameters

In Table 9.1, we present the ZnO parameters used for defining their valence and conduction bands. We also present the lattice constant along different solid directions, as well as the Kane parameters. Here, we denote through $i = x, y, z$ the solid bands $\Gamma - M$, $\Gamma - K$ and $\Gamma - A$, respectively. These data have been extracted from Refs. [2–4].

9.4 A Wannier-Bloch Picture of HHG in Solids

Instead of employing Bloch states for both the valence and conduction bands, this subsection characterizes light-solid interactions considering an hybrid approach, where valence band states are characterized through Wannier states while conduction band states with Bloch states. This representation using localized states in the valence band offers an atomistic-like interpretation of the HHG process in solids [5]. Within the single-active electron approximation, one can conceptualize the electron as being initially localized in a specific Wannier site and subsequently recombining with a potentially different site, resembling to some extent the simple man's model picture of ionization and recombination in atomic systems.

Table 9.1 Solid parameters used to model crystalline ZnO. The data we used have been extracted from Refs. [2–4]

Valence band	Conduction band	Lattice constants (in a.u.)		
$\alpha^0_{v,x} = -0.0928$	$\alpha^0_{c,x} = 0.0898$	$a_x = 5.32$	$a_y = 6.14$	$a_z = 9.83$
$\alpha^1_{v,x} = 0.0705$	$\alpha^1_{c,x} = -0.0814$	Kane parameters (in a.u.)		
$\alpha^2_{v,x} = 0.0200$	$\alpha^2_{c,x} = -0.0024$	$\mathfrak{E}_{p,x} = 0.479$	$\mathfrak{E}_{p,y} = 0.355$	$\mathfrak{E}_{p,z} = 0.355$
$\alpha^3_{v,x} = -0.0012$	$\alpha^3_{c,x} = -0.0048$			
$\alpha^4_{v,x} = 0.0029$	$\alpha^4_{c,x} = -0.0003$			
$\alpha^5_{v,x} = 0.0006$	$\alpha^5_{c,x} = -0.0009$			
$\alpha^0_{v,y} = -0.0307$	$\alpha^0_{c,y} = 0.1147$			
$\alpha^1_{v,y} = 0.0307$	$\alpha^1_{c,y} = -0.1147$			
$\alpha^0_{v,z} = -0.0059$	$\alpha^0_{c,z} = 0.0435$			
$\alpha^1_{v,z} = 0.0059$	$\alpha^1_{c,z} = -0.0435$			

9.4.1 Semiclassical Description

In this section, we introduce a semiclassical depiction of solid dynamics utilizing the Wannier-Bloch formalism. Wannier states, unlike Bloch states, deviate from being eigenstates of the crystal Hamiltonian. This distinction underscores an additional electron degree of freedom within solid-state systems in contrast to atomic assemblies: electrons exhibit the capacity to transition from one Wannier site to another –a phenomenon absent among distinct atomic nuclei, which, due to their substantial separation, experience minimal interaction. Specifically, the matrix elements of the crystal Hamiltonian within the Wannier basis, under the tight-binding approximation, are expressed as

$$\langle w_{j,v}|\hat{H}_{\mathrm{cr}}|w_{j',v}\rangle = -I_v \delta_{|j-j'|,1}, \tag{9.13}$$

where $I_v < 0$ represents the hopping parameter.

Similar to the analysis presented in Sect. 4.2.3, we adopt a single-active electron description. Within this framework and utilizing the mixed Wannier-Bloch basis representation, we introduce the following ansatz to address the Schrödinger equation, elucidating the dynamics of light interacting with the solid

$$|\psi(t)\rangle = \sum_j a_j(t)|w_{j,v}\rangle + \int \mathrm{d}\boldsymbol{k} a_c(\boldsymbol{k}, t)|\phi_{\boldsymbol{k},c}\rangle, \tag{9.14}$$

with the initial condition specified as $a_j(0) = \delta_{j,j_0}$. Upon substituting this ansatz into Eq. (4.90) and performing a projection onto both a Wannier state $|w_{j,v}\rangle$ and a Bloch state $|\phi_{\boldsymbol{k},c}\rangle$, we derive the following set of coupled differential equations

$$\begin{aligned} i\hbar\frac{\partial a_j(t)}{\partial t} =& - I_v a_{j-1}(t) - I_v a_{j+1}(t) + \mathrm{e} r_j E_{\mathrm{cl}}(t) a_j(t) \\ &+ E_{\mathrm{cl}}(t)\int \mathrm{d}\boldsymbol{k}\, d_{jc}(\boldsymbol{k}) a_c(\boldsymbol{k}, t), \end{aligned} \tag{9.15}$$

$$i\hbar\frac{\partial a_c(\boldsymbol{k}, t)}{\partial t} = E_c(\boldsymbol{k}) a_c(\boldsymbol{k}, t) + i\hbar E_{\mathrm{cl}}(t)\frac{\partial a_c(\boldsymbol{k}, t)}{\partial k_i} + E_{\mathrm{cl}}(t)\sum_j d^*_{jc}(\boldsymbol{k}) a_j(t), \tag{9.16}$$

where r_j represents the position of the jth site, and $d_{jc}(\boldsymbol{k}) \equiv \mathrm{e}\langle w_{j,v}|\hat{r}_i|\phi_{\boldsymbol{k},v}\rangle$. The latter denotes the transition matrix element between an electron located in site j of the valence band, to a Bloch state characterized by crystal momentum $\boldsymbol{k}$ placed in the conduction band.

It is important to note that both interband and intraband dynamics are described in Eqs. (9.15) and (9.16). Specifically, intraband transitions within the valence band are delineated in Eq. (9.15) by the electron's movement between adjacent sites. Similarly, intraband transitions within the conduction band are characterized by the crystal momentum derivatives. Finally, interband transitions between the valence and conduction bands manifest through terms incorporating the transition dipole matrix elements $d_{jc}(\boldsymbol{k})$.

For the first equation, and in accordance with the SFA, we assume that the depletion caused by excitations to the conduction band is significantly weaker compared to that resulting from intraband transitions. Consequently, we simplify Eq. (9.15) as

$$i\hbar\frac{\partial a_j(t)}{\partial t} \approx -I_v a_{j-1}(t) - I_v a_{j+1}(t) + \mathsf{e} r_j E_{\text{cl}}(t) a_j(t), \tag{9.17}$$

and perform a discrete Fourier transform over the site label j by multiplying the entire equation by e^{-iqj} and summing over all sites j. Hence, if we define

$$a_q(t) = \sum_j a_j(t) e^{-iqj}, \tag{9.18}$$

the Schrödinger equation in Eq. (9.17) transforms into

$$i\hbar\frac{\partial a_q(t)}{\partial t} = -2I_v \cos(q) a_q(t) + i\hbar \mathsf{ea} E_{\text{cl}}(t) \frac{\partial a_q(t)}{\partial q}, \tag{9.19}$$

with a denoting the lattice constant. This differential equation can also be written as

$$i\hbar\frac{\mathrm{d}a_{\tilde{q}}(t)}{\mathrm{d}t} = -2I_v \cos(\tilde{q} + \mathsf{ea} E_{\text{cl}}(t)) a_{\tilde{q}}(t) \tag{9.20}$$

where $\tilde{q} = q - \mathsf{ea} A_{\text{cl}}(t)$, and whose solution is straightforwardly given by

$$a_{\tilde{q}}(t) = \exp[\frac{i}{\hbar} 2I_v \int_{t_0}^{t} \mathrm{d}\tau \cos(\tilde{q} + \mathsf{ea} A_{\text{cl}}(\tau))] a_{\tilde{q}}(t_0). \tag{9.21}$$

Consequently, a solution to Eq. (9.17) can be obtained by inverting the discrete Fourier transform. Thus, we arrive at

$$a_j(t) = \sum_q e^{iq(j-j_0)} \exp[\frac{i}{\hbar} 2I_v \int_{t_0}^{t} \mathrm{d}\tau \cos(\tilde{q} + \mathsf{ea} A_{\text{cl}}(\tau))], \tag{9.22}$$

where the initial condition $a_j(t_0) = \delta_{j-j_0}$ has already been implemented.

Regarding Eq. (9.16), this expression corresponds to a first-order inhomogeneous differential equation with a well defined homogeneous and inhomogeneous parts. As has been discussed along the main text, its solution can be written as a linear combination of the solution to the homogeneous part and a solution to the inhomogeneous component. In this case, after introducing the initial conditions, we arrive at [5, 6]

$$a_c(\boldsymbol{K},t) = -\frac{i}{\hbar}\sum_j \int_{t_0}^{t} \mathrm{d}t_1 \exp[-\frac{i}{\hbar}\int_{t_0}^{t_1} \mathrm{d}\tau E_c\big(\boldsymbol{K} + \mathrm{e}\epsilon_i A_{\mathrm{cl}}(\tau)\big)] \\ \times E_{\mathrm{cl}}(t_1) d^*_{jc}\big(\boldsymbol{K} + \mathrm{e}\epsilon_i A_{\mathrm{cl}}(t_1)\big) a_j(t_1), \tag{9.23}$$

which bears some resemblance to the atomic-ionization process. Specifically, at time t_1 the electron is found at site j, which is in general different from the initial j_0. Afterward, it transitions to the conduction band driven by the field at time t_1, these dynamics represented through the dipole matrix element $d_{jc}(\boldsymbol{k})$. As a last step, it accelerates in the conduction band until time t, with the state gaining an additional phase factor accounting for this process.

Having in mind that [5]

$$d_{jc}(\boldsymbol{k}) = d_{vc}(\boldsymbol{k})\tilde{w}_v e^{ikr_j/\hbar}, \tag{9.24}$$

where $\tilde{w}_m$ is a normalization constant for Wannier states which for 1D lattices is k-independent [7], we can express the transition dipole matrix element as

$$\begin{aligned} D(t) &= \langle\psi(t)|\hat{r}|\psi(t)\rangle \\ &\approx -\frac{i}{\hbar}|\tilde{w}_v|^2 \sum_{j,j'} \int_{t_0}^{t} \mathrm{d}t_1 \int \mathrm{d}\boldsymbol{K} a^*_j(t) d_{vc}\big(\boldsymbol{K} + \mathrm{e}\epsilon_i A_{\mathrm{cl}}(t)\big) e^{i(\boldsymbol{K}+\mathrm{e}\epsilon_i A_{\mathrm{cl}}(t))r_j/\hbar} \\ &\qquad \times e^{-i\varphi(p,t,t_1)/\hbar} E_{\mathrm{cl}}(t_1) d^*_{vc}\big(\boldsymbol{K} + \mathrm{e}\epsilon_i A_{\mathrm{cl}}(t_1)\big) \\ &\qquad + e^{-i(\boldsymbol{K}+\mathrm{e}\epsilon_i A_{\mathrm{cl}}(t_1))r_{j'}/\hbar} a_{j'}(t_1) + \text{c.c.}, \end{aligned} \tag{9.25}$$

where we have restricted ourselves to interband transitions, of interest since they dominate over the non-perturbative part of the HHG spectrum. In this expression, we defined $\varphi(p,t,t_1) \equiv \int_{t_1}^{t} \mathrm{d}\tau E_c(\boldsymbol{K} + \mathrm{e}\epsilon_i A_{\mathrm{cl}}(\tau))$. The obtained time-dependent dipole moment, whose Fourier transform leads to the HHG spectrum, yields similar dynamics to those occurring in atomic systems. At time t_1 the electron gets ionized from the Wannier site j' to the conduction band. Within the conduction band, it accelerates gaining a phase $\varphi(p,t,t_1)$, and finally recombines at time t with the jth site.

Since in general $j \neq j'$,we observe one of the main distinctions between HHG in atoms and solid-state systems, where electrons do not need to begin and finalize the HHG dynamics in the same state. This could explain the reduced dependence of the resultant harmonics on the field's ellipticity found in solid systems [8]. However, depending on the material considered and how the excitation process takes place (for instance, the crystal direction along which the laser is polarized), the influence of increasing values $\Delta j = |j - j'|$ on the HHG spectra might differ. In Fig. 9.2, this is shown for a ZnO material excited by a linearly polarized field parallel to the $\Gamma - A$ direction, with field strength $E_0 \approx 0.2$ V/Å, central wavelength $\lambda_L = 3.0$ μm and duration of $\Delta t \approx 100$ fs (≈ 10 optical cycles). As observed, in this case, the contribution to the HHG spectra starts to decay already for $\Delta j \geq 2$, meaning that for this particular case, the HHG process is highly localized. The HHG spectra obtained

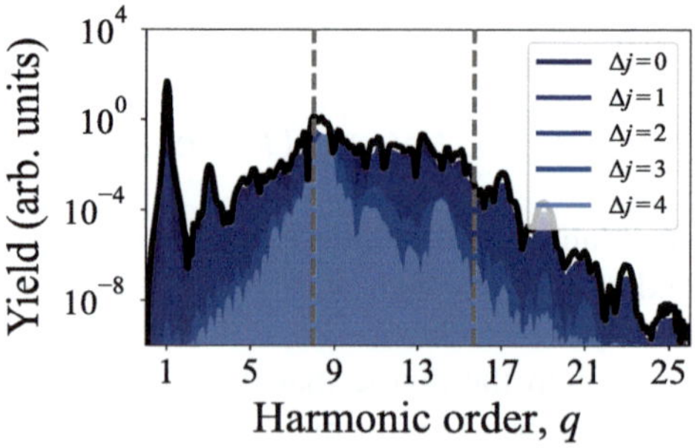

Fig. 9.2 Interband contribution to the HHG spectrum employing the Wannier-Bloch approach. The figure illustrates the computed HHG spectrum contribution employing the Wannier-Bloch formalism. The material under consideration is ZnO, subjected to excitation by a linearly polarized field along the $\Gamma - A$ direction. The applied field strength is $E_0 \approx 0.2$ V/Å, with a central wavelength of $\lambda_{\mathbf{k}_L} = 3.0\,\mu$m and a duration of $\Delta t \approx 100$ fs (equivalent to ≈ 10 optical cycles). Various colors denote distinct contributions to the spectra based on the separation between initial and final Wannier sites. The black curve represents the cumulative contribution. This figure is adapted from Fig. 3 in Ref. [5]

here recovers similar features to those found in the Bloch-based analysis in the main text (shown in Fig. 4.15 (b) with the darkest curve): a plateau structure defined within the lowest and maximum values of the bandgap, as well as the presence of poorly resolved harmonic peaks in the absence of dephasing.

9.4.2 Quantum Optical Description

As mentioned earlier, the discreteness of Wannier states provides a more manageable framework for exploring the entanglement between light and matter. In this subsection, we address the single-electron quantum optical version by considering the hybrid Wannier-Bloch approach presented earlier

$$i\hbar\frac{\partial|\Psi(t)\rangle}{\partial t} = \big[\hat{H}_{\text{cr}} + \mathsf{e}\hat{r}\hat{E}(t) + \hat{H}_{\text{field}}\big]. \tag{9.26}$$

with the initial state given in this case by

$$|\Psi(t_0)\rangle = |w_{j_0,v}\rangle \otimes |\alpha_{\mathbf{k}_L}\rangle \bigotimes_{q=2}^{q_c} |0_{\mathbf{k}_q}\rangle, \tag{9.27}$$

where the single-active electron initially occupies the j_0th Wannier site.

By applying the same unitary transformations as in the atomic HHG analysis, we arrive at

$$i\hbar\frac{\partial|\tilde{\Psi}'(t)\rangle}{\partial t} = \mathsf{e}\hat{r}(t)\hat{E}(t)|\tilde{\Psi}'(t)\rangle, \tag{9.28}$$

where $\hat{r}(t) \equiv \hat{U}_{\text{sc}}^{\dagger}(t)\hat{r}\hat{U}_{\text{sc}}(t)$. Next, we introduce the identity in the Wannier-Bloch representation, namely

$$\mathbb{1} = \sum_j |w_{j,v}\rangle\langle w_{j,v}| + \int \mathrm{d}\boldsymbol{k}\ |\phi_{\boldsymbol{k},c}\rangle\langle\phi_{\boldsymbol{k},c}|, \tag{9.29}$$

which inputted in Eq. (9.28) leads to

$$\begin{aligned} i\hbar\frac{\partial|\tilde{\Psi}'(t)\rangle}{\partial t} &= \mathsf{e}\sum_j \hat{r}(t)\hat{E}(t)\langle w_{j,v}|\tilde{\Psi}'(t)\rangle|w_{j,v}\rangle \\ &\quad + \mathsf{e}\int \mathrm{d}\boldsymbol{k}\ \hat{r}(t)\hat{E}(t)\langle\phi_{\boldsymbol{k},c}|\tilde{\Psi}'(t)\rangle|\phi_{\boldsymbol{k},c}\rangle. \end{aligned} \tag{9.30}$$

In the following, we assume the probability of the electron residing in the conduction band to be minimal at all times, which we expect to be valid in the regime of weak depletion of the valence band [5]. Consequently, we approximate Eq. (9.30) as

$$i\hbar\frac{\partial|\tilde{\Psi}'(t)\rangle}{\partial t} \approx \mathsf{e}\sum_j \hat{r}(t)\hat{E}(t)\langle w_{j,v}|\tilde{\Psi}'(t)\rangle|w_{j,v}\rangle, \tag{9.31}$$

and by projecting with respect to the set of Wannier states, we arrive at the following system of differential equations

$$i\hbar\frac{\partial|\Phi_i\rangle(t)}{\partial t} = \sum_j M_{i,j}(t)\hat{E}(t)|\Phi_j(t)\rangle, \quad \forall j \in \mathbb{W}, \tag{9.32}$$

where we defined $|\Phi_i(t)\rangle \equiv \langle w_{i,v}|\tilde{\Psi}(t)\rangle$ and $M_{i,j}(t) \equiv \mathsf{e}\langle w_{i,v}|\hat{r}(t)|w_{j,v}\rangle$. It is important to note that $M_{i,j}(t)$ is a non-diagonal matrix since an electron initially on site i can end up in site j either through an intraband transition, described by off-diagonal elements, or via an excitation-recombination mechanism, mainly described by the diagonal elements.

In what follows, we consider HHG processes under the regime of weak electron delocalization, that is, $|M_{i,j}(t)| \ll |M_{i,i}(t)|$ with $i \neq j$. For simplicity, we solely consider nearest neighbor contributions, restricting our focus to the non-diagonal elements of $M_{i,j}(t)$ for which $j = i \pm 1$. Treating these as a perturbation parameter, we expand our state $|\Phi_i(t)\rangle$ up to first order as

$$|\Phi_i(t)\rangle \approx |\Phi_i^{(0)}(t)\rangle + |\Phi_i^{(1)}(t)\rangle, \tag{9.33}$$

such that the zeroth-order term satisfies

$$i\hbar\frac{\mathrm{d}|\Phi_i^{(0)}(t)\rangle}{\mathrm{d}t} = M_{i,i}(t)\hat{E}(t)|\Phi_i^{(0)}(t)\rangle, \tag{9.34}$$

and whose solution can be written as

$$\begin{aligned}|\Phi_i^{(0)}(t)\rangle &= \hat{\mathcal{D}}\big(\chi_{i,i}(t,t_0)\big)|\Phi_i^{(0)}(t_0)\rangle \\ &\equiv \prod_q e^{i\varphi_{i,i}^{(q)}(t,t_0)}\hat{D}_q\big(\chi_{i,i}^{(q)}(t,t_0)\big)|\Phi_i^{(0)}(t_0)\rangle,\end{aligned} \tag{9.35}$$

where $\hat{D}_q(\alpha) = \exp[\alpha\hat{a}_q^\dagger - \alpha^*\hat{a}_q]$ is the displacement operator acting on mode q, $\varphi_{i,i}^{(q)}(t)$ is a prefactor arising from applying the BCH relation, and $\chi_{i,i}^{(q)}(t,t_0)$ is a coherent state amplitude given by

$$\chi_{i,i}^{(q)}(t,t_0) = -\frac{1}{\hbar}\int_{t_0}^{t}\mathrm{d}\tau e^{i\omega_{\mathbf{k}_q}\tau}g(\omega_{\mathbf{k}_q})M_{i,i}(\tau), \tag{9.36}$$

where it is worth noting that, although we have explicitly depicted the dependence of $\chi_{i,i}^{(q)}(t,t_0)$ on the Wannier site, specifically the ith site, the dynamics across all sites are equivalent. Consequently, they result in identical HHG spectra, which enables us to disregard the i index in $\chi_{i,i}^{(q)}(t,t_0)$.

Regarding the first-order term, we get the following differential equation

$$i\hbar\frac{\mathrm{d}|\Phi_i^{(1)}(t)\rangle}{\mathrm{d}t} = M_{i,i}(t)\hat{E}(t)|\Phi_i^{(1)}(t)\rangle + \sum_{j\in\{i\pm1\}} M_{i,j}(t)\hat{E}(t)|\Phi_j^{(0)}(t)\rangle, \tag{9.37}$$

whose solution can be generally written as

$$\begin{aligned}|\Phi_i^{(1)}(t)\rangle =&\hat{\mathcal{D}}\big(\chi_{i,i}(t,t_0)\big)|\Phi_i^{(1)}(t_0)\rangle \\ &-\frac{i}{\hbar}\sum_{j\in\{i\pm1\}}\int_{t_0}^{t}\mathrm{d}t'\hat{\mathcal{D}}\big(\chi_{i,i}(t,t')\big)M_{i,j}(t')\hat{E}(t') \\ &\qquad\times\hat{\mathcal{D}}\big(\chi_{j,j}(t',t_0)\big)|\Phi_j^{(0)}(t_0)\rangle.\end{aligned} \tag{9.38}$$

Introducing the initial conditions at this point, we observe two distinct contributions based on the final site, j, where the electron ends up:

- If the electron concludes its dynamics in the initial Wannier site, j_0, then the final state is given by

$$|\Phi_{j_0}(t)\rangle = \hat{\mathcal{D}}\big(\chi_{i,i}(t,t_0)\big)\bigotimes_q|0_{\mathbf{k}_q}\rangle; \tag{9.39}$$

- Conversely, if the electron concludes its dynamics in a different Wannier site from where it started, j, the final state becomes

$$|\Phi_j(t)\rangle = -\frac{i}{\hbar}\int_{t_0}^{t} dt'\hat{\mathcal{D}}\big(\chi_{j,j}(t,t')\big)M_{j,j_0}(t')\hat{E}(t') \times \hat{\mathcal{D}}\big(\chi_{j_0,j_0}(t',t_0)\big)\bigotimes_q|0_{\mathbf{k}_q}\rangle. \tag{9.40}$$

Taking into account that $\chi_{i,i}^{(q)}(t) = \chi_{j,j}^{(q)}(t) = \chi_{\mathbf{k}_q}(t)$, we can rewrite Eq. (9.40) as

$$|\Phi_j(t)\rangle = -\frac{i}{\hbar}e^{i\varphi(t)}\hat{\mathcal{D}}\big(\chi(t,t_0)\big)\int_{t_0}^{t} dt' e^{-i\theta(t')}\hat{\mathcal{D}}^{\dagger}\big(\chi(t',t_0)\big) \times \big(M_{j,j_0}(t')\hat{E}(t')\big)\hat{\mathcal{D}}\big(\chi(t',t_0)\big)\bigotimes_{q=1}|0_{\mathbf{k}_q}\rangle, \tag{9.41}$$

where we defined $\theta(t') \equiv \sum_q \Im\{\chi_{\mathbf{k}_q}(t,t_0)\chi^*_{\mathbf{k}_q}(t',t_0)\}$. Numerical analysis evidenced that the integral above differ in the fifth significant decimal when $j = \pm 1$. Hence, in the following we approximate $|\Phi_{+1}(t)\rangle \approx |\Phi_{-1}(t)\rangle \equiv |\Phi_{\text{NN}}(t)\rangle$.

For a 1D lattice, the joint state of the system can be written as

$$|\tilde{\Psi}'(t)\rangle = |w_{j_0,v}\rangle \otimes |\Phi_{j_0}(t)\rangle + \sqrt{2}|w_{\text{NN},v}\rangle \otimes |\Phi_{\text{NN}}(t)\rangle, \tag{9.42}$$

where we defined $|w_{\text{NN},v}\rangle \equiv (1/\sqrt{2})(|w_{j_0+1,v}\rangle + |w_{j_0-1,v}\rangle)$. If we denote the coordinate number of the lattice by f_c (i.e. $f_c = 2d$ in hypercubic lattices in d-dimensions, $f_c = 3$ for a honeycomb lattice in 2D, $f_c = 6$ for a triangular lattice in 2D), we can provide a generalization of our state to arbitrary dimensions

$$|\tilde{\Psi}'(t)\rangle = |w_{j_0,v}\rangle \otimes |\Phi_{j_0}(t)\rangle + \sqrt{f_c}|w_{\text{NN},v}\rangle \otimes |\Phi_{\text{NN}}(t)\rangle, \tag{9.43}$$

where now $|w_{\text{NN},v}\rangle \equiv (1/\sqrt{f_c})(\sum_{j\in\text{NN}}|w_{j_0+j,v}\rangle)$.

9.4.2.1 An Extension to the Many-Electron Regime

Until now we have worked under the single-active electron approximation. However, the initial state of the solid corresponds to a completely filled Fermi sea, where all Wannier sites are occupied. As electron-electron correlations are weak in typical semiconductor materials, we can treat the interaction of all electrons contributing to the HHG process with the laser field as being independent. Consequently, after introducing the same unitary transformations and approximations as in the single-electron analysis, the Schrödinger equation reads

$$i\hbar\frac{\mathrm{d}|\tilde{\Psi}'(t)\rangle}{\mathrm{d}t} \approx \mathsf{e}\sum_{\mathbf{m}} \hat{\tilde{r}}(t)\hat{E}(t)\langle w_{\mathbf{m},v}|\tilde{\Psi}'(t)\rangle|w_{\mathbf{m},v}\rangle, \tag{9.44}$$

where $|w_{\mathbf{m},v}\rangle \equiv |w_{m_0,v}\rangle \otimes |w_{m_1,v}\rangle \otimes \cdots \otimes |w_{m_{N_\mathsf{e}},v}\rangle$; and $\hat{\tilde{r}} = \sum_{j=0}^{N_\mathsf{e}-1} \hat{r}_{i,j}$, with N_e being the number of electrons contributing in a phase-matched way to the HHG process, and $r_{i,j}$ the position operator along direction i acting on the jth electron.

Akin to our previous analysis, we assume the regime of weak-delocalization and focus on situations where the single-electron wavepacket does not significantly spread across different Wannier sites. Therefore, we differentiate the contribution of $\tilde{r}(t)$ into their diagonal and off-diagonal elements, allowing us to treat the latter using perturbation theory. By projecting Eq. (9.44) with respect to $|w_{\mathbf{i},v}\rangle$, we arrive at

$$i\hbar\frac{\mathrm{d}|\Phi_\mathbf{i}(t)\rangle}{\mathrm{d}t} = \mathsf{e}\sum_{\mathbf{m}} \langle w_{\mathbf{i},v}|\tilde{r}(t)|w_{\mathbf{m},v}\rangle\hat{E}(t)|\Phi_\mathbf{m}(t)\rangle, \tag{9.45}$$

where we defined $|\Phi_\mathbf{i}(t)\rangle \equiv \langle w_{\mathbf{i},v}|\tilde{\psi}(t)\rangle$, and where the matrix elements of $\tilde{r}(t)$ are given by

$$\langle w_{\mathbf{n},v}|\tilde{r}(t)|w_{\mathbf{m},v}\rangle = \sum_{j=0}^{N_\mathsf{e}-1} \langle w_{n_j,v}|\hat{r}_{i,j}(t)|w_{m_j,v}\rangle \prod_{l\in O_j} \delta_{n_l,m_l}, \tag{9.46}$$

wherein the previous expression $O_j := \{l : l \in \mathbb{N} - \{j\}\}$.

Up to first order perturbation theory, the equations we get are identical to the ones found in the single-electron regime. Thus, taking into account the initial conditions, we find two kinds of contributions depending on the final sites j where the electrons end up in:

- If the electrons end up the dynamics in the initial Wannier sites where they were initially located, then the final state reads

$$|\Phi_\mathbf{i}(t)\rangle = \hat{\mathcal{D}}\big(N_\mathsf{e}\chi(t,t_0)\big)\bigotimes_{q=0}^{q_c}|0_{\mathbf{k}_q}\rangle; \tag{9.47}$$

- If one of the electrons ends up the dynamics in a different Wannier site j from which it initially started the dynamics, then the final state reads

$$\begin{aligned}|\Phi_\mathbf{j}(t)\rangle = -\frac{i}{\hbar}\hat{\mathcal{D}}\big(N_\mathsf{e}\chi(t,t_0)\big)\int_{t_0}^{t}\mathrm{d}t' e^{-i\theta(t,t',t_0)}M_{j_n,j_{n,0}}(t')\\ \times\hat{\mathcal{D}}^\dagger\big(N_\mathsf{e}\chi(t',t_0)\big)\hat{E}(t')\hat{\mathcal{D}}\big(N_\mathsf{e}\chi(t',t_0)\big)\bigotimes_{q=1}^{q_c}|0_{\mathbf{k}_q}\rangle,\end{aligned} \tag{9.48}$$

  where we considered that one of the electrons has transitioned from its initial site $j_{n,0}$ to site j_n. The first-order perturbation theory term describes events where only one of the electrons ends up in a different Wannier site from which it initially started the dynamics. In order to consider events where we find n transitions, we would have to perform a perturbation theory expansion up to the nth order term. Neverheless, in the many-electron scenario, we restrict ourselves to solids for which the HHG process is highly localized, as it happens for instance with solid argon [9], or for specific crystal directions of certain solids like ZnO (see Fig. 9.2).

With all this, we extend Eq. (9.43) to the many-electron scenario as

$$|\tilde{\Psi}'(t)\rangle = |w_{\boldsymbol{j_0},v}\rangle \otimes |\Phi_{\boldsymbol{j_0}}(t)\rangle + \sqrt{f_c N_{\mathrm{e}}}|w_{\overline{\mathrm{NN}},v}\rangle \otimes |\Phi_{\overline{\mathrm{NN}}}(t)\rangle, \tag{9.49}$$

such that we arrive at an entangled state between light and matter, where the main advantage in this case is that we have characterized the electronic states through a discrete basis set. In this expression, we defined $|w_{\overline{\mathrm{NN}},v}\rangle \equiv (1/\sqrt{f_c N_{\mathrm{e}}}) \sum_{\boldsymbol{j}\in\overline{\mathrm{NN}}}|w_{v,\boldsymbol{j}}\rangle$, with $\overline{\mathrm{NN}} := \{\boldsymbol{j} : \boldsymbol{j}_0 \pm (0_1, \ldots, 1_i, \ldots, 0_{N_{\mathrm{e}}})\ \forall i\}$.

9.4.2.2 Conditioning on HHG and Its Consequences

In the following, we consider the use of the conditioning on HHG operation presented in Sect. 3.3.2.1. By applying this operation to Eq. (9.49) and subsequently projecting it onto the coherent state $|\gamma\rangle \equiv \bigotimes_{q\geq 2}|\chi_{\mathbf{k}_q}(t, t_0)\rangle$, representing the measured harmonics, we obtain

$$|\Psi_{\mathrm{cond}}(t)\rangle = \langle\gamma|\Psi(t)\rangle - \xi(\gamma, t)|\alpha_{\mathbf{k}_L}\rangle, \tag{9.50}$$

with $\xi(\gamma, t) = \langle\alpha_{\mathbf{k}_L}, 0_{\mathrm{HH}}|\Psi(t)\rangle\langle\gamma|0_{\mathrm{HH}}\rangle$.

The discreteness of the employed Wannier basis significantly simplifies the characterization of entanglement compared to continuous-variable bases sets, as it happens with the Bloch-based approach outlined in Sect. 4.2.3. Moreover, within the regime of weak delocalization, we can focus on nearest-neighbor interactions, further simplifying the entanglement analysis as we can treat the electronic degrees of freedom in Eq. (9.49) as if they were qubits. Therefore, in this context the entanglement entropy in Eq. (2.123) serves as a suitable entanglement measure. Additionally, we examine the fidelity of the state with respect to the $|w_{\overline{\mathrm{NN}},v}\rangle$, which provides insights into the likelihood of finding electrons in their nearest neighbors. In this case, the fidelity can be expressed as

$$\mathcal{F}_{\mathrm{del}}(\hat{\sigma}) = \mathrm{tr}_{\mathrm{field}}\Big(\langle w_{\overline{\mathrm{NN}},v}|\hat{\sigma}|w_{\overline{\mathrm{NN}},v}\rangle\Big). \tag{9.51}$$

In Fig. 9.3, we present the two aforementioned quantities when considering different lattice geometries, and their dependency on the number of electrons N_{e}. Panel (a) demonstrates the fidelity, while panel (b) showcases the entanglement entropy. The dashed curves represent the state without conditioning, i.e., when considering Eq. (9.49). Conversely, solid curves showcase Eq. (9.50), where the conditioning on

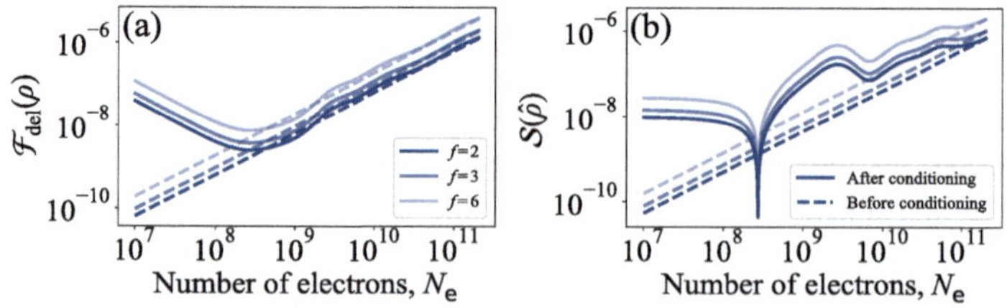

Fig. 9.3 Entropy of entanglement and fidelity computed under a Wannier-Bloch approach. In (**a**) we display the fidelity measure as defined in Eq. (9.51) plotted against the number of electrons independently contributing to the HHG process. In (**b**) we present the entanglement entropy. Dashed lines indicate scenarios without conditioning on HHG operation, while solid curves depict cases where this conditioning has been applied. Each color represents a distinct coordination number. The analysis involves the interaction of ZnO with a linearly polarized electric field along the $\Gamma - A$ crystal direction, featuring a peak field strength of $E_0 \approx 0.2$ V/Å, $\lambda_{\mathbf{k}_L} = 3.25\ \mu$m, and a duration of $\Delta t \approx 40$ fs (equivalent to ≈ 4 optical cycles)

HHG operation has been applied. For both cases, the observed values of these measures generally remain within the order of 10^{-6} at maximum suggesting that, under the considered excitation conditions, the entanglement features are nearly negligible. This underscores the validity of the approximations utilized in Sect. 4.2.3. Nevertheless, there is a noticeable trend: the amount of entanglement tends to increase with the number of electrons involved in the interaction region. However, for larger values of N_e, considering more than nearest neighbors in the final quantum optical state becomes mandatory.

Despite the reduced values observed for the degree of entanglement and the fidelity in Fig. 9.3, other noteworthy features can be observed. Firstly, for $N_e \lesssim 2 \times 10^8$, both the fidelity and the entanglement after the conditioning (solid curves) exhibit higher values compared to those before the conditioning (dashed curves). This enhancement is primarily attributed to $|\chi(t, t_0)|$, which remains a small quantity ($\sim 10^{-2}$), allowing the conditioning to accentuate the contributions from electron recombination in the nearest neighbors. This increase in fidelity goes along with a growth in the value of $S(\hat{\rho})$. Secondly, an intriguing observation occurs around $N_e \simeq 2.5 \times 10^8$: an abrupt drop in $S(\hat{\rho})$ after the conditioning, which is absent for $\mathcal{F}_{\text{del}}(\hat{\rho})$. This phenomenon stems from the cancellation of quantum optical contributions arising from the $|w_{\overline{\text{NN}},v}\rangle$ term in Eq. (9.50), i.e., $\langle\gamma|\Phi_{\overline{\text{NN}}}(t)\rangle - \xi(\gamma, t)|\alpha_{\mathbf{k}_L}\rangle$. As N_e increases, $\xi(\gamma, t)$ diminishes with the increment of $|\chi(t, t_0)|$, resulting in a minimal overlap with the initial state. Therefore, in this regime, where the conditioning on HHG operation leads to negligible effects, $\mathcal{F}_{\text{del}}(\hat{\rho})$ and $S(\hat{\rho})$ for the state before and after conditioning coincide. Furthermore, these two quantities demonstrate a marginal increment with the coordinate number f_c.

References

1. Atkins P, Friedman R (2015) Techniques of approximation. In: Molecular quantum mechanics. Oxford University Press, Oxford, pp 168–205. Chap. 6
2. Vampa G, McDonald C, Orlando G, Klug D, Corkum P, Brabec T (2014) Phys Rev Lett 113:073901. https://doi.org/10.1103/PhysRevLett.113.073901
3. Vampa G, McDonald CR, Orlando G, Corkum PB, Brabec T (2015) Phys Rev B 91:64302. https://doi.org/10.1103/PhysRevB.91.064302
4. Vampa G, McDonald C, Fraser A, Brabec T (2015) IEEE J Sel Top Quantum Electron 21:1. https://doi.org/10.1109/JSTQE.2015.2402636
5. Osika EN, Chacón A, Ortmann L, Suárez N, Pérez-Hernández JA, Szafran B, Ciappina MF, Sols F, Landsman AS, Lewenstein M (2017) Phys Rev X 7:021017. https://doi.org/10.1103/PhysRevX.7.021017
6. Amini K, Biegert J, Calegari F, Chacón A, Ciappina MF, Dauphin A, Efimov DK, Figueira de Morisson Faria C, Giergiel K, Gniewek P, Landsman AS, Lesiuk M, Mandrysz M, Maxwell AS, Moszyński R, Ortmann L, Pérez-Hernández JA, Picón A, Pisanty E, Prauzner-Bechcicki J, Sacha K, Suárez N, Zaïr A, Zakrzewski J, Lewenstein M (2019) Rep Prog Phys 82:116001. https://doi.org/10.1088/1361-6633/ab2bb1
7. Kohn W (1959) Phys Rev 115:809. https://doi.org/10.1103/PhysRev.115.809
8. Ghimire S, DiChiara AD, Sistrunk E, Agostini P, DiMauro LF, Reis DA (2011) Nat Phys 7:138. https://doi.org/10.1038/nphys1847
9. Ndabashimiye G, Ghimire S, Wu M, Browne DA, Schafer KJ, Gaarde MB, Reis DA (2016) Nature 534:520. https://doi.org/10.1038/nature17660

Chapter 10
Additional Material of Chapter 5

10.1 Optimization of the Asymptotic Key Rate Using the Convergent Hierarchy on the Von Neumann Entropy

In Sect. 5.2.3, we described how the numerical optimization of the key rate was performed based on the use of Eq. (5.14) to bound the $H(A|x = m, E)$. Here, we provide more specific details about the SDP implementation and the numerical optimization.

10.1.1 SDP Relaxation

The expression presented in Eq. (5.14) corresponds to a SDP relaxation of the exact expression outlined in Theorem 2.11 of Ref. [1], utilizing the NPA hierarchy [2, 3]. In our numerical implementation, we confine ourselves to the second level of the hierarchy, encompassing solely up to second-order moments in our certificate. However, we introduce additional third-order moments, specifically terms such as $\langle \hat{M}_{a|x} \hat{M}_{b|y} \hat{Z} \rangle$ and $\langle \hat{M}_{a|x} \hat{Z}^{\dagger} \hat{Z} \rangle$ (where the $\hat{Z}$ operators are those included in Eq. (5.14)).

The SDP relaxation depicted in Eq. (5.14) is reliant on M, representing the maximum number of terms considered in the expansion of integrals defined within the quantum relative entropy in terms of the Gauss-Radau quadratures method (refer to Theorem 2.11 in Ref. [1] for comprehensive details). Naturally, as M tends to infinity ($M \to \infty$), we converge towards the exact definition of integral terms, consequently yielding improved bounds.

J. Rivera-Dean, *Non-classical States of Light*, Springer Theses,
https://doi.org/10.1007/978-3-031-73769-5_10

10.1.2 Parallelization of the Optimization

After generating several points outside the local set $\mathcal{L}$ that fulfill the criteria of our implementation in Fig. 5.8 and the subsequent creation of the SDP relaxation problem, we proceed with optimizing the measurement angles to maximize the key rate. This optimization task is computationally challenging, demanding the execution of the SDP relaxation with each evaluation of the optimization function.

To address this, we implement a parallelized approach using the MPI standard [4], allowing simultaneous optimization of the non-local points. Notably, higher values of M in our SDP relaxation entail increased computational times. To strike a balance between computation and results, we set $M = 2$ for each optimization iteration. This initial step filters out non-local points that evidently yield unfavorable (negative) key rate values. Subsequently, a subset of *good* points emerges, which are then evaluated using a larger M value ($M = 8$). This process leads to the plots showcased in Sect. 5.2.3.

10.2 Specifics About the Finite Size Analysis

In this section, we present the Theorem 1 of Ref. [5] and provide additional information about how it was computed in the finite size analysis presented in Sect. 5.2.4.

10.2.1 Theorem 1 in E. Y.-Z. Tan Et Al, Quantum 5, 880 (2022)

This Theorem relies on several parameters, some of which we optimize while keeping others constant. Initially, we have γ, w_{exp}, and δ_{tol}. These parameters respectively denote: the probability of a test round; the expected winning probability of a CHSH game, optimized for scenarios where the CHSH inequality is maximally violated; and the permissible deviation from the anticipated winning probability w_{exp}. Among these, we conduct a brute force optimization over γ, determine w_{exp} based on the characteristics of the source and local detectors, and maintain δ_{tol} fixed at 10^{-3}.

Next, we elucidate additional, more intricate parameters associated with the parameter estimation, error correction, and privacy amplification steps executed by Alice and Bob to distill a secure key. The parameters detailed here remain constant throughout our numerical implementation, while the remaining unmentioned variables will persist as optimization variables. The forthcoming parameters involve the inference Bob makes about Alice's outcome, denoted as $\tilde{A}$ for the former and A for the latter

- $\epsilon_{\rm EC}^{\rm com}$: bound on the probability that Bob's guess $\tilde{A}$ for A to be wrong.
- $\epsilon_{\rm PE}^{\rm com}$: bound on the probability of Bob's guess $\tilde{A}$ for A to be correct, but not satisfying the parameter estimation checks.
- $\epsilon_{\rm EA}$: bound on the probability that a virtual parameter estimation step accepts in a scenario where the devices are not sufficiently secure.
- $\epsilon_{\rm PA}$: bound on the secrecy parameter of the keys after privacy amplification in a scenario where the devices are not sufficiently secure.
- ϵ_h: bound on the probability that $\texttt{hash}(\tilde{A})$ matches $\texttt{hash}(A)$ even when Bob's guess $\tilde{A}$ for A is wrong.
- $\epsilon_s, \epsilon_s', \epsilon_s''$: smoothing parameters for several entropic terms.

With all this, Theorem 1 in Ref. [5] reads as follows:

Theorem *Take any* $\epsilon_{EC}^{com}, \epsilon_{PE}^{com}, \epsilon_{EA}, \epsilon_{PA}, \epsilon_h, \epsilon_s, \epsilon_s', \epsilon_s'' \in [0,1]$ *such that* $\epsilon_s > \epsilon_s' + 2\epsilon_s''$, *and any* $\alpha \in (1,2)$, $\alpha' \in (1,2)$, $\beta \in [g(0), g(1)], \gamma \in (0,1)$, *where* $V' := 2\log_2 5$. *The protocol used in Sect. 5.2 is* $(\epsilon_{EC}^{com} + \epsilon_{PE}^{com})$*-complete and* $(\max(\epsilon_{EA}, \epsilon_{PA} + 2\epsilon_s) + 2\epsilon_h)$*-sound when performed with any choice of* EC_{max} *such that* $P(\tilde{A} \neq A) \leq \epsilon_{EC}^{com}$ *holds and* δ_{tol} *and* l_{key} *satisfying*

$$\epsilon_{EC}^{com} \geq B_{n,\gamma w_{exp}}\big(\lfloor (w_{exp} - \delta_{tol})\gamma n \rfloor\big) + B_{n,1-\gamma+\gamma w_{exp}}\big(\lfloor (1 - \gamma + \gamma w_{exp} - \gamma\delta_{tol})n \rfloor\big), \tag{10.1}$$

$$\begin{aligned} l_{key} \leq\ & ng(w_{exp} - \delta_{tol}) - n\frac{(\alpha-1)\ln 2}{2}V^2 - n(\alpha-1)^2 K_\alpha - n\gamma - n\left(\frac{\alpha'-1}{4}\right)V'^2 \\ & - \frac{\vartheta_{\epsilon_s'}}{\alpha-1} - \frac{\vartheta_{\epsilon_s''}}{\alpha'-1} - \left(\frac{\alpha}{\alpha-1} + \frac{\alpha'}{\alpha'-1} - 2\right)\log_2\frac{1}{\epsilon_{EA}} - 3\vartheta_{\epsilon_s - \epsilon_s' - 2\epsilon_s''} \\ & - EC_{max} - \left\lceil \log_2\left(\frac{1}{\epsilon_h}\right)\right\rceil - 2\log_2\frac{1}{\epsilon_{PA}} + 2, \end{aligned} \tag{10.2}$$

where $B_{n,p}(k)$ *is the cumulative distribution function of a binomial distribution, and*

$$\vartheta_\epsilon := \log_2\frac{1}{1-\sqrt{1-\epsilon^2}} \leq \log\frac{2}{\epsilon^2}, \tag{10.3}$$

$$V := \sqrt{Var_{Q_f}(f_{min}) + 2} + \log_2 33, \tag{10.4}$$

$$K_\alpha := \frac{2^{(\alpha-1)(2\log_2 4 + Max(f_{min}) - Min_{Q_f}(f_{min}))}}{6(2-\alpha)^3\ln 2}\ln^3\left(2^{2\log_2 4 + Max(f_{min}) - Min_{Q_f}(f_{min})} + e^2\right), \tag{10.5}$$

with f_{min} and $\mathcal{Q}_f$ being a function and a set that satisfy

$$Max(f_{min}) = \frac{1}{\gamma}g(1) + \left(1 - \frac{1}{\gamma}\right)\beta, \quad Min_{\mathcal{Q}_f}(f_{min}) = g\left(\frac{2-\sqrt{2}}{4}\right), \tag{10.6}$$

$$Var_{\mathcal{Q}_f}(f_{min}) \leq \frac{2-\sqrt{2}}{4\gamma}\min\{\Delta_0^2, \Delta_1^2\} + \frac{2+\sqrt{2}}{4\gamma}\max\{\Delta_0^2, \Delta_1^2\}, \tag{10.7}$$

where $\Delta_w := \beta - g(w)$.

The Theorem above provides an upper bound for the length of the key l_{key}. In our scenario, the aforementioned parameters were chosen such that Eq. (10.1) holds, ensuring a protocol completeness of 10^{-3} and a protocol soundness of 10^{-6}. These parameters are standard in such analyses (for more details, refer to Ref. [5]) and respectively represent the probability of aborting the protocol and the probability of generating an insecure key.

Additionally, the function $g(\cdot)$ has not been previously defined and is typically referred to as the *min-tradeoff* function. This function establishes a lower bound on the entropy accumulated in a single round of the protocol. In our case, we construct the min-tradeoff function from Eq. (5.10) as follows

$$g_t(w) = \eta(t) + (w - t)\eta'(t), \tag{10.8}$$

where the $\eta(t)$ function is defined as

$$\eta(t) = \begin{cases} 0, \text{ if } t \leq 3/4 \\ 1 - H_t(A|x = m, E), \text{ otherwise} \end{cases}, \tag{10.9}$$

where $H_t(A|x = m, E)$ is defined in Eq. (5.10) with $t = (S/8) + (1/2)$. The function presented in Eq. (10.8) establishes a lower bound for Eq. (5.10) across all values of w, as depicted in Fig. 10.1.

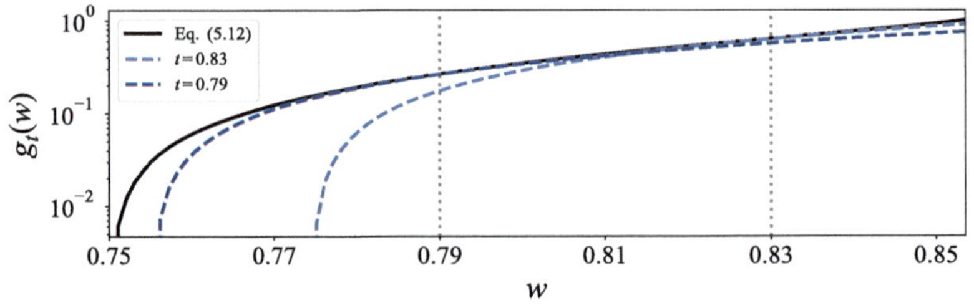

Fig. 10.1 Min-tradeoff function computed from Eq. (5.12). The black solid curve showcases Eq. (5.10) after the transformation $w = (S/8) + (1/2)$. The lighter and darker blue dashed curves represent $g_t(w)$ for $t = 0.79$ and $t = 0.83$, respectively. These values correspond to the leftmost and rightmost grey dotted lines, respectively

10.3 Parametrization of the POVM Set

The parametrization utilized for the measurements discussed in Sect. 5.3 is fundamentally reliant on the dimensions attributed to the state shared between Alice and Bob. For the case of $d = 2$, the projective operators considered for characterizing Alice's measurements (similarly for Bob) are spanned by

$$\left\{\hat{\Pi}_{0|x} = |\psi_2\rangle\langle\psi_2|\,,\, \hat{\Pi}_{1|x} = \mathbb{1} - \hat{\Pi}_{0|x}\right\} \tag{10.10}$$

where we parameterize the state $|\psi_2\rangle$ in the equation above as

$$|\psi_2\rangle = \cos\theta^{(x)}\,|0\rangle + e^{i\phi^{(x)}} \sin\theta^{(x)}\,|1\rangle\,, \tag{10.11}$$

which relies on two parameters: $(\theta^{(x)}, \phi^{(x)})$. Hence, the total count of parameters utilized when Alice implements m measurements and Bob implements $m+1$ measurements amounts to $2m(m+1)$. It is essential to note that this parametrization must be performed for each value of the input x.

As for $d = 3$, the set of projective operators spanning the measurements applied by both parties is given by

$$\left\{\hat{\Pi}_{0|x} = |\psi_3\rangle\langle\psi_3|\,,\, \hat{\Pi}_{1|x} = |\psi_3^{\perp}\rangle\langle\psi_3^{\perp}|,\, \hat{\Pi}_{2|x} = \mathbb{1} - \hat{\Pi}_{0|x} - \hat{\Pi}_{1|x}\right\}, \tag{10.12}$$

where, in this case, we express the state $|\psi_3\rangle$ as

$$|\psi_3\rangle = \cos\phi_0^{(x)} \sin\theta_0^{(x)}\,|0\rangle + e^{-\frac{2}{3}i\alpha_0^{(x)}\pi} \sin\phi_0^{(x)} \sin\theta_0^{(x)}\,|1\rangle + e^{-\frac{4}{3}i\beta_0^{(x)}\pi} \cos\theta_0^{(x)}\,|2\rangle\,. \tag{10.13}$$

Subsequently, employing the Gram-Schmidt process allows us to derive a parameterized state orthogonal to $|\psi_3\rangle$, represented as

$$\begin{aligned}
|\psi_3^{\perp}\rangle = \frac{1}{\mathcal{N}}\Bigg\{\Bigg[&\cos\phi_1^{(x)} \sin\theta_1^{(x)} \left(\sin^2\phi_0^{(x)} \sin^2\theta_0^{(x)} + \cos^2\theta_0^{(x)}\right) \\
&- \cos\phi_0^{(x)} \sin\theta_0^{(x)} \Big(e^{\frac{2}{3}\pi i(\alpha_0^{(x)}-\alpha_1^{(x)}+1)} \sin\phi_0^{(x)} \sin\phi_1^{(x)} \sin\theta_0^{(x)} \sin\theta_1^{(x)} \\
&\qquad + e^{\frac{4}{3}\pi i(\beta_0^{(x)}-\beta_1^{(x)}+1)} \cos\theta_0^{(x)} \cos\theta_1^{(x)}\Big)\Bigg]|0\rangle \\
&+ e^{-\frac{2}{3}\pi i\alpha_0^{(x)}}\Big[\cos\phi_0^{(x)} \sin^2\theta_0^{(x)} \sin\theta_1^{(x)} \\
&\qquad \times \left(-\sin\phi_0^{(x)} \cos\phi_1^{(x)} + (-1)^{2/3} e^{\frac{2}{3}\pi i(\alpha_0^{(x)}-\alpha_1^{(x)})} \cos\phi_0^{(x)} \sin\phi_1^{(x)}\right) \\
&\qquad + (-1)^{2/3} e^{\frac{2}{3}\pi i(\alpha_0^{(x)}-\alpha_1^{(x)})} \sin\phi_1^{(x)} \cos^2\theta_0^{(x)} \sin\theta_1^{(x)} \\
&\qquad + (-1)^{1/3} e^{\frac{4}{3}\pi i(\beta_0^{(x)}-\beta_1^{(x)})} \sin\phi_0^{(x)} \sin\theta_0^{(x)} \cos\theta_0^{(x)} \cos\theta_1^{(x)}\Big]|1\rangle \\
&+ e^{-\frac{4}{3}\pi i\beta_0^{(x)}} \sin\theta_0^{(x)}\Big[\cos\theta_0^{(x)} \sin\theta_1^{(x)} \\
&\qquad \times \left(\cos\phi_0^{(x)} \cos\phi_1^{(x)} + (-1)^{2/3} e^{\frac{2}{3}\pi i(\alpha_0^{(x)}-\alpha_1^{(x)})} \sin\phi_0^{(x)} \sin\phi_1^{(x)}\right) \\
&\qquad + (-1)^{1/3} e^{\frac{4}{3}\pi i(\beta_0^{(x)}-\beta_1^{(x)})} \sin\theta_0^{(x)} \cos\theta_1^{(x)}\Big]|2\rangle\Bigg\},
\end{aligned} \tag{10.14}$$

where $\mathcal{N}$ is a normalization constant. Thus, utilizing this parametrization, the set described in Eq. (10.12) is contingent on a total of 8 parameters. This implies that the overall count of parameters, considering that Alice and Bob apply m and $m + 1$ measurements respectively, amounts to $8m(m + 1)$.

References

1. Brown P, Fawzi H, Fawzi O (2024) Quantum 8:1445. https://doi.org/10.22331/q-2024-08-27-1445
2. Navascués M, Pironio S, Acín A (2007) Phys Rev Lett 98:010401. https://doi.org/10.1103/PhysRevLett.98.010401
3. Navascués M, Pironio S, Acín A (2008) New J Phys 10:073013. https://doi.org/10.1088/1367-2630/10/7/073013
4. Message Passing Interface Forum (2021) MPI: a message-passing interface standard version 4.0. https://www.mpi-forum.org/docs/mpi-4.0/mpi40-report.pdf
5. Tan EY-Z, Sekatski P, Bancal J-D, Schwonnek R, Renner R, Sangouard N, Lim CC-W (2022) Quantum 6:880. https://doi.org/10.22331/q-2022-12-22-880

GPSR Compliance

The European Union's (EU) General Product Safety Regulation (GPSR) is a set of rules that requires consumer products to be safe and our obligations to ensure this.

If you have any concerns about our products, you can contact us on ProductSafety@springernature.com

In case Publisher is established outside the EU, the EU authorized representative is:

Springer Nature Customer Service Center GmbH
Europaplatz 3
69115 Heidelberg, Germany

Batch number: 10755227

Printed by Printforce, the Netherlands